SIGMUND EXNER   HANSJOCHEM AUTRUM

D.G. Stavenga   R.C. Hardie (Eds.)

Facets of Vision

With 171 Figures

Springer-Verlag
Berlin Heidelberg New York
London Paris Tokyo

Dr. DOEKELE GERBEN STAVENGA
Laboratorium voor Algemene Natuurkunde
Rijksuniversiteit Groningen
Department of Biophysics
Westersingel 34
NL-9718 CM Groningen
The Netherlands

Dr. ROGER CLAYTON HARDIE
Cambridge University
Department of Zoology
Downing Street
GB-Cambridge CB2 3EJ
England

Cover illustration: The cover shows a reconstruction of the so-called H1 neuron, from the lobula plate of the blowfly, *Calliphora*. Recordings from this widefield motion sensitive neuron have made major contributions to our understanding of the mechanisms of motion detection (see Chs. 17 & 18). Figure courtesy of Professor Klaus Hausen.

Frontispiece: Professor Sigmund Exner. Photograph reproduced with permission from the Bildarchiv of the Austrian National Library. N° 516.604. Professor Hansjochem Autrum in his laboratory. Picture taken in 1957.

ISBN 3-540-50306-4 Springer-Verlag Berlin Heidelberg New York
ISBN 0-387-50306-4 Springer-Verlag New York Berlin Heidelberg

Typesetting: International Typesetters Inc., Makati, Philippines
Printing: Druckhaus Beltz, Hemsbach
Binding: J. Schäffer, Grünstadt
2131/3145-543210 – Printed on acid-free paper

Preface

The papers published in this Volume are the fruits of a symposium held in Regensburg in April 1987. The meeting was held to commemorate two most significant events in the development of compound eye research. In chronological order these are firstly, Sigmund Exner's seminal monograph on the physiology of compound eyes of crustaceans and insects, which was first published in Vienna in 1891, and is now shortly to appear for the first time in the English translation [Exner, S. (1989) The Physiology of the Compound Eyes of Insects and Crustaceans. Springer Berlin Heidelberg New York Tokyo]. Secondly, the meeting was also held in honour of Professor Hansjochem Autrum's 80th birthday. Professor Autrum, who is justly acknowledged as one of the pioneers of modern compound eye research, attended the meeting as the guest of honour.

In keeping with these historical occasions, it has been our intention in this volume to present a comprehensive collection of short reviews covering the major aspects of compound eye research. Whilst the most up-to-date developments have been included in every field from optics, through photochemistry, phototransduction, integrative processes and behavior, an attempt has also been made to provide a historical perspective.

Scientists represent a truly international community with no respect for borders, and as such, we are very grateful to our various sponsors: the Bavarian State Government, the Deutsche Forschungsgemeinschaft, the Friends of Regensburg University and the Springer publishers, whose generous financial support allowed us to invite leading researchers from all over the world.

Of the colleagues who were unable to attend we would particularly like to mention Carl Gustav Bernhard who organized the first symposium ever held on the compound eye. He sent his best wishes for a successful conference. Thanks in particular to the efforts of Professor Burkhardt and other members of the Zoology department of Regensburg University, who organized and arranged a suitable setting for the meeting, we hope this has indeed been achieved.

Groningen/Cambridge, Winter 1988/89   D.G. STAVENGA
R.C. HARDIE

Contents

Contributors

BACKHAUS, W., Freie Universität Berlin, Neurobiologie,
Königin-Luise-Str. 28–30, 1000 Berlin 33, FRG

BARLOW, R. B. jr., Syracuse University, Institute for Sensory
Research, Merrill Lane, Syracuse, NY 13244-5290, USA

BURKHARDT, D., Institut für Zoologie der Universität
Regensburg, Universitätsstr. 31, 8400 Regensburg, FRG

CHAMBERLAIN, S. C., Syracuse University, Institute for Sensory
Research, Merrill Lane, Syracuse, NY 13244-5290, USA

EGELHAAF, M., Max-Planck-Institut für biologische Kybernetik,
Spemannstr. 38, 7400 Tübingen, FRG

FEIN, A., University of Connecticut Health Center, School of
Medicine, Farmington, CT 06032, USA

FRANCESCHINI, N., Laboratoire de Neurobiologie, C.N.R.S.,
31, Chemin J. Aiguier, 13402 Marseille, France

GOLDSMITH, T. H., Department of Biology, Kline Tower Building,
Yale University, PO Box 6666, New Haven, CT 06511-8112, USA

HARDIE, R. C., Cambridge University, Department of Zoology,
Downing St., Cambridge CB2 3EJ, England

VAN HATEREN, J. H., Rijksuniversiteit Groningen, Department of
Biophysics, Westersingel 34, 9718 CM Groningen,
The Netherlands

HAUSEN, K., Zoologisches Institut, Universität zu Köln,
Weyertal 119, 5000 Köln 41, FRG

LAND, M. F., University of Sussex, School of Biological Sciences,
Falmer, Brighton BN1 9QG, England

LAUGHLIN, S. B., Cambridge University, Department of Zoology, Downing St., Cambridge CB2 3EJ, England

LEHMAN, H. K., Syracuse University, Institute for Sensory Research, Merrill Lane, Syracuse, NY 13244-5290, USA

MENZEL, R., Freie Universität Berlin, Neurobiologie, Königin-Luise-Str. 28–30, 1000 Berlin 33, FRG

LE NESTOUR, A., Laboratoire de Neurobiologie, C.N.R.S., 31, Chemin J. Aiguier, 13402 Marseille, France

NILSSON, D.-E., University of Lund, Department of Zoology, Helgonavägen 3, 223 62 Lund, Sweden

PAYNE, R., Department of Zoology, University of Maryland College Park, College Park, MD 20742, USA

RIEHLE, A., Laboratoire de Neurobiologie, C.N.R.S., 31, Chemin J. Aiguier, 13402 Marseille, France

ROSSEL, S., Albert-Ludwigs-Universität, Institut für Biologie I, Albertstr. 21a, 7800 Freiburg, FRG

SCHWEMER, J., Ruhr-Universität Bochum, Lehrstuhl für Tierphysiologie, Universitätsstr. 150, 4630 Bochum 1, FRG

SCHWIND, R., Institut für Zoologie der Universität Regensburg, Universitätsstr. 31, 8400 Regensburg, FRG

SHAW, S. R., Life Sciences Centre, Department of Psychology, Dalhousie University, Halifax, Nova Scotia, Canada B3H 4J1

STAVENGA, D. G., Rijksuniversiteit Groningen, Department of Biophysics, Westersingel 34, 9718 CM Groningen, The Netherlands

STRAUSFELD, N. J., Arizona Research Laboratories, Division of Neurobiology, 611 Gould-Simpson Science Building, The University of Arizona, Tucson, AZ 85721, USA

VOGT, K., Albert-Ludwigs-Universität, Institut für Biologie I, Albertstr. 21a, 7800 Freiburg, FRG

Chapter 1

Compound Eyes and the World of Vision Research

TIMOTHY H. GOLDSMITH, New Haven, Connecticut, USA

The invitation to provide this introductory chapter suggested that I present my views on the past, present and future of compound eye research, including short side steps into related work on vertebrate eyes. The result is a contribution with two themes. The opening section provides an overview of the research on compound eyes that has been accomplished since the time of Sigmund Exner's landmark book on arthropod optics and hints at the more detailed reviews that follow. Although it provides little more than a historical outline, for those new to the field it may help to anchor seminal events in time and to show the relationship to other currents in vision research. The subsequent sections illustrate on a limited front how work on arthropods relates both to vision in other animals as well as to much broader biological issues, while suggesting some questions for the future.

1 Compound Eyes Since Exner

Figure 1 is an effort to capture a sense of how investigations of arthropod vision have developed during the last 100 years. This is an idiosyncratic view of events; any two people can take the same journey and see different things, and in this instance others would doubtless have noted other landmarks. I have arbitrarily divided the flow of work in this table into five parallel streams, but in reality they intersect repeatedly. And although I have omitted names of people in order to keep the visual clutter under control, appropriate references are present in the text.

For want of a better name, the first column is called Physiological Optics. Exner's 1891 pioneering work on the dioptrics of compound eyes, which was the catalyst for this meeting, was accepted for 70 years, during which time nothing of significance happened, save for Horace Barlow (1952) calling attention to the importance of diffraction in a context that encouraged closer comparison of simple and compound eyes. (Actually, Mallock [1894] had pointed to the role of diffraction at the time of Exner, but his contribution was overlooked until much later.) Then, starting in the mid 1960's, there was a rush of important discoveries and theoretical insights: neural superposition eyes of Diptera (Kirschfeld 1967), a revival of the credibility of superposition optics after a brief but well-founded period of skepticism (e.g., Kunze 1969; see Kunze 1979 for review), the extension of waveguide theory to fused rhabdoms (e.g., Snyder et al. 1973; but see Snyder 1979), and the discovery of mirror optics in crustacean ommatidia (Vogt 1975; Land 1976).

Stavenga/Hardie (Eds.) Facets of Vision
© Springer-Verlag Berlin Heidelberg 1989

A CENTURY OF ARTHROPOD VISION

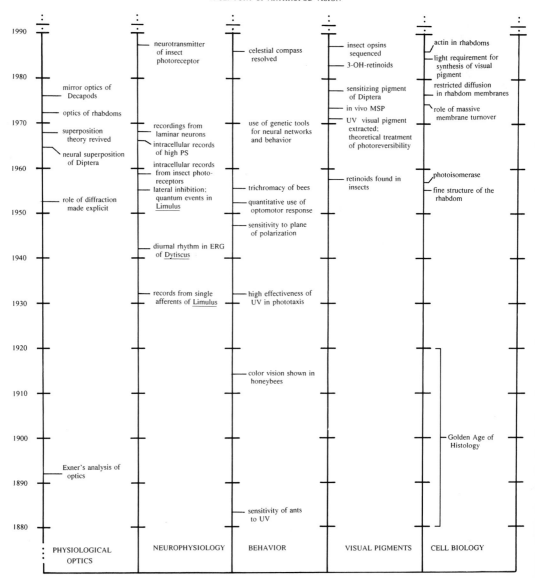

Fig. 1. A synopsis of some major events in the study of arthropod vision since the time of Exner

Major advances in neurophysiology have roughly paralleled similar work on vertebrates; for technical reasons leading the way with recordings from photo-receptors but lagging with intracellular recordings from visual interneurons. This work started with the supposition by Hartline and Graham (1932) that they were able to record from the axons of single photoreceptor cells. Although the eccentric cells subsequently turned out to be second-order neurons electrotonically coupled to retinular cells, *Limulus* continued to play an important role with the discovery of lateral inhibition (Hartline and Ratliff 1957) and quantum bumps (Yeandle, 1958; see Scholes, 1964 for insects). Moreover, the ventral photoreceptor of *Limulus* (Millecchia and Mauro 1969a,b) has become a particularly important preparation for the study of phototransduction (see Fein and Payne this Vol.).

Jahn and Wulff (1943) made an early effort to study diurnal changes in sensitivity by electrophysiological means. Intracellular recordings from insect photoreceptors were accomplished by Kuwabara and Naka (1959), Burkhardt and Autrum (1960) and Naka (1961), and were quickly used to measure spectral sensitivity by Burkhardt (1962). Other events such as the discovery of high polarization sensitivity of photoreceptors (Shaw 1969), records from laminar neurons (Autrum et al. 1970), and identification of neurotransmitters (Hardie 1987) have followed.

In the column labeled Behavior the chronology begins with Exner's contemporary Sir John Lubbock (1882) and his discovery that the visible spectrum of ants extends into the ultraviolet. Sensitivity to UV light has been a matter of recurrent interest and surprise during the ensuing century, and has had a pervasive influence on the field. Moreover, Lubbock (Lord Avebury) was an important figure in his own time. Charles Darwin once wrote him ". . . I settled some time ago that I should think more of Huxley's and your opinion, from the course of your studies and the clearness of your minds, than of any other men in England."

Behavior has had a noble history, providing a synergistic interplay with both physiology and anatomy. And along this stream we find a special role played by Karl von Frisch and his intellectual descendants: in particular, the demonstration of color vision (von Frisch 1914) and its trichromacy (Daumer 1956), and the use of polarized light (von Frisch 1949) and the resolution of the underlying sensory mechanism (Rossel and Wehner 1984). Other benchmarks were the demonstration of the high efficiency of UV in phototaxis (Bertholf 1931), a forerunner of the concept of wavelength-specific behaviors; the quantification of optomotor behavior (Hassenstein 1951); and the introduction of genetic techniques (Heisenberg 1979).

Unlike the neurophysiology of arthropod photoreceptors, the biochemistry and photochemistry of arthropod visual pigments has tended to follow work on vertebrates, primarily because the arthropods do not have as much pigment in their retinas and therefore present more difficulties to the biochemist. Retinal was not found in insects until the late 1950's (Goldsmith 1958), and the stability of invertebrate metarhodopsins necessitated some different approaches to the way the photochemistry was studied (Hamdorf et al. 1973; Stavenga 1975; Hamdorf 1979; Bernard 1982, 1983a). But the isolation of the UV-sensitive visual pigment by the Bochum group (Hamdorf et al. 1971) and the discoveries in Tübingen of a sensitizing pigment in Diptera (Kirschfeld et al. 1977) and of a new chromophore (Vogt 1983) are all novel contributions.

After a spate of meticulous comparative histological work (references in Goldsmith 1964), events in cell biology awaited the introduction of electron microscopy. The fine structure of the rhabdom was first described at this time and attention drawn to its likely role in the detection of polarized light (Miller 1957; Goldsmith and Philpott 1957; Danneel and Zeutschel 1957; Wolken et al. 1957). In addition to a large body of elegant neuroanatomy that was also made possible (and which is not explicitly recognized in this column, but see Strausfeld and Nässel 1981 as well as Strausfeld this Vol.), principal discoveries include the relative lack of fluidity of rhabdomeric membranes (Goldsmith and Wehner 1977), the widespread nature of membrane turnover and its relation to the regeneration of visual pigment (White and Lord 1975; Stein et al. 1979), the early discovery of a soluble retinal photoisomerase (Goldsmith 1958) — correctly interpreted only later (Schwemer et al. 1984), the light requirement (in some species) for pigment renewal (Schwemer 1983), bleaching in the rhabdoms of butterflies (Bernard 1983a,b), and the discovery of actin in rhabdomeric microvilli (e.g., de Couet et al. 1984).

1.1 Whither Now?

However one might choose the examples, it is clear that the eyes of arthropods have posed a number of fundamental biological problems that have been attacked with resourcefulness and skill. The problem of transduction seems on the verge of solution, there is active work on the cell biology of photoreceptors, and the opportunities made possible by molecular biology lie largely in front of us.

But what additional lessons can we draw from this survey? One of the trends in biology, at least in America, is that as the funding process focuses resources in an ever tighter way on the study of cellular and subcellular phenomena, it imposes a value system on the entire educational process. Simply put, organismal and evolutionary biology and the study of diversity for its own sake ('systematics' in George Gaylord Simpson's broad sense of the word) have become unfashionable in a number of research universities, and in consequence there are important intellectual issues that suffer neglect.

But what has this to do with Exner and eyes? The record shows that this seemingly specialized area of vision reaches into the farthest corners of biology and illustrates how broadly based efforts to understand nature produce unexpected results. One could illustrate this thesis narrowly by pointing to a number of functional discoveries in Fig. 1, but my wider point is better made by developing a couple of evolutionary issues that are currently not at the forefront of thinking in the vision research community. The second part of the argument is developed at greater length elsewhere (Goldsmith 1989).

1.2 Optical Design and Adaptation by Natural Selection

One of the current major research approaches in behavioral ecology is an effort to apply optimization models. My impression is that in behavior and ecology the hypotheses are difficult to formulate in ways that cover all the variables over which adaptation integrates, and when successes seem to be achieved, the boundary conditions are frequently so restricted that it is difficult to generalize the conclusions. Yet the effort continues, with the theoretically reasonable hope that adaptation is working towards an optimization of something, at least in a restricted time frame.

One of the great successes in arthropod vision in the last 15 years has been a much clearer understanding of the physical parameters driving the optical design of eyes. From the work of a number of people — Bernard, Horridge, Kirschfeld, Kunze, Land, Laughlin, Miller, Nilsson, Snyder, Stavenga, Vogt — has come not only an understanding of the evolutionary variety of optical systems and how they work, but a better general insight into the interplay of contrast sensitivity and spatial resolution and how they jointly influence the evolution of eyes. This has, in short, been a very successful exercise in optimization thinking, and it is appropriate to ask why it has done so well.

The answer is that it was possible to identify most of the variables that have influenced natural selection in the evolution of image-forming eyes. Furthermore, because these factors involved some straightforward physics, the problem lent itself readily to quantitative analysis. The conclusions, moreover, are not trivial. Perhaps evolution does tend to optimize, if we are just keen enough to see what the compromises are that need to be made, what represents 'good enough,' and what restrictions may be imposed at particular times and on particular phyletic lines. Moreover, it is clear that if multiple solutions are possible, they are likely to occur, even if some are manifestly better than others. The pinhole eye of the *Nautilus*, Mike Land (1981) has reminded us, is a poor device without a lens, but there it is. In short, this body of work represents a paradigmatic example that should be a stimulus to those whose primary interest is the consideration of optimization criteria in evolutionary theory and behavioral ecology, as well as a reminder to those who are inclined toward vertebrate chauvinism, that important general principles are where you find them.

2 The Evolution of Visual Proteins

The techniques of molecular biology have opened new vistas. In the past 2 years the primary structures of ten visual pigments have been determined. From the amphipathic character of the side chains, it has been inferred that all fold in seven membrane-spanning helices (for references and review see Applebury and Hargrave 1986). When the proteins are aligned for optimal matching one can see regions of high similarity, particularly the cytoplasmic loop between helix 1 and helix 2, as well as helix 7, where the chromophore binding site is located (Fig. 2).

The results of pair-wise comparisons between the ten pigments can be expressed as a matrix of percent identity of amino acid composition (Fig. 3); by this measure the three vertebrate rhodopsins (rod pigments) are very similar, as are the red-sensitive and green-sensitive cone pigments of the human eye. The blue-sensitive cone pigment is as different from the other cone pigment(s) as each is from rhodopsin. And the pigments of *Drosophila* show more remote affinity.

The data on which this table is based can also be expressed as a difference matrix (not shown), from which it is possible to calculate the minimally linked tree of (presumably) shortest total distance, the so-called Wagner network (Fig. 4, lower part). The contemporary pigments are the leaves, the internal nodes are hypo-

loop 1-2   helix 7

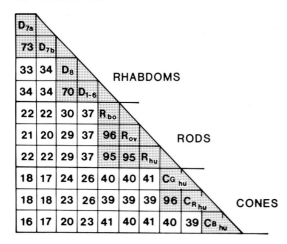

```
f a t t k s l r t p a n    l n - t i w g a c f a k s a a c y n p i v y g i s h    D1
f s a a k s l r t p s n    g a - t m i p a c a c k m v a c i d p f v y a i s h    D7a
f s t s k s l r t p s n    g a - t m i p a c t c k l v a c i d p f v y a i s h    D7b
f g g t k s l r t p a n    l t - t i w g a t f a k t s a v y n p i v y g i s h    D8

t v q h k k l r t p l n    i f m t i - p a f f a k t s a v y n p v i y - i m m    Rbo
t v q h k k l r t p l n    i f m t i - p a f f a k s s s v y n p v i y - i m m    Rov
t v q h k k l r t p l n    i f m t i - p a f f a k s a a i y n p v i y - i m m    Rhu

t m k f k k l r h p l n    l m a a l - p a y f a k s a t i y n p v i y - v f m    Crh
t m k f k k l r h p l n    l m a a l - p a f f a k s a t i y n p v i y - v f m    Cgh
t l r y k k l r q p l n    r l v t i - p s f f s k s a c i y n p i i y - c b m    Cbh
```

☐ complete homology   ⌐¬ extensive homology   ⋯ conservative substitutions

Fig. 2. Two regions of high similarity in ten opsins that have been sequenced — the cytoplasmic loop between helices 1 and 2; and helix 7, in which the chromophore binds at the lys residue (k) that is present in all the opsins. The amino acids are indicated by the single-letter code. Amino acid identities and functionally conservative substitutions are shown by *boxed areas.* D_{1-6}, D_{7a}, D_{7b}, and D_8 are the pigments of retinular cells 1-6, cell 7, and cell 8 of *Drosophila.* Cell 7 is polymorphic with regard to pigment composition and one of the pigments has λ_{max} in the near UV. R_{bo}, R_{ov}, R_{hu} are mammalian rod pigments from cattle, sheep, and human retinas; C_{Bh}, C_{Gh}, and C_{Rh} are the human cone pigments. Original references are O'Tousa et al., 1985 (D_{1-6}); Zuker et al., 1987 (D_{7a}), Montell et al., 1987 (D_{7b}); Cowman et al., 1986 (D_8); Nathans and Hogness, 1983 (R_{bo}); Findlay, 1986 (R_{ov}); Nathans and Hogness, 1984 (R_{hu}); and Nathans et al. 1986 (human cone pigments). (Note added in proof: recent evidence (Feiler et al. 1988, Pollock and Benzer 1988) indicates that the so-called R8 opsin in fact represents the opsin expressed in the ocelli)

PERCENT IDENTITY OF VISUAL PIGMENTS

D7a									
73	D7b								
33	34	D8							
34	34	70	D1-6						
22	22	30	37	Rbo					
21	20	29	37	96	Rov				
22	22	29	37	95	95	Rhu			
18	17	24	26	40	40	41	CGhu		
18	18	23	26	39	39	39	96	CRhu	
16	17	20	23	41	40	41	40	39	CBhu

RHABDOMS

RODS

CONES

Fig. 3. Matrix of identities of amino acids at corresponding positions for ten visual pigments. The closest relationships are shown by the *shaded triangular areas.* Symbols and sources as described for Fig. 2

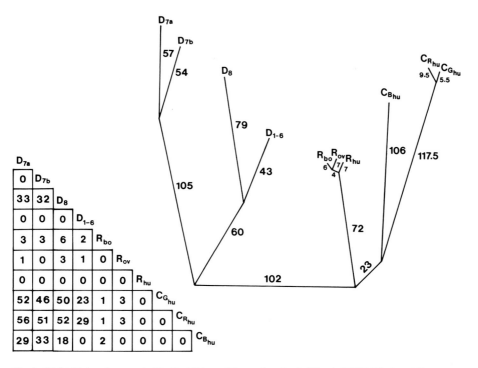

Fig. 4. Right: Network computed by the 'distance Wagner' method of Farris (1972). The input distance matrix was derived from all possible pairwise comparisons of the ten opsins referred to in Figs. 2–3. (The original difference matrix is not shown, but it is an alternative way of tabulating the data in Fig. 3.) The tree is unrooted and the branches are drawn to scale. It is not guaranteed to be of minimal length nor to replace a full maximum parsimony analysis of sequences. A total of 533 extra steps was needed to fit the data to the tree, and the total length of the tree is 857.

Left: Apparent evolutionary convergence. This matrix was compiled by subtracting the observed number of amino acid differences between each pair of proteins (the input distance in the computation of the Wagner network) from the corresponding calculated internode lengths in the tree. The numbers in this matrix indicate the extent to which observed differences are fewer than expected from a simple model of divergent evolution.

thetical pigments which would exist in a parsimonious series of stepwise transitions between leaves, and the lengths of the internodes are numbers of amino acid changes required to make the transitions. In most cases, the sums of the internode lengths between pairs of pigments are not significantly different from the observed numbers of differences from which the network was calculated, but in some cases the internodes are longer. The excess lengths, which are tabulated in the matrix in the upper half of Fig. 4, are a measure of the degree to which the proteins seem to exhibit convergent evolution. The pattern is surprising. For example, most of the apparent convergence involves several of the *Drosophila* genes and the human R/G pigment genes. These apparently convergent substitutions are not, however, clustered in any particular part of the molecule, so we gain no insight from that quarter in understanding their significance.

The Evolution of Absorption Maxima

The tools of molecular biology not only offer many opportunities for exploring the relation between structure and function in the molecules of visual pigment, but they reopen some fascinating evolutionary questions as well. For example, why do visual pigments absorb where they do? The λ_{max} of vertebrate rhodopsins (for which there is a large sample) is tightly clustered around 500 nm; very few pigments have λ_{max} longer than 512 nm, but animals inhabiting deep water characteristically have hypsochromically shifted rhodopsins (Fig. 5). It was originally suggested that a λ_{max} at 500 nm provides a match to the solar spectrum, but as Dartnall (1975) pointed out, when the solar energy distribution is expressed in quanta per wavelength interval, this hypothesis is not supported by the numbers. The quantum flux per wavelength interval is greatest at much longer wavelengths. The blue shift of the rhodopsins of deep sea fish and deep-diving mammals is accepted as an adaptation (Lythgoe 1984), but if one computes the gain in absolute sensitivity achieved by moving the λ_{max} to 480 from 500 nm, it seems inconsequential (Fig. 5). Perhaps this small advantage was all that was required, but if that is so, why is the λ_{max} not found at longer wavelengths in nocturnal vertebrates, as the spectrum of moonlight, like that of the sun, suggests it ought to be?

The adaptational explanations that have been proposed for this phenomenon seem tenuous (Goldsmith 1989). The most interesting is due to Barlow (1957), who hypothesized that if the position of the λ_{max} can be taken as an indication of the energy barrier that must be overcome to isomerize, long-wavelength pigments should be subject to enormously greater rates of thermal isomer-

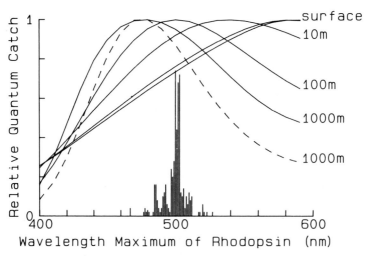

Fig. 5. Histogram: distribution of λ_{max} of 274 vertebrate rhodopsins, as tabulated by Lythgoe (1972). Full scale height corresponds to 50 pigments. *Curves* relative sensitivity of rhodopsin as a function of λ_{max}, calculated for sunlight and at various depths of water. The *solid curves* assume an axial absorbance of rhodopsin of 1 at the λ_{max}; the *broken curve*. 0.1. Further details in Goldsmith (1989)

ization and consequently their receptors should be plagued with much greater dark noise. On his assumption, the effect of λ_{max} on relative rate of dark isomerization is huge.

There are several reasons for doubting this adaptational explanation of the spectral position of most rhodopsins. Baylor et al. (1980) have observed quantum-like fluctuations in the photocurrents of toad rods in darkness. The activation energy is about 22 kcal/mol, which is similar to the activation energy for thermal isomerization of 11-cis to all-trans retinal in solution. But this value corresponds to a wavelength in the near infrared, and therefore does not suggest an involvement of the first excited singlet states of either retinal (λ_{max} 380 nm) or rhodopsin (λ_{max} 500 nm). Second, thermal isomerizations around carbon-carbon double bonds are believed on theoretical grounds to involve elevated vibrational modes in the ground state and perhaps a low-lying triplet state whose energy lies well below the first excited singlet state (Cundall 1964). There thus appears to be little reason to expect that the λ_{max} of the visual pigment is an indication of the energy barrier for thermal isomerization of the chromophore. Third, from the entropies of activation, the rates of both processes (shots of dark current in rods and thermal isomerization of retinal in solution) seem to be governed by conformational factors and not simply an energy barrier. And finally, despite the suggestive value for the activation energy, it is conceivable that the spontaneous pulses of dark current could be caused by thermally induced interactions of rhodopsin with G-protein that occur without isomerization of the chromophore.

Perhaps we need to seek another kind of explanation. Perhaps there is no simple mutational path available to the genes for vertebrate rod pigments that both preserves function and leads to a substantial bathochromic shift in spectrum. In other words, in the genes coding for rod pigments the mutational changes that would be required to shift the λ_{max} to longer wavelengths may not be within easy evolutionary reach, and the distribution of λ_{max} in the family of rod pigments is as much a result of evolutionary inertia as it is of adaptation by natural selection. In this context it is well to remember that λ_{max} is only one of many attributes of the molecule that is sensitive to natural selection.

This idea of evolutionary constraint is speculative, but in principle it can soon be tested. If a single amino acid substitution would suffice to move the spectrum to longer wavelengths without loss of function (such as transducer action), then this hypothesis almost certainly fails. But if a sequence of events is required, with loss of function at intermediate stages, then the transition may be improbable. The fact that some cone and invertebrate pigments absorb at longer wavelengths is not particularly relevant, for those subfamilies of opsins diverged from rod pigments more than 500 million years ago. Further discussion of this conundrum can be found in Goldsmith (1989).

We will understand more about the answer to this evolutionary riddle — and for that matter, a host of other matters — when we know the roles of specific amino acids in determining the tertiary structure and the functional properties of rhodopsin. In this context it would be very useful to have available the primary structures of additional visual pigments, both vertebrate and invertebrate. A rich base of comparative data will not only help to clarify evolutionary relationships, but should provide other important insights by revealing what features of the molecule have been conserved for particular functional purposes.

This kind of biological thinking is not constrained by taxonomic boundaries and even now has implications for the human visual system. The strong overlap of the R and G cone pigments of human color vision generates a color triangle that is compressed and therefore would seem to be suboptimal (Barlow 1982). A complement of pigments spaced more evenly in the spectrum should be a better design, and that is the way the honeybee has arranged its color vision (see Menzel and Backhaus this Vol.). This feature of human color vision has perplexed those who have thought solely in terms of adaptation by natural selection, and plausible adaptational explanations have to do with optimizing spatial acuity rather than wavelength resolution (Barlow 1982).

Without denying the plausibility of this adaptational argument, one can still make a case that this pair of pigments represents the best that primates have had time to do, given the dichromatic heritage with which their ancestral stock emerged from 50–100 million years of Mesozoic nocturnality (Goldsmith 1989). The molecular biology is certainly consistent with this idea. Not only has the R/G gene pair seemingly arisen from a relatively recent gene duplication (Nathans et al. 1986), but all the gene polymorphisms at these loci in both Old and New World primates code for visual pigments with λ_{max} between about 525 and 565 nm (Bowmaker et al. 1983, 1987; Dartnall et al. 1983; MacNichol et al. 1983).

3 Neural Integration

Which brings us back to arthropods and to a final and very general point about the CNS. One of the central problems in neurobiology is understanding the relationship between the simple forms of neural processing that can be studied at early stages of the afferent pathway on the one hand, and on the other, the generation of behavior or an understanding of perception, memory, and those higher mental functions like imagination that have traditionally been considered uniquely human. We stand now in a position somewhat analogous to that occupied by geneticists in 1916, when William Bateson wrote a skeptical review of a book summarizing the evidence that genes are arranged in linear sequence on chromosomes. What was really troubling him becomes apparent in a single sentence:

> "The supposition that particles of chromatin, indistinguishable from each other and indeed almost homogeneous under any known test, can by their material nature confer all the properties of life surpasses the range of even the most convinced materialism."

Some may argue that the problem that so perplexed Bateson is different in kind; that understanding particles of chromatin was an exercise in reductionism that required only time and the maturation of the science of chemistry for the appearance of a solution. Understanding how the nervous system generates behavior — behavior in the widest sense of the word — this reasoning goes, is a horse of a different color, for behavior is an emergent property characteristic of ensembles of neurons.

Bateson's dilemma was that in the scientific framework of his day, the solution to the problem he articulated required reaches of imagination so divorced from what was then known about the natural world as to court rejection as serious science. In light of subsequent history, an optimist might argue by analogy that our prospects for understanding the biological bases of behavior are really quite promising. And although the road ahead may be hard to see, on the basis of our passage to date, those who travel it will certainly keep company with numerous arthropods and other invertebrates.

Acknowledgments. Supported by NIH grant EY-00222. I am indebted to colleagues at the University of Cambridge for hospitality during a sabbatical stay, and in particular for numerous helpfully critical conversations. I am especially grateful to Adrian Friday for providing the calculation of the Wagner network in Fig. 4.

References

Applebury ML, Hargrave PA (1986) Molecular biology of the visual pigments. Vision Res 26:1881–1895

Autrum H, Zettler F, Järvilehto M (1970) Post-synaptic potentials from a single monopolar neuron of the ganglion opticum I of the blowfly *Calliphora*. Z Vergl Physiol 70:414–424

Barlow HB (1952) The size of ommatidia in apposition eyes. J Exp Biol 29:667–674

Barlow HB (1957) Purkinje shift and retinal noise. Nature (London) 179:255–256

Barlow HB (1982) What causes trichromacy? A theoretical analysis using comb-filtered spectra. Vision Res 22:635–643

Bateson W (1916) review of The mechanism of Mendelian heredity, by TH Morgan, AH Sturtevant, HJ Muller and CB Bridges. Science 44:536–543

Baylor DA, Mathews G, Yau K-W (1980) Two components of electrical dark noise in toad retinal rod outer segments. J Physiol 309:591–621

Bernard GD (1982) Noninvasive optical techniques for probing insect photoreceptors. In: Packer L, Part H (eds) Visual pigments and purple membranes, vol 81. Methods in enzymology. Academic Press, New York London, pp 752–759

Bernard GD (1983a) Dark-processes following photoconversion of butterfly rhodopsin. Biophys Struct Mech 9:277–286

Bernard GD (1983b) Bleaching of rhabdoms in eyes of intact butterflies. Science 219:69–71

Bertholf LM (1931) The distribution of stimulative efficiency in the ultraviolet spectrum for the honey bee. J Agr Res 43:703–713

Bowmaker JD, Mollon JD, Jacobs GH (1983) Microspectrophotometric results for old and new world primates. In: Mollon JD, Sharpe LT (eds) Color vision. Physiology and psychophysics. Academic Press, New York London, pp 57–68

Bowmaker JK, Jacobs GH, Mollon JD (1987) Polymorphism of photopigment in the squirrel monkey: a sixth phenotype. Proc R Soc Lond Ser B 231:383–390

Burkhardt D (1962) Spectral sensitivity and other response characteristics of single visual cells in the arthropod eye. Symp Soc Exp Biol 16:86–109

Burkhardt D, Autrum H (1960) Die Belichtungspotentiale einzelner Sehzellen von *Calliphora erythrocephala* Meig. Z Naturforsch 15b:612–616

de Couet HG, Stowe S, Blest AD (1984) Membrane-associated actin in the rhabdomeral microvilli of crayfish photoreceptors. J Cell Biol 98:834–846

Cowman AF, Zucker CS, Rubin GM (1986) An opsin gene expressed in only one photoreceptor cell type of the *Drosophila* eye. Cell 44:705–710

Cundall RB (1964) The kinetics of cis-trans isomerizations. Prog Reaction Kinet 2:165-215

Danneel R, Zeutschel B (1957) Uber den Feinbau der Retinula bei *Drosophila melanogaster.* Z Naturforsch 12b:580-583

Dartnall HJA (1975) Assessing the fitness of visual pigments for their photic environment. In: Ali MA (ed) Vision in fishes. Plenum Press, New York, pp 543-563

Dartnall HJA, Bowmaker JK, Mollon JD (1983) Microspectrophotometry of human photoreceptors. In: Mollon JD, Sharpe LT (eds) Color vision. Physiology and psychophysics. Academic Press, New York London, pp 69-80

Daumer K (1956) Reizmetrische Untersuchungen des Farbensehens der Bienen. Z Vergl Physiol 38:413-478

Exner S (1891) Die Physiologie der facettirten Augen von Krebsen und Insecten. Franz-Deuticke, Leipzig Wien

Farris JS (1972) Estimating phylogenetic trees from distance matrices. Am Nat 106:645-668

Feiler R, Harris WA, Kirschfeld K, Wehrahn C, Zuker CS (1988) Targeted misexpression of a *Drosophila* gene leads to altered visual function. Nature (London) 333:737-741

Findlay JBC (1986) The biosynthetic, functional and evolutionary implications of the structure of rhodopsin. In: Stieve H (ed) The molecular mechanism of photoreception. Dahlem Konferenzen. Springer, Berlin Heidelberg New York Tokyo, pp 11-30

Frisch K von (1914) Der Farbensinn und Formensinn der Biene. Zool J Physiol 37:1-238

Frisch K von (1949) Die Polarisation des Himmelslichtes als orientierender Faktor bei den Tänzen der Bienen. Experientia 5:142-148

Goldsmith TH (1958) The visual system of the honeybee. Proc Natl Acad Sci 44:123-126

Goldsmith TH (1964) The visual system of insects. In: Rockstein M (ed) The physiology of insecta, vol 1, chap 10, 1st edn. Academic Press, New York London, pp 397-462

Goldsmith TH (1989) The evolution of visual pigments and colour vision. In: Gouras P (ed) The perception of color, vol 7. Vision and visual dysfunction. MacMillan (in press)

Goldsmith TH, Bernard GD (1985) Visual pigments of invertebrates. Photochem Photobiol 42:805-809

Goldsmith TH, Philpott DE (1957) The microstructure of the compound eyes of insects. J Biophys Biochem Cytol 3:429-440

Goldsmith TH, Wehner R (1977) Restriction of rotational and translational diffusion of pigment in the membranes of a rhabdomeric photoreceptor. J Gen Physiol 70:453-490

Hamdorf K (1979) The physiology of invertebrate visual pigments. In: Autrum H (ed) Handbook of sensory physiology, vol VII/6A. Springer, Berlin Heidelberg New York, pp 145-224

Hamdorf K, Schwemer J, Gogala M (1971) Insect visual pigment sensitive to ultraviolet light. Nature (London) 231:458-459

Hamdorf K, Paulsen R, Schwemer J (1973) Photoregeneration and sensitivity control of photoreceptors of invertebrates. In: Langer H (ed) Biochemistry and physiology of visual pigments. Springer, Berlin Heidelberg New York, pp 155-174

Hardie RC (1987) Is histamine a neurotransmitter in insect photoreceptors? J Comp Physiol A 161:201-213

Hartline HK, Graham CH (1932) Nerve impulses from single receptors in the eye. J Cell Comp Physiol 1:277-295

Hartline HK, Ratliff F (1957) Inhibitory interactions of receptor units in the eyes of *Limulus.* J Gen Physiol 40:357-376

Hassenstein B (1951) Ommatidienraster und afferente Bewegungsintegration (Versuche an dem Russelkäfer Chlorophanus miridis). Z Vergl Physiol 33:301-326

Heisenberg M (1979) Genetic approach to a visual system. In: Autrum H (ed) Handbook of sensory physiology, vol VII/6A. Springer, Berlin Heidelberg New York, pp 665-679

Jahn TL, Wulff VJ (1943) Electrical aspects of a diurnal rhythm in the eye of *Dytiscus fasciventris.* Physiol Zool 16:101-109

Kirschfeld K (1967) Die Projektion der optischen Umwelt auf das Raster der Rhabdomere im Komplexauge von *Musca.* Exp Brain Res 3:248-270

Kirschfeld K, Franceschini N, Minke B (1977) Evidence for a sensitizing pigment in fly photoreceptors. Nature (London) 269:386-390

Kuwabara M, Naka K-I (1959) Response of a single retinula cell to polarized light. Nature (London) 184:455-456

Kunze P (1969) Eye glow in the moth and superposition theory. Nature (London) 223:1172-1174

Kunze P (1979) Apposition and superposition eyes. In: Autrum H (ed) Handbook of sensory physiology, vol VII/6A. Springer, Berlin Heidelberg New York, pp 441–502

Land M (1976) Superposition images are formed by reflection in the eyes of some oceanic decapod crustacea. Nature (London) 263:764–765

Land M (1981) Optics and vision in invertebrates. In: Autrum H (ed) Handbook of sensory physiology, vol VII/6B. Springer, Berlin Heidelberg New York, pp 472–592

Lubbock J (1882) Ants, bees, and wasps. A record of observations on the habits of the social Hymenoptera, 2nd edn. Kegan Paul, Trench

Lythgoe JN (1972) List of vertebrate visual pigments. In: Dartnall HJA (ed) Handbook of sensory physiology, vol VII/1. Springer, Berlin Heidelberg New York, pp 604–624

Lythgoe JN (1984) Visual pigments and environmental light. Vision Res 24:1539–1550

MacNichol EF, Jr., Levine JS, Mansfield RJW, Lipetz LE, Collins BA (1983) Microspectrophotometry of visual pigments in primate photoreceptors. In: Mollon JD, Sharpe LT (eds) Color vision. Physiology and psychophysics. Academic Press, New York London, pp 13–38

Mallock A (1894) Insect sight and the defining power of composite eyes. Proc R Soc London Ser B 55:85–90

Millecchia R, Mauro A (1969a) The ventral photoreceptor cells of *Limulus*. II. The basic photoresponse. J Gen Physiol 54:310–330

Millecchia R, Mauro A (1969b) The ventral photoreceptor cells of *Limulus*. III. A voltage-clamp study. J Gen Physiol 54:331–351

Miller WH (1957) Morphology of the ommatidia of the compound eye of *Limulus*. J Biophys Biochem Cytol 3:421–428

Montell CK, Jones C, Zuker C, Rubin G (1987) A second opsin gene expressed in the ultraviolet sensitive R7 photoreceptor cells of *Drosophila melanogaster*. J Neurosci 7:1558–1566

Naka K-I (1961) Recording of retinal action potentials from single cells of the insect compound eye. J Gen Physiol 44:571–584

Nathans J, Hogness DS (1983) Isolation, sequence analysis and intron-exon arrangement of the gene encoding bovine rhodopsin. Cell 34:807–814

Nathans J, Hogness DS (1984) Isolation and nucleotide sequence of the gene encoding human rhodopsin. Proc Natl Acad Sci USA 81:4851–4855

Nathans J, Thomas D, Hogness DS (1986) Molecular genetics of human color vision: the genes encoding blue, green, and red pigments. Science 232:193–202

O'Tousa JE, Baehr W, Martin RL, Hirsh I, Pak WL, Applebury ML (1985) The *Drosophila nina E* gene encodes an opsin. Cell 40:839–850

Pollock JA, Benzer S (1988) Transcript localization of four opsin genes in the three visual organs in *Drosophila*: RH2 is ocellus specific. Nature (London, 333:779–782)

Rossel S, Wehner R (1984) How bees analyze the polarization patterns in the sky. J Comp Physiol A 154:607–615

Scholes JH (1964) Discrete subthreshold potentials from the dimly lit insect eye. Nature (London) 202:572–573

Schwemer J (1983) Pathways of visual pigment regeneration of fly photoreceptor cells. Biophys Struct Mech 9:287–298

Schwemer J, Pepe IM, Paulsen R, Cugnoli C (1984) Light-induced *trans-cis* isomerization of retinal by a protein from honeybee retina. J Comp Physiol A 154:549–554

Shaw SR (1969) Sense-cell structure and interspecies comparisons of polarized light absorption in arthropod compound eyes. Vision Res 9:1031–1041

Snyder AW (1979) Physics of vision in compound eyes. In: Autrum H (ed) Handbook of sensory physiology, vol VII/6A. Springer, Berlin Heidelberg New York, pp 441–502

Snyder AW, Menzel R, Laughlin SB (1973) Structure and function of the fused rhabdom. J Comp Physiol 87:99–135

Stavenga DG (1975) Derivation of photochrome absorption spectra from absorbance difference measurements. Photochem Photobiol 21:105–110

Stein PJ, Brammer JD, Ostroy SE (1979) Renewal of opsin in the photoreceptor cells of the mosquito. J Gen Physiol 74:565–582

Strausfeld NJ, Nässel DR (1981) Neuroarchitecture of brain regions that subserve the compound eyes of crustacea and insects. In: Autrum H (ed) Handbook of sensory physiology, vol VII/6B. Springer, Berlin Heidelberg New York, pp 1–132

Vogt K (1975) Zur Optik des Flusskrebsauges. Z Naturforsch 30c:691

Vogt K (1983) Is the fly visual pigment a rhodopsin? Z Naturforsch 38c:329–333

White RH, Lord E (1975) Diminution and enlargement of the mosquito rhabdom in light and darkness. J Gen Physiol 65:583–598

Wolken JJ, Capenos J, Turano A (1957) Photoreceptor structures. III Drosophila melanogaster. J Biophys Biochem Cytol 3:441–448

Yeandle S (1958) Electrophysiology of the visual system. Discussion. Am J Ophthalmol 46:82–87

Zuker CS, Montell C, Jones K, Laverty T, Rubin GM (1987) A rhodopsin gene expressed in photoreceptor cell R7 of the *Drosophila eye:* Homologies with other signal-transducing molecules. J Neurosci 7:1550–1556

Chapter 2

Autrum's Impact on Compound Eye Research in Insects

DIETRICH BURKHARDT, Regensburg, FRG

1 The Origins of Compound Eye Research

To appreciate the contributions of a scientist in any particular field of research one
has to follow the course of history: As compound eyes are rather small objects, the
history of compound eye research is closely linked with the development of the
microscope. Singer (1953) tells how Galileo, after constructing his telescope in
1609, already used it a year later in a reversed manner to observe small objects like
fleas and eyes of insects at close range. A Frenchman, not named in that source,
visited Galileo in Florence in 1614 and complained afterwards, Galileo was still
using his old-fashioned device despite Kepler having already developed an improved construction in 1610.

Next Hooke's Micrographia (1665; see Fig. 1), deserves mention. Reading
carefully his Chapter XXXIX: "Of the Eyes and Head of a Grey Drone-Fly, and
of several other Creatures", one becomes aware how many of the ideas discussed
later are already stated in this book with great clarity. From this period, work of
Swammerdam and Leeuwenhoek should also be recalled (quoted in Bernhard
1966).

Finally in the first half of the 19th century Johannes Müller presented his
mosaic theory in the famous book: "Zur vergleichenden Physiologie des Gesichtssinnes des Menschen und der Tiere" (1826). Johannes Müller had many
students, but it seems that only one of them, Max Schultze, worked on the
compound eye (Schultze 1868). Other students of Johannes Müller, like Henle,
Schwann, du Bois-Reymond, Brücke, Helmholtz, Virchow and Haeckel became
still more famous and in turn had famous students (Fig. 2). Obviously, even though
they themselves did no major work on the compound eye, the interest in the subject was kept and resurrected in subsequent scientific generations. Among the
"grandsons" of Johannes Müller I name only three: W. Kühne, famous for his work
on visual purple; to his pupils belong von Uexküll and Holmgren, who detected the
ERG in 1865. Another grandson was Koelliker, who became an anatomist at
Würzburg. His students included Leydig, and Grenacher, both of whom made
major contributions to the field (Leydig 1864; Grenacher 1879). Finally, among the
grandsons of Johannes Müller we find Sigmund Exner, whose work on the
arthropod eye: "Die Physiologie der facettirten Augen von Krebsen und Insecten",
published in 1891, is the basis of modern compound eye research. The English
translation of his work will soon be published, and one of the purposes of these
proceedings is to survey the progress made during the century since the first edition
appeared.

Stavenga/Hardie (Eds.) Facets of Vision
© Springer-Verlag Berlin Heidelberg 1989

MICROGRAPHIA:

OR SOME

Physiological Descriptions

OF

MINUTE BODIES

MADE BY

MAGNIFYING GLASSES.

WITH

OBSERVATIONS and INQUIRIES thereupon.

By R. *HOOKE*, Fellow of the ROYAL SOCIETY.

Non possis oculo quantum contendere Linceus,
Non tamen idcirco contemnas Lippus inungi. Horat. Ep. Lib. 1.

LONDON, Printed for *John Martyn*, Printer to the ROYAL SOCIETY, and are to be sold at his Shop at the *Bell* a little without *Temple Barr*. M DC LXVII.

Fig. 1. Front page of Hooke's Micrographia (1665) (*left*), and scheme 24 (*right*), depicting the head of a "grey Drone-Fly"

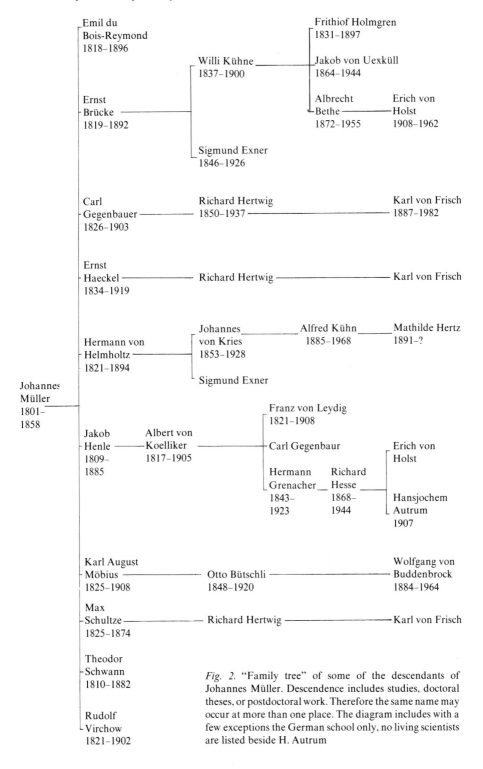

Fig. 2. "Family tree" of some of the descendants of Johannes Müller. Descendence includes studies, doctoral theses, or postdoctoral work. Therefore the same name may occur at more than one place. The diagram includes with a few exceptions the German school only, no living scientists are listed beside H. Autrum

2 From the Late 19th to the Early 20th Century

Whilst microscopy was almost the only approach towards an understanding of the compound eye up to the end of the 19th century, it was Exner's outstanding contribution that he applied physical optics as another major approach. At the end of that century and in the beginning of the 20th century, a first summit was reached in light microscopy. At the end of the last century, microscopy had already advanced to the physical limits of resolution and new staining techniques were made available. In this context two most important contributions, from Grenacher's student, R. Hesse, should be noted (Hesse 1901, 1908). In these papers he dealt with the "Stäbchensäume", the rhabdomeric structures of invertebrate visual cells. Only much later were the details of the fine structure of these rhabdomeres revealed by electron microscopy (Danneel and Zeutzschel 1957; Goldsmith and Philpott 1957). Between 1907 and 1925 detailed analysis of the retina and of the neuronal network connecting the retina with the brain was performed by Vigier, Cajal, Zawarzin and others (literature cited in Strausfeld 1976).

At the end of the 19th century and the beginning of the 20th century, the observation of animal behavior became another powerful tool in the study of the performance of the compound eye. Fabre (1879, 1882) published his first observations on insect behavior, Lubbock (1882) found that ants reacted to ultraviolet light, and Santschi (1911) worked on compass orientation. Most impressive was von Frisch's (1914) work on color vision in bees, which we regard today also as the very beginning of modern invertebrate ethology. Quantifying the animal's reaction combined with careful analysis of the stimulus situation, and profound knowledge of the underlying structures, contributed new facets to an exact understanding of the compound eye's function. Von Frisch never lost his interest in the compound eye. Beside his discovery of the polarization sensitivity in bees in 1949, one should remember the contributions made by many of his scholars, for example Baumgärtner (1928), del Portillo (1936), and Stockhammer's work (1956) with the polarization microscope, and the beautiful contributions of Daumer (1956, 1958) towards an understanding of color vision of bees.

An important step was made in 1921 when Kühn and Pohl introduced the use of pure spectral lights into the study of the compound eye. Mathilde Hertz, a pupil of Kühn, published her famous work on pattern discrimination in bees in 1933. Using the optomotor response and phototactic orientation, von Buddenbrock and his students gained insight into the angular resolution of the insect eye (von Buddenbrock and Schulz 1932/1933; Fig. 3), and tried to tackle the question of whether color vision is present in various arthropods (Schlieper 1927). A picture constructed by von Uexküll and Brock (1927) originated in these times and was widely used in textbooks to give an idea what the world might look like for an insect (Fig. 4).

While such anatomical studies and behavioral work marked the progress made on the continent, in England and the United States a quite different approach began. The development of modern electrophysiological methods provided a powerful tool for physiologists to study receptor organs. Very soon these methods, first used to study vertebrate receptors, were also applied to the compound eye. The famous work of Hartline (1928) and Hartline and Graham (1932) on the lateral

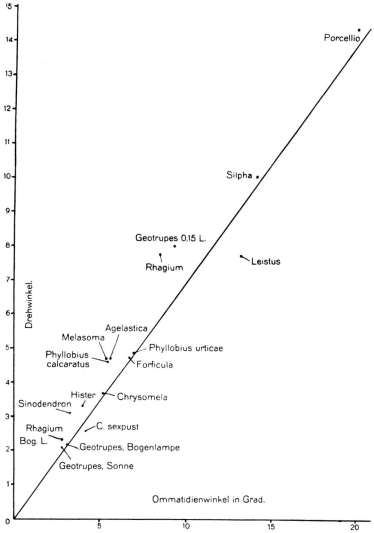

Fig. 3. Correlation between interommatidial inclination as found by histological investigation and the angular movement of a point light source necessary to elicit in 70% of phototactic runs a change of the animals' course. (von Buddenbrock and Schulz 1932/33)

eyes of *Limulus* should be mentioned, as well as the early work of Adrian (1932, 1937) on the eyes of the water beetle *Dytiscus*. Others to apply these methods successfully in the 1930's and 1940's included Crescitelli and Jahn (1939), Jahn and Crescitelli (1938, 1939), Roeder (1939, 1940), Wulff (1943) and Wulff and Jahn (1943). Finally, I should mention C. G. Bernhard, who succeeded in separating the generator response of the retina from the impulse traffic in the optic nerve in the water beetle (Bernhard 1942). Also in 1942 Bernhard formulated the concept of the generator potential together with Granit and Skoglund. In 1965 Bernhard organized the first symposium on the compound eye.

Fig. 4. View of an old village. *a* photography; *b, c* the same view depicted in increasingly more coarse screens; *d* scheme of what the view may look like to an insect. (von Uexküll and Brock 1927)

3 Autrum's Contributions

Returning to R. Hesse, one of his last students was Autrum. Autrum's interest in the compound eye begun in 1943, but due to the turbulent times, the first short papers were published in 1948; a thorough report of his early studies on the compound eye followed in 1950. Summarizing his early findings in a review article Autrum (1952) begins with the following sentences (translated from the original): "An analysis of the action potentials offers information about the physiology of an eye in two respects. First, from the range of stimuli which elicit responses it is possible to gain information about the capability of the eye. Second, analyzing the action potentials offers a tool to investigate the underlying response mechanisms. The conclusions thus drawn must then be confirmed by behavioral studies on the one hand, by biophysical and biochemical analysis at the cellular level on the other". Viewed in retrospect this sounds like the program absolved by Autrum, his school, and other groups during the decades to follow. In evaluating the impact of Autrum's work on compound eye research, I wish to emphasize first that Autrum also made major contributions in many fields of comparative physiology other than vision and secondly that I can select only parts of his work on insect vision. For the benefit of the younger colleagues, I concentrate particularly on the earlier studies of Autrum and his coworkers.

One of Autrum's first important discoveries was the extremely high flicker fusion frequency found in some insects, in particular in those which are rapid fliers (Autrum 1950). In this comparative study, it was evident that a low flicker fusion frequency was found in insects with a monophasic and negative ERG, while a high flicker fusion frequency was present in insects which had a polyphasic ERG with pronounced transients after onset and cessation of light. Finding fusion frequencies up to 300 flashes per second (Fig. 5) was unexpected and raised a lot of questions. As typical for Autrum's successful habit of double checking each result, behavioral experiments followed (Autrum and Stöcker 1950, 1952). Using the optomotor responses and responses to stroboscopic motion, the high temporal resolution of these eyes was confirmed. These first experiments were performed with flies which had their wings cut. Only a few years later G. Schneider (1953, 1955), a pupil of Autrum, had the idea that flies might react much more sensitively to

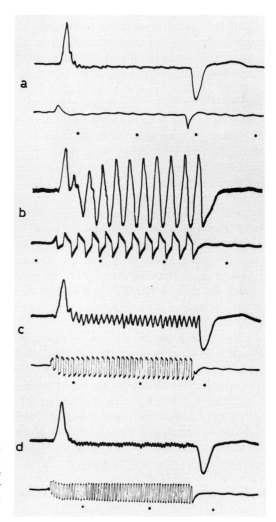

Fig. 5. ERG of a blowfly stimulated with intermittent light. *Upper traces* ERG; *lower traces* light stimulus; *dots* time marker: 0.1 s. *a* Constant light; *b* flicker frequency 47 flashes per second; *c* 133 f/s; *d* 246 f/s. (Autrum 1952)

moving patterns while flying. Hence he used tethered flying flies to study their tendency to follow the rotation of the drum. Although this method has been widely applied since, G. Schneider has hardly been credited in the literature for this important contribution.

To return to the question of high flicker fusion frequencies: this finding implicated tremendous consequences for the analysis of movement and pattern detection and pattern discrimination in insects and completely changed our view. Therefore Autrum's 1950 paper is still most frequently cited. At the cellular level, Autrum (1952) proposed that in eyes with such high fusion frequency the biochemical events following quantum absorption should differ greatly from those in the vertebrate visual cells. Interestingly, much later it was shown that in all insects and many other invertebrates the biochemical events following the rhodopsin-metarhodopsin conversion are different from those in vertebrate visual cells (see Fein and Payne this Vol.). Autrum had a second idea, which, however, was never published as far as I know. He favored a model where the visual pigment molecules would be arranged in the rhabdomere in the manner of a crystal lattice. These ideas proved to have been a prophecy: Today we know that the visual pigments compose a major fraction of the membrane proteins, that they are more rigidly implanted than those of the membrane of vertebrate visual cells and that they are strongly oriented with respect to the microvillar axes; cf. Goldsmith and Wehner (1977), Hardie (1985), and el Gammal et al. (1987). These differences are in part responsible for the polarization sensitivity of the compound eyes. At that time, in 1950, Autrum's model led to the first theme of my thesis. I was to test this proposed structural arrangement by trying to elicit piezoelectric effects in eyes treated with ultrasonic waves. But all we found was heating of the eye caused by absorption of the ultrasonic energy.

Autrum proposed the hypothesis that the transients of the flies' ERG might originate as a response of some lamina neurons, while the response of the retina itself should be monophasic and negative. Ursula Gallwitz skilfully severed the ganglia, and indeed the monophasic response was found in the remaining retina (Autrum and Gallwitz 1951, Fig. 6). A second demonstration was provided by studying the development of eyes of dragonflies. In the succession of the larval instars, the ganglia move from a remote to a close position towards the retina. Simultaneously, the ERG changes from a monophasic one to the polyphasic type with strongly expressed transients.

Recently, a paper was published by Coombe (1986) giving proof that in *Drosophila* the transients of the ERG are due to the response of the lamina L1 and L2 cells: In a mutant with degenerating L1 and L2 cells the transients vanish (Fig. 6). I do not know whether to blame the author or the chief editor of the Journal of Comparative Physiology that Autrum's work is not cited at all in that paper.

When I joined Autrum's group in 1950, von Frisch (1949, 1950) had just published a description of the incredible ability of bees to orientate with respect to the direction of the E-vector of polarized light. Autrum and Stumpf (1950) tried an electrophysiological analysis using the ERG. Stimulating single ommatidia with a point light source, they found the maximal response to polarized light higher than that to unpolarized light. This effect was more pronounced with shorter wavelengths than with longer wavelengths. Autrum and Stumpf concluded there-

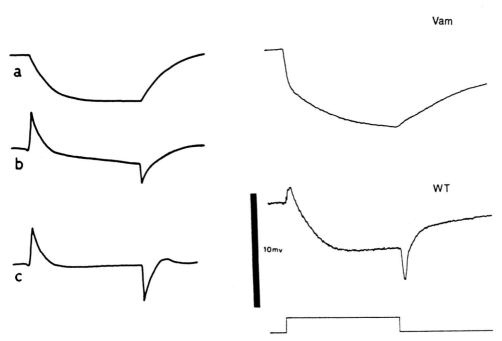

Fig. 6. *Left* Changes of recorded ERG after removal of the optic ganglia by microsurgery in *Calliphora*. *c* recorded response of the intact eye; *b* after removal of the medulla; *a* after removal of medulla and lamina (Autrum 1952). *Right* Changes of recorded ERG in *Drosophila: WT* Wild type; *Vam* mutant vacuolar medulla, investigated when the lamina neurons L1 and L2 are degenerated. (Coombe 1986)

fore that the analyzer must lie within the visual cells themselves. Again today we have final proof that the visual cells themselves are indeed polarization sensitive and that especially the ultraviolet receptors are involved in the E-vector discrimination (cf. Rossel this Vol.).

During the next years, Autrum's interest turned first towards oxygen consumption and metabolism of the insect eye (Autrum and Hoffmann 1960; Autrum and Tscharntke 1962; and Autrum and Hamdorf 1964), a line followed later by Ch. Hoffman (1960), Langer (1962), and Hamdorf and Kaschef (1964). Secondly, Autrum became interested in the spectral sensitivity of insect eyes. Together with Stumpf (1953), he had found a strong peak at about 620 nm in the spectral response curve of the flies' ERG. He considered this peak due to lack of screening action of the accessory pigments. Using the mutant white apricot of *Calliphora erythrocephala* supplied by Tate (1947), he showed this idea to be correct (Autrum 1955).

I believe that this represented the first use of mutants as a tool in physiological research on insect eyes. Obviously, it was an elegant way to study the contributions of different cells and tissues towards the reactions of the organ as a whole. About the same time, Walther (1958) in Autrum's laboratory studied the spectral sensitivity of the cockroach eye and later Walther, together with Dodt (1959), published one of the first reliable electrophysiological recordings of the spectral sensitivity of an insect eye in the near ultraviolet.

In 1958 Autrum succeeded von Frisch as a chairman and director of the Department of Zoology in Munich. About that time I had begun to work on the crayfish stretch receptor and I had made myself familiar with intracellular recording electrodes. A year later in Munich, I was tempted to try recording from the much smaller visual cells of flies, and it proved possible to record resting and action potentials sometimes even for a quarter of an hour (Burkhardt and Wendler 1960; Burkhardt and Autrum 1960); see Fig. 7. This promised to allow one run through the spectrum and a second as a control. Thus Autrum and I assembled our set-ups. My recording device and his equipment for equal quanta stimuli in the range between 300 and 700 nm were combined. The first intracellular records of the spectral sensitivities of insect visual cells resulted (Fig. 8; Autrum and Burkhardt 1960, 1961).

The atmosphere in the Munich department at that time was extremely stimulating. Von Frisch still came to the seminars and his questions were always exactly to the point.

While we did the intracellular recordings, Langer and Hamdorf were busy studying the insect eye's metabolism, the properties of the screening pigments and of the visual pigments (Langer 1962, 1967; Hamdorf et al. 1973). Furthermore, Autrum gave Ingrid Wiedemann the task of investigating in which direction the optical axes in the flies' ommatidia point. Using the eyecap preparation under the light microscope, she found that in each ommatidium the optical axes of the visual cells diverge, while on the other hand within a group of neighboring ommatidia, visual cells could be found with optical axes in parallel (Autrum and Wiedemann 1962). This question was also investigated in Tübingen with more sophisticated methods and finally led to the theory of neural superposition (Kirschfeld 1967, 1973). In my group, Washizu investigated the directional sensitivity of visual cells using intracellular techniques (Washizu et al. 1964), and Seitz (1968, 1969) examined the refractive power of the dioptric apparatus using the interference microscope. From Ljubljana, first Michieli and later Gogala came as visiting research workers to Autrum's laboratory, shortly after they discovered the UV-specialized eyes of *Ascalaphus* (Gogala and Michieli 1965).

Undoubtedly, the most important step in these years in Munich was the work of Autrum and von Zwehl (1962, 1963, 1964), investigating the spectral sensitivities of the bee's visual cells. In the very same department, where nearly 40 years earlier

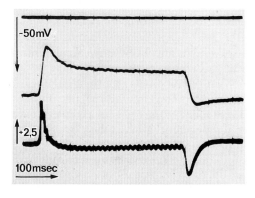

Fig. 7. One of the first records of intracellular action potentials from a blowfly visual cell. *Top trace* reference beam; *middle trace* intracellular recorded resting and action potential; *lower trace* simultaneous record of the ERG

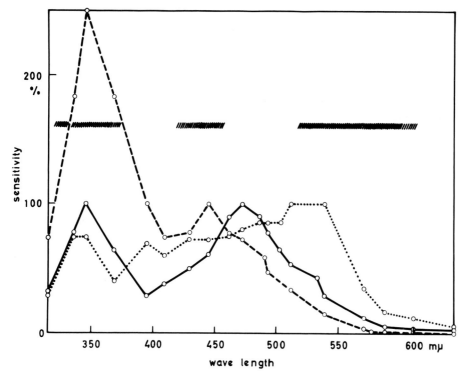

Fig. 8. Spectral sensitivity curves of visual cells of *Calliphora erythrocephala. (Burkhardt 1962)*

von Frisch had given his famous proof of color vision in the honeybee, and later his student Daumer had succeeded in finding that color vision in bees is trichromatic and color metrics follow about the same rules as in man, now the single cell technique gave clear evidence of the spectral response curves of three differing receptor cell types (see Menzel and Backhaus this Vol.). These data fitted all the ealier predictions (Fig. 9). The existence of three receptors mediating color vision had been postulated in the 19th century by Young and Helmholtz. Now proof was presented. In this instance insect vision research was ahead of vertebrate eye research, where the single cell approach was achieved only later (Marks 1965; Tomita et al. 1967).

Many years of fruitful research on the insect eye were to follow under Autrum's guidance in the department at Munich. Certainly, our knowledge has increased tremendously in the meantime and is much more detailed. Some of the early hypotheses needed correction, others had to be rejected, but all of them were useful in stimulating research. Also some of the older data had to be revised when more refined techniques became available. But one should bear in mind that exploring an unknown region means that those who try to find the way will not necessarily find the shortest one, and sometimes do not even reach their goal. But they are the pioneers who direct the next generation. Much of Autrum's work was pioneer work for today's insect eye reseach. We are also indebted to him for the enormous task

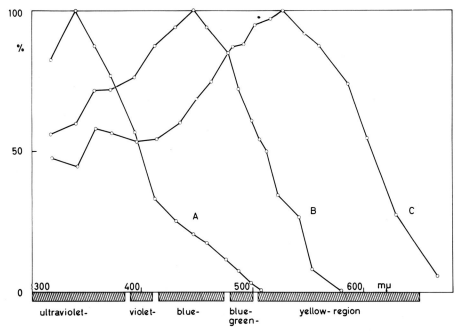

Fig. 9. Spectral sensitivity curves of single visual cells in the bee. The curves shown here are from a drone (Autrum and von Zwehl 1963), those found in the workers are much the same (Autrum and von Zwehl 1964). Along the abscissa the color regions of the bees' chromaticity diagram are indicated. Note that at those parts of the spectrum where the receptor curves intersect with their flanks, the hue changes much more rapidly with wavelength than in other parts of the spectrum

he undertook during the 1970's in editing the Handbook of Sensory Physiology (1971-1981). In many of these volumes our current knowledge of insect vision was accumulated and presented in a digestible manner for future research workers. I trust all of us are aware of Autrum's contributions towards this knowledge, of his impact on today's research as well as on future developments which draw on his ideas. As our common interest is the compound eye, we are all very grateful to him for his leading example.

References

Adrian ED (1932) The activity of the optic ganglion of *Dytiscus marginalis*. J Physiol 75:26–27 P
Adrian ED (1937) Synchronous reactions in the optic ganglion of *Dytiscus*. J Physiol 91:66–89
Autrum H (1948a) Über das zeitliche Auflösungsvermögen des Insektenauges. Nachr Akad Wiss Göttingen Math Phys Kl: 8–12
Autrum H (1948b) Zur Analyse des zeitlichen Auflösungsvermögens des Insektenauges. Nachr Akad Wiss Göttingen Math Phys Kl: 13–18
Autrum H (1950) Die Belichtungspotentiale und das Sehen der Insekten (Untersuchungen an *Calliphora* und *Dixippus*). Z Vergl Physiol 32:176–227
Autrum H (1952) Über zeitliches Auflösungsvermögen und Primärvorgänge im Insektenauge. Naturwissenschaften 39:290–297
Autrum H (1955) Die spektrale Empfindlichkeit der Augenmutation white-apricot von *Calliphora erythrocephala*. Biol Zentralbl 74:515–524

Autrum H, Burkhardt D (1960) Die spektrale Empfindlichkeit einzelner Sehzellen. Naturwissenschaften 47:527

Autrum H, Burkhardt D (1961) Spectral sensitivity of single visual cells. Nature (London) 190:639

Autrum H, Gallwitz U (1951) Zur Analyse der Belichtungspotentiale des Insektenauges. Z Vergl Physiol 33:407–435

Autrum H, Hamdorf K (1964) Der Sauerstoffverbrauch des Bienenauges in Abhängigkeit von der Temperatur bei Belichtung und im Dunkeln. Z Vergl Physiol 48:266–269

Autrum H, Hoffmann C (1960) Diphasic and monophasic responses in the compound eye of *Calliphora*. J Insect Physiol 4:122–127

Autrum H, Stöcker M (1950) Die Verschmelzungsfrequenzen des Bienenauges. Z Naturforsch 5b:38–43

Autrum H, Stöcker M (1952) Über optische Verschmelzungsfrequenzen und stroboskopisches Sehen bei Insekten. Biol Zentralbl 71:129–152

Autrum H, Stumpf H (1950) Das Bienenauge als Analysator für polarisiertes Licht. Z Naturforsch 5b:116–122

Autrum H, Stumpf H (1953) Elektrophysiologische Untersuchungen über das Farbensehen von *Calliphora*. Z Vergl Physiol 35:71–104

Autrum H, Tscharntke H (1962) Der Sauerstoffverbrauch der Insektenretina im Licht und im Dunkeln. Z Vergl Physiol 45:695–710

Autrum H, Zwehl V von (1962) Zur spektralen Empfindlichkeit einzelner Sehzellen der Drohne (*Apis mellifica* ♂). Z Vergl Physiol 46:8–12

Autrum H, Zwehl V von (1963) Ein Grünrezeptor im Drohnenauge (*Apis mellifica* ♂). Naturwissenschaften 50:698

Autrum H, Zwehl V von (1964) Die spektrale Empfindlichkeit einzelner Sehzellen des Bienenauges. Z Vergl Physiol 48:357–384

Autrum H, Wiedemann I (1962) Versuche über den Strahlengang im Insektenauge (Appositionsauge). Z Naturforsch 17:480–482

Autrum H, Jung R, Loewenstein WR, MacKay DM, Teuber HL (eds) (1971–1981) Handbook of sensory physiology, vol I–IX. Springer , Berlin Heidelberg New York

Baumgärtner H (1928) Der Formensinn und die Sehschärfe der Bienen. Z Vergl Physiol 7:56–143

Bernhard CG (1942) Isolation of retinal and optic ganglion response in the eye of *Dytiscus*. J Neurophysiol 5:32–48

Bernhard CG (ed) (1966) The functional organization of the compound eye. Pergamon, Oxford New York

Bernhard CG, Granit R, Skoglund CR (1942) The breakdown of accommodation. Nerve as a model sense organ. J Neurophysiol 5:55–68

Buddenbrock W von, Schulz E (1932/33) Beiträge zur Lichtkompaßbewegung und der Adaptation des Insektenauges. Zool Jahrb Allg Zool Physiol 52:513–536

Burkhardt D (1962) Spectral sensitivity and other response characteristics of single visual cells in the arthropod eye. Symp Soc Exp Biol 16:86–109

Burkhardt D, Autrum H (1960) Die Belichtungspotentiale einzelner Sehzellen von *Calliphora erythrocephala* Meig. Z Naturforsch 15b:612–616

Burkhardt D, Wendler L (1960) Ein direkter Beweis für die Fähigkeit einzelner Sehzellen des Insektenauges, die Schwingungsrichtung polarisierten Lichtes zu analysieren. Z Vergl Physiol 43:687–692

Coombe PE (1986) The large monopolar cells L1 and L2 are responsible for ERG transients in *Drosophila*. J Comp Physiol A 159:655–665

Crescitelli F, Jahn TL (1939) The electrical response of the dark-adapted grasshopper eye to various intensities of illumination and to different qualities of light. J Cell Comp Physiol 13:105–112

Danneel R, Zeutschel B (1957) Über den Feinbau der Retinula bei *Drosophila melanogaster*. Z Naturforsch 12b:580–583

Daumer K (1956) Reizmetrische Untersuchung des Farbensehens der Bienen. Z Vergl Physiol 38:413–478

Daumer K (1958) Blumenfarben, wie sie die Bienen sehen. Z Vergl Physiol 41:49–110

el Gammal St., Hamdorf K, Henning U (1987) The paracrystalline structure of an insect rhabdomere (*Calliphora erythrocephala*). Cell Tissue Res 248:511–518

Exner S (1891) Die Physiologie der facettirten Augen von Krebsen und Insecten. Deuticke, Leipzig Wien

Fabre, JH (1879, 1882) Souvenirs entomologiques. Etudes sur l'instinct et les moeurs des insectes. Ser 1 et 2. Delagrave, Paris (German Transl, 2nd edn, 1987, Artemis, Zürich München)

Frisch K von (1914) Demonstration von Versuchen zum Nachweis des Farbensinnes bei angeblich total farbenblinden Tieren. Verh Dtsch Zool Ges Freiburg 24:50–58

Frisch K von (1914/15) Der Farbensinn und Formensinn der Bienen. Zool Jahrb Physiol 35:1–188

Frisch K von (1949) Die Polarisation des Himmelslichtes als orientierender Faktor bei den Tänzen der Bienen. Experientia 5:142–148

Frisch K von (1950) Die Sonne als Kompaß im Leben der Bienen. Experientia 6:21–221

Gogala M, Michieli S (1965) Das Komplexauge von Ascalaphus, ein spezialisiertes Sinnesorgan für kurzwelliges Licht. Naturwissenschaften 52:217–218

Goldsmith TH, Philpott DE (1957) The microstructure of the compound eyes of insects. J Biophys Biochem Cytol 3:429–438

Goldsmith TH, Wehner R (1977) Restrictions on rotational and translational diffusion of pigment in the membranes of a rhabdomeric photoreceptor. J Gen Physiol 70:453–490

Grenacher H (1879) Untersuchungen über das Sehorgan der Arthropoden, insbesondere der Spinnen, Insekten und Crustaceen. Vandenhoeck & Ruprecht, Göttingen

Hamdorf K (1979) The physiology of invertebrate visual pigments. In: Autrum H (ed) Handbook of sensory physiology, vol VII/6A. Springer, Berlin Heidelberg New York, pp 145–224

Hamdorf K, Kaschef AH (1964) Der Sauerstoffverbrauch des Facettenauges von Calliphora erythrocephala in Abhängigkeit von der Temperatur und dem Ionenmilieu. Z Vergl Physiol 48:251–265

Hamdorf K, Paulsen R, Schwemer J (1973) Photoregeneration and sensitivity control of photoreceptors of invertebrates. In: Langer H (ed) Biochemistry and physiology of visual pigments. Springer, Berlin Heidelberg New York, pp 155–166

Hardie RC (1985) Functional organization of the fly retina. In: Ottoson D (ed) Progress in sensory physiology, vol V. Springer, Berlin Heidelberg New York Tokyo, pp 1–79

Hartline HK (1928) A quantitative and descriptive study of the electric response to illumination of the arthropod eye. Am J Physiol 83:466–483

Hartline HK, Graham CG (1932) Nerve impulses from single receptors in the eye. J Cell Comp Physiol 1:277–295

Hertz M (1933) Über figurale Intensitäten und Qualitäten in der optischen Wahrnehmung der Biene. Biol Zentralbl 53:10–40

Hesse R (1901) Untersuchungen über die Organe der Lichtempfindung bei niederen Tieren. VII. Von den Arthropoden-Augen. Z Wiss Zool 70:347–473

Hesse R (1908) Das Sehen der niederen Tiere. Fischer, Jena

Hoffmann CH (1960) Belichtungspotentiale der Insekten und Sauerstoffdruck. Verh Dtsch Zool Ges 53:220–225

Holmgren F (1856/66) Method att objectivera effecten av ljusintryck pa retina. Upsala Läkar Förh 1:177–191

Hooke R (1665) Micrographia or some physiological descriptions of minute bodies made by magnifying glasses with observations and inquiries thereupon. Martyn & Allestry, London (reprint by Dover, New York 1961)

Jahn TL, Crescitelli F (1938) The electrical response of the grasshopper eye under conditions of light and dark adaptation. J Cell Comp Physiol 12:39–55

Jahn TL, Crescitelli F (1939) The electrical response of the Cecropia moth eye. J Cell Comp Physiol 13:113–119

Kirschfeld K (1967) Die Projektion der optischen Umwelt auf das Raster der Rhabdomere im Komplexauge von Musca. Exp Brain Res 3:248–270

Kirschfeld K (1973) Das neurale Superpositionsauge. Fortschr Zool 21:229–257

Kühn A, Pohl R (1921) Dressurfähigkeit der Bienen auf Spektrallinien. Naturwissenschaften 9:738–740

Langer H (1962) Untersuchungen über die Größe des Stoffwechsels isolierter Augen von Calliphora erythrocephala Meigen. Biol Zentralbl 81:691–720

Langer H (1967) Grundlagen der Wahrnehmung von Wellenlänge und Schwingungsebene des Lichtes. Verh Dtsch Zool Ges 60:195–233

Langer H, Patat U (1962) Über die Bedeutung einer neuen Augenfarbenmutante von Calliphora erythrocephala Meig. für die Untersuchung der Funktion des Facettenauges. Verh Dtsch Zool Ges 55:174–180

Leydig F (1864) Das Auge der Gliederthiere. Laupp, Tübingen

Lubbock J (1882) Ants, bees and wasps. Appleton, New York

Marks WB (1965) Visual pigments of single goldfish cones. J Physiol 178:14–32

Müller J (1826) Zur vergleichenden Physiologie des Gesichtssinnes des Menschen und der Tiere. Cnobloch, Leipzig

Portillo J del (1936) Beziehungen zwischen den Öffnungswinkeln der Ommatidien, Krümmung und Gestalt der Insektenaugen und ihrer funktionellen Aufgabe. Z Vergl Physiol 23:100–145

Roeder KD (1939) Synchronized activity in the optic and protocerebral ganglia of the grasshopper *Melanoplus femur-rubrum*. J Cell Comp Physiol 14:299–307

Roeder KD (1964) The origin of visual rhythms in the grasshopper *Melanoplus femur-rubrum*. J Cell Comp Physiol 15:399–401

Rossel S (1987) Das Polarisationssehen der Biene. Naturwissenschaften 74:53–62

Santschi F (1911) Observations et remarques critiques sur le mécanisme de l'orientation chez les fourmis. Rev Suisse Zool 19:303–338

Schlieper C (1927) Farbensinn der Tiere und optomotorische Reaktionen. Z Vergl Physiol 6:453–472

Schneider G (1953) Die Halteren der Schmeissfliege (*Calliphora*) als Sinnesorgane und als mechanische Flugstabilisatoren. Z Vergl Physiol 35:416–458

Schneider G (1955) Die spektrale Empfindlichkeit des Rezeptors für das "Dämmerungssehen" bei *Calliphora*. Verh Dtsch Zool Ges 48:346–351

Schultze M (1868) Untersuchungen über die zusammengesetzten Augen der Krebse und Insecten. Cohen, Bonn

Seitz G (1968) Der Strahlengang im Appositionsauge von *Calliphora erythrocephala* (Meig.). Z Vergl Physiol 59:205–231

Seitz G (1969) Untersuchungen am dioptrischen Apparat des Leuchtkäferauges. Z Vergl Physiol 62:61–74

Singer Ch (1953) Die ältesten Abbildungen mikroskopischer Objekte. Endeavour 12:197–201

Stockhammer K (1956) Zur Wahrnehmung der Schwingungsrichtung linear polarisierten Lichtes bei Insekten. Z Vergl Physiol 38:30–83

Strausfeld NJ (1976) Atlas of an insect brain. Springer, Berlin Heidelberg New York

Tate P (1947) A sex-linked and sex-limited white-eyed mutation of the blow-fly (*Calliphora erythrocephala*). J Genet 48:176–191

Tomita T, Kaneko A, Murakami M, Pautler EL (1967) Spectral response curves of single cones in the carp. Vision Res 7:519–531

Uexküll J von, Brock F (1927) Atlas zur Bestimmung der Orte in den Sehräumen der Tiere. Z Vergl Physiol 5:167–178

Walther JB (1958) Untersuchungen am Belichtungspotential des Komplexauges von *Periplaneta* mit farbigen Reizen und selektiver Adaptation. Biol Zentralbl 77:63–104

Walther JB, Dodt E (1959) Die Spektralsensitivität von Insekten-Komplexaugen im Ultraviolett bis 290 mμ. Z Naturforsch 14b:273–278

Washizu Y, Burkhardt D, Streck P (1964) Visual field of single retinula cells and interommatidial inclination in the compound eye of the blowfly *Calliphora erythrocephala*. Z Vergl Physiol 48:413–428

Wulff VJ (1943) Correlation of photochemical events with the action potential of the retina. J Cell Comp Physiol 21:319–326

Wulff VJ, Jahn TL (1943) Intensity-EMF relationships of the electroretinogram of beetles possessing a visual diurnal rhythm. J Cell Comp Physiol 22:89–94

Annotation:

A short biography of Autrum and a bibliography complete to 1976 is given in the J. Comp. Physiol., *120* 101–107 (1977). Also a special issue in honour of H. Autrum has been published: Senses and environment. J. Comp. Physiol., *161* A, Number 4 (1987)

Historical data for this review have been extracted mainly from two sources:
K.E. Rothschuh: Geschichte der Physiologie. Springer, Berlin Göttingen Heidelberg 1953
Jahn I., Löther R., and K. Senglaub: Geschichte der Biologie. VEB Fischer, Jena 1985

Chapter 3

Optics and Evolution of the Compound Eye

DAN-ERIC NILSSON, Lund, Sweden

1 Introduction

During the last 100 years, the number of known optical types of compound eye has
grown from one to at least seven. With this increasing knowledge we have learned
that there are many ways to construct image-forming optical systems and the
similarity of compound eyes of different animals is indeed only superficial. The
radically different optical systems found in the eyes of closely related groups of both
insects and crustaceans present a serious problem concerning the evolution of
optical mechanisms: some compound eyes form multiple inverted images whereas
others form a single erect image. The problem arises from the fact that hypothetical
intermediate designs may seem nonfunctional.

Charles Darwin, although ignorant about compound eye optics, realized the
problem in more general terms and stated the following in his famous 1859
monograph on the origin of species: "If it could be demonstrated that any complex
organ existed, which could not possibly have been formed by numerous, successive, slight modifications, my theory would absolutely break down". Interestingly,
the opponents to Darwin's theory argued that an eye is exactly the kind of highly
perfected organ that would disrupt the evolutionary theory. Their mistake was to
believe that a "well-developed" eye is the only functional design, and thus would
have to be created by "accident" in a single major step. By the advancement of
comparative anatomy and embryology, this classical issue was soon resolved for
single-lens eyes, like our own (for a summary see Salvini-Plaven and Mayr 1977).
But the problem remained almost unrecognized for compound eyes whose designs
are even more remarkable.

To fully realize the design problems in a compound eye, we may take the
skipper butterflies as an example. Their eyes are of the superposition type where the
lens systems of many ommatidia cooperate to form a superimposed image on the
retina. To accomplish this, the crystalline cone in each ommatidium has a powerful
refractive index gradient that makes the ommatidial optics behave as an astronomical telescope. Such a system will work only when the design of the crystalline
cones and the gradients within them are perfect. Moreover, the position and size of
the photoreceptors have to match the overall geometry of the array of lens systems.
Also the size of the facet lenses, the angles between the ommatidia and the visual
fields of the receptors have to be matched to each other in a way that is determined
by the light intensity to which the eye is optimized. As if this were not enough, there
are regional differences requiring that the optics of each ommatidium be fitted

Stavenga/Hardie (Eds.) Facets of Vision
© Springer-Verlag Berlin Heidelberg 1989

according to the particular position of the facet. The demands of mutual matching of parameters in such an eye are enormous and a mismatch of any single parameter would prevent the formation of a crisp image on the retina. It is this optical perfection that leads to the geometrical beauty of compound eyes, but there seems to be little freedom for evolution to invent new and radically different optical solutions. This last conclusion is, however, quite wrong, and we shall see that although eye structure and function may be very conservative in some groups, it certainly changes easily in others.

The principal considerations in the design of an eye are the comparatively simple laws of optics and, because of this, eyes provide an unparalleled opportunity to 'understand' biological structures. There are, however, several questions involved in this understanding of compound eyes: First, how is the eye organized structurally? Second, how do these structures function? Third, for what reason have particular functions developed? In this chapter I compare the different types of compound eye and attempt to give an answer to all three questions. But first, I will shortly summarize the history of studies of compound eye optics, since this is almost as interesting as the topic itself.

2 The History of Compound Eye Optics

The facetted nature of insect eyes was discovered shortly after the invention of the light microscope. In 1644 the Italian scientist Hodierna published a work entitled *L'occhio della mosca*, where he described the fly's eye as looking like a raspberry. This first account of the compound eye was uncovered by Seitz (1971) but seems, apart from that, to have remained unnoticed. Hodierna's work certainly deserves more attention because he made remarkably accurate observations for the time and he also presented the first theory of compound eye function, which in fact pre-empted Müller's mosaic theory. "Each little-eye possesses a *cornea tunica*, followed inwards by a crystalline body and then a dark layer. In each little-eye, only the light that enters parallel to its axis will be effective." (translated into English from the German translation of Seitz 1971).

Beautiful and famous illustrations of compound eyes include those by Robert Hooke (1665) and Jan Swammerdam (1737), but their insight into the function of the eyes was no more advanced than that of Hodierna. The first optical experiments on compound eyes were performed in the late 17th century by Antoni van Leeuwenhoek, who found small inverted images behind the corneal lenses. Interesting accounts of Leeuwenhoek's work are given by Bernhard (1966) and Wehner (1981).

In the 18th century there seems to have been little progress in understanding the compound eye, but a new era of research was initiated by Müller (1826). He argued that each ommatidium of the compound eye was sensitive only to light parallel to its own axis. His theory came to be known as the *mosaic theory*, where each ommatidium measures only the intensity in its own direction. The resolution of the eye would then be determined by the number of facets and, in the image, each ommatidium would correspond to one tile in the mosaic. Although he knew about the corneal lenses, he did not properly consider them in his theory, and he thought of each facet as a pigment-lined tube with a nerve in the bottom. This view that there was no image perception within the ommatidium led to a serious controversy when Grüel (1844) and Gottsche (1852) rediscovered the tiny inverted images that Leeuwenhoek found almost two centuries before. Their finding became even more controversial when Leydig (1855, 1864) considered the crystalline cone to be the site of photoreception, thus enabling a perception of the tiny images behind each lens. This position was also taken by Schultze (1868) and by Patten as late as 1886. The conceptual enigma came from the fact that the individual images were inverted whereas the global image in the eye is erect.

The idea of image vision within an ommatidium was gradually abandoned as Boll (1871) and Grenacher (1877, 1879) provided more accurate information about the retinal anatomy and it was established that the rod-shaped rhabdom was the site of photoreception. The above two authors shared with Sigmund Exner (1876, 1891) the view that there was no analysis of the image within a single ommatidium, and Müller's theory became generally accepted.

Exner also demonstrated the existence of two main types of image formed by compound eyes: the *apposition image*, which relied essentially on the mosaic theory of Müller but with the corneal lenses also taken into account, and the *superposition image* which is formed by the cooperation of many corneal lenses and crystalline cones (Fig. 1). Both images are erect but the superposition image is brighter and found mainly in nocturnal animals. Exner chose the horseshoe crab *Limulus* and the firefly *Lampyris* as representative examples of animals that employ apposition and superposition imaging, respectively. Partly in collaboration with his physicist brother, Exner presented the idea of the lens cylinder, an inhomogeneous re-fracting structure, which seemed necessary in the formation of superposition images. A summary of Exner's views is given in Fig. 1. Despite the technical limitations of the time, Exner's observations and predictions were remarkably accurate, and his monograph of 1891 was so much ahead of its time that it took the scientific community 80 years to catch up with his state of knowledge.

Many of Exner's contemporaries were excellent anatomists and a broad comparative knowledge of compound eye structure was provided by Schultze (1868), Grenacher (1879), Patten (1886), Parker (1891), Chun (1896) and Hesse (1901), just to mention a few. The amount of fine detail described by these authors is astonishing. Sometimes, however, a few structures too many were discovered and, rightly, Chun (1896) castigated Patten for paying too much attention to histological details below the resolution limit of the microscope!

The question of image resolution in a single ommatidium was brought up again by Vigier (1907, 1909) and Dietrich (1909), because the eye of dipteran flies has an open rhabdom, i.e., the rhabdomeres of the receptor cells are separate and do not fuse to a single lightguide (fused rhabdom) as in the majority of compound eyes. Each receptor of an ommatidium will thus sample its own part of the inverted image produced by the corneal lens. Vigier (1909) found that each receptor shared the same direction of view as one of the receptors in each of six adjacent ommatidia, and that the axons of these receptors converged so that the neural image was no different from that of any other compound eye. At that time, the reason for this peculiar arrangement was not understood and Vigier's remarkable finding became forgotten.

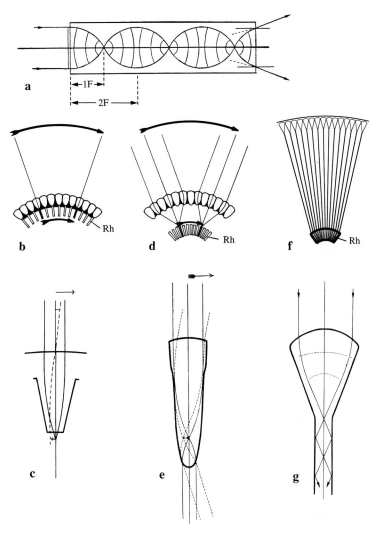

Fig. 1a-g. A summary of Exner's view of compound eye optics. *a* Ray path in a lens cylinder. The refractive index is highest along the axis and decreases towards the periphery. This will slow down the central part of the wavefront so that rays will continuously bend towards the axis. The curvature of both rays and wavefronts are shown in the figure. A lens cylinder of one focal length (*1F*) produces a focus at its rear surface, whereas, the double length (*2F*) behaves as a telescope. *b* Formation of an apposition image, where the ommatidia operate as optically isolated units: *Rh* rhabdom. *c* Ray path in the corneal cone of a *Limulus* apposition eye, which is an example of a lens cylinder of one focal length. *d* Formation of a superposition image by cooperation of many ommatidia. *e* Ray path in the corneal cone of a *Lampyris* superposition eye. Here the optical system behaves as a lens cylinder of double focal length. *f* The lightguide eye of *Phronima*, which Exner called a "catoptric" eye because light is transmitted to the rhabdom by reflections in the long lightguiding extensions of the cones. *g* Ray path in the distal part of a crystalline cone of *Phronima*. *a, c, e* and *g* after Exner (1891); *b* and *d* after Land (1981a)

The 50 years following Vigier's discovery were dominated by Exner's theories (e.g., Demoll 1917, Eltringham 1933). More anatomical work was performed, but in compound eye optics nothing conceptually new was discovered. The progress during these years was instead the development of electrophysiology, electron microscopy, and new techniques in light microscopy, all of which were to become powerful tools in the further unraveling of the function of compound eyes.

The work done in the 1960's and early 1970's offers a cascade of rival new theories on compound eye optics. This caused much confusion, which is now reflected as incomprehensible accounts in many modern textbooks. Electrophysiological recordings indicated an extensive overlap of visual fields from neighboring ommatidia and at first this was taken as a disproof of the mosaic theory. Although the reason for this overlap, i.e., diffraction by the corneal lenses, was anticipated as early as 1894 by Mallock and discussed in detail by Barlow (1952), it seemed in conflict with the measured ability of the eyes to resolve fine striped patterns. This led Burtt and Catton (1961, 1966) to propose an alternative mechanism for image formation, which was based on interference images formed by the regular array of the facets. However, both the reasons for developing this theory and the theory itself were so seriously wrong that the whole idea was soon abandoned.

Another challenge to Exner's theories was initiated when Kuiper (1962) found that the crystalline cones of crayfish are optically homogeneous. According to Exner, these eyes are of the superposition type, which would require a radial gradient of refractive index (a lens cylinder) in the crystalline cone. The eyes of many decapod crustaceans do indeed form superposition images but, as discovered much later by Vogt (1975) and independently by Land (1976), it is in this case accomplished by orthogonal sets of mirrors and not by lens cylinders. In the early 1960's, however, the homogeneity of crayfish cones caused serious doubt as to the validity of the superposition principle in general. Horridge (1968), Miller et al. (1968) and Døving and Miller (1969) believed that superposition images were not formed at all in living eyes (specifically in fireflies and moths). Their argument was that certain structures in the eyes, acting as lightguides, would maintain optical isolation of the ommatidia. The lack of optical isolation is however easily demonstrated in these eyes and Horridge (1975) considered superposition as being one of several candidates for image formation in the eyes of neuropterans, beetles and moths, although he believed many of these eyes to be only partly focused. After a convincing chain of evidence (Seitz 1969; Kunze 1969, 1970, 1972; Kunze and Hausen 1972; Hausen 1973; Horridge et al. 1972, 1977; Cleary et al. 1977) it turned out that Exner was indeed right and the superposition principle survived after more than a decade of doubt.

Another highlight of the 1960's came from the renewed interest in the open rhabdom of dipteran flies (Autrum and Wiedemann 1962), which culminated in the discovery of the *neural superposition* principle (Kirschfeld 1967). This was actually a rediscovery, since Vigier, who had been forgotten, largely solved the problem in 1909. The new results, however, were more precise and the principle not only got a proper name — it was also better understood.

In retrospect it may seem that this period in the 1960's and 1970's resulted in nothing but a return to the theories that were in vogue 50 years earlier. But such an interpretation fails to account for the many definitive proofs and the considerably improved understanding that came from this long overdue testing of the old theories. In recent years the field of compound eye optics has developed further. It now embraces optimization theory, wave optics and evolutionary theory. Conceptually new optical principles have also been discovered and these, together with previously known principles, will be described in detail in this chapter.

3 The Building Blocks of Compound Eyes

Compound eyes are the most prominent and most important visual organs in both insects and crustaceans. A comparable case of general occurrence of compound eyes can be found only in the extinct trilobites. Compound eyes are also known from a handful of other animals, but these are always isolated occurrences in groups where other types of eye are dominant. Among the chelicerates, *Limulus* constitutes such an isolated case and similarly, *Scutigera* is the only chilopod with a true compound eye. Outside the Arthropoda, compound eyes are only known from some tube-living polychaetes and a small group of clams, but in both cases the eyes are rather small and of simple design (for further information and references on eye types see Land 1981a).

We will here concentrate on the compound eyes of arthropods, which all have the same functional components in their ommatidia: the cornea, a cone, a rhabdom and a more or less complicated pigment screen. The cellular organization of ommatidia shows a much larger variation, and representative cases are illustrated in Fig. 2.

The cornea is a specialized part of the cuticle, which in most animals is formed into corneal lenses. Exceptions to this are the amphipods and the lower Crustacea where the cornea is not facetted and thus of little optical significance. The opposite case is seen in *Limulus* and some beetles, where the corneal lenses bulge into the eye as a substitute for the crystalline cones that are lacking in these animals. The number of cells that form the crystalline cone is four in insects and crustaceans, but exceptions occur in the latter group. Especially in insects, these cells are called "Semper cells". In *Scutigera,* the most common number of cone cells seems to be six (Nilsson unpublished) and in *Limulus*, where no crystalline cone is formed, there may be as many as 100 "cone cells" (Fahrenbach 1975). The terms exocone, eucone, and acone are sometimes used to describe eyes with corneal cones, hard crystalline cones, and soft watery cones respectively, but for simplicity we will here only use the terms corneal cone and crystalline cone, depending on whether it is a cuticular structure or not.

Each ommatidium has its own group of photoreceptor cells. These are called retinula cells and the group, a retinula. The number of these cells per ommatidium is basically eight in insects and higher crustaceans but exceptions are common in both groups. Lower crustaceans have fewer retinula cells, typically 5 or 6. In *Scutigera* and *Limulus* the number varies, commonly between 12–15 in *Scutigera* and 10–13 in *Limulus*. The rhabdom, which is the photosensitive structure, is formed by microvillar contributions (rhabdomeres) of the retinula cells. The way

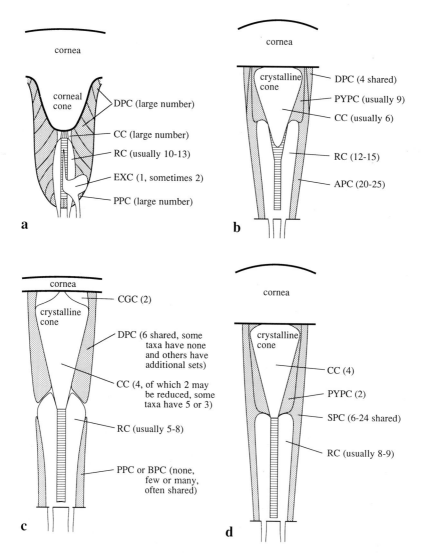

Fig. 2a-d. Schematic diagrams of the cellular organization in ommatidia of *a Limulus: b Scutigera; c* a crustacean; *d* an insect. For each type of cell, the usual number per ommatidium is indicated. Abbreviations: *DPC* distal pigment cells; *CC* cone cells; *RC* retinula cells; *EXC* eccentric cell; *PPC* proximal pigment cells; *PYPC* primary pigment cells; *APC* accessory pigment cells; *CGC* corneagenous cells; *BPC* basal pigment cells; *SPC* secondary pigment cells

that rhabdomeres are organized into a rhabdom is a matter of great variation and depends on the optical type of eye as well as on other factors including spectral discrimination and polarization sensitivity. A characteristic feature of the *Limulus* ommatidium is the eccentric cell which provides the main nervous output, although it does not itself form a rhabdomere.

Most compound eyes have several types of screening pigment cells (Fig. 2). The number of these cells, their position and pigment content are so variable that a common plan cannot be found. Depending on their position it is usually possible to define distal, proximal, and basal cells, but the terminology is far from consistent. Only a few definable cell types are found repeatedly. In insects these are the two primary pigment cells and in the malacostracan crustaceans there are six distal pigment cells, which are shared by neighboring ommatidia, so that there are twice as many pigment cells as there are ommatidia. The crustacean ommatidium is also characterized by a pair of corneagenous cells which, although devoid of pigment, show similarities with the primary pigment cells of insects. The pigment screen in *Scutigera* shows vague similarities with both insects and crustaceans, but in *Limulus* the arrangement is more reminiscent of simple ocelli.

Of great functional and experimental importance is the presence of a reflecting tapetum at the base of many insect and crustacean eyes. In crustaceans this is formed by proximal or basal pigment cells containing a reflecting pigment. For the same purpose, insects have instead made good use of their tracheal system and beautifully arranged tracheolar interference reflectors are found in the eyes of Neuroptera, Lepidoptera, Trichoptera, and some Coleoptera.

4 Optical Types of Compound Eye

Before we discuss and evaluate the different optical designs of compound eye we must ask the nature of the specific task that the eye is designed to solve. There are two modalities of visual information, spectral sensitivity and polarization sensitivity, that are attributed to properties of the receptor cells and not to the mechanisms of image formation (see Menzel and Backhaus; Stavenga; Rossel, this Vol.). Rather, it is the demand for spatial resolution that dictates the optical design of an eye. In compound eyes, just as in our own eyes, spatial information is conveyed by an array of parallel channels such that each channel corresponds to a point in the image, and apart from differences in resolution, a compound eye and a human eye look at the world in essentially the same way. What thereafter is done to the spatial information in the nervous system, is quite another matter. It seems obvious that the more spatial channels an eye has, the better is the resolution, but the signal the channels provide must be reliable. This brings us to the limitations of the eye.

Light is detected by the eye as single particles (photons) and their arrival is a random event. Intensity, in these terms, is the average number of photons per unit time, and the fewer the photons, the larger is the uncertainty about the true average. This means that for two image channels to be discriminated, there must be a sufficient number of photons per integration time for the uncertainty (noise) to become smaller than the actual difference in intensity. Photon noise is reduced by collecting more photons. If this is done, for example by a larger lens, the information capacity of the eye is increased. In an arbitrary eye, such an increase in information capacity can be exploited in several ways: it can be used to see at lower intensities; it can be used to detect objects of less contrast; it can be spent on more image channels so that resolution is improved or it can be used for sampling at shorter periods so that faster events are seen. We see that, in a very general way, the light-gathering capacity of an eye determines the amount of information it can provide, and it must therefore be important for an eye to collect as much light as possible from each point in space.

Another limitation to vision is due to the wave nature of light which results in the phenomenon of diffraction. When parallel light is passed through an aperture it will suffer an angular spread that is inversely related to the size of the aperture. If the aperture contains a lens, parallel light will not be imaged as a point but as a blur spot, the Airy disc. Diffraction, therefore, leads to an uncertainty about the direction of origin of light. The only way to reduce diffraction is to increase the size of the lens. If the receptors (rhabdoms) are larger than the Airy disc, diffraction will have little effect, but the visual field of a single receptor can never become smaller than the limit set by diffraction. Animals that are active in bright sunlight are often diffraction-limited whereas nocturnal animals are limited mainly by photon noise (see Land this Vol.).

A third limitation to vision comes from aberrations in the optical system. For the performance of an eye these will have the same effect as diffraction. The amount of aberration depends on which mechanisms are employed for image formation, but as a general rule, attempts to increase the brightness of the image will also lead to more serious aberrations.

To summarize the above, a well-performing eye should gather as many photons as possible from the environment and it should be as sure as possible about their direction of origin. We must bear this basic purpose of an image-seeing eye in mind as we compare the different optical types of compound eye in the following discussion. For further information on operational principles, information capacity and the limits of vision, the reader is referred to Snyder (1979), Land (1981a and this Vol.), van Hateren (this Vol.) and Laughlin (this Vol.).

4.1 Apposition Eyes

An apposition eye is defined as a compound eye where the rhabdoms receive light only from their 'own' corneal facets. This is the same as saying that the ommatidia work as optically isolated units. Anatomically, these eyes are characterized by the rhabdoms which are usually long and slender and always directly apposed to the crystalline cones.

4.1.1 Simple Apposition

The most straightforward way to construct a compound eye is to have the distal tip of a fused rhabdom in the focal plane of the corneal lens, and to isolate the ommatidia by an extensive shield of screening pigment. Since the rhabdom only measures the intensity in the part of the image that falls on its distal tip (without resolving it), the ommatidium becomes the functional unit for spatial vision. As in all compound eyes the global image is erect whereas the little, unresolved image behind each lens is inverted. Historically, this is the kind of compound eye described by Müller and which Exner called "an eye that forms an apposition image". Today we know that apposition optics can be realized in more than one way and I have here chosen to call this first type "simple apposition" (see Fig. 3). A more descriptive but longer term would be "focal apposition with a fused rhabdom."

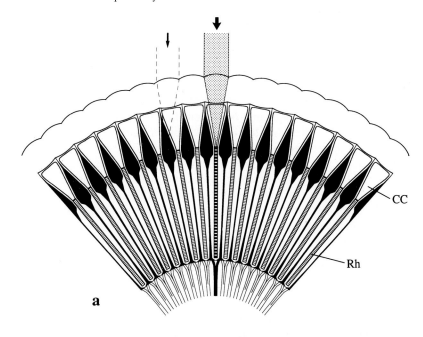

Fig. 3a-d. The simple apposition principle. *a* Schematic section of an eye cut along the ommatidia, showing the ray path from a point source: *CC* crystalline cone; *Rh* rhabdom. Off-axis light (*dashed lines*) is absorbed in the pigment around the crystalline cone. *b* The distal parts of ommatidia in the simple apposition eye of a bumble bee (*Scale bar* 25 μm). *c* Similar view of ommatidia from the primitive crustacean *Lepidurus* (*Scale bar* 25 μm). *d* Cross-section of the fused rhabdom from a bumble-bee eye, as seen in the electron microscope (Scale bar 1 μm)

The principle requires only two components: a positive lens and a single sensitive lightguide (rhabdom). Exner (1891) was the first to predict the rhabdom's lightguide properties. These rely on the higher refractive index of the rhabdom material relative to that of the surrounding cytoplasm. From recent work on insects we know that the refractive index of the rhabdom is very close to 1.36 compared with 1.34 for the surrounding medium (Stavenga 1974a,b; Beersma et al. 1982; Nilsson et al. 1988). Light that is guided by the rhabdom will be continuously scrambled so that all the rhabdomeres of one ommatidium share the same visual field. Rhabdoms of apposition eyes are frequently so narrow that geometrical optics fails to account for their behavior. In these cases waveguide optics is appropriate, but that goes beyond the scope of this chapter and the interested reader is referred to Snyder (1979), Nilsson et al. (1988) and van Hateren (this Vol.). We shall go no further than stating that light propagates as discrete interference patterns called waveguide modes, and that these modes make the waveguide behave as if it had diffuse borders, i.e., some of the power propagates outside the rhabdom and the distal tip can also catch some light outside its diameter.

Together with diffraction in the lens, these waveguide effects make the ommatidium's angular sensitivity function bell-shaped, and neighboring ommatidia will therefore have some overlap of visual fields. The optimum amount of overlap depends on the trade-off between contrast, resolution and sensitivity. This is true for all types of compound eye and a comprehensive treatment is given by Snyder (1977) and Land (this Vol.).

The simple apposition principle is in many respects a poor solution to spatial vision. As we have already concluded, the two most important limitations to vision are photon noise and diffraction. Both increase with decreasing lens diameter and if the corneal surface of the eye is to be divided into one "private" lens per image channel, it is indeed used in the least economical way.

Despite the good theoretical understanding of the simple apposition type of eye (see Snyder 1975; Pask and Barrell 1980a,b; van Hateren this Vol.), there is relatively little experimental work done on the actual optics. Judging from the anatomy, this optical principle seems to be by far the most common and it is found, amongst others, in bees, locusts, mantids, cockroaches, dragonflies, many crabs, stomatopods, amphipods, isopods, all lower crustaceans, *Limulus* and *Scutigera*. It is possible that this list will lose one or two members after further experimental examination, because anatomically, this type of eye is indistinguishable from afocal apposition which exists in butterflies and will be described later.

In insects, crabs, stomatopods and *Scutigera*, the ommatidial lens is formed by the cornea, either by its curved outer surface, by graded refractive index within it, or by a combination of both. In these cases the crystalline cone is a mere optical spacer that allows the cornea to focus on the distal tip of the rhabdom (see Fig. 3b). In *Limulus* and some beetle apposition eyes (Caveney 1986) the facets project into the ommatidia as corneal cones and no crystalline cones are formed at all. The opposite case is found in amphipods and all lower crustaceans where the crystalline cone contains the entire ommatidial lens (Fig. 3c). Here the cornea seems to be nothing but a protective cover over the eye. The isopod crustaceans take an intermediate position because here the cornea and crystalline cone may both take part in the focusing (Nilsson and Nilsson 1981, 1983). In general, the curved outer surface of the cornea is a useful lens in air, but for aquatic animals, graded refractive index lenses (lens cylinders) within the cornea or the crystalline cone provide both the most sensible and the most common solution. Amphibious animals in particular have a flat cornea so that the focal distance stays the same in air and water.

The rhabdoms of insect apposition eyes are usually one or a few μm in diameter (Fig. 3d) and 100 μm or more in length. In many small crustacean apposition eyes, where the interommatidial angles are large, the rhabdoms are typically wider and much shorter (Fig. 3c). In the extreme case (e.g., the isopod *Cirolana*, Nilsson and Nilsson 1981) the width equals the length. It has been repeatedly found that, in these simple apposition eyes with very short rhabdoms, the ommatidial lens is not focused on the distal tip of the rhabdom but much deeper (Young and Downing 1976; Nilsson and Odselius 1981, 1983) and sometimes even below the rhabdom as in *Cirolana*. The reason for this underfocusing is as yet unknown.

4.1.2 The Open Rhabdom and Neural Superposition

Quite a few insects have apposition eyes with rhabdomeres that are not fused into a single lightguide (Fig. 4). These so-called "open-rhabdom eyes" are found in dipteran flies, hemipteran bugs, earwigs, and many beetles. Among the crustaceans, this kind of eye is known only from the isopod genus *Ligia* (Edwards 1969; Hariyama et al. 1986). The ommatidial optics of eyes with open rhabdoms rely on the same basic principle as the previous type: a lens — usually the cornea — is focused on the distal tip of the rhabdom (this may not always be valid in the light-adapted state, see Sect 4.1.5). The crystalline cone is frequently not crystalline at all (nor is it cone-shaped) but forms a watery space between cornea and the rhabdom distal tip. The unifying feature is the open rhabdom and we may define this type as "focal apposition with an open rhabdom." The separation of the rhabdomeres has a drastic effect on spatial vision: the retinula cells of an ommatidium will not share the same visual field.

In higher dipteran flies (Brachycera), the rhabdomeres act as seven separate lightguides, all "looking" in different directions. In each ommatidium there is a narrow central lightguide surrounded by six wider ones in an asymmetric trapezoidal pattern (Fig. 4d). The central lightguide is composed of two rhabdomeres, one on top of the other, whereas the peripheral lightguides are single rhabdomeres. As shown by Vigier (1909) and later worked out in detail by Kirschfeld (1967), the central pair of rhabdomeres "look" in the same direction in space as one of the peripheral rhabdomeres in each of six neighboring ommatidia (Fig. 4a,b). Thus the same spatial information reaches all these rhabdomeres, and below the basement membrane the axons of their cells converge to one cartridge in the lamina (Fig. 4a). Kirschfeld (1967) termed this principle "neural superposition" because the information from one point in space is collected through several facets and then superimposed neurally.

Such a convergence of seven equivalent signals gives the same result as if light was collected through a lens with a seven times larger area. This will reduce photon noise and thus increase sensitivity but without a sacrifice in resolution. Diffraction will, however, not be reduced. As already mentioned, the central lightguide, which

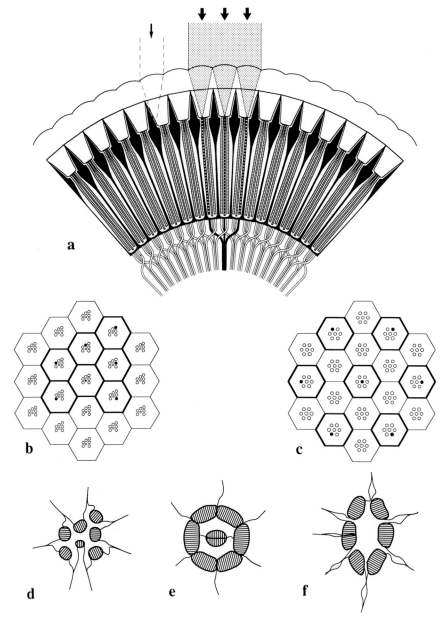

Fig. 4a-f. Apposition eyes with ipen rhabdoms. *a* Schematic section of an eye cut along the ommatidia, illustrating the neural superposition principle. Off-axis rays (*dashed lines*) are handled in the same way as in a simple apposition eye. *b* Cross-section geometry of rhabdomeres in ommatidia of a brachyceran neural superposition eye.*c* The corresponding arrangement in the dorsal eye of a bibionid fly. *Filled circles* in *b* and *c* indicate rhabdomeres that share the same visual axis. *d-f* Semi-schematic drawings of open rhabdoms from a brachyceran fly (*d*), an insect without neural superposition (*e*), and the isopod crustacean *Ligia* (*f*). In all cases *d-f*, the rhabdomeres come closer to each other at the distal rhabdom tip

is composed of rhabdomeres from two cells, is narrower than the six surrounding ones (Fig. 4d). It turns out that these two central cells do not take part in the neural superposition mechanism but send their axons straight through the lamina cartridge without making synaptic connections (an informative description of this neural network is given by Strausfeld 1976). The sensitivity gain by neural superposition is therefore six times instead of seven times, and the central rhabdomeres make a seperate system which does not have the increased sensitivity but instead a slightly better resolution. We see that the eye of brachyrean flies is indeed sophisticated, not only because it has neural superposition, but also because it contains two separate systems for spatial vision that operate simultaneously and are optimized to different intensities. Interesting and more complete accounts are given by Franceschini (1975), van Hateren (1984), Hardie (1985, 1986) and Shaw (this Vol.).

In the other dipteran suborder, the Nematocera, the open rhabdoms do not have the trapezoid arrangement typical for brachyceran flies (see Shaw, this Vol.). Instead, the typical nematoceran ommatidium has a radially symmetric arrangement with six rhabdomeres equally spaced from a central two-rhabdomere lightguide. In bibionid flies, such an arrangement is combined with neural superposition (Fig. 4c), but in this case with the next-but-one ommatidia and not with the immediate neighbors as in higher flies (Zeil 1979, 1983).

In order to demonstrate the presence of a neural superposition mechanism, the necessary arrangement of rhabdomere visual axes and the appropriate neural connections both have to be revealed and shown to match. This has only been done in brachyceran flies and in the nematoceran group Bibionidae. In other nematoceran flies and in bugs (Hemiptera: Heteroptera), many beetles (Coleoptera: Cucujiformia) and earwigs (Dermaptera), the rhabdomeres also show an open arrangement, but in most cases probably without the presence of any neural superposition. Although there are variations in rhabdom arrangement, a central pair of rhabdomeres surrounded by six peripheral ones seems to be the most common (Diptera: Brammer 1970; Meyer-Rochow and Waldvogel 1979; Zeil 1979; Williams 1980; Hemiptera: Burton and Stockhammer 1969; Schneider and Langer 1969; Walcott 1971; Ioannides and Horridge 1975; Coleoptera: Chu et al. 1975; Wachmann 1977, 1979; Gokan and Hosobuchi 1979a,b; Schmitt et al. 1982; Dermaptera: McLean and Horridge 1977). What then is the functional significance of such a rhabdom if it is not combined with neural superposition? A clue to this question comes from the fact that the peripheral rhabdomeres are often not separate from each other but form a ring around the two central ones (Fig. 4e). This kind of open rhabdom cannot resolve much spatial information within the ommatidium, and the peripheral rhabdomeres would simply have a larger acceptance angle than the two central ones. If this is true, then these eyes would have two types of image channel, one with high sensitivity and poor resolution (the peripheral rhabdomeres) and one with low sensitivity and high resolution (the two central rhabdomeres). In contrast to neural superposition eyes, the gain in sensitivity, conveyed by the peripheral rhabdomeres, must be at the expense of resolution. This mechanism with one photopic and one scotopic system has been proposed before (e.g., Ioannides and Horridge 1975), and is supported experimentally and by the way these eyes adapt to changes in light intensity (see Sect. 4.1.5). The above hypothesis, however, cannot apply to the crustacean isopod *Ligia* because here there are no central rhabdomeres (Fig. 4f).

Although the fly's eye constitutes the most thoroughly analyzed and best understood visual system of any arthropod, there is obviously much left to be learned about the open rhabdom eyes of other insects and indeed also of *Ligia*.

4.1.3 Afocal Apposition

The afocal apposition eye (Nilsson et al. 1984) is a recent addition to the number of known optical types of compound eye. As yet, it is only known from butterflies where it seems to be of universal occurrence (Nilsson et al. 1988; the only exception being the skipper butterflies). Anatomically, this type is indistinguishable from a simple apposition eye (Sect. 4.1.1). The ommatidia are optically isolated and the fused rhabdom behaves as a single lightguide. The secret to its function is concealed in the optics of the crystalline cone. The corneal lens alone — in this case its curved outer surface — is sufficient to focus an image at the distal rhabdom tip. However, a cross-section of the proximal part of a fresh crystalline cone behaves as a powerful lens (Fig. 5b) and obviously the optical system is more complicated than in a simple apposition eye. In fact, the combined action of the crystalline cone and the corneal lens, is that of an astronomical telescope. Parallel light incident on the cornea is focused to an image inside the crystalline cone, well above the rhabdom and, in the proximal stalk of the cone, the beam is recollimated into a parallel pencil of rays which enters the rhabdom (Fig. 5c). Such a system is termed *afocal* in contrast to the *focal* systems we have discussed earlier. Afocal optics has long been known to exist in superposition eyes (Sect. 4.2.1), but it is not so obvious how it can be combined with apposition imaging.

Using geometrical optics, we find a number of peculiarities in the behavior of the butterfly ommatidium. Light from a distant point source will reach the rhabdom as a narrow parallel beam. Tilting the incident beam will cause a six to seven times larger tilt to the parallel beam at the rhabdom. From this angular magnification, which is the essential feature of man-made telescopes, it follows that the diameter of the incident beam (set by the facet diameter) becomes correspondingly minified in the cone stalk so that the exit beam fits the size of the rhabdom and pivots around the rhabdom tip when the incident beam is tilted (Fig. 5d). This in turn leads to the remarkable conclusion that the corneal facet is imaged onto the rhabdom tip and vice versa. The field of view of the rhabdom will then be determined by the critical angle for total internal reflection in the rhabdom. We see that the addition of the tiny lens in the cone stalk makes radical changes to the optical system, but geometrical optics does not on its own offer an explanation to why butterflies have this system.

From the dimensions of the rhabdom and the cone stalk lens it is clear that wave-optics phenomena must be important (Fig. 5e). Van Hateren and Nilsson (1987) treated the cone-stalk lens as part of the waveguide and were able to show, experimentally and with a theoretical model, that such a system is superior to a conventional focal system (see also Nilsson et al. 1988, van Hateren this Vol.). To avoid a lengthy introduction to wave-optics, we shall here only summarize the main conclusions, namely: (1) the butterfly system is about 10% more efficient in

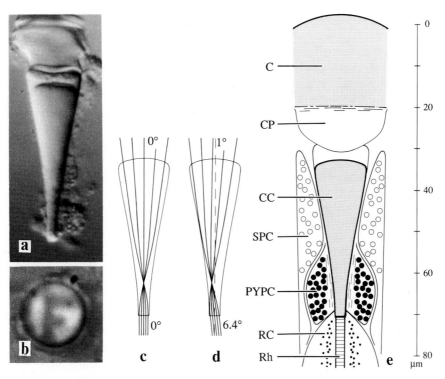

Fig. 5a-e. The afocal apposition eye of butterflies. A Nomarski micrograph of an isolated crystalline cone (*a*) from a butterfly (*Heteronympha*) reveals nothing unusual. But a parallel-sided cross-section from the proximal tip is capable of forming an image (*b*) of a distant object (the letter 'F'). *c* and *d* show the ray path in the crystalline cone for parallel rays incident on the cornea at 0° and 1° respectively. The cellular structure of the distal part of the ommatidium is shown in e: *C* cornea; *CP* corneal process; *CC* crystalline cone; *SPC* secondary pigment cell; *PYPC* primary pigment cell; *RC* retinula cell; *Rh* rhabdom. All at the same scale (*right*) except for *b* which is magnified x5 relative to the other

accepting on-axis light; (ii) the resolution can reach very close to the diffraction limit set by the corneal lenses; (iii) the superiority of the sytem can be exploited only with a diurnal mode of life; (iv) the optical system can be interpreted as a refinement of a simple focal system. The significance of these results will be considered in Section 5.2.

Since anatomical criteria are not sufficient to demonstrate the presence or absence of an afocal optical system of this kind, it is possible that it occurs in several other diurnal insects.

4.1.4 Transparent Apposition

Pelagic and planktonic crustaceans are often transparent because they lack body pigmentation. The transparency makes these animals difficult to spot when they swim in open water, and it can be thought of as an odd example of camouflage: to be concealed in open water, one of the few possibilities is to look like water. But such a strategy is hampered by the presence of eyes which need pigmentation for their function. This reasoning explains why many pelagic and planktonic crustaceans seem to have gone to great length to reduce the extent of the pigment screen in their apposition eyes. A typical eye of a transparent planktonic crustacean contains unusually long crystalline cones without pigment shields, and a small compressed retina which is the only pigmented part of the eye (Fig. 6a-e). As a result, the pigmented part appears as a small core in an otherwise much larger transparent eye and the animal can avoid being spotted by predators or prey.

Transparent apposition eyes are found in onychopod and haplopod waterfleas, hyperiid amphipods, most larval euphausiids, all larval stomatopods, all larval decapods and also in a few adult decapods. Thus, it seems that this type of eye is widespread in crustaceans and, owing to the very special optical mechanisms involved, there are good reasons to define transparent eyes as a special optical type of apposition eye. In fact, this has been done at least twice in the old literature. Exner (1891) examined the transparent eye of the hyperiid amphipod *Phronima* and called it the "catoptric eye" as opposed to "dioptric eyes" which were all the other compound eyes. On purely anatomical grounds, Chun (1896) made a similar division between "retinopigmented eyes" (transparent eyes) and "iridopigmented eyes" (normally pigmented eyes).

The combination of very long crystalline cones and the lack of screening pigment between them would mean that optical isolation is lost if there were no alternative mechanisms to maintain it. Such mechanisms are indeed present and we shall see that there are three principally different optical designs of the crystalline cones, which all serve to maintain optical isolation without the need for screening pigment between the cones (Fig. 6b-d).

The first type of transparent eye, found in the dorsal eye of *Phronima*, is characterized by extremely elongated crystalline cones (Figs. 1f,g, 6b) with a distal cone-shaped part and an extended lightguide of considerable length (Exner 1891, Ball 1977, Land 1981b). The entire cone has a higher refractive index than the surrounding fluid in the eye. As anticipated by Exner (1891) and later shown by Land (1981b), the distal-most part contains a graded-index lens which forms a focus in the bottom of the cone-shaped part. Light is then guided in the threadlike part, which penetrates the pigmented retina and joins to a rhabdom. Due to the enormous length of the lightguides in the eye of *Phronima*, the small and compact retina lies about 5 mm below the optical array of the eye, and the spatial information is conveyed to the retina in a way not dissimilar to that used in manufactured image-transmitting fiber-bundles. In other eyes where this optical design is found, the lightguide part of the cone is shorter but still much longer than the rhabdom. So far, there are no conclusive results about the mechanism of optical isolation in these eyes, but a simple mechanism can be inferred from the shape of the cones. Due to the lack of pigment, light will enter the side of the cone-shaped part as well as the lightguide. The criterion for optical isolation is that this "unwanted" light should not become guided. The criterion is fulfilled if the refractive index of the lightguide is such that the critical angle for total internal reflection fits with the tapering angle of the cone-shaped part. From the refractive index values presented by Land (1981b) this mechanism seems plausible. A remarkable parallel to the lightguide eye of *Phronima* is present in the predatory water-flea *Bythotrephes* (Miltz 1899).

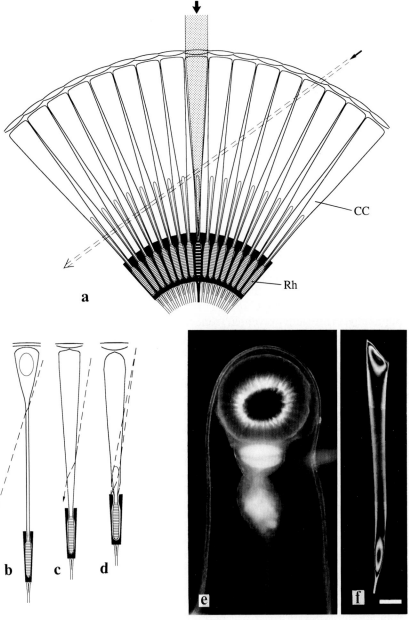

Fig. 6a-e. Transparent apposition eyes of pelagic or planktonic crustaceans. *a* Schematic drawing and ray path of an eye cut along the ommatidia: *CC* crystalline cone; *Rh* rhabdom. Light that is not used for vision (*dashed lines*) is passed through the eye. Three types of transparent eyes are known (*b-d*), which all use different strategies to prevent light from neighboring facets (*dashed rays*) from entering the rhabdom: the schematic drawings are of *b*, lightguide type; *c* axial gradient type; *d* radial gradient type. The dark-field micrograph (*e*) shows the transparent eye of the water flea *Leptodora* (from Nilsson et al. 1983a). The proximal gradients – here of the radial type – shine brightly in dark-field illumination because they deviate the transmitted light. *f* is an interference micrograph of a crystalline cone from the hyperiid amphipod, *Hyperia* (*Scale bars* 50 μm). The cone reveals two graded index lenses: one distal and one proximal.

Another design of the crystalline cone is found in the larvae of decapods and stomatopods (Nilsson 1983a), a few adult decapods (Nilsson unpubl.) and the isopod *Astacilla* (Nilsson and Nilsson 1983). Here the crystalline cones have a more normal shape and do not form lightguides (Fig. 6c). The corneal lenses are focused at the rhabdom distal tip, and the long crystalline cones maintain optical isolation simply by their high refractive index. Light entering a cone from a neighboring corneal facet is bent and deflected at the border of the cone so that it is forced to emerge at the other side. Light that enters the side of the cone at a proximal position will have to be bent more and these cones typically have an axial gradient of refractive index with values increasing towards the proximal tip (Nilsson 1983a; Nilsson and Nilsson 1983).

The third mechanism for optical isolation in transparent eyes relies on the presence of an additional proximal lens in the crystalline cone. The ommatidia thus contain two lenses: the first one is the corneal facet or sometimes a graded index lens in the distal end of the cone, and the second one is a graded index lens in the proximal part of the cone (Fig. 6d,f). Such an ommatidium shows similarities with the afocal system of butterfly ommatidia. The proximal lens serves to image the corneal facet, or distal gradient, at the proximal tip of the cone. The image is minified and fits the pigment aperture through which the cone penetrates the retina (Nilsson unpubl.). Light from neighboring corneal facets will thus be imaged outside the pigment aperture and prevented from entering the rhabdom. Eyes containing this double-lens system are found in onychopod and haplopod water-fleas (Nilsson et al. 1983a), many hyperiid amphipods (Nilsson 1982), and most euphaussiid larvae (Nilsson 1983a).

It is important to note that these designs do not compromise vision. Only in a few deep sea hyperiids (Land 1981b), where screening pigment is absent from the retinas as well, must there be a trade-off between vision and visibility. The three different mechanisms described, employing lightguides, axial gradients or radial gradients are adaptations found only in crustaceans. It is advantageous for camouflage in water and would thus not be expected in any terrestrial animals.

4.1.5 Adaptational Changes

Most animals have to use their eyes under a variety of illumination conditions. Although the receptor cells can adapt to changes in intensity, they have a limited dynamic range. Optical mechanisms that control the light flux in the receptors will extend the dynamic range of the eye and thus extend its use. Several such adaptational mechanisms are known, and many are present in connection only with special types of eye. I will here briefly summarize the adaptational mechanisms of apposition eyes.

A mechanism of the same type as the pupil of our own eyes would be quite disastrous in a compound eye. The facet lenses are already small and a further reduction by a corneal pupil would lead to a dramatic increase in diffraction. There are, however, other factors determining image brightness that can be successfully exploited by a compound eye. One such factor is the focal length which is altered in response to changes in light intensity in the eye of the brine shrimp *Artemia* (Nilsson and Odselius 1981). Changing the focal length requires a flexible lens

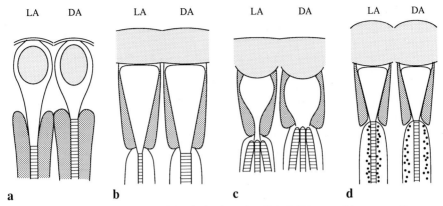

Fig. 7a-d. Four different strategies for light/dark adaptation (*LA, DA*) in apposition eyes. *a* Changes of focal length in *Artemia* (Nilsson and Odselius 1981). *b* Changes of rhabdom diameter as in grapsid crabs and many insects (after Nässel and Waterman 1979). *c* A variable pigment pupil in front of the rhabdom. *d* Radial migration of screening pigment granules in the retinula cells

which is present in the crystalline cone of *Artemia* (Fig. 7a). This mechanism affects not only rhabdom light flux but also the acceptance angle which enlarges on dark adaptation. An identical effect is reached by changing the diameter of the rhabdom (Fig. 7b). In grapsid crabs, the rhabdom may be up to three times wider at night (Nässel and Waterman 1979) which should lead to a ninefold increase in sensitivity. These changes in rhabdom diameter are a very sensible way to exploit the daily cycle that renews the microvillar membrane. The same mechanism has also been reported for locusts and mantids (Rossel 1979; Horridge et al. 1981; Williams 1982).

Another mechanism that results in a coupled change of sensitivity and acceptance angle is to vary a pigment aperture, or iris, in the focal plane of the lens in front of the rhabdom distal tip. This is employed almost universally in the open rhabdom eyes that do not have neural superposition (see Autrum 1981; Walcott 1975; Ioannides and Horridge 1975). Also *Limulus*, a few ants and some beetles, that have fused rhabdoms, seem to have adopted this principle (Behrens 1974; Home 1976; Menzi 1987; Barlow et al. this Vol.). In insects it is invariably the primary pigment cells and the soft cone cells that bring about the change in aperture (Fig. 7c). In the dark-adapted state, the pigment cells are withdrawn to a peripheral position and the dilated cone leaves all the rhabdomeres exposed. During light adaptation, the pigment cells contract, the rhabdom is pressed down and the cone cells form a narrow thread transmitting light to the rhabdom. This mechanism allows a very direct control of the acceptance angle and hence the rhabdom light flux.

A mechanism that is faster, and has much less effect on the acceptance angle, is a change of the medium around the rhabdom. As mentioned before, light propagating along a narrow waveguide will have part of its power outside the diameter of the waveguide. This light is accessible for absorption by screening pigment granules in the retinula cells. In flies, butterflies and bees these pigment granules migrate radially in response to illumination (Kirschfeld and Franceschini 1969; Stavenga and Kuiper 1977; Stavenga et al. 1977; Ribi 1978; Stavenga 1979;

Nilsson et al. 1988). In the light-adapted state the pigment granules aggregate close to the rhabdom and thus attenuate the light flux (Fig. 7d). In the dark-adapted state the pigment granules are dispersed peripherally so that light is no longer extracted from the rhabdom. The pigment granules only have to move a few microns to create a dramatic effect and the process is therefore very fast, typically 5–20 s. In locusts it is mitochondria that migrate radially instead (Horridge and Barnard 1965). This will primarily affect the refractive index around the rhabdom, and this in turn alters the efficiency of the rhabdom as a waveguide. Rhabdoms of apposition eyes are usually surrounded by empty intracellular cisternae (Fig. 3d), or extracellular invaginations. These form a palisade around the rhabdom and improve its lightguide properties. A change in the extent of the palisade is a common complement to movements of mitochondria or pigment granules.

Eyes that possess a reflecting tapetum in the bottom of the retina may be able to control its efficiency. Although tapeta are rare in apposition eyes, a number of amphipods and isopods have tapeta that are exposed in the dark-adapted state but obscured or even moved below the basement membrane during light adaptation (Debaisieux 1944).

4.2 Superposition Eyes

The term "superposition" was coined by Exner (1891) to describe image formation in compound eyes where the ommatidia are not optically isolated. Anatomically, superposition eyes are characterized by the presence of an unpigmented space, the clear zone, separating the optical array from a much deeper-lying retina. The rhabdoms are typically short and wide, often not separated by any screening pigment. The clear zone permits light to cross between ommatidia so that each rhabdom will receive light from a large number of facets. Alternatively it can be said that each facet produces an image that extends over many rhabdoms. Since the overall image in a compound eye is always erect, it requires the optics of each facet to produce an erect image. The single lens of an apposition eye produces an inverted image, and a superposition eye must therefore have a mechanism that re-inverts these primary images. This is the equivalent of saying that rays entering from one side of the optical axis must be bent across the axis and directed back to the same side. This can be realized by refraction, reflection or a combination of both, and these three possibilities have led to the existence of three different types of superposition eye.

Superposition optics has evolved for one single reason: the need for an improved photon catch. Quite independently of the specific optical mechanism involved, the superposition design will greatly improve sensitivity but it will not reduce diffraction. The gain in sensitivity is roughly proportional to the number of facets contributing to the superposition pupil. This collective pupil, however, breaks up the wavefront and diffraction is thus determined by the size of individual facets — just as in the apposition eye.

The majority of superposition eyes display pigment migrations into the clear zone during light adaptation. Since these will prevent the formation of superposition images, the following descriptions apply to the dark-adapted state where the superposition mechanism is in operation.

4.2.1 Refracting Superposition

An astronomical telescope of the Kepler type contains two lenses separated by the sum of their focal lengths. A parallel bundle of rays incident on the front lens converges to a focus between the two lenses and is then recollimated into a parallel bundle by the second lens. This ray path has the properties required by an ommatidium of a superposition eye because the second lens will re-invert the ray-path from the first lens, and thus bend light across the axis. In order to handle rays entering at large angles, the second lens would have to be very large. This problem can be avoided if light is instead continuously bent in a system with a radial gradient of refractive index. The system outlined here is that of a lens cylinder of double focal length (see Fig. 1), and it is also an accurate description of the ommatidial optics in a refracting superposition eye (Fig. 8a,b). The system is afocal and thus closely resembles that in an afocal apposition eye (compare Figs. 5d and 8b). We recall that in the apposition eyes of butterflies the ommatidial optics provides an angular magnification of about 6–7. In refracting superposition eyes this is much smaller, typically between 1 and 2 (see Kunze 1979). If the angular magnification is 1 — incident and emerging rays make the same angle to the optical axis — then the focal length of the lens cylinder must be equal on both sides of the intermediate focus. If we consider the whole eye, an angular magnification of one leads to the formation of a superposition image halfway between the optical array and its center of curvature. As already mentioned, the angular magnification may be slightly higher, and this puts the superposition image somewhat closer to the layer of crystalline cones.

Even if the ommatidial optics behave as ideal telescopes, the superposition image will suffer from spherical aberration. If the concentric symmetry of the eye is abandoned, as in many euphausiids, aberrations can be reduced to a minimum, but such a design only works for a very restricted field of view (Land et al. 1979). For eyes used in air, refraction by the outer curvature of the entire eye will counteract spherical aberration so that it is largely cancelled (Bryceson and McIntyre 1983; Land 1985). Some aberrations will always persist and as a general rule, the more ommatidia that contribute to the superposition image, the more will it suffer from aberrations. It is a common misconception, however, that superposition eyes are not as good as apposition eyes in resolving images. In fact, diurnal animals, with apposition and superposition eyes respectively, have about the same resolving power (Land 1984). A similar comparison between animals from dim environments would certainly show the superposition eyes to resolve better. The reason is that apposition eyes have to pay for every increase in sensitivity with a corresponding loss in resolution (reduction in number of ommatidia), whereas superposition eyes can gain sensitivity by a larger superposition pupil which results in only a modest degradation of the image (Caveney and McIntyre 1981; McIntyre and Caveney 1985). The spherical geometry of superposition eyes (Fig. 8c) is not compatible with large regional differences over the eye, and the acute zones encountered in most apposition eyes are therefore absent in superposition eyes in general (see Land, this Vol.).

The refracting type of superposition eye is the most common and found in both insects and crustaceans, where it evolved independently in several orders. Among the insects, refracting superposition eyes occur in some mayflies, lacewings and

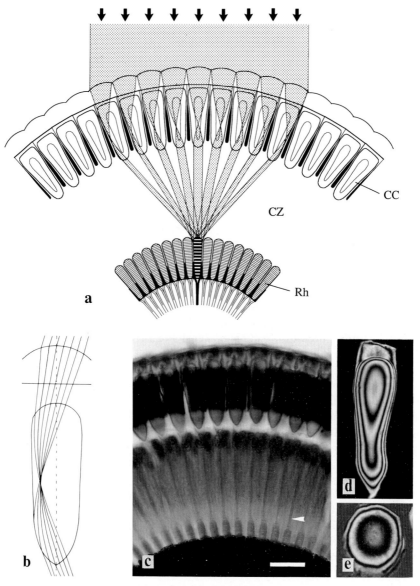

Fig. 8a-e. Refracting superposition eyes. *a* Schematic drawing of an eye cut along the ommatidia, showing superposition of rays from many ommatidia:

Fig. 8a-e. Refracting superposition eyes. *a* Schematic drawing of an eye cut along the ommatidia, showing superposition of rays from many ommatidia: *CC* crystalline cone; *CZ* clear zone; *Rh* rhabdom. *b* The afocal ray path in a moth crystalline cone (after Cleary et al. 1977). *c* Micrograph of a longitudinal section through a Mysid (*Neomysis*) eye: the *arrow* points at an epirhabdom (*Scale bar* 25 μm). *d, e* Interference micrographs of an intact and a cross-sectioned cone of *Neomysis*, showing the characteristic profile and the lens-cylinder gradient

most other neuropterans, several groups of beetles, most moths and most caddis flies (for a comparison see Horridge 1975). A corresponding list for the crustaceans includes euphausiid and mysid shrimps (Fig. 8c-e, Land and Burton 1979; Nilsson et al. 1983b) and, as recently found (Nilsson unpubl.), also the Anaspideacea, a few decapod shrimps (e.g., the genus *Gennadas*) and one isolated case in a hermit crab (Anomura).

Although the eyes of these animals all share the same basic design of a refracting superposition eye, there are some variations in how it is realized structurally. In crustaceans, the crystalline cone seems to contain the entire afocal system, whereas in insects, the corneal lens probably constitutes an important part of the first half of the telescope. An extreme case is seen in some of the beetles — including fireflies. click beetles and related groups — where the lens cylinder is formed by an inward projection of the cornea, and the cone cells do not produce a proper crystalline cone (see Caveney 1986). There are also large variations in the number of ommatidia contributing to the superposition pupil. In the skipper butterflies only some ten ommatidia may contribute but in nocturnal moths it may be several hundred.

Up till now we have assumed that the clear zone is optically homogeneous. In many refracting superposition eyes, however, this is obviously not the case. There have been numerous reports of light-guiding structures crossing the clear zone and joining rhabdoms to cones. In some eyes it is actually a narrow part of the rhabdom that reaches all the way up to the crystalline cone (Fischer and Horstmann 1971; Horridge and Giddings 1971; Horridge 1972). It is also common that entirely separate accessory rhabdoms are formed immediately below the cone (Horridge 1969, 1975; Kuster 1979; Caveney 1986). Apart from rhabdoms, there may be crystalline tracts, formed by the cone cells or retinula cells, that serve as light-guide connections across the clear zone (Horridge 1975, 1976; Horridge and Henderson 1976; Wolburg-Buchholz 1976; Welsch 1977; Schneider et al. 1978; Meinecke 1981). These structures do not prevent the formation of a superposition image but may take care of axial light (Kunze 1979). In mysids and euphausiids, a sharp peg-shaped structure (Fig. 8c), the epirhabdom, connects to the rhabdom and extends through part of the clear zone (Hallberg 1977; Hallberg and Nilsson 1983; Denys et al. 1983). The epirhabdom is not rhabdomeric but made of a clear material of high refractive index. Its functional significance is still obscure.

4.2.2 Reflecting Superposition

The action of a lens cylinder of double focal length — such as that in refracting superposition eyes — shows some similarities with a mirror. In fact, mirrors can also be used successfully in superposition imaging (Fig. 9). Many decapod crustaceans have superposition eyes with the typical clear zone, but the crystalline cones are optically homogeneous and could by no means act as lens cylinders. Characteristic of these eyes are the square facets and the square cross-section of the crystalline cones (Fig. 9d). These features are due to the reflecting superposition mechanism which relies on mirrors instead of lenses (Land 1976; Vogt 1975, 1980). The mirrors that form the superposition image are the walls of the crystalline cone (see Fig.

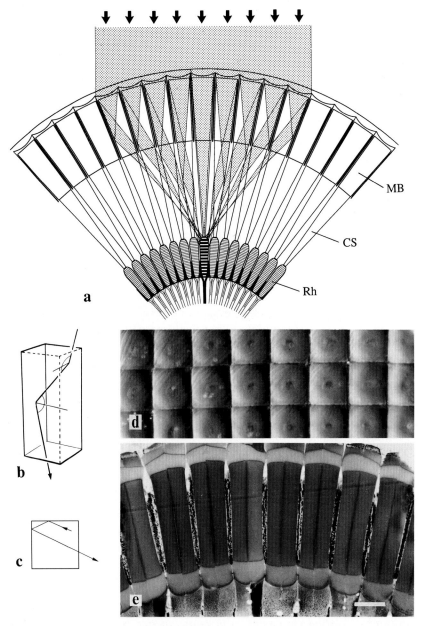

Fig. 9a-e. Reflecting superposition eyes. *a* Schematic drawing of the ray path in an eye cut along the ommatidia: *MB* mirror box; *CS* cone stalk; *Rh* rhabdom. *b* The two reflections redirecting an oblique ray in the orthogonal mirror box (after Vogt 1977). *c* The same ray path seen from a point on the optical axis ("cross-section"). *d* The square facets of a shrimp (*Leander*) as seen in the Nomarski microscope. *e* Micrograph of a longitudinal section through the mirror boxes of the same shrimp (*Scale bar* 25 μm)

9a,e). The mechanism is easily conceivable when looking at the ray path in a section cut along the ommatidia. It is, however, less obvious when one looks at the ray path from a point on the optical axis (Fig. 9b,c). It is here that the square cross-section of the cone comes in. In this view from above the ommatidium, superposition imaging requires the rays to be redirected so that they emerge in the same direction from the optical axis as they enter. It is perhaps easier to think of it as rays that are paraxial with an ommatidium to one side, and thus should be bent towards the rhabdom of that ommatidium. A single mirror would do this if it was perpendicular to the ray as seen from above the ommatidium, but for an arbitrary direction of incidence, four orthogonally arranged plane mirrors are required. An oblique ray will then be reflected once in each of two orthogonal sides of the mirror box. Such an arrangement has the same effect as a single mirror that is always perpendicular to the ray as seen from above (Fig. 9c). Rays that are parallel with one of the sides of the crystalline cone are only reflected once, and these are the rays seen in the longitudinal section (Fig. 9a). All other oblique rays must undergo two reflections to be redirected correctly. In some shrimps the mirrors act solely by total internal reflection but in eyes of other animals like, for instance, *Astacus*, the reflections in the mirror box are augmented by a multilayer coating of reflecting pigment (Land 1976, 1981c; Vogt 1977; Ball et al. 1986).

Rays that are only slightly oblique may bounce at only one of the sides of the mirror box before entering the clear zone. Such rays could also pass diagonally through the mirror box without being reflected at all. These slightly oblique rays would be wrongly directed and thus degrade the image, but they are taken care of by the cone stalk, which is a tapering extension of the cone reaching across the entire clear zone down to the rhabdom. At least the distal half of the cone stalk is square in cross-section and operates in the same way as the outer part of the cone. The slightly oblique rays will thus undergo one or both the reflections in the cone stalk rather than in the distal mirror box itself. The refractive index of the cone stalk has to be very precisely tuned to allow the already redirected rays to emerge from the cone to form the superposition image. If the refractive index is too high, many rays will undergo more than two reflections and if it is too low, the slightly oblique rays will miss out on one or both of the reflections. The optimum compromise that takes care of most of the rays involves an axial gradient in the cone stalk, with a proximally decreasing refractive index. The angle for total internal reflection will thus increase towards the distal part of the stalk. This gradient was predicted by Vogt (1977) and has also been demonstrated by measurements (Vogt 1977; Nilsson 1983b).

We see that the reflecting superposition eye is capable of forming images without any lenses at all. But these eyes do in fact have corneal lenses that cannot be disregarded (Bryceson 1981). The focal length is long and corresponds approximately to the distance between cornea and rhabdom layer. Although the superposition mechanism works without the lenses, the bundle of rays from each facet will converge to a true focus rather than having the width of a corneal facet. The result will thus be a slightly improved quality of the superposition focus.

The superposition image and the rhabdoms are located halfway between the spherical array of mirror boxes and its centre of curvature. As in the refracting superposition eye, the image will suffer from spherical aberration (Bryceson and McIntyre 1983). In addition to that, light will have to cross several cone stalks on its way to the superposition focus. This will deviate the rays and further impair the quality of the image.

Despite these aberrations, the reflecting superposition design seems successful because it is almost universally present in decapod shrimps (Natantia), crayfish and lobsters (Macrura) and the squat lobsters (Anomura, Galatheoidea). It is perhaps clearer to say that all decapods have reflecting superposition eyes with the main exceptions being the true crabs (Brachyura) and the hermit crabs (Anomura, Paguroidea; see Fincham 1980; Cronin 1986). For a yet unknown reason, the eyes of squat lobsters differ from the other reflecting superposition eyes in lacking the characteristic stalks of the crystalline cones (Kampa 1963; Bursey 1975). Neither in insects nor in any crustacean outside the decapod group are there any examples of a reflecting superposition eye. The reason for this will be obvious when we discuss the evolution of optical mechanisms.

4.2.3 Parabolic Superposition

This type of eye is the most recent addition to the list of optical principles in compound eyes (Nilsson 1988). The first animals with parabolic superposition eyes that I encountered were swimming crabs of the genus *Macropipus* (previously *Portunus*). The general features of their eyes include corneal lenses with a short focal length, crystalline cones that look very much like those of simple apposition eyes, a wide clear zone below the cone tips of the dark-adapted eye and a deep-lying retina typical for superposition eyes (Fig. 10a). The crystalline cones do not contain lens-cylinder gradients, nor are they square in cross-section, but it can be demonstrated experimentally that the dark-adapted eyes produce a relatively well-focused superposition image.

To explain how this image is formed, we again have to consider the ray path first in a longitudinal section and then in an ommatidial cross-section. The corneal lens is focused at the proximal tip of the crystalline cone, and oblique rays that hit the side of the cone are reflected at its border (Fig. 10c,d) by a multilayer reflector of regularly arranged endoplasmic reticulum. A profile of the cone reveals an inward parabolic curvature along most of its length, hence the name parabolic superposition. This parabolic mirror is necessary to cancel the convergence of rays caused by the corneal lens. The combination of lens and parabolic mirror is actually equivalent to a positive lens closely followed by a negative one that recollimates the rays. Axial rays will not hit the mirror and would diverge strongly after emerging from the cone tip. This is prevented by lightguides connecting cone tips with rhabdoms, and thus guiding axial light as in an apposition eye (Fig. 10b).

The reflecting border of the cone is circular in cross-section and obviously it cannot operate by the same principles as the reflecting superposition eye. Instead, the cone contains a homogeneous internal structure that acts as a cylindrical lens (not to be confused with a lens cylinder). Since the corneal lens is focused at the tip of the cone, oblique rays would reach the reflecting border of the cone before they come to a focus. This condition is altered by the cylindrical lens which shortens the focal length so that the focus falls on the reflecting surface (Fig. 10e). An oblique bundle of initially parallel rays will thus be focused on the mirror surface at the cone border. The diverging fan of reflected rays will then pass the cylindrical surface

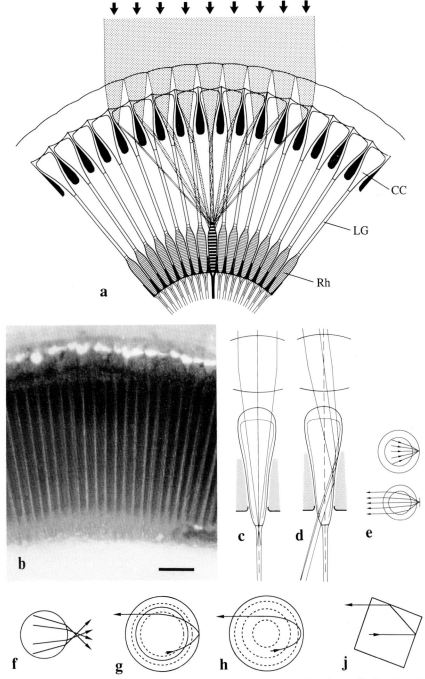

Fig. 10a-j. Parabolic superposition eyes. *a* Schematic drawing of the ray path in a longitudinal section of the eye: *CC* crystalline cone; *LG* lightguide; *Rh* rhabdom. *b* The lightguides seen in a fresh hemisected eye of *Macropipus* (*Scale bar* 100 μm). *c, d* Side view of the ray path and reflections at the parabolic surface inside the cone of *Macropipus*. *e* Schematic ray path in a cone seen from above, of rays before and after reflection. Aberrations in the cylindrical lens (*f*) can be cancelled by a radial refractive index gradient (*g*), whereby the optics comes very close to that of a refracting superposition eye which is shown for comparison in *h*. As an alternative to circular cross-section with an internal cylindrical lens, a square cross-section (*j*), as in xanthid crabs, can be used in the same way as in reflecting superposition eyes once on its re-entrance into the internal structure and once on its way out on the other side. When the rays emerge from the cone they will also have to pass the cylindrical outer wall of the cone (Fig. 10e). These three cylindrical surfaces will together recollimate the beam so that it emerges parallel from the cone. Note that the internal lens is *cylindrical*, and thus acts as a lens only in the cross-section plane. Consequently, the cylindrical lens does not add any dioptric power to the ray path seen in the longitudinal section. An interesting result of this is that an oblique point-source will be imaged as a line along the cone border and then recollimated into a parallel beam that is directed towards the superposition focus.

A cylindrical lens will display a phenomenon that corresponds to spherical abberation, i.e., peripheral rays will be bent too much (Fig. 10f). To solve this problem, a graded index cylindrical lens could be used. This would be a two-dimensional analog of a fish lens, and in fact, this mechanism for cancelling aberration exists in the parabolic superposition eye of a hermit crab. Interestingly, an eye using this mechanism handles light in a way that resembles the ray path in a refracting superposition eye (Fig. 10g,h).

There are many variations on the basic theme of parabolic superposition. The xanthid crabs, for example, have no cylindrical lenses inside the crystalline cone, but use a square cross-section in the same way as reflecting superposition eyes (Fig. 10j). The lightguides that are required to take care of axial light are often a narrower distal part of the rhabdom, and in some species the parabolic superposition mechanism is so little developed that the eyes may anatomically look like apposition eyes.

Despite this large variation, there are four distinctive features that are characteristic for parabolic superposition eyes: (1) the cyrstalline cones have an inward parabolic profile along part of the length; (2) corneal lenses are focused at the distal end of the clear zone; (3) the clear zone is crossed by lightguides; (4) the crystalline cones are either circular in cross-section with an internal cylindrical lens, or square in cross-section without internal structures.

In one form or another, the parabolic superposition type of eye is now found in numerous brachyuran crabs and hermit crabs. Using the above characteristics, it seems that parabolic superposition has also been invented in some mayflies (see Horridge et al. 1982).

4.2.4 Adaptational Changes

Light-dark adaptation mechanisms are of two basic types in superposition eyes: those that regulate the size of the superposition pupil and those that control tapetal reflection. The cellular components involved in adaptation vary greatly between species, but frequently, most of the cells in the ommatidium take part in one way or another (see Autrum 1981; Walcott 1975). Since the superposition principle relies on the cooperation of many ommatidia to form a common image, a natural way to control image brightness is to regulate the light flux across ommatidia in the clear zone (Exner 1891; Kunze 1979; Land 1981a). This is accomplished by longitudinal migrations of screening pigment (Fig. 11a). It may be a movement of entire cells or just a redistribution of pigment granules within the cells. In the

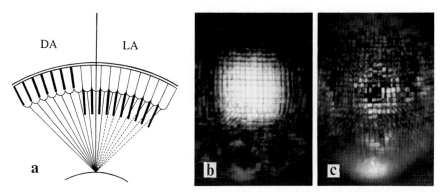

Fig. 11a-c. Light/dark adaptation (*LA, DA*) in superposition eyes. *a* Schematic drawing illustrating the principle of longitudinal pigment migrations (after Exner 1891). *b, c* Micrographs of dark- and light-adapted live eyes of a decapod shrimp (*Leander*). In the dark-adapted state (*b*) the wide superposition pupil displays an eyeshine because the tapetum is exposed. In the light-adapted eye (*c*) superposition rays are cut off and the tapetum is obscured, resulting in a small dark pseudopupil

dark-adapted condition, the screening pigment is compressed into the narrow spaces between the cones, so that a wide clear zone separates the optical array from the rhabdoms. During light adaptation, many superposition eyes effectively convert to apposition optics by moving down the pigment curtains into the clear zone to intercept the rays crossing between ommatidia. It is in this state that the enigmatic lightguides probably become important in refracting superposition eyes. Quite commonly, the crystalline cones also become more pointed (see Horridge 1975), perhaps optimizing the ray path to the apposition mode.

There are a few examples, like some skipper butterflies (Horridge et al. 1972) and some mysid shrimps (Hallberg et al. 1980a), where adaptation involves only minor changes in pigment position around the cone. It seems that these animals make only small changes in size of the superposition pupil, without ever turning into pure apposition. Yet other examples are found where no changes at all can be detected between light- and dark-adapted eyes (Exner 1891; Horridge et al. 1972, 1977; Meyer-Rochow and Walsh 1978; Land 1984). There are probably two different reasons for this. First, animals that are active mainly at high light intensities, like agaristid moths and again some skipper butterflies, may have optimized their superposition mechanism accordingly. When light intensity drops, they simply retain the maximally open pupil. Second, animals that inhabit constantly rather dark environments, like deep sea shrimps (euphausiids and decapods), do not need a light-adaptation mechanism because naturally they never encounter any high light intensities. Many migrate vertically in the sea, maintaining similar dim conditions day and night.

An additional type of adaptation strategy in superposition eyes is to expose or obscure a reflecting tapetum. This was mentioned in connection with apposition eyes, but the mechanism is more common in superposition eyes. In the dark-adapted state these eyes show an "eyeshine" or "eye glow" if they are illuminated and observed from the same direction (see Fig. 11b and Stavenga this Vol.). During light adaptation (Fig. 11c), the tapetum is either obscured by dark pigment or, as in many crustaceans, the tapetum itself moves to below the basement membrane of the eye.

Apart from pigment migrations, it is not uncommon for the rhabdoms to change position during adaptation. Such phenomena are reported from many insects (see Horridge 1975; Burghause 1976; Autrum 1981).

4.3 Other Types of Design

A number of insects, including apterygotes, coccids, lice, and fleas, have single-lens eyes or ocelli replacing the compound eyes (Paulus 1979; Land 1981a). Whether these ocelli are derived from compound eyes is impossible to say. There is no structural evidence that they have been formed by fusion of ommatidia into single-lens eyes. Such an example exists, however, in the ampeliscid amphipods. These crustaceans are tube dwellers that inhabit soft deposits at moderate depths in the sea. Instead of compound eyes, they have three pairs of single-lens eyes that are clearly derived by fusion of ommatidia (Elofsson et al. 1980; Hallberg et al. 1980b). In each eye, the cuticle forms a common corneal lens which is separated from the retina by a vitreous body (Fig. 12a). The scalloped retina is composed of closely packed ommatidia, where cone cells, retinula cells, rhabdoms and pigment cells can be identified. The cellular composition of the ommatidia is actually not unlike that of other amphipod compound eyes. The existence of this "single-lens compound eye" has some interesting evolutionary implications that will be discussed later.

Another aberrant development of compound eyes is found in some deep-sea crustaceans, including representatives from decapods, mysids, and amphipods. Instead of reducing their eyes, which seem of little use in the darkness of the deep sea, these crustaceans have eyes of normal size or larger, but without any trace of optics left (Welsh and Chace 1937; Zharkova 1975; Elofsson and Hallberg 1977; Hallberg et al. 1980b; Nilsson unpuslibhed). The volume of their eyes is occupied by enormous hypertrophied rhabdoms and reflecting pigment (Fig. 12b). It seems

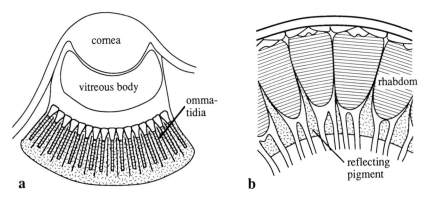

Fig. 12a,b. Modified compound eyes. *a* The "single-lens compound eye" of the amphipod crustacean *Ampelisca. b* Part of a longitudinal section through the compound eye of the deep-sea decapod *Hymenodora glacialis* which lacks all traces of dioptric apparatus

that these animals have given up spatial resolution altogether and put all their cards on sensitivity. Although these eyes do not have any "optical mechanism", we shall see below that they may be important when discussing evolution of optical mechanisms.

5 Evolution of Optical Mechanisms

Although the preceding presentations of optical types of eye are generalized and simplified, we have provided the basic information necessary to make a few conclusions about the evolutionary paths that produced the enormous diversity of compound eyes. At a first glance it may seem that similarities and differences in design are the only facts on which we can base evolutionary arguments. This way of thinking, however, is unlikely to lead to any solid conclusions. Similarities, for example, may indicate common ancestry, or parallel but independent evolution in related groups, or true convergence from different origins. Likewise, differences in design may occur between closely related groups as a result of divergent evolution. Therefore, considerations based on structural characteristics alone are of little value if we do not know the reason for their presence. In this respect, compound eyes may be one of the most promising organs for an evolutionary analysis because the design is very well understood in terms of optics and performance.

A prerequisite for accurate conclusions is a realistic view of the evolutionary process: gradual changes will take place if there are functional reasons for them *and* if the conditions are present that make the changes possible. Depending on life style and habitat, an eye design that is advantageous for one species is not necessarily so for another. Evolution is also limited by possibilities of the moment, meaning that better solutions are often out of reach because less efficient intermediate solutions may "block the way".

5.1 The Origin of Compound Eyes

There are three hypothetical ways by which a compound eye may be derived. First, a small number of nonresolving ocelli or eye spots may fuse into a cluster where each eye spot has a different field of view. Second, a number of image-resolving ocelli may fuse after which the ocelli reduce into nonresolving ommatidia. Third, a single image-resolving ocellus may split into separate ommatidia.

The last possibility is highly unlikely, but let us first take a known example of the reverse transformation. The single-lens eye of the amphipod *Ampelisca* is clearly derived from a standard amphipod compound eye (Sect. 4.3). Although it is easy to argue that a single-lens eye is better than a compound eye in collecting photons, the evolution from compound to single-lens eye is far from straightforward. The problem lies in the overall retinal image which is erect in a compound eye but inverted in a single lens eye. Not only will intermediates be nonfunctional, but the nervous system will also have to interpret the spatial information in the opposite way. How then is it possible that the ampeliscids have managed to do the transformation? The most obvious possibility is that they once gave up image vision altogether because they exploited dark habitats or developed a burrowing life, where their compound eyes were not sensitive enough to be useful. If this led to a loss of the optical apparatus so that only light sensitivity was left — as in many deep sea crustaceans — then the successive build-up of a common lens would lead to an improved sensitivity and also a potential recovery of spatial resolution. Going the other way and deriving a compound eye from a single-lens eye would require a similar unique route via nonresolving eyes, and this is in fact a re-invention of optics rather than a change in optical design.

The fusion of a cluster of image-resolving ocelli is equally unlikely because the process would lead to a continuous loss in photon capture rate and image resolution. We are therefore left with the first alternative which is the only one that results in a continuous improvement in performance. It was also pointed out by Land (1981a) that a fusion of nonresolving eye spots seems to be the way that the primitive compound eyes of sabellid polychaetes came about.

The development of some kind of ommatidial lens system follows naturally because this involves an inrease in both sensitivity and resolution compared to just pigmented cups or pigment-lined tubes. If this was the way it happened, then the first compound eyes must have been of the simple apposition type. The compound eyes of presumably "primitive" animals like the xiphosurans, scutigeromorphs, lower crustaceans, and apterygote insects are all simple apposition and the in-evitable conclusion is that such an eye is the primitive and ancestral type of compound eye. This does not mean that apposition eyes of recent species are primitive in all respects, and certainly, almost all species have specialized their eyes in accordance with their need for spatial resolution. The impressive eyes of dragonflies, for example, are of the simple apposition type.

Since the derivation of a primitive compound eye is a relatively simple and straightforward process, it means that a common ancestry of compound eyes is indeed not necessary. Instead it seems likely that the major arthropod groups — insects, crustaceans, chilopods, xiphosurans and trilobites — have evolved their compound eyes independently, and similarities of the eyes from different groups may be of functional rather than phylogenetic significance. Paulus (1979) argued for the monophyly of arthropods on the basis of purely structural similarities of compound eye ommatidia, but this way of thinking seems less adequate the more we learn about the eye's functional morphology.

5.2 Transition Between Optical Types

Perhaps the most intriguing question in compound eye evolution is how super-position eyes can evolve from apposition eyes. Table 1 contains a summary of the eye types found in crustaceans and insects. The diversity within these two groups is impressive, and to a large extent due to the need for improved sensitivity. As we have concluded before, the simple apposition design is a most uneconomical way to use the space available for an eye. It is only a small exaggeration to say that evolution seems to be fighting a desperate battle to improve a basically disastrous

Table 1. Summary of the distribution of eye types in the major groups of insects and crustaceans. Question marks indicate possible but unconfirmed occurrence

Group	Taxon	Simple apposition	Open rhabdom apposition	Neural superposition	Afocal apposition	Transparent apposition — Lightguide	Transparent apposition — Axial gradient	Transparent apposition — Radial gradient	Refracting superposition	Reflecting superposition	Parabolic superposition
Insecta	Apterygota	×									
	Ephemerida	×							×		?
	Plecoptera	×									
	Odonata	×									
	Dictyoptera	×									
	Orthoptera	×									
	Dermaptera		×								
	Isoptera	×									
	Hemiptera: Homoptera	×									
	Hemiptera: Heteroptera	×		?							
	Mecoptera	×							×		
	Neuroptera	×							×		
	Coleoptera	×	×						×		
	Diptera		×	×							
	Lepidoptera				×				×		
	Trichoptera								×		
	Hymenoptera	×									
Crustacea	Branchiopoda	×				×		×			
	Ostracoda	×									
	Maxillopoda	×									
	Anaspidacea								×		
	Amphipoda	×				×		×			
	Isopoda	×	×				×				
	Mysidacea*							(×)	×		
	Euphausiacea							×	×		
	Decapoda: Natantia						×		×	×	
	Decapoda: Macrura						×			×	
	Decapoda: Anomura	×					×		×	×	×
	Decapoda: Brachyura	×					×				×
	Stomatopoda	×					×				

*Transparent apposition design only in postembryonic development (before hatch from the female marsupium).

design. If an animal has to cope with moderate or low light intensities then a superposition eye is the most efficient solution within the basic framework of a facetted eye. There is thus no doubt that for many insects and crustaceans, a superposition eye involves a significant selective advantage. Although there is no difficulty in seeing the reasons for developing superposition optics, it is indeed difficult to find the conditions that make it possible. The problem again lies in the type of imaging. Superposition eyes produce a single erect image whereas simple apposition eyes produce multiple inverted images. If a superimposed erect image is to occur, each ommatidium has to produce an erect image at some distance below the cornea. It seems that all attempts to go from inverted to erect images have to pass through unfocused systems and, needless to say, there would be no selective pressure for such nonfunctional intermediates.

The seriousness of this problem is obvious, especially since superposition eyes must have evolved many times in both insects and crustaceans, but no attention has been paid to this question until very recently. However, investigations of the transparent apposition eyes of pelagic crustaceans and the discovery of two new optical types, afocal apposition and parabolic superposition, have now provided at least three possible paths along which superposition eyes can evolve by a series of small successive improvements.

1. The afocal apposition eye of butterflies has the same kind of telescopic ommatidial optics as refracting superposition eyes. If butterflies had a clear zone in their eyes, it is possible that their crystalline cones would be able to form a superposition image. Butterflies are diurnal and the majority of species are active only in bright sunlight. Under these conditions, a superposition eye would not be better and there is thus no need for butterflies to change their optical system. The butterfly optical system is in fact the best choice of known optical types when light intensities are so high that diffraction is the main limitation to vision (Sect.4.1.3). The proximal lens in the crystalline cones can be interpreted as a distal specialization of the waveguide, that improves the transfer of light from the Airy disc of the lens to the rhabdom (van Hateren and Nilsson 1988). Such a specialization ca be developed gradually from a simple apposition system via continuously improved designs. Once the butterfly optical system is developed, the potential for superposition imaging is present (Fig. 13a). If an animal with such a system changes to crepuscular or nocturnal habits, a smooth transformation to refracting superposition is both possible and advantageous (Nilsson et al. 1987). Apart from butterflies, most lepidopterans, i.e., the moths, are more or less nocturnal and they do have refracting superposition eyes. It therefore seems likely that the evolutionary path outlined here has actually taken place in the Lepidoptera, perhaps more than once.

This does not mean that moths are descended from butterflies, but it does suggest an ancestral lepidopteran with diurnal habits, from which both butterflies and moths are derived. Of course, the reverse transformation from refracting superposition to afocal apposition is also possible, but not actually necessary, if a nocturnal animal converts into a strictly diurnal life.

2. The transparent apposition eyes of pelagic crustaceans have optical specializations in the crystalline cones, that replace screening pigment between the cones. Nilsson (1983a) showed that some transparent apposition eyes produce

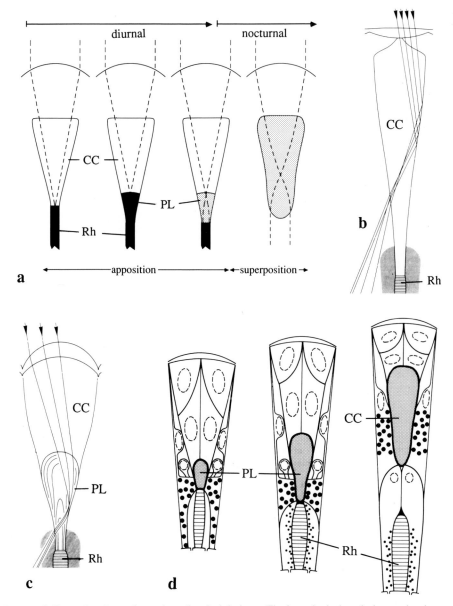

Fig. 13a-d. Examples of transformation of optical design. *a* The hypothetical evolutionary development from simple apposition, via afocal apposition, to refracting superposition (from Nilsson et al. 1987). *b* Unused "superposition rays" formed by reflection in the transparent apposition eye of a decapod larva. *c* A similar phenomenon caused by refraction in the transparent apposition eye of a euphausiid larva. *b* and *c* are from Nilsson (1983a). *d* The ontogenetic development from transparent apposition to refracting superposition design in euphausiids and mysids (from Nilsson et al. 1986). *CC* crystalline cone; *Rh* rhabdom; *PL* proximal lens

"superposition rays" as a by-product of their specialized mechanisms to obtain optical isolation without pigment. These superposition rays are formed by reflection inside the cone in the transparent eyes of larval decapods (Fig. 13b), which all rely on axial refractive index gradients. As adults many of these decapods transform their eyes into the reflecting superposition type. Larval euphausiids instead use a proximal cone-lens for optical isolation and, at least for oblique rays, this system produces an afocal ray path (Fig. 13c) of the same kind as that used in the refracting superposition eyes of adult euphausiids (see Fig. 8b). Thus, it seems that the two different strategies for obtaining transparent eyes in larval decapods and euphausiids have pre-adapted these eyes for reflecting and refracting super-position optics respectively. There are no nonfunctional intermediates involved, which is also demonstrated by the gradual way in which the larval apposition eyes turn into adult superposition (Land 1981c; Fincham 1984; Nilsson et al. 1986).

Although this path explains the evolution of reflecting superposition in decapods and refracting superposition in euphausiids, it does not seem to incor-porate the refracting superposition eyes of mysids, as these animals do not have free-living planktonic larvae. Instead, the whole postembryonic development occurs in the female marsupium. This problem is now resolved (Nilsson et al. 1986) because the postembryonic development of the mysid eye is shown to be identical to the larval development of the eye in euphausiids (Fig. 13d). Although not used, the mysid embryos possess a typical transparent apposition eye of the same kind as that of euphausiid larvae. The conclusion of this must be that mysids once used the transparent design, either as planktonic larvae or as adults, and that this has led to the refracting superposition eyes of present day adult mysids. It also shows that the ontogenetic development can be a useful tool in revealing the evolutionary processes.

3. The parabolic superposition eyes are perhaps the easiest to derive from an apposition type. Even a fully dark-adapted parabolic superposition eye employs an apposition mechanism for axial rays. Provided the crystalline cones have a re-fractive index that is slightly higher than that of the surrounding medium, reflections will occur, and if the number of ommatidia contributing to the super-position pupil is small, specializations in the crystalline cones are not necessary. The lightguides (rhabdoms) and corneal lenses are already present in the appo-sition eye. The parabolic profile of the cone, and the cylindrical lens or square cross-section are features that gradually become important as the clear zone widens and the superposition pupil becomes larger. In fact, almost every imaginable stage of development of a parabolic superposition eye is present today in various species of true crabs and hermit crabs (Nilsson 1988).

We have also seen in Section 4.2.3 that the correction of aberrations in the cylindrical lenses of some hermit-crab parabolic superposition eyes results in a design that is very close to a refracting superposition eye. This is possibly the reason why other hermit crabs have refracting superposition eyes. Since refracting superposition is the most aberration-free type of superposition eye, this trans-formation certainly makes sense.

A mixed occurrence of apposition eyes, presumed parabolic superposition eyes and refracting superposition eyes is also present in the mayflies (see Table 1). This parallel between crabs and mayflies could very well explain the presence of refracting superposition eyes in mayflies, and it is possible that a similar development accounts for superposition eyes in other insects as well.

We see that there are numerous ways by which apposition eyes can turn into superposition, and indeed, the reverse is also possible (for a summary see Fig. 14). The constraints put upon an eye that is optimized for a specific ecological niche may prevent a principal change in eye function, and clearly, a change in habitat provides the best opportunity to get around evolutionary obstacles. Although we are yet far from a complete understanding, the diversity of optical mechanisms is no longer impenetrable problem. There may be more yet unsuspected evolutionary transformations, but it is already obvious that evolution of optical types is still very much an on-going process.

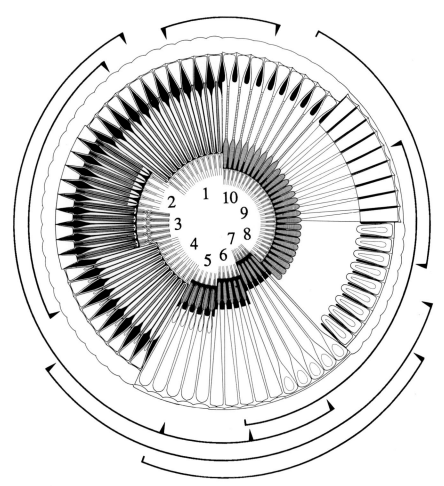

Fig. 14. A summary of the known optical types of compound eye, and the functionally possible evolutionary transitions that are discussed in the text. The eye types are: *1* simple apposition; *2* open rhabdom apposition; *3* neural superposition; *4* afocal apposition; *5, 6, 7* transparent apposition with radial gradient, axial gradient and with lightguides; *8* refracting superposition; *9* reflecting superposition; *10* parabolic superposition

Acknowledgment. My thanks are due to M.F. Land for critical and helpful discussion of the manuscript.

References

Autrum H (ed) (1981) Light and dark adaptation in invertebrates. In: Handbook of sensory physiology, vol VII/6C. Springer, Berlin Heidelberg New York, pp 1–91

Autrum H, Wiedemann I (1962) Versuche über den Strahlengang im Insektenauge (Appositionsauge). Z Naturforsch 17:480–482

Ball EE (1977) Fine structure of the compound eyes of the midwater amphipod *Phronima* in relation to behaviour and habitat. Tissue Cell 9:521–536

Ball EE, Kao LC, Stone RC, Land MF (1986) Eye structure and optics in the pelagic shrimp *Acetes sibogae* (Decapoda, Natantia, Sergestidae) in relation to light-dark adaptation and natural history. Philos Trans R Soc London Ser B 313:251–270

Barlow HB (1952) The size of ommatidia in apposition eyes. J Exp Biol 29:667–674

Beersma DGM, Hoenders BJ, Huiser AMJ, Toorn P van (1982) Refractive index of the fly rhabdomere. J Opt Soc Am 72:583–588

Behrens ME (1974) Photomechanical changes in the ommatidia of the *Limulus* lateral eye during light and dark adaptation. J Comp Physiol 89:45–57

Bernhard CG (ed) (1966) Opening address. In: The functional organization of the compound eye. Pergamon, Oxford New York, pp. 1–11

Boll F (1871) Beiträge zur physiologischen Optik. Arch Anat Physiol Wiss Med 1:530–549

Brammer JD (1970) The ultrastructure of the compound eye of a mosquito *Aedes aegypti* L. J Exp Zool 175:181–196

Bryceson KP (1981) Focusing of light by corneal lenses in a reflecting superposition eye. J Exp Biol 90:347–350

Bryceson KP, McIntyre P (1983) Image quality and acceptance angle in a reflecting superposition eye. J Comp Physiol A 151:367–380

Burghause F (1976) Adaptationserscheinungen in den Komplexaugen von *Gyrinus natator* L. (Coleoptera: Gyrinidae). Int J Insect Morphol Embryol 5:335–348

Bursey CR (1975) The microanatomy of the compound eye of *Munida irrasa* (Decapoda: Galatheidae). Cell Tissue Res 160:505–514

Burton PR, Stockhammer KA (1969) Electron microscopic studies of the compound eye of the toadbug, *Gelastocoris osulatus.* J Morphol 127:233–258

Burtt ET, Catton WT (1961) Is the mosaic theory of insect vision true? Int Congr Entanol 11:670–673

Burtt ET, Catton WT (1966) The role of diffraction in compound eye vision. In: Bernhard CG (ed) The functional organization of the compound eye. Pergamon, Oxford New York, pp 63–76

Caveney S (1986) The phylogenetic significance of ommatidium structure in the compound eyes of polyphagan beetles. Can J Zool 64:1787–1819

Caveney S, McIntyre P (1981) Design of graded-index lenses in the superposition eyes of scarab beetles. Phil Trans R Soc London Ser B 294:589–635

Chu H, Norris DM, Carlson SD (1975) Ultrastructure of the compound eye of the diploid female beetle, *Xyleborus ferrugineus.* Cell Tissue Res 165:23–26

Chun C (1896) Leuchtorgane und Facettenaugen. Bibl Zool 19:191–262

Cleary P, Deichsel G, Kunze P (1977) The superposition image in the eye of *Ephestia kühniella.* J Comp Physiol A 119:73–84

Cronin TW (1986) Optical design and evolutionary adaptation in crustacean eyes. J Crustacean Biol 6:1–23

Darwin C (1859) On the origin of species by means of natural selection. Oxford Univ Press

Debaisieux P (1944) Les yeux des Crustacés: structure, développement, réactions à l'éclairement. Cellule 50:9–122

Demoll R (1917) Die Sinnesorgane der Arthropoden, ihr Bau und ihre Funktion. Vieweg, Braunschweig

Denys CJ, Adamian M, Brown PK (1983) Ultrastructure of eye of a euphausiid crustacean. Tissue Cell 15:77–95

Dietrich W (1909) Die Facettenaugen der Dipteren Z Wiss Zool 92:465–539

Dóving KB, Miller WH (1969) Function of insect compound eyes containing crystalline tracts. J Gen Physiol 54:250–267

Edwards AS (1969) The structure of the eye of the *Ligia oceanica* L. Tissue Cell 1:217–228

Elofsson R, Hallberg E (1977) Compound eyes of some deep-sea and fiord mysid crustaceans. Acta Zool (Stockholm) 58:169–177

Elofsson R, Hallberg E, Nilsson HL (1980) The juxtaposed compound eye and organ of Bellonci in *Haploops tubicola* (Crustacea: Amphipoda) — The fine structure of the organ of Bellonci. Zoomorphologie 96:255–262

Eltringham H (1933) The senses of insects. Methuen's Biol Monogr, London

Exner S (1876) Uber das Sehen von Bewegungen und die Theorie des zusammengesetzten Auges. Sitz Ber Kaiserl Akad Wiss Math Nat Wiss 72:156–191

Exner S (1891) Die Physiologie der facettirten Augen von Krebsen und Insecten. Deuticke, Leipzig Wien

Fahrenbach WH (1975) The visual system of the horseshoe crab *Limulus polyphemus*. Int Rev Cytol 41:285–349

Fincham AA (1980) Eyes and classification of malacostracan crustaceans. Nature (London) 287:729–731

Fincham AA (1984) Ontogeny and optics of the eyes of the common prawn *Palaemon (Palaemon) serratus* (Pennant, 1777). Zool J Linnean Soc 81:89–113

Fischer A, Horstmann G (1971) Der Feinbau des Auges der Mehlmotte, *Ephestia kühniella* Zeller (Lepidoptera, Pyralididae). Z Zellforsch 116:275–304

Franceschini N (1975) Sampling of the visual environment by the compound eye of the fly: Fundamentals and applications. In: Snyder AW, Menzel R (eds) Photoreceptor optics. Springer, Berlin Heidelberg New York, pp 97–125

Gokan N, Hosobuchi K (1979a) Fine structure of the compound eyes of the longicorn beetles (Coleoptera: Cerambycidae). Appl Entomol Zool 14:12–27

Gokan N, Hosobuchi K (1979b) Ultrastructure of the compound eye of the Longicorn beetle, *Prionus insularis* Motschulsky (Coleoptera: Cerambycidae). Kontyû 47:105–116

Gottsche A (1852) Beitrag zur Anatomie und Physiologie des Auges der Krebse und Fliegen. Arch Anat Physiol Wiss Med 483–492

Grenacher H (1877) Untersuchungen über das Arthropodenauge. Beil Klin Monatsbl Augenheilkd 15

Grenacher H (1879) Untersuchungen über das Sehorgan der Arthropoden, inbesondere der Spinnen, Insecten und Crustaceen. Vandenhoek & Ruprecht, Göttingen

Grüel C (1844) Mikroskopische Beobachtungen. Ann Phys Chem 61:220–222

Hallberg E (1977) The fine structure of the compound eyes of mysids (Crustacea: Mysidacea). Cell Tissue Res 184:45–65

Hallberg E, Nilsson D-E (1983) The euphausiid (Crustacea: Euphausiacea) compound eye — a morphological re-investigation. Zoomorphology 103:59–66

Hallberg E, Andersson M, Nilsson D-E (1980a) Responses of the screening pigment in the compound eye of *Neomysis integer* (Crustacea: Mysidacea). J Exp Zool 212:397–402

Hallberg E, Nilsson HL, Elofsson R (1980b) Classification of amphipod compound eyes — the fine structure of the ommatidial units (Crustacea, Amphipoda). Zoomorphologie 94:279–306

Hardie RC (1985) Functional organization of the fly retina. Progr Sens Physiol 5:1–79

Hardie RC (1986) The photoreceptor array of the dipteran retina. Trends Neuro Sci 9:419–423

Hariyama T, Meyer-Rochow VB, Eguchi E (1986) Diurnal changes in structure and function of the compound eye of *Ligia exotica* (Crustacea, Isopoda). J Exp Biol 123:1–26

Hateren JH van (1984) Waveguide theory applied to optically measured angular sensitivities of fly photoreceptors. J Comp Physiol A 154:761–771

Hateren JH van, Nilsson D-E (1987) Butterfly optics exceed the theoretical limits of conventional apposition eyes. Biol Cybernet 57:159–168

Hausen K (1973) Die Brechungsindices im Kristallkegel der Mehlmotte *Ephestia kühniella*. J Comp Physiol 82:365–378

Hesse R (1901) Untersuchungen über die Organe der Lichtempfindung bei niederen Thieren. VII. Von den Arthropoden-Augen. Z Wiss Zool 70:347–473

Hodierna DG (1644) L'occhio della mosca. Decio Cirillo, Palermo

Home EM (1976) The fine structure of some carabid beetle eyes, with particular reference to ciliary structures in the retinula cells. Tissue Cell 8:311–333

Hooke R (1665) Micrographia or some physiological descriptions of minute bodies made by magnifying glasses. Martin & Allestry, London

Horridge GA (1968) Pigment movement and the crystalline threads of the firefly eye. Nature (London) 218:778–779

Horridge GA (1969) The eye of *Dytiscus* (Coleoptera). Tissue Cell 1:425–442

Horridge GA (1972) Further observations on the clear zone eye of *Ephestia*. Proc R Soc London Ser B 181:157–173

Horridge GA (ed) (1975) Optical mechanisms of clear zone eyes. In: The compound eye and vision of insects. Oxford Univ Press, pp 295–298

Horridge GA (1976) The ommatidium of the dorsal eye of *Cloeon* as a specialization for photoreisomerization. Proc R Soc London Ser B 193:17–29

Horridge GA, Barnard PBT (1965) Movement of palisade in locust retinula cells when illuminated. Q J Microsc Sci 106:131–135

Horridge GA, Giddings C (1971) The retina of *Ephestia* (Lepidoptera). Proc R Soc London Ser B 179:87–95

Horridge GA, Henderson I (1976) The ommatidium of the lacewing *Chrysopa* (Neuroptera). Proc R Soc London Ser B 192:259–271

Horridge GA, Giddings C, Stange G (1972) The superposition eye of skipper butterflies. Proc R Soc London Ser B 182:457–495

Horridge GA, McLean M, Stange G, Lillywhite PG (1977) A diurnal moth superposition eye with high resolution *Phalaenoides tristifica* (Agaristidae). Proc R Soc London Ser B 196:233–250

Horridge GA, Duniec J, Marçelja L (1981) A 24-hour cycle in single locust and mantis photoreceptors. J Exp Biol 91:307–322

Horridge GA, Marçelja L, Jahnke R (1982) Light guides in the dorsal eye of the male mayfly. Proc R Soc London Ser B 216:25–51

Ioannides AC, Horridge GA (1975) The organization of visual fields in the hemipteran acone eye. Proc R Soc London Ser B 190:373–391

Kampa EM (1963) The structure of the eye of a galatheid crustacean, *Pleuroncodes planipes*. Crustaceana 6:69–80

Kirschfeld K (1967) Die Projektion der optischen Umwelt auf das Raster der Rhabdomere im Komplexauge von *Musca*. Exp Brain Res 3:248–270

Kirschfeld K, Franceschini N (1969) Ein Mechanismus zur Steuerung des Lichtflusses in den Rhabdomeren des Komplexauges von *Musca*. Kybernetik 6:13–22

Kuiper JW (1962) The optics of the compound eye. Symp Soc Exp Biol 16:58–71

Kunze P (1969) Eye glow in the moth and superposition theory. Nature (London) 223:1172–1174

Kunze P (1970) Verhaltensphysiologische und optische Experimente zur Superpositionstheorie der Bildentstehung in Komplexaugen. Verh Dtsch Zool Ges 64:234–238

Kunze P (1972) Comparative studies of arthropod superposition eyes. Z Vergl Physiol 76:347–357

Kunze P (1979) Apposition and superposition eyes. In: Autrum H (ed) Handbook of sensory physiology, vol VII/6A. Springer, Berlin Heidelberg New York, pp 441–502

Kunze P, Hausen K (1972) Inhomogeneous refractive index in the crystalline cone of a moth eye. Nature (London) 231:392–393

Kuster JE (1979) Comparative structure of compound eyes of Cicindelidae and Carabidae (Coleoptera): evolution of scotopy and photopy. Quaest Entomol 15:297–334

Land MF (1976) Superposition images are formed by reflection in the eyes of some oceanic decapod crustacea. Nature (London) 263:764–765

Land MF (1981a) Optics and vision in invertebrates. In: Autrum H (ed) Handbook of sensory physiology, vol VII/6B. Springer, Berlin Heidelberg New York, pp 471–592

Land MF (1981b) Optics of the eyes of *Phronima* and other deep-sea amphipods. J Comp Physiol A 145:209–226

Land MF (1981c) Optical mechanisms in the higher Crustacea with a comment on their evolutionary origins. In: Laverack MS, Cosens DJ (eds) Sense organs. Blackie, Glasgow, pp 31–48

Land MF (1984) The resolving power of diurnal superposition eyes measured with an ophthalmoscope. J Comp Physiol A 154:515–533

Land MF (1985) The eye: optics. In: Kerkut GA, Gilbert LI (eds) Comprehensive insect physiology biochemistry and pharmacology. Pergamon Press, Oxford New York, pp 225–275

Land MF, Burton FA (1979) The refractive index gradient in the crystalline cones of the eyes of a euphausiid crustacean. J Exp Biol 82:395–398

Land MF, Burton FA, Meyer-Rochow VB (1979) The optical geometry of euphausiid eyes. J Comp Physiol A 130:49–62

Leydig F (1855) Zum feineren Bau der Arthropoden. Müller's Arch Anat Physiol 22:406–444

Leydig F (1864) Das Auge der Gliedertiere. Tübinger Universitätsschr, Tübingen

Mallock A (1894) Insect sight and the defining power of composite eyes. Proc R Soc London Ser B 55:85–90

McIntyre P, Caveney S (1985) Graded-index optics are matched to optical geometry in the superposition eyes of scarab beetles. Philos Trans R Soc London Ser B 311:237–269

McLean M, Horridge GA (1977) Structural changes in light-and dark-adapted compound eyes of the Australian earwig *Labidura riparia truncata* (Dermaptera). Tissue Cell 9:653–666

Meinecke CC (1981) The fine structure of the compound eye of the African armyworm moth, *Spodoptera exempta* Walk. (Lepidoptera, Noctuidae). Cell Tissue Res 216:333–347

Menzi U (1987) Visual adaptation in nocturnal and diurnal ants. J Comp Physiol A 160:11–22

Meyer-Rochow VB, Waldvogel H (1979) Visual behaviour and the structure of dark and light-adapted larval and adult eyes of the New Zealand glowworm *Arachnocampa luminosa* (Mycetophilidae: Diptera). J Insect Physiol 25:601–613

Meyer-Rochow VB, Walsh S (1978) The eyes of mesopelagic crustaceans: III. *Thysanopoda tricuspidata* (Euphausiacea). Cell Tissue Res 195:59–79

Miller WH, Bernard GD, Allen JL (1968) The optics of insect compound eyes. Science 162:760–767

Miltz O (1899) Das Auge der Polyphemiden. Zoologica (Stuttgart) 28:1–61

Müller J (1826) Zur vergleichenden Physiologie des Gesichtsinnes. Cnobloch, Leipzig

Nässel DR, Waterman TH (1979) Massive diurnally modulated photoreceptor membrane turnover in crab light and dark adaptation. J Comp Physiol A 131:205–216

Nilsson D-E (1982) The transparent compound eye of *Hyperia* (Crustacea): Examination with a new method for analysis of refractive index gradients. J Comp Physiol A 147:339–349

Nilsson D-E (1983a) Evolutionary links between apposition and superposition optics in crustacean eyes. Nature (London) 302:818–821

Nilsson D-E (1983b) Refractive index gradients subserve optical isolation in a light-adapted reflecting superposition eye. J Exp Zool 225:161–165

Nilsson D-E (1988) A new type of imaging optics in compound eyes. Nature (London) 332:76–78

Nilsson D-E, Nilsson HL (1981) A crustacean compound eye adapted for low light intensities (Isopoda). J Comp Physiol A 143:503–510

Nilsson D-E, Nilsson HL (1983) Eye camouflage in the isopod crustacean *Astacilla longicornis* (Sowerby). J Exp Mar Biol Ecol 68:105–110

Nilsson D-E, Odselius R (1981) A new mechanism for light-dark adaptation in the *Artemmia* compound eye (Anostraca, Crustacea). J Comp Physiol A 143:389–399

Nilsson D-E, Odselius R (1983) Regionally different optical systems in the compound eye of the water-flea *Polyphemus* (Cladocera, Crustacea). Proc R Soc London Ser B 217:163–175

Nilsson D-E, Odselius R, Elofsson R (1983a) The compound eye of *Leptodora kindtii* (Cladocera): An adaptation to planktonic life. Cell Tissue Res 230:401–410

Nilsson D-E, Andersson M, Hallberg E, McIntyre P (1983b) A micro-interferometric method for analysis of rotation-symmetric refractive-index gradients in intact objects. J Microsc 132:21–29

Nilsson D-E, Land MF, Howard J (1984) Afocal apposition optics in butterfly eyes. Nature (London) 312:561–563

Nilsson D-E, Hallberg E, Elofsson R (1986) The ontogenetic development of refracting superposition eyes in crustaceans: Transformation of optical design. Tissue Cell 18:509–519

Nilsson D-E, Land MF, Howard J (1988) Optics of the butterfly eye. J Comp Physiol A 162:341–366

Parker A (1891) The compound eyes in crustaceans. Bull Mus Comp Zool 21:45–140

Pask C, Barrell KF (1980a) Photoreceptor optics I: Introduction to formalism and excitation in a lens-photoreceptor system. Biol Cybernet 36:1–8

Pask C, Barrell KF (1980b) Photoreceptor optics II: Application to angular sensitivity and other properties of a lens-photoreceptor system. Biol Cybernet 36:9–18

Patten W (1886) Eyes of molluscs and arthropods. Mitt Zool Stn Neapel VI:542–756

Paulus HF (1979) Eye structure and the monophyly of the Arthropoda. In: Gupta AP (ed) Arthropod phylogeny. Van Nostrand Reinhold, New York, pp 299–383

Ribi WA (1978) Ultrastructure and migration of screening pigments in the retina of *Pieris rapae* L. (Lepidoptera, Pieridae). Cell Tissue Res 191:57–73

Rossel S (1979) Regional differences in photoreceptor performance in the eye of the praying mantis. J Comp Physiol A 131:95–112

Salvini-Plawen L v, Mayr E (1977) On the evolution of photoreceptors and eyes. In: Hecht MK, Sterre WC, Wallace B (eds) Evolutionary biology, vol 10. Plenum, New York, pp 207–263

Schmitt M, Mischke U, Wachmann E (1982) Phylogenetic and functional implications of the rhabdom patterns in the eyes of Chrysomeloidea (Coleoptera). Zool Scr 11:31–44

Schneider L, Langer H (1969) Die Struktur des Rhabdoms im "Doppelauge" des Wasserläufers *Gerris lacustris.* Z Zellforsch 99:538–559

Schneider L, Gogala M, Draslar K, Langer H, Schlecht P (1978) Feinstruktur und Schirmpigment-Eigenschaften der Ommatidien des Doppelauges von *Ascalaphus* (Insecta, Neuroptera). Cytobiology 16:274–307

Schultze M (1868) Untersuchungen über die zusammengesezten Augen der Krebse und Insekten. Cohen, Bonn

Seitz G (1969) Untersuchungen am dioptrischen Apparat des Leuchtkäferauges. Z Vergl Physiol 62:61–74

Seitz G (1971) Bau und Funktion des Komplexauges der Schmeißfliege. Naturwissenschaften 58:258–265

Snyder AW (1975) Photoreceptor optics – theoretical principles. In: Snyder AW, Menzel R (eds) Photoreceptor optics. Springer, Berlin Heidelberg New York, pp 38–55

Snyder AW (1977) Acuity of compound eyes: physical limitations and design. J Comp Physiol A 116:161–182

Snyder AW (1979) Physics of vision in compound eyes. In: Autrum H (ed) Handbook of sensory physiology, vol VII/6A. Springer, Berlin Heidelberg New York, pp 225–313

Stavenga DG (1974a) Refractive index of fly rhabdomeres J Comp Physiol A 91:417–426

Stavenga DG (1974b) Waveguide modes and refractive index in photoreceptors of invertebrates. Vision Res 15:323–330

Stavenga DG (1979) Pseudopupils of compound eyes. In: Autrum H (ed) Handbook of sensory physiology, vol VII/6A. Springer, Berlin Heidelberg, New York, pp 357–439

Stavenga DG, Kuiper JW (1977) Insect pupil mechanisms I. On the pigment migration in the retinula cells of Hymenoptera (Suborder Apocrita). J Comp Physiol A 113:55–72

Stavenga DG, Numan JAJ, Tinbergen J, Kuiper JW (1977) Insect pupil mechanisms II. Pigment migration in retinula cells of butterflies. J Comp Physiol A 113:73–93

Strausfeld NJ (1976) Atlas of an insect brain. Springer, Berlin Heidelberg New York

Swammerdam J (1737) Bibia Naturae sive Historia Insectorum. Boerhaave H (ed) Severinus, Leyden

Vigier P (1907) Sur la reception de l'exitant lumineux dans les yeux composés des insectes, en particulier chez les muscides. C R Acad Sci Paris 63:633–636

Vigier P (1909) Mecanisme de la synthèse des impressions lumineuses par les yeux composés des Diptères. C R Acad Sci Paris 148:1221–1223

Vogt K (1975) Zur Optik des Flußkrebsauges. Z naturforsch 30:691

Vogt K (1977) Ray path and reflection mechanisms in crayfish eyes. Z Naturforsch 32:466–468

Vogt K (1980) Die Spiegeloptik des Flußkrebsauges. J Comp Physiol A 135:1–19

Wachmann E (1977) Vergleichende Analyse der feinstrukturellen Organisation offener Rhabdome in den Augen der Cucujiformia (Insecta, Coleoptera), unter besonderer Berücksichtigung der Chrysomelidae. Zoomorphologie 88:95–131

Wachmann E (1979) Untersuchungen zur Feinstruktur der Augen von Bockkäfern (Coleoptera, Cerambycidae). Zoomorphologie 92:19–48

Walcott B (1971) Cell movement on light adaptation in the retina of *Lethocerus* (Belostomatidae, Hemiptera). Z Vergl Physiol 74:1–16

Walcott B (1975) Anatomical changes during light adaptation in insect compound eyes. In: Horridge GA (ed) The compound eye and vision of insects. Clarendon, Oxford, pp 20–36

Wehner R (1981) Spatial vision in arthropods. In: Autrum H (ed) Handbook of sensory physiology, vol VII/6C. Springer, Berlin Heidelberg New York, pp 287–616

Welsch B (1977) Ultrastruktur und funktionelle Morphologie der Augen des Nachtfalters *Deilephila elpenor* (Lepidoptera, Sphingidae). Cytobiology 14:378–400

Welsh JH, Chace FA (1937) Eyes of deep sea crustaceans. I Acanthephyridae. Biol Bull 72:57–74

Williams DS (1980) Organisation of the compound eye of a tipulid fly during the day and night. Zoomorphologie 95:85–104

Williams DS (1982) Ommatidial structure in relation to turnover of photoreceptor membrane in the locust. Cell Tissue Res 225:595–617

Wolburg-Buchholz K (1976) The dorsal eye of *Cloëon* dipterum (Ephemeroptera): A light and electronmicroscopical study. Z Naturforsch 31c:335–336

Young S, Downing AC (1976) The receptive fields of *Daphnia* ommatidia. J Exp Biol 64:185–202

Zeil J (1979) A new kind of neural superposition eye: the compound eye of male Bibionidae. Nature (London) 278:249–250

Zeil J (1983) Sexual dimorphism in the visual system of flies: the compound eyes and neural superposition in Bibionidae (Diptera). J Comp Physiol A 150:379–393

Zharkova IS (1975) Reduction of organs of sight in deep-water Isopoda, Amphipoda and Decapoda. Zool Zh 54:200–208

Chapter 4

Photoreceptor Optics, Theory and Practice

J. HANS VAN HATEREN, Groningen, The Netherlands

1 Introduction

According to Snyder and Menzel (1975b), the goal of photoreceptor optics is to explain the structural basis of a photoreceptor's absolute, spectral, directional, and polarization sensitivities. This review will concentrate, however, on the directional sensitivity, as this is the quality, together with the absolute sensitivity, most influenced by photoreceptor optics and most thoroughly investigated both theoretically and experimentally. For other reviews on related topics see Snyder 1975, 1979; Snyder and Love 1983; and Horowitz 1981.

The optics of well-developed eyes in general consists of two parts: the first part, forming an image of the surroundings, funnels the light collected into the second part of the optical system, the light-sensitive structures. These structures usually consist of lightguides containing visual pigment. Absorption of light by this pigment eventually leads to electrical signals transmitted to the brain. The most common design, and the one theoretically best understood, is the combination of lens and waveguide. Many variations on this theme are found, though (see the reviews of Land and Nilsson this Vol.). The lens-waveguide system is found in many compound eyes, but also in camera-type eyes, such as those of vertebrates.

Although physiological optics is not a new discipline (e.g., Exner 1891), only quite recently has it received an impetus from wave optics. Several investigators noted the fact that the size of the lens(es) imposes a constraint on the resolution an eye can reach, due to diffraction (Mallock 1922, Barlow 1952, de Vries 1956). Toraldo di Francia (1949) was the first to compare the directional sensitivity of a visual lightguide to the receptive field of an antenna, and suggested that the visual lightguide functions as a dielectric waveguide. Snitzer and Osterberg (1961) developed mathematical expressions for waveguide modes, and observed them in small glass fibers. Enoch (1961, 1963) observed modes in retinae of vertebrates.

Photoreceptor optics received a great impetus through the theoretical work of Snyder (1969a,b), who developed approximations for waveguide modes that were more convenient to handle, and applied concepts from waveguide optics to eyes (e.g., Snyder and Pask 1973). The beginning of the 1970's witnessed a blooming of this area, culminating in a conference in 1974 entirely devoted to photoreceptor optics (Snyder and Menzel 1975a). As sometimes happens, theory was ahead of experiment for some time. Only recently has experimental work partly caught up through work on fly and butterfly (van Hateren 1984, 1985; Nilsson et al. 1984, 1988; van Hateren and Nilsson 1987).

Stavenga/Hardie (Eds.) Facets of Vision
© Springer-Verlag Berlin Heidelberg 1989

2 Methods

Physiological optics is blessed with a wide variety of theoretical and experimental methods. Below we will see how these methods can be applied to the best-investigated system, a lens with a waveguide in its focal plane, but many of the results are valid for other designs as well.

2.1 Theoretical Methods

In the next sections we will use several of the following theoretical approaches. *Geometrical optics* uses light rays to explain optical phenomena, such as imaging by a lens and guiding of light in a lightguide. It can deal with more complicated structures through the method of ray tracing, but it fails on a scale smaller than a few wavelengths of light, i.e., about a micrometer. Then we enter the realm of *(vectorial) wave optics*, which is closely related to Maxwell's equations for the electromagnetic field. Problems tend to be difficult to handle, due to the vectorial nature of the electromagnetic field. Fortunately, vectorial wave optics has a simpler approximation, *scalar wave optics*, where a single scalar quantity, called the amplitude of the field, replaces the electric and magnetic field vectors. The intensity of the light is proportional to the squared modulus of this amplitude. Scalar wave optics gives accurate results if the refractive indices vary only slightly over distances in the order of the wavelength of light. It explains diffraction at a lens, and the emergence of waveguide modes in a waveguide, but it cannot always deal successfully with polarization. A most useful approach based on scalar wave optics is *Fourier optics* (Goodman 1968). This approach decomposes amplitude distributions into plane waves, which explains diffraction in a very natural way. Furthermore, it leads to a simple and basic understanding of how lenses work.

2.1.1 The Lens

Depending on the viewpoint a lens is: (1) a device that focuses light rays (geometrical optics, Fig. 1a), (2) that converts an incoming plane wave into part of a spherical wave, propagating into a diffraction pattern (wave optics, Fig. 1b), or (3) that yields Fourier transforms of amplitude distributions (Fourier optics, Fig. 1b and c).

As the geometrical optics of a lens (Fig. 1a) will be familiar to all readers, we will turn right away to wave optics. Figure 1b shows that a lens converts an incoming plane wave into part of a spherical wave. If this spherical wave were complete, it would produce a small spot of maximum intensity in the center with roughly the size of a wavelength. But information is lacking because there is only part of a spherical wave, with the result that the small spot in its center is enlarged to a diffraction pattern. The smaller the diameter of the lens (with the same focal distance), the smaller the part of the spherical wave, and the larger the diffraction pattern that results. A circular lens thus produces the well-known Airy diffraction pattern.

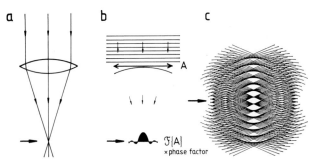

Fig. 1a-c. Optics of a lens. *a* Geometrical optics: a lens focuses light rays. *b* Wave optics: a lens converts a plane wave into part of a spherical wave. The resulting Airy diffraction pattern is the Fourier transform of the lens aperture, times a phase factor. *c* Fourier optics: a diffraction pattern can be decomposed into a set of plane waves with an angular distribution filling the lens aperture exactly. *Arrows* indicate the focal plane

Fourier optics considers lenses as devices that yield Fourier transforms. The Fourier transform of the aperture of a lens (e.g., circular or square) yields the amplitude of the diffraction pattern, apart from a phase factor (see Goodman 1968). Another way of putting this Fourier relationship is that the diffraction pattern can be thought of as being composed of a distribution of plane waves with directions filling the lens aperture exactly. Figure 1c illustrates how a superposition of plane waves produces a spherical wave front at the lens and a diffraction pattern at the focus.

2.1.2 The Waveguide

Geometrical optics explains the guiding of light in a lightguide by total reflection of light rays (Fig. 2a, left); a light ray is totally reflected if $n_1 > n_2$ and the angle α is sufficiently small [$\alpha \leq \arccos(n_2/n_1)$, with n_1 and n_2 the refractive indices inside and outside the waveguide respectively]. This explanation fails, however, for lightguides with diameters in the order of the wavelength of light (about 0.5 μm), which is often the case for visual lightguides. Wave optics must be used for describing light propagation in these small lightguides, then called waveguides. Interference of the waves inside a waveguide leads to waveguide modes, stable patterns of light travelling along the waveguide. They are the only way light can propagate through it.

The modes of a cylindrical waveguide can be obtained directly from the Maxwell equations (vectorial wave optics). This leads to modes designated as HE_{11}, TM_{01}, etc. (Snitzer and Osterberg 1961). These modes are exact solutions for a waveguide, fully describing the vectorial nature of the electromagnetic field. Snyder (1969a, b) introduced a more convenient approximation, assuming that the refractive indices inside and outside the waveguide are only slightly different — which is usually the case in visual waveguides. This approximation leads to a scalar wave equation, and thus belongs to scalar wave optics (see Marcuse 1974). This yields modes which are designated as LP_{01}, LP_{11}, etc. LP means Linearly Polarized:

the modes are linearly polarized superpositions of the exact modes, which are not linearly polarized (apart from HE_{11}, which is identical to LP_{01}). As the scalar approximation is much simpler to handle and understand, we will use it in the following.

Figure 2a (right) shows an LP_{01} travelling in a waveguide. When it leaves the waveguide, it will radiate away with a certain angular amplitude distribution, which is, far away from the waveguide aperture, called the far field radiation

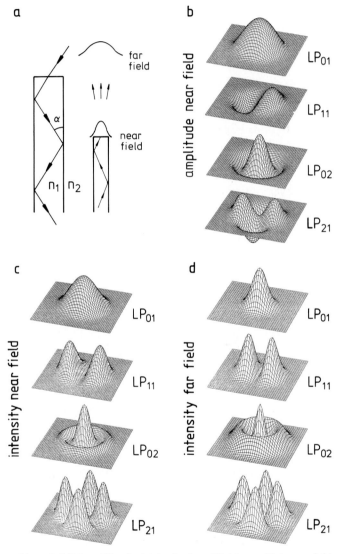

Fig. 2a-d. Optics of a waveguide. *a Left* light guiding by total reflection of light rays. *Right* near field pattern of a mode LP_{01} radiates away to its far field pattern. *b* Amplitude distributions of the near field patterns of the first four modes. *c* Near field intensity distributions of the first four modes *d* Far field intensity distributions of the first four modes

pattern. Figure 2b shows the amplitude distributions of the first four waveguide modes (called the near field to distinguish them from the far field distributions), Figure 2c shows their intensity distributions, and Figure 2d the intensity distributions of their far field radiation patterns. The indices designating each mode have a simple interpretation: the first digit is half the number of maxima in the intensity distribution encountered azimuthally, the second digit the number of maxima encountered going from the center toward infinity. How many modes can propagate in a waveguide is determined by an important parameter, the V-number:

$$V = \frac{2\pi b}{\lambda} \sqrt{n_2^1 - n_2^2},$$

with b the radius of the waveguide, λ the wavelength of the light, n_1 the refractive index inside, and n_2 outside the waveguide. If $V \leq 2.4$, only LP_{01} can propagate, if $V \leq 3.8$ both LP_{01} and LP_{11}, if $V \leq 5.1$ all four modes shown in Fig. 2b, and if $V > 5.1$ other modes as well. As an example, fly photoreceptors (R1–6) have $n_1 \approx 1.36$, $n_2 \approx 1.34$ (Stavenga 1974, 1975; Beersma et al. 1982), and b ≈ 0.9 μm (Horridge et al. 1976), which yield $V = 2.2$ for $\lambda = 600$ nm, and $V = 3.3$ for $\lambda = 400$ nm (Kirschfeld and Snyder 1976). Thus at 600 nm only LP_{01} can exist in fly photoreceptors, whereas at 400 nm LP_{11} also exists.

Fourier optics is useful for waveguides because it relates any amplitude distribution to its far field radiation pattern: these are Fourier transforms of each other. Thus the near field of a mode is related to its far field in a simple way: their amplitudes are each other's Fourier transforms. We will show in the next section that this notion can give us insight into how a lens-waveguide system works.

2.1.3 The Lens-Waveguide System

Figure 3a shows how geometrical optics explains a lens-waveguide system. Light rays coming from the lens and entering the lightguide will be trapped inside the fiber through total reflection if the angle α is sufficiently small. Ideally, the lens aperture should be matched to the maximum α allowing total reflection. The angular sensitivity of the photoreceptor, i.e., its receptive field, is the projection of the fiber aperture to infinity. Thus the fiber aperture determines the width of the angular sensitivity. This shows the weakness of this model: geometrical optics does not take diffraction at the lens into account, and this is only accurate if the fiber aperture is much larger than the diffraction pattern of the lens.

Figure 3b shows a more realistic picture of how light enters the waveguide: an incoming plane wave is first diffracted by the lens, and the resulting diffraction pattern subsequently excites waveguide modes in the waveguide, which are finally absorbed by the visual pigment (see Sect. 2.1.4). How strong the various modes are excited depends on how well their amplitude distributions fit with the exciting amplitude distribution, in this case the Airy diffraction pattern produced by a distant point light source. This goodness of fit depends, of course, on the relative position of diffraction pattern and waveguide aperture, and thus on the angular position of the stimulus. For example, on-axis illumination will only excite

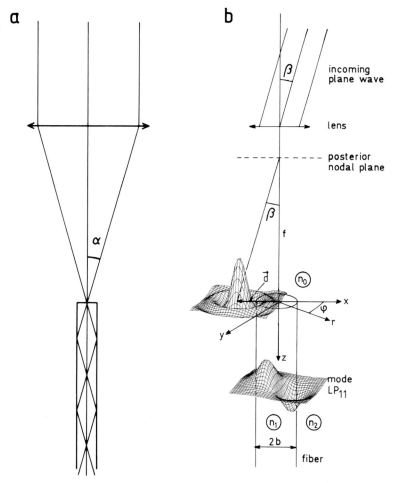

Fig. 3a,b. Optics of a lens-waveguide system. *a* Light rays focused by the lens are totally reflected inside the lightguide. *b* Wave optics: a diffraction pattern excites modes in a waveguide. The waveguide shown is large enough to support LP_{11} (thus LP_{01} is also present, not shown)

azimuthally symmetrical modes (LP_{01}, and LP_{02} etc.) because the positive and negative parts of the azimuthally asymmetrical modes cancel each other with symmetrical illumination. Off-axis illumination, on the other hand, may also excite asymmetrical modes like LP_{11} — if the waveguide dimensions allow their existence. The excitation of the modes as a function of the stimulus position thus leads to the angular sensitivity of the system. In mathematical terms it is for each mode the convolution of mode and diffraction pattern.

Fourier optics is most illuminating when used with an extremely useful method, both theoretically and experimentally, namely reversal of the light direction. From the reciprocity theorem of Helmholtz (see, e.g., Born and Wolf 1965) it follows that the sensitivity of a photoreceptor to light travelling in the normal direction (orthodromic light) is similar in shape to the intensity distribution obtained when the light direction is reversed (antidromic light). Some care when applying this reciprocity theorem is necessary, however, because it only applies to each mode separately if the modes are absorbed independently (see Sect. 2.1.4). We will now illustrate the power of this method.

Suppose an LP_{01} is travelling through the waveguide in the antidromic direction. After leaving the fiber aperture it transforms gradually from near field pattern to far field pattern (the Fourier transform the mode). At the lens we thus find in good approximation the far field radiation pattern of LP_{01}, with its phase front spherical with the fiber aperture as its center. Just as a lens converts a plane wave front into a spherical one, it will convert a spherical wave front into a plane one. Therefore, the light emerging from the facet lens has a plane phase front, and an amplitude distribution equal to the far field pattern of the waveguide, but limited (cut off) by the lens aperture. This light will subsequently radiate toward infinity, which means another Fourier transform (see end of Sect. 2.1.2). Consequently, the pattern we find at infinity is the Fourier transform of the Fourier-transformed mode (the far field of the mode) restricted by the lens aperture. A mathematical theorem states that the Fourier transform (actually the inverse Fourier transform, but this leads only to a reflection of coordinates which we will ignore) of the product of two functions, in this case the Fourier-transformed mode and the lens aperture, equals the convolution of their (inverse) Fourier transforms, in this case the mode and the diffraction pattern. This is the same result as we mentioned above for the angular sensitivity. Thus

$$R = F\{F\{M\} \times L\} = F\{F\{M\} \times F\{D\}\} = M * D = S,$$

where R is the far field radiation pattern of the system (or, from Helmholtz's reciprocity theorem, the angular sensitivity S), F denotes a Fourier transform, M the mode pattern, L the lens aperture, D the diffraction pattern, and * convolution. Efficient excitation of a mode by the diffraction pattern depends on how well they fit, and we have just shown that this is equivalent to how well the far field pattern of the mode matches the lens aperture. Details and formulas can be found in van Hateren 1984 and 1985.

Van Hateren and Nilsson (1987) show that in butterflies this matching of mode pattern and diffraction pattern is even better than in conventional lens-waveguide systems, resulting in an improved resolution and sensitivity. They suggest that this is accomplished by the cone stalk of the butterfly through matching the diffraction pattern with a combination of LP_{01} and LP_{02} and subsequent coupling of these modes. The cone stalk is an optically very complicated structure, showing the properties of a strong lens (Nilsson et al. 1984, 1988, van Hateren and Nilsson 1987).

2.1.4 Absorption of Modes

When modes are excited in a waveguide, they will travel along it, and be absorbed by the visual pigment it contains. An important notion is the following: although the modes are coherent, their absorption is independent, which means that the intensity of each mode determines how strongly it is absorbed (see van Hateren and Nilsson 1987). This independent absorption is caused by the fact that the modes beat with each other due to different wave velocities (see, e.g., Bernard 1975). Because the beat period is typically in the order of 10–20 μm, thus much shorter than most visual waveguides, the modes appear to be absorbed independently. The reverse is true for another kind of beating, namely of the vectorial modes that compose the LP modes. These vectorial modes also have slightly different wave velocities, but their beat period is typically in the order of 5–10 mm, thus much longer than a typical visual waveguide. Therefore, the absorption of these modes is in effect not independent, which is the reason why the LP modes describe absorption accurately. An experimental technique introduced by Nilsson et al. (1984) for the butterfly, however, looks at light that is depolarized during passage through the waveguide. For explaining this effect one needs the vectorial modes, and the fact that they are beating (Nilsson and Howard in prep.).

Modes can also be absorbed by the longitudinal pupil of the photoreceptors. In bright light this pupil closes, i.e., pigment granules in the photoreceptor cells move close to the waveguide (Kirschfeld and Franceschini 1969). As only part of the power of a mode is propagated inside the waveguide (this fraction is called η, see Snyder 1975) the remaining part, travelling alongside the waveguide, can be absorbed by the pigment granules. The fraction η is different for different modes (higher order modes have a smaller η than lower order ones), thus the pupil affects higher order modes more. One consequence of this is that the pupil can influence the angular sensitivity of the photoreceptors (Smakman et al. 1984; Nilsson et al. 1988). For a theoretical discussion of absorption by the pupil see Snyder 1975.

2.2 Experimental Methods

2.2.1 Physiological Methods

Physiological methods for studying physiological optics are defined here as those methods that use the sensitivity of the photoreceptor cells, either directly or indirectly. Indirect methods use higher-order neurons or the behavior of the animal. For example, the halfwidth of the angular sensitivity of housefly photo-receptors was inferred from behavioral experiments (Götz 1965; Buchner 1976). Another example is the psychophysical determination of the sensitivity profile of human photoreceptors at the level of the cornea (Stiles-Crawford effect, review: Enoch and Bedell 1981).

Direct methods use the response of the photoreceptor cells, either their membrane potential (e.g., Washizu et al. 1964) or their pupillary response (Beersma 1979; Bernard and Wehner 1980). Angular sensitivities are usually obtained from responses to flashes of constant intensity at different positions in the receptive field of the cell. Afterward a correction is then necessary for the (usually nonlinear) stimulus-response curve of the cell. Methods not needing this correction are counting responses to single photons, and clamping of the response to a constant level by feedback to the intensity of the stimulus (Franceschini 1979; Smakman and Pijpker 1983). Finally, angular sensitivities can also be inferred from the response of the visual sense cells to sinusoidal gratings of varying spatial frequency, by Fourier transforming the so-obtained optical transfer function (Dubs 1982; see Goodman 1968 for an explanation of the method).

2.2.2 Optical Methods

Optical methods for studying physiological optics are defined here as those methods that aim at determining directly optical properties of the investigated systems. One approach is to determine structure and properties, e.g., refractive indices, of the optical components through anatomy and interference microscopy. An early example of this is Exner's work on *Limulus* cones (Exner 1891). Recent examples are the work of Seitz (1968) on the fly optical system (see also Stavenga 1974), of Varela and Wiitanen (1970) on the honeybee optical system, and of Nilsson and Odselius (1981), Nilsson et al. (1983), and Nilsson (1983) on crustaceans, combining anatomy, interference microscopy, and ray tracing. A limit to this approach is the wavelength of light: it is next to impossible to obtain reliable refractive indices about structures varying at the scale of wavelengths. Unfortunately, this is also a range where theorizing is difficult and where interesting properties are likely.

It is also possible to directly obtain optical properties of the whole system, like its angular sensitivity. This approach can be subdivided into orthodromic and antidromic methods. An example of an orthodromic method is the technique using eye slices, where the light transmitted through the waveguides is monitored at their cut end, with the remaining optics intact (Kuiper 1966; Eheim and Wehner 1972). This method works well if only LP_{01} is present. If more modes are propagated, however, it can give erroneous results, because the modes are absorbed independently (see Sect. 2.1.4). This absorption will be different for each mode, thus the weighting of the various modes will not be the same for the light transmitted by the waveguide (used in this method) and the light absorbed by the waveguide (determining the angular sensitivity of the photoreceptor cell).

Antidromic methods have the same problem (discussed in van Hateren 1984, 1985, and van Hateren and Nilsson 1987), but can be highly accurate at longer wavelengths, where the waveguide is monomodal. This approach has been mainly used on two insect species, flies (Kirschfeld and Franceschini 1968; Franceschini 1975; van Hateren 1984) and butterflies (Land 1984; Nilsson et al. 1984, 1988). Although the latter studies use orthodromic light to begin with, they use in fact antidromic light reflected back from the tapeta as their main diagnostic tool.

3 Results

3.1 Observations of Modes

Observations of mode patterns are interesting, because they prove that photoreceptors behave as dielectric waveguides. Various mode patterns were observed by Enoch (1963). These observations were done in relatively wide photoreceptors, often leading to quite complicated patterns. More recently, observations in fly (far field: Franceschini and Kirschfeld 1971; Pick 1977; van Hateren 1984; cornea: van Hateren 1985) and butterfly (cornea: Nilsson et al. 1984, far field: Nilsson et al. 1988) showed clear LP_{01} and LP_{11} patterns. Examples are shown in Fig. 4; in Fig. 4a the far field radiation pattern from a single ommatidium of the fly at several

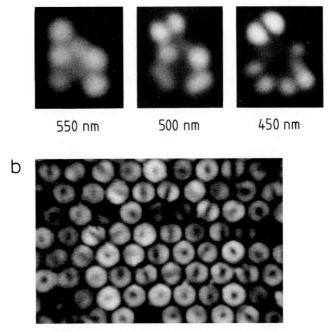

550 nm 500 nm 450 nm

b

Fig. 4.a Modes seen in the far field radiation pattern of a single ommatidium of the fly, containing seven waveguides. With a wavelength of 550 nm, only LP_{01} is observed, whereas LP_{11} is present at 500 nm, and strongly so at 450 nm. *b* Cornea of a butterfly using a technique with polarized light introduced by Nilsson et al. (1984). Mainly LP_{11} is observed (*b* courtesy of Dr. D.G. Stavenga and Mr. H.L. Leertouwer)

wavelengths (550 nm, yielding only LP_{01}; 500 nm, yielding also some LP_{11}; and 450 nm, yielding predominantly LP_{11}), and in Fig. 4b patterns seen at the cornea of a butterfly, mainly showing LP_{11}.

3.2 Angular Sensitivities

Angular sensitivities have been measured in visual sense cells by many investigators (e.g., Washizu et al. 1964; Järvilehto and Zettler 1973; Laughlin 1974; Hardie 1979). Explaining the measurements with models derived from photoreceptor optics also has a long history. One model (Kuiper 1966; Horridge et al. 1976) assumed diffraction at the lens, and subsequent convolution of the intensity of the diffraction pattern with a rectangular acceptance function for the waveguide. This procedure is accurate when the waveguide behaves according to geometrical optics (approximately correct if four or more modes are supported, thus for $V >$ 3.8), but gives erroneous results if only one or two modes are supported. Smakman et al. (1984) performed accurate measurements of angular sensitivities of fly photoreceptors by a feedback method, and succeeded in interpreting these data with a model using a complete wave-optical analysis. An example of these measurements and theoretical fits is shown in Fig. 5. In this study it was found that

J. Hans van Hateren

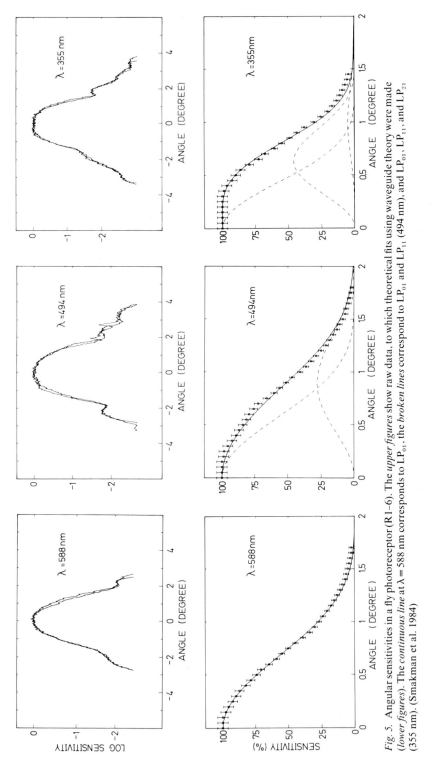

Fig. 5. Angular sensitivities in a fly photoreceptor (R1–6). The *upper figures* show raw data, to which theoretical fits using waveguide theory were made (*lower figures*). The *continuous line* at $\lambda = 588$ nm corresponds to LP_{01}, and the *broken lines* correspond to LP_{01}, and LP_{11} (494 nm), and LP_{01}, LP_{11}, and LP_{21} (355 nm). (Smakman et al. 1984)

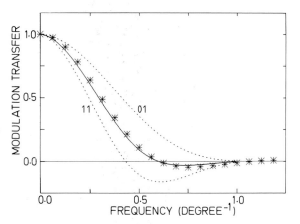

Fig. 6. Optical transfer function of a fly photoreceptor (R1–6). *Asterisks* Fourier transform of a measured angular sensitivity ($\lambda = 450$ nm); *solid line* theoretical fit, consisting of OTFs from LP_{01} and LP_{11} (*broken lines*). The fit was obtained by adjusting the relative weighting of the modes. All other parameters, including the lens diameter (27 μm), were known or measured

closure of the longitudinal pupil narrows the angular sensitivity slightly, presumably because the pupil absorbs more light from higher-order modes like LP_{11} than from the more narrow LP_{01}. Optical measurements of angular sensitivities were performed by Eheim and Wehner (1972), Land (1984), and van Hateren (1984).

Optical transfer functions (OTF's) were measured by Dubs (1982). Fourier transforming the OTF's yields the angular sensitivity (see, e.g., Goodman 1968). On the other hand, the transfer function may be obtained from the angular sensitivity by Fourier transforming it. Figure 6 shows an example of the latter procedure of an angular sensitivity at $\lambda = 450$ nm in the fly, thus involving LP_{11} as well as LP_{01}. We see that both theory and experiment yield a transfer function with a negative part, caused by LP_{11}. In this spatial frequency range the cell gives a maximum response while in reality looking at a minimum in the stimulus (a sinusoidal grating). This contrast reversal is also known from defocussed lenses (Goodman 1968), and may lead the experimenter astray: he should be aware of the possibility of contrast reversal, and check the phase of the response as well as the modulation transfer.

3.3 Corneal, Spectral, and Polarization Sensitivities

Corneal sensitivities, i.e., the directional sensitivity of the photoreceptors measured at the level of the cornea, were investigated electrophysiologically and optically in the fly (van Hateren 1985) and optically in the butterfly (Nilsson et al. 1984, 1988). Most work has been done on this subject, however, in the vertebrate lens eye, where it is called the Stiles-Crawford effect (Stiles and Crawford 1933).

Although the spectral sensitivity is in principle influenced by the optics, e.g., through the wavelength-dependent emergence of waveguide modes, the influence seems to be small in practice. An important point to note is the fact that in fly photoreceptors the angular sensitivity has approximately the same half-width over the entire wavelength range: the emergence of higher order modes at shorter wavelengths compensates for the narrowing of the diffraction pattern at these

wavelengths (see Fig. 5). In effect, the influence of waveguide effects on spectral
sensitivity is only slight (Smakman and Stavenga 1986). Much more important for
the spectral sensitivity are effects caused by absorption by the visual pigment, such
as self-screening (Snyder and Pask 1973; Hardie 1985), lateral filtering (Snyder et
al. 1973), and the spectral properties of the longitudinal pupil (Hardie 1979; Vogt
et al. 1982).

Polarization sensitivity in visual systems is mainly determined by the way the
visual pigment is packed in the rhabdoms. A discussion of polarization sensitivity
is beyond the scope of this review, but can be found in the review of Hardie (1985)
on the fly retina.

4 Discussion

4.1 How Accurately Must a Lens-Waveguide System be Built?

An important question, particularly for the owner, is how accurately a lens-
waveguide system must be built in order to function properly. We will discuss in the
following the effects of skewness of the waveguide, and of defocusing and aberrations of the lens.

Ideally, the waveguide must be pointing to the center of the lens. If it is not, it
does not trap as much light as it could, and the angular sensitivity is broadened.
Both effects can be understood intuitively by realizing that the acceptance profile
of the waveguide must match the lens aperture. A mismatch leads to a smaller
effective lens aperture, reducing the light capture and leading to stronger
diffraction. Wave-optical calculations show that for fly photoreceptors a 2.5°
misalignment leads to a 10% reduction in sensitivity (van Hateren 1985).

Similar effects occur when the waveguide is not in the focal plane of the lens:
this also causes a mismatch (see McIntyre and Kirschfeld 1982). Finally, lens
aberrations also cause these problems, because they lead to phase mismatches: a
lens with aberrations converts a plane wave into a wave front which is not perfectly
spherical. This means that the match with the waveguide acceptance profile
becomes worse, or, equivalently, that the phase of the resulting diffraction pattern
is less well matched to the (plane) phase front of the mode. Lenses in compound
eyes, however, seem to be effectively aberration-free (Stavenga et al. in prep.), thus
lens aberrations are not really a limiting factor.

4.2 The Future of Photoreceptor Optics

Although the paradigmatic lens-waveguide system is now reasonably well understood, photoreceptor optics is far from being a finished subject. Many optical
structures in eyes need a more detailed study, e.g., oil droplets just in front of
photoreceptors of birds, and other specializations found in front of the photoreceptors elsewhere. Furthermore, the shape of many visual waveguides is far from
circular, but rather elliptical, rectangular, or even more complicated. We do not at

all understand the consequences of this for vision. Moreover, many eyes deviate from the simple lens-waveguide scheme. Often, eyes seem to use their cone as well as their facet lenses to funnel the light into the waveguide, and more complicated designs using mirrors or optical superposition may yield surprising results when studied in great (optical) detail (see Nilsson, this Vol.).

A potentially very powerful method not yet applied to waveguides in eyes is the use of sinusoidal gratings consisting of coherent light (instead of incoherent as for determining the OTF). With this method the role of the various modes can be studied directly, because the modes show different phase shifts for coherent light. Also, transfer functions for coherent light easily yield many other interesting properties of the system. The potential of Fourier methods in physiological optics is still vast.

Photoreceptor optics may also be expected to play an important role in the addressing of ecological and evolutionary problems (see Land and Nilsson this Vol.). Furthermore, it provides a solid basis for understanding higher-order processing, including problems like hyperacuity and problems related to sampling and flow fields. Last but not least, the thorough understanding of the optics we now have in several insects allows the investigator of higher order neurons to make use of the acquired techniques, e.g., for methods of stimulating individual photo-receptors very precisely (Franceschini 1975 and this Vol.; Riehle and Franceschini 1984; Lenting 1985; van Hateren 1986, 1987).

References

Barlow H (1952) The size of ommatidia in apposition eyes. J Exp Biol 29:667–674

Beersma DGM (1979) Spatial characteristics of the visual fields of flies. PhD Thesis, Univ Groningen, NL

Beersma DGM, Hoenders BJ, Huiser AMJ, Toorn P van (1982) Refractive index of the fly rhabdomere. J Opt Soc Am 72:583–588

Bernard GD (1975) Physiological optics of the fused rhabdom. In: Snyder AW, Menzel R (eds) Photoreceptor optics. Springer, Berlin Heidelberg New York, pp 78–97

Bernard GD, Wehner R (1980) Intracellular optical physiology of the bee's eye. I. Spectral sensitivity. J Comp Physiol A 137:193–203

Born M, Wolf E (1965) Principles of optics. Pergamon, Oxford New York

Buchner E (1976) Elementary movement detectors in an insect visual system. Biol Cybernet 24:85–101

de Vries HL (1956) Physical aspects of the sense organs. Prog Biophys 6:208–264

Dubs A (1982) The spatial integration of signals in the retina and lamina of the fly compound eye under different conditions of luminance. J Comp Physiol A 146:321–343

Eheim WP, Wehner R (1972) Die Sehfelder der zentralen Ommatidien in den Appositionsaugen von *Apis mellifica* and *Cataglyphis bicolor* (Apidae, Formicidae; Hymenoptera). Kybernetik 10:168–179

Enoch JM (1961) Nature of transmission of energy in the retinal receptors. J Opt Soc Am 51:1122–1126

Enoch JM (1963) Optical properties of retinal photoreceptors. J Opt Soc Am 53:71–85

Enoch JM, Bedell HE (1981) The Stiles-Crawford effects. In: Enoch JM, Tobey FL, Jr. (eds) Vertebrate photoreceptor optics. Springer, Berlin Heidelberg New York, pp 83–126

Exner S (1891) Die Physiologie der facettierten Augen von Krebsen und Insecten. Deuticke, Leipzig

Franceschini N (1975) Sampling of the visual environment by the compound eye of the fly: fundamentals and applications. In: Snyder AW, Menzel R (eds) Photoreceptor optics. Springer, Berlin Heidelberg New York, pp 98–125

Franceschini N (1979) Voltage clamp by light: rapid measurement of the spectral and polarization sensitivity of receptor cells. Invest Ophthalmol Vis Sci Suppl Apr:5

Franceschini N, Kirschfeld K (1971) Etude optique in vivo des éléments photorécepteurs dans l'oeil composé de *Drosophila*. Kybernetik 8:1–13

Goodman JW (1968) Introduction to Fourier optics. McGraw-Hill, New York

Götz KG (1965) Die optischen Übertragungseigenschaften der Komplexaugen von *Drosophila*. Kybernetik 2:215–221

Hardie RC (1979) Electrophysiological analysis of the fly retina. I. Comparative properties of R1–6 and R7 and R8. J Comp Physiol A 129:19–33

Hardie RC (1985) Functional organization of the fly retina. In: Ottoson D (ed) Progress in sensory physiology, vol 5. Springer Berlin Heidelberg New York, pp 1–79

Hateren JH van (1984) Waveguide theory applied to optically measured angular sensitivities of fly photoreceptors. J Comp Physiol A 154:761–771

Hateren JH van (1985) The Stiles-Crawford effect in the eye of the blowfly, *Calliphora erythrocephala*. Vision Res 25:1305–1315

Hateren JH van (1986) Electrical coupling of neuro-ommatidial photoreceptor cells in the blowfly. J Comp Physiol A 158:795–811

Hateren JH van (1987) Neural superposition and oscillations in the eye of the blowfly. J Comp Physiol A 161:849–856

Hateren JH van, Nilsson D-E (1987) Butterfly optics exceed the theoretical limits of conventional apposition eyes. Biol Cybernet 57:159–168

Horowitz BR (1981) Theoretical considerations of the retinal receptor as a waveguide. In: Enoch JM, Tobey FL, Jr. (eds) Vertebrate photoreceptor optics. Springer, Berlin Heidelberg New York, pp 219–300

Horridge GA, Mimura K, Hardie RC (1976) Fly photoreceptors. III. Angular sensitivity as a function of wavelength and the limits of resolution. Proc R Soc London Ser B 194:151–177

Järvilehto M, Zettler F (1973) Electrophysiological-histological studies on some functional properties of visual cells and second order neurons of an insect retina. Z Zellforsch 136:291–306

Kirschfeld K, Franceschini N (1968) Die optischen Eigenschaften der Ommatidien im Komplexauge von *Musca*. Kybernetik 5:47–52

Kirschfeld K, Franceschini N (1969) Ein Mechanismus zur Steuerung des Lichtflusses in den Rhabdomeren des Komplexauges von *Musca*. Kybernetik 6:13–22

Kirschfeld K, Snyder AW (1976) Measurements of a photoreceptor's characteristic waveguide parameter. Vision Res 16:775–778

Kuiper JW (1966) On the image formation in a single ommatidium of the compound eye in Diptera. In:Bernhard CG (ed) The functional organization of the compound eye. Pergamon, Oxford New York, pp 35–50

Land MF (1984) The resolving power of diurnal superposition eyes measured with an ophthalmoscope. J Comp Physiol A 154:515–533

Laughlin SB (1974) Neural integration in the first optic neuropile of dragonflies. III. The transfer of angular information. J Comp Physiol A 92:377–396

Lenting BPM (1985) Functional characteristics of a wide-field movement processing neuron in the blowfly visual system. Thesis, Univ Groningen, NL

Mallock A (1922) Divided composite eyes. Nature (London) 110:770–771

Marcuse D (1974) Theory of dielectric optical waveguides. Academic Press, New York London

McIntyre P, Kirschfeld K (1982) Chromatic aberration of a dipteran corneal lens. J Comp Physiol 146:493–500

Nilsson D-E (1983) Evolutionary links between apposition and superposition optics in crustacean eyes. Nature (London) 302:818–821

Nilsson D-E, Odselius R (1981) A new mechanism for light-dark adaptation in the *Artemia* compound eye (Anostraca, Crustacea). J Comp Physiol A 143:389–399

Nilsson D-E, Odselius R, Elofsson R (1983) The compound eye of *Leptodora kindtii* (Cladocera). Cell Tissue Res 230:401–410

Nilsson D-E, Land MF, Howard J (1984) Afocal apposition optics in butterfly eyes. Nature (London) 312:561–563

Nilsson D-E, Land MF, Howard J (1988) Optics of the butterfly eye. J Comp Physiol A 162:341–366

Pick B (1977) Specific misalignments of rhabdomere visual axes in the neural superposition eyes of dipteran flies. Biol Cybernet 26:215–224

Riehle A, Franceschini N (1984) Motion detection in flies: parametric control over on-off pathways. Exp Brain Res 54:390–394

Seitz G (1968) Der Strahlengang im Appositionsauge von *Calliphora erythrocephala* (Meig). Z Vergl Physiol 62:61–74

Smakman JGJ, Pijpker BA (1983) An analog-digital feedback system for measuring photoreceptor properties with an equal response method. J Neurosci Meth 8:365–373

Smakman JGJ, Stavenga DG (1986) Spectral sensitivity of blowfly photoreceptors: dependence on waveguide effects and pigment concentration. Vision Res 26:1019–1025

Smakman JGJ, Hateren JH van, Stavenga DG (1984) Angular sensitivity of blowfly photoreceptors: intracellular measurements and wave-optical predictions. J Comp Physiol A 155:239–247

Snitzer E, Osterberg H (1961) Observed dielectric waveguide modes in the visible spectrum. J Opt Soc Am 51:499–505

Snyder AW (1969a) Asymptotic expressions for eigenfunctions and eigenvalues of a dielectric or optical waveguide. IEEE Trans Microwave Theor Tech 17:1130–1138

Snyder AW (1969b) Excitation and scattering of modes on a dielectric or optical fiber. IEEE Trans Microwave Theor Tech 17:1138–1144

Snyder AW (1975) Photoreceptor optics — theoretical principles. In: Snyder AW, Menzel R (eds) Photoreceptor optics. Springer, Berlin Heidelberg New York, pp 38–55

Snyder AW (1979) Physics of vision in compound eyes. In: Autrum H (ed) Handbook of sensory physiology, vol VII/6A. Springer, Berlin Heidelberg New York, pp 225–313

Snyder AW, Love DJ (1983) Optical waveguide theory. Chapman & Hall, London New York

Snyder AW, Menzel R (eds) (1975a) Photoreceptor optics. Springer, Berlin Heidelberg New York

Snyder AW, Menzel R (eds) (1975b) Introduction to photoreceptor optics — an overview. In: Photoreceptor optics. Springer, Berlin Heidelberg New York, pp 1–13

Snyder AW, Pask C (1973) Spectral sensitivity of dipteran retinula cells. J Comp Physiol A 84:59–76

Snyder AW, Menzel R, Laughlin SB (1973) Structure and function of the fused rhabdom. J Comp Physiol 87:99–135

Stavenga DG (1974) Refractive index of fly rhabdomeres. J Comp Physiol A 91:417–426

Stavenga DG (1975) Optical qualities of the fly eye — an approach from the side of geometrical, physical and waveguide optics. In: Snyder AW, Menzel R (eds) Photoreceptor optics. Springer, Berlin Heidelberg New York, pp 126–144

Stiles WS, Crawford BH (1933) The luminous efficiency of rays entering the eye pupil at different points. Proc R Soc London Ser B 112:428–450

Toraldo di Francia G (1949) Retinal cones as dielectric antennas. J Opt Soc Am 39:324

Varela FG, Wiitanen W (1970) The optics of the compound eye of the honeybee *(Apis mellifera)*. J Gen Physiol 55:336–358

Vogt K, Kirschfeld K, Stavenga DG (1982) Spectral effects of the pupil in fly photoreceptors. J Comp Physiol A 146:145–152

Washizu Y, Burkhardt D, Streck P (1964) Visual field of single retinula cells and interommatidial inclination in the compound eye of the blowfly *Calliphora erythrocephala*. Z Vergl Physiol 48:413–428

Chapter 5

Variations in the Structure and Design of Compound Eyes

MICHAEL F. LAND, Brighton, England

1 Introduction

At first glance the compound eyes of insects and crustaceans seem to be spherical structures with a more or less uniform distribution of facets. A second look, however, often reveals that the eye is really not symmetrical. There may be regional variations in its curvature, in the sizes of the facet lenses and consequently their packing density. These are reflected in the size and shape of the pseudopupil – the dark spot which marks the part of the eye that images the observer (Stavenga 1979). Even closer observation shows that as the eye is rotated the pseudopupil moves at different speeds in different parts, showing that in some regions the same angle in space occupies more of the eye surface than it does in others. Gordon Walls (1942) wrote that "everything in the vertebrate eye means something", and although our subjects here are exclusively invertebrate, the same dictum is just as valid. The small variations I have mentioned are not haphazard developmental oddities; they all reflect the way in which the eye samples its environment. Properly interpreted, they tell us a great deal about the role of vision in the animal's life.

The reason we can be confident that we are not trying to read too much into these small variations of structure and symmetry is that compound eyes are always cramped for space, and, like a business with a tight budget, expansions in one part must be accompanied by contractions in another. It is not just that eyes are expensive to run, however; the cramping has an optical cause that amounts almost to a design weakness of compound eyes. Because of the small diameter of each lens, the spatial resolution of a compound eye is limited by diffraction, to what by human standards is a poor performance. The 25 μm facets of a bee eye can resolve only about 1 degree, as opposed to 1 minute in our own eyes. Let us suppose that the requirements of a predatory life-style, courtship, distance estimation, or something else make a doubling of a compound eye's resolution desirable; what happens? The diameter of each facet must double, because the angular resolution of a diffraction limited lens is inversely proportional to its diameter. However, because the resolution is now greater, the number of ommatidia can also be increased to exploit the improved image, and the net effect is that the diameter of the eye must increase as the square of the required resolution, and its surface area as the fourth power. (In a camera-type eye like our own the effects are much less drastic. The lens must increase in size, but obviously not also in number, and resolution is directly proportional to eye size, not to its square.) Supposing, then, that a fly needs a

Stavenga/Hardie (Eds.) Facets of Vision
© Springer-Verlag Berlin Heidelberg 1989

better-resolved image. Rather than quadrupling the size of the entire eye from 2 mm to an unliftable 8 mm, what it can do instead is to insert a small region of improved resolution in a part of the eye where it will do most good. This "acute zone" must still be paid for, in increased eye size or lowered resolution elsewhere, but unlike a general increase in resolution, it does not make the eye unmanageable. Because of these constraints, any variation in the way one part of the visual field needs to be treated relative to the rest tends to be reflected in the local structure of the eye, and not the eye as a whole. Conversely, these local structural differences provide an interpretable physical statement of the nature of the animal's visual priorities.

Exner wrote his monograph of 1891 just too early for the importance of diffraction to be appreciated. The Abbe theory of the microscope dates from 1873, but Rayleigh's application of diffraction theory to telescope resolution only came in 1896. It was Mallock (1894) who first pointed out that the small lenses of insect eyes limited their performance, although his work was forgotten until recently (de Vries 1956). Mallock's work was not known to Barlow, whose paper in 1952 on the size of ommatidia really began the modern revival of interest in compound eye optics. Exner had an interest in the optical asymmetries of compound eyes, but his explanations for them were in terms of improved motion perception rather than resolution.

2 The Limits of Visual Performance in Compound Eyes

2.1 Resolution

The ability of any eye to resolve detail depends on two factors: the fineness of the mosaic of receptive elements that sample the image, and the optical quality of the image that those elements receive. The first of these is most conveniently described in terms of the spatial sampling frequency (v_s) of the mosaic, and the second by the spatial cut-off frequency (v_{co}) of the optics. The highest spatial frequency that the mosaic can deal with (v_s) is that for which there is one element for each half cycle of the grating — one for each dark and one for each light stripe — so that $v_s = 1/(2\Delta\Phi)$, were $\Delta\Phi$ is the interreceptor (or in a compound eye the interommatidial) angle (Fig. 1). In an optical system that is otherwise free of defects, the optical limit (v_{co}) is set by diffraction at the aperture of the lens, and it is given by $v_{co} = D/\lambda$, where D is the lens diameter, and λ the wavelength of light in vacuum. In general, one would expect these two limits to be roughly matched, so that the receptor mosaic is just fine enough to sample adequately the highest spatial frequency that the optics provides, but is no finer than that because there is no further information to be gained. In practice we find that v_s is a little lower than v_{co}, probably because at the optical cut-off the contrast of the image vanishes, and retinae need some contrast to work with. Hence:

$$v_{co} \geqslant v_s, \text{ or } D/\lambda \geqslant 1/(2\Delta\Phi), \text{ or}$$
$$D\Delta\Phi \geqslant \lambda/2. \tag{1}$$

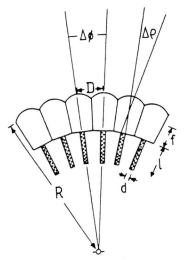

Fig. 1. The geometry of an apposition eye. Inter-ommatidial angle $\Delta\Phi = D/R$ where D is the facet diameter and R the local radius of curvature. The acceptance angle $\Delta\rho$ by geometrical optics is d/f, where d is the rhabdom diameter and f the focal length of the facet lens. l is the rhabdom length

This expression shows that in a diffraction-limited eye the interreceptor angle is inversely related to the size of the aperture. The bigger the lens the finer — in angular terms — the mosaic. If we take the examples of our own eye and that of a bee, we can see how important this limit really is. In a bee with 25-μm diameter facets, and taking the wavelength of light as 0.5 μm, we find from Eq. (1) that $\Delta\Phi$ is at least 1/100th rad, or 0.57°. In our own eye, with a daylight pupil 100 times bigger (2.5 mm), $\Delta\Phi$ is 100 times smaller, with a minimum value of 0.0057° or 0.34 minutes. These values are close to, though slightly smaller than, those actually measured, and they eloquently make the point that compound eyes are *intrinsically* low resolution structures, *because of the small diameter of the individual facets.*

The product $D\Delta\Phi$ in Eq. (1) is a measure of how close the eye comes to the diffraction limit, and is known as the eye parameter p (Snyder et al. 1977). From (1) its minimum value is 0.25 (μm) if λ is 0.5 μm, and indeed insects show p values down to about 0.3, although even for insects active in bright sunlight 0.5 is a more common value (see Howard and Snyder 1983). At the other extreme, the king-crab *Limulus* has facets 300 μm wide, and $\Delta\Phi$ of about 6° (0.105 radians), which makes p about 31. Clearly, this can have nothing to do with diffraction, and here other factors, such as the need for a bright image at night, are much more important. We will return to this question later.

The key problem that faces animals with compound eyes is that it is very difficult to increase the resolution of the eye as a whole, because it simply becomes too big. As we have already seen, this is because the ommatidia must increase in both number and size. More formally, the interommatidial angle in a spherical apposition compound eye must be equal to the subtense of one facet at the centre of curvature; thus $\Delta\Phi = D/R$, where R is the eye radius (Fig. 1). Substituting for D in (1) gives:

$$R\Delta\Phi^2 \geqslant \lambda/2 \qquad\qquad\qquad (2)$$

and, since $v_s = 1/(2\Delta\Phi)$

$$R \geqslant 2\lambda v_s^2 \tag{3}$$

That is, the radius of the eye is proportional to the *square* of the required resolution. Simple eyes like our own are not subject to this fierce constraint because they have only one lens, not many, and size increases linearly with resolution. Were we to have compound eyes, however, the consequences would be dire. With a value for $\Delta\Phi$ of 0.5 minutes (approximately the value in our fovea) Eq. (2) predicts an eye radius of 5.3 metres! Although it was Mallock (1894) who first pointed out the immense size of such eyes, Kirschfeld (1976) was the first to illustrate it, and I have borrowed his delightful picture (Fig. 2). Kirschfeld has reduced the size of the eye greatly by taking into account the decrease in functional resolution of the human eye away from the fovea, but the eye remains huge nonetheless.

The impossibility of providing high overall resolution has led to the insertion of small regions of higher acuity into eyes with modest overall resolution, resulting in interesting and often easily visible asymmetries in the structure of the eyes. These acute zones are the main subject of this chapter. A feature of these zones, which is a common source of confusion, is that the facets are usually bigger than elsewhere in the eye (see Fig. 9); this has to be the case if diffraction is to be overcome, as Eq. (1) states. Nevertheless, the coarse appearance of these regions seems to imply coarse resolution, as indeed it would be if we were dealing with the retina of a camera-type eye like our own. This is a mistake; in compound eyes regions of large facets are almost invariably associated with small interommatidial angles, and they correspond to the high resolution "foveas", "areas" or "visual streaks" of the vertebrate retina (Hughes 1977).

Fig. 2. A compound eye with the same distribution of resolution as the human eye (Kirschfeld 1976). Each facet in the drawing represents 10^4 ommatidia. The dependence of eye size on the square of resolution in apposition eyes [Eq. (3)] makes them impossibly large where resolution better than about 1° is needed

2.2 Sensitivity

The other feature that affects the size of a compound eye is its sensitivity, which can be defined in terms of the number of photons that each receptive element receives when the eye is looking at a surface of a standard luminance. The reason that this is important is that small numbers of photons, arriving randomly, produce a very "noisy" signal, and all visual judgments of position or timing or contrast improve as higher rates of photon capture are procured. It is easy to show that in all conditions other than bright sunlight eyes are effectively "photon-starved", so one might well expect to see adaptations directed towards increasing image brightness incorporated into the design of compound eyes. The factors that affect sensitivity are indicated in equation (4):

$$\text{Sensitivity (S)} = \frac{\text{Flux absorbed/receptor}}{\text{Luminance of Source}}$$
$$= (\pi/4)^2 \, (D/f)^2 \, d^2 \, (1 - e^{-kl}), \tag{4}$$

where D is the diameter of the lens and f its focal length; d is the receptor diameter, l the receptor (or rhabdom) length and k the natural extinction coefficient of photopigment in the receptor (see Fig. 1). A derivation of Eq. (4) is given in Land (1981a), and valuable discussions are given by Barlow (1964), Kirschfeld (1974) and Snyder (1979). Equation (4) is only approximate, principally because it is based entirely on geometrical optics and ignores waveguide mode effects and uncertainties of focus (see van Hateren 1984), but it is accurate enough for comparing the performances of different eyes.

The terms that most affect eye design are those involving D, f and d. Receptor length (l) is also important, as it requires a rhabdom several hundred μm in length for 90% of the entering light to be absorbed, and in most arthropods the rhabdoms are not as long as this. Rhabdom diameter d and focal length f are not independent of each other, because the ratio d/f is actually the geometrical acceptance angle of a receptor (Fig. 1). Except at the very lowest light levels one would expect this to be closely similar to the interreceptor angle $\Delta\Phi$ (otherwise either resolution or sensitivity would be lost). Thus, once the resolution of an eye is set, any variations in d must be matched by changes in f, and the sensitivity will not be altered. This effectively only leaves changes to the lens diameter D as a mechanism for varying an eye's sensitivity. This in turn means that if there are variations in the need for sensitivity across an eye, these may show up as differences in facet diameter, *just as do changes in resolution*. Thus, on its own, facet size is an ambiguous indicator of function.

The eye parameter (p = D$\Delta\Phi$) is useful here. As mentioned earlier, if it approaches its minimum value of 0.25 μm, then resolution is the paramount consideration; if on the other hand, p is much greater than about 0.6 μm, it is likely that large facet size is being used to obtain greater sensitivity. If we assume that d/f is equal to $\Delta\Phi$, then Eq. (4) shows that sensitivity is proportional to the square of p; an eye with p = 2 is 16 times as sensitive as one where p = 0.5. Thus p gives an indication of the light environment to which the eye is adapted, high values implying that the eye is used in dim conditions; Fig. 3 shows how p would be

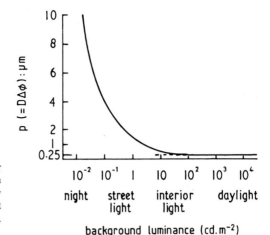

Fig. 3. Dependence of the eye parameter ($D\Delta\Phi$) in an optimized apposition eye on the light level at which it operates. Below about 10 cd/m² the facet diameter D must increase sharply if $\Delta\Phi$ is kept constant. (After Snyder 1979)

expected to vary in eyes designed to operate at different background luminances. (The curve in Fig. 3 was developed by Snyder et al. (1977) based on the proposition that there is an optimum value of p which maximizes the amount of information in the image at each light level.) In practice, if an eye shows regional variations in facet size linked to changes in $\Delta\Phi$ but not in p, then these are concerned with resolution, not sensitivity. However, variations in p, whether or not accompanied by changes in D or $\Delta\Phi$, mean that there are likely to be local differences in sensitivity.

It was implied that higher p values mean that the eyes are used in dim conditions. That may be true, but it is not the only possible reason. The higher photon numbers made available can be used in two other ways. The sampling time could be cut down, to enable the eye to respond faster; Snyder (1979, p. 250) argues that the relatively high value of p in houseflies (about 1) is related to the short sampling times associated with high speed aerobatics. Alternatively, the high photon count could be used to improve the statistics of detection, and so permit very low contrasts to be detected. This is likely to be particularly important in insects which detect potential mates against the sky (Kirschfeld 1979).

2.3 Apposition and Superposition Eyes

The discussion so far applies mainly to apposition eyes, and with very minor modification to the neural superposition eyes of dipterans. The other principal type of compound eye, the superposition type that Exner first elucidated, is different in a number of ways. Its large absolute pupil size means that it is intrinsically more sensitive than the apposition type, by 2 or even 3 orders of magnitude (Kirschfeld 1974). The resolution is limited by diffraction at single facets, just as in apposition eyes (Land 1984). However, the option of inserting small higher resolution regions does not seem to be available, because the image-forming system will not work if there are local variations. There are superposition eyes with regions of different resolution – in euphausiid crustaceans for example (Land et al. 1979) – but here there are really two eyes joined together rather than a single one with an optical gradient.

3 Regional Variations in the Organization of Compound Eyes

3.1 Acute Zones Concerned with Forward Flight

Bees, butterflies, and acridid grasshoppers are all flying herbivores in which the female, at least, does not pursue the other sex. Thus they have no need for a special eye region concerned with capture, but nevertheless they all have consistent and pronounced variations in interommatidial angle across the eye. There are two separate gradients: one is a front-to-back increase in horizontal $\Delta\Phi$ (see footnote), and the other an increase in vertical $\Delta\Phi$ from the equator of the eye to both dorsal and ventral poles (Figs. 4–6). The result is a forward-pointing acute zone and a band around the equator with enhanced vertical (but not horizontal) resolution. This pattern was first described by Baumgärtner (1928) for bees, and del Portillo (1936) for bees and butterflies; it was documented for locusts by Autrum and Wiedemann (1962), and for various bees and wasps by Horridge (1978). The pattern is well demonstrated in the Australian butterfly, *Heteronympha merope*, illustrated in Fig. 4. At the front both vertical and horizontal $\Delta\Phi$'s are small, and the result is a tight pattern with considerable overlap between neighboring receptive fields. Going backwards around the eye, the horizontal $\Delta\Phi$ values increase so that the vertical rows of fields separate, and going vertically from the equator in either direction results in larger vertical $\Delta\Phi$'s, so that at the poles the field pattern is again hexagonal, but with a much looser spacing than at the front. Interestingly, the dark-adapted acceptance angles are the same across the eye (1.9° in *Heteronympha*), and there is very little variation in facet diameter (21–26 μm). Thus although p will vary as $\Delta\Phi$ varies, the sensitivity will not as this is proportional to the square of the acceptance angle [in Eq.(4) d/f can be approximated by $\Delta\rho^1$, as in Fig. 1].

A very similar pattern is seen in bees. Figure 5 shows the Australian bee *Amegilla* viewed from a series of directions, 20° apart, in the horizontal plane. The dark pseudopupil is clearly visible; it is vertically elongated, and whereas its vertical height changes little from front to back, the horizontal extent decreases markedly. In general, the smaller the interommatidial angle, the larger the pseudopupil (see Stavenga 1979), so that the vertical elongation can be taken to mean that vertical $\Delta\Phi$'s are smaller than horizontal, and the decreasing width means that the horizontal $\Delta\Phi$'s increase from front to back, very much as in butterflies (Fig. 4). Indeed, butterfly pseudopupils show very similar though less

---

[1]$\Delta\Phi$ on its own is used throughout to mean the angle between ommatidial axes along a facet row. However, $\Delta\Phi_h$ and $\Delta\Phi_v$, with suffixes indicating horizontal and vertical, are the partial values in the sense used by Stavenga (1979); that is, either ½ or $\sqrt{3}/2$ times the angle between neighboring horizontal or vertical ommatidia, depending on the orientation of the hexagonal lattice

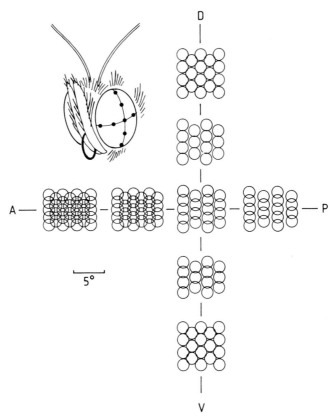

Fig. 4. Regional variations in the way different parts of the eye of a butterfly sample the visual surroundings. The acceptance angle is nearly constant across the eye, and the value used here ($\Delta\rho = 1.9°$) is that of *Heteronympha merope* with a fully open pupil. The horizontal inter-ommatidial angles increase from front to back, and vertical inter-ommatidial angles increase from the equator to the dorsal and ventral poles. The vertical/horizontal difference is greatest at the side on the eye's equator

Fig. 5. Appearance of the eye of the Australian bee *A megilla*, viewed from the front (*left*) and at 20° intervals round to the side, showing how the pseudopupil changes. In general, the dimensions of the pseudopupil are largest when the inter-ommatidial angles are small (Stavenga 1979). Thus all views show that the vertical inter-ommatidial angles are smaller than the horizontal, but whereas the vertical angles stay much the same from front to side, the horizontal angles increase greatly

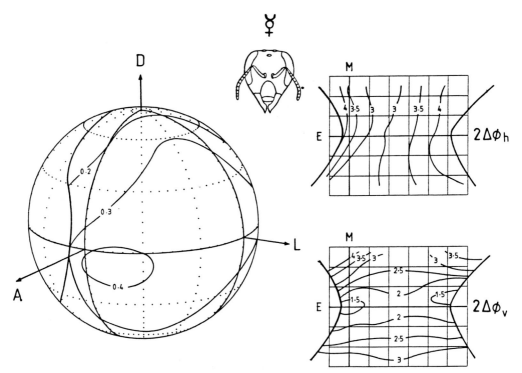

Fig. 6. Variation of resolution across the eye of the worker honey bee *Apis mellifica. Left* orthographic projection of the visual field of the left eye showing the densities of ommatidial axes (axes per square degree). There is a weak forward-pointing acute zone. *Right* distribution of horizontal and vertical inter-ommatidial angles in the visual field, plotted onto a square grid (30° squares); *M* is the midline and *E* the equator. As in the butterfly (Fig. 4) horizontal angles ($\Delta\Phi_h$) increase from near the front to the rear, whereas $\Delta\Phi_v$ increases from the equator to the poles. (Compiled from data in Seidl 1982)

dramatic transformations. The situation in the honey bee (*Apis mellifica*) is similar, and has been documented thoroughly by Seidl (1982), and Figs. 6 and 7 are based on his data. In the worker (Fig. 6) there is a modest frontally directed acute zone, with a maximum of 0.4 ommatidial axes per square degree of the visual field. The plots on the right show that this acute zone is actually the result of quite separate gradients of horizontal and vertical interommatidial angle. $\Delta\Phi_h$ increases from a minimum about 40° from the front, but with little vertical variation, whereas $\Delta\Phi_v$ has a minimum on the equator and maxima at the dorsal and ventral extremes, and shows only slight horizontal variation.

It seems likely that both the front-to-back gradient and the vertical/horizontal asymmetry can be explained as adaptations to forward flight through a textured environment. They tend not to be present in nonpredatory animals that seldom or never fly, phasmids and tettigoniid grasshoppers for example. During flight the visual world moves past the eyes in a "flow field" whose velocity is zero in front and is maximal at 90°. Motion of the image will cause blurring, to the extent that each rhabdom no longer samples a spot, but a horizontal streak, whose length depends on the integration time of the receptors and the angular velocity of the image. As there is no information in the blur streaks, it makes sense for the ommatidial axes

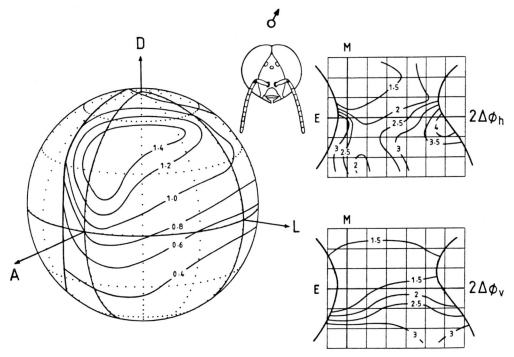

Fig. 7. Resolution across the eye of the drone honey bee. All conventions as in Fig. 6. There is a pronounced acute zone directed forwards and upwards. In the lower half of the eye the pattern of inter-ommatidial angles is similar to that of the worker, but in the upper half it is greatly distorted. (Data from Seidl 1982)

to be spaced more widely at the side where the image degradation is greatest. Further, since the degradation is confined to the horizontal, there is no reason for the vertical $\Delta\Phi$'s to decrease in line with the horizontal $\Delta\Phi$'s, so that one would expect to see the kind of difference that is in fact observed. More quantitatively, suppose a butterfly whose receptors have a sampling time of 10 ms flies at 2 m/s past a spot at a distance of 0.5 m, then the corresponding angular length of the blur streak will be 2.3°, and so at the side of the eye we might expect the horizontal value for $\Delta\Phi$ to be similar to this. In *Heteronympha* $\Delta\Phi$ at 90° is 2.6° as opposed to 1.4° at the front. Although the calculation is only illustrative, it shows that the argument is quantitatively plausible. What it does not fully explain is the increase in vertical $\Delta\Phi$ from the equator to the poles. It may simply be that the region around the horizon contains more useful information than the upper and lower parts of the field, and that this is a gradient which reflects the interest of the animal and has nothing to do with geometry.

In female houseflies and blowflies there is an acute zone which points directly forwards, and this too is probably related to forward flight (Land and Eckert 1985). In these dipterans the eyes are round rather than oval, the difference between vertical and horizontal interommatidial angles is smaller than in bees, and differences between the equator and the poles are also fairly small. However, the overall pattern is similar to that in bees and butterflies. Dipteran acute zones are discussed in more detail later in connexion with sexual pursuit.

3.2 Acute Zones Concerned with the Capture of Prey and Females

A great many insects are equipped with a region of eye pointing forwards or upwards which serves for the detection of other small insects at greater distances than would be possible with other parts of the eye. Often this region is only present in the male, which immediately suggests a role in sexual pursuit (simuliid midges, hoverflies, drone bees, mayflies) whereas in other, predatory insects it is found in both sexes (mantids, robberflies, dragonflies). These acute zones take a great variety of forms, but we will consider them here in two broad categories: (1) forward-pointing regions used in detecting targets against a background, and (2) upward-pointing regions for detecting targets against the sky or ocean surface. This functional differentiation seems to go together with an anatomical one; there is usually a gentle gradation of facet size out from the centre of the acute-zone in (1), but in (2) the two parts of the eye are nearly separate with a clear demarcation between the large facets on top and the smaller ones elsewhere.

1. Frontal acute zones. The eye of the drone honey bee samples the visual field in a strikingly different way from the eye of a worker (Figs. 6, 7). The most notable feature of the drone eye is the large region in the dorso-frontal quadrant where the density of ommatidial axes is three to four times greater than anywhere in the female eye (Seidl 1982). The lenses in this region are also enlarged. Van Praagh et al. (1980) found that when drones chase the queen — or a dummy queen on a string — they always keep her in this region of the eye, so the function seems to be clear.

Amongst the dipteran flies there are many examples of forward-pointing acute zones. As we have seen already, some are probably concerned only with forward flight, but even in species where the female has such a region, the corresponding zone in the male usually has larger facets, smaller $\Delta\Phi$'s, and it may be shifted to a more dorsal position. This is shown for *Calliphora* in Fig. 8. Whilst the female acute zone points directly forwards, that of the male points about 20° more dorsally and it is partly contralateral. The facets reach a maximum of 37 μm compared with 29 μm in the female, and the minimum value of $\Delta\Phi$ is 1.07° compared with 1.28° (Land and Eckert 1985). The value of p is more or less constant across the eye, between 0.7 and 0.8 μm. This is a little high for a basically diurnal fly, but not as high as in *Musca domestica* where it can reach 1.3 μm (Snyder 1979) — suggesting that houseflies were indeed pre-adapted for life indoors!

The function of the male acute zone is to allow the males to chase and catch females. In houseflies the males keep the females in the dorso-frontal region while chasing, just as in drone bees, and they continuously adjust their path so as to keep the female ahead (Land and Collett 1974; Wehrhahn 1979; Wagner 1986). The details of the tracking tasks involved have been best studied in a small hoverfly *Syritta pipiens* (Fig. 9; Collett and Land 1975). Here the male has an acute zone where $\Delta\Phi$ is nearly three times smaller than in the female, enabling him to track her

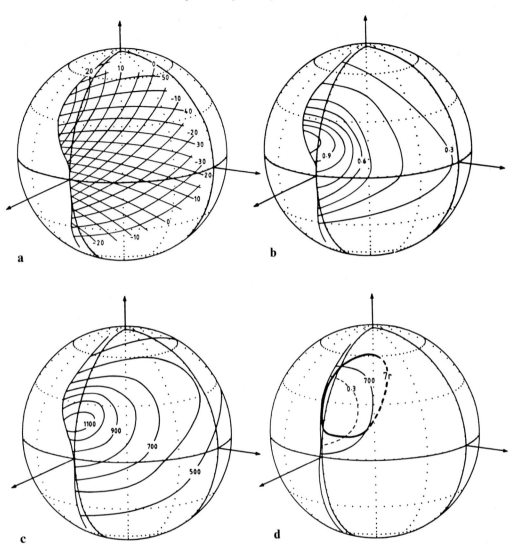

Fig. 8a-d. The acute zone in male houseflies and blowflies. *a* Orthographic map of the left visual field of the male blow-fly *Calliphora erythrocephala,* showing the direction of view of every 5th facet row. The rows concentrate near the midline, above the equator. *b* Similar projection showing the densities of ommatidial axes (axes per square degree). *c* The distribution of facet areas (μm^2) on a similar plot. Note that large facets correspond closely with high resolution, measured as axis density. *d* Similar plot of the male housefly *Musca domestica,* showing the correspondence between contours representing 0.3 axes/deg^2, 700 μm facet area, and the region of sex-specific "7r" photoreceptors (the "love spot") described by Hardie et al. (1981)

Fig. 9. Eyes of four dipteran flies showing different types of acute zone. *Upper left* male *Syritta,* a small hover fly with a fronto-dorsal zone of enlarged facets not present in the female. *Upper right* fully divided eye of a male bibionid fly (*Dilophus,* facing left). The female completely lacks the large-facetted dorsally directed eye. *Lower left* a large robber fly (*Neoaratus hercules*). Both sexes have a frontal acute zone with facets as large as 80 μm. *Lower right* an empid fly which catches insects trapped in the water surface. There is a band of enlarged facets just below the eye's equator. (Empid photograph courtesy of Dr. Jochen Zeil)

at a large enough distance for him to be effectively invisible. When she moves slowly he can keep her within 5° of the acute zone centre using a continuous feedback loop that minimises position error. Outside this region — which means outside the acute zone — saccades are used to recenter the female, as in fixation by mantids (see below). *Syritta* males can maintain a constant distance while tracking, but probably by a method not involving triangulation, as there is very little binocular overlap (see also Schwind this Vol.).

The male acute zone in the fly is anatomically specialised as well. Hardie et al. (1981) showed that the receptors in this region were unusual in that one of the central pair, R7, resembled the surrounding R1–6 receptors in spectral and angular sensitivity, and in sending its axon to the lamina rather than the medulla (Fig. 8d). R7 receptors like this are not found in females, and the suggestion is that by adding their contribution to the signal in the lamina they improve the sensitivity in this region — though only by about 8%. There are special neurons associated with the male acute zone centrally as well as peripherally. Hausen and Strausfeld (1980) found a number of "male-specific" neurons in the lobula, all of which have fields that include the acute zone and presumably have a function in chasing.

Amongst the Diptera there are many other examples of sexually dimorphic acute zones, and the single most comprehensive account is still that of Dietrich (1909), who illustrates almost all the known types. Not all are concerned with sex, and some of the predatory flies like the asilids and empids have impressive acute zones in both sexes (Fig. 9). The asilids in particular have nearly flat frontal acute zones with very large facets, sometimes up to 80 μm in diameter and rivalling those of dragonflies.

Frontal acute zones are not confined to flying insects. The relationship between prey-capture and eye structure was first recognized in that most sinister of insects, the praying mantis. Barrós-Pita and Maldonado (1970) referred to the large flattened acute zone as a "fovea" by analogy with the human eye, although the literal meaning of fovea (pit) makes it an inappropriate term for compound eyes. Since then there have been many studies on these animals but the most comprehensive are those of Rossel (1979, 1980). His optical findings in *Tenodera australasiae* showed that $\Delta\Phi$ varied from 0.6° in the forward-pointing direction to 2.5° laterally; that throughout the eye electrophysiologically measured values of acceptance angle ($\Delta\rho$) were virtually identical to the local value of $\Delta\Phi$; and that facet diameters did decrease from the center of the acute zone outwards, but only from 50 to 35 μm, meaning that p increases somewhat away from the acute zone center. In his second paper, Rossel showed that the saccadic turn the mantis makes to a target located peripherally always brings that target to within a few degrees of the center of the acute zone; and that the probability of a saccade being made decreases with distance from the center of the acute zone. Thus the acute zone of mantids is the place where potential prey are fixated and their position established with sufficient accuracy for a successful strike. The role of these regions in binocular distance estimation was examined by Rossel (1983), and is discussed by Schwind (this Vol.). Mantis shrimps (Stomatopoda) show a remarkable convergence with praying mantids in the way they catch their prey, and they too have eyes with pronounced acute zones (Horridge 1978; Schiff et al. 1986). They differ from all other compound eyes in that *each* eye contains as many as three regions which point in the same forward direction, one on the specialized central band, and one on each side of it; when the eye is viewed from the appropriate direction three pseudopupils are visible. Since both eyes can view the same object, vision may be sextuple! Exner (1891) himself suggested that each eye should be capable of judging object distance by triangulation, and this remains the most likely explanation for the multiple images (see Schwind this Vol.).

2. Dorsal acute zones. The detection tasks of frontal and dorsal acute zones are somewhat different. In the frontal type, the presence of a cluttered background means that it is the spatial resolution of the eye (sampling and cut-off spatial frequencies) that sets the limit to performance in moderate to bright light, whereas the task of detecting a black dot on a light background is more a matter of small signal detection, and absolute photon numbers are as important as small sampling angles (Kirschfeld 1979; Land 1981b). This may be the reason why dorsal acute zones tend to be different, and even separate, from the rest of the eye.

The males of swarming insects have some of the most impressive dorsal acute zones, and the most dramatic sexual dimorphism. In male simuliid flies the upper and lower parts of the eye touch, but the dorsal ommatidia are massive compared with the others, with much larger facets (25–40 compared with 10–15 μm) and rhabdomeres reaching down almost to the roof of the mouth (Dietrich 1909; Kirschfeld 1979). Kirschfeld and Wenk (1976) found that at noon these flies could detect a female at a distance of 50 cm, when it subtended only 0.2°, a small fraction of an ommatidial acceptance angle. Another group of nematoceran dipterans with divided eyes are the bibionids (Fig. 9), which include the large black "St Mark's fly" *Bibio marci*. These were studied by Zeil (1983a,b) who found that the $\Delta\Phi$ values in the upper eye were less than half those in the lower (1.6 compared with 3.7°), and that facet diameters were much greater in the upper eye (33 compared with 21 μm), giving p values in the range 0.9–1.4, similar to houseflies. Upper and lower eyes had a slightly different rhabdomere arrangement, and the eyes of females were similar in most respects to the lower eyes of the males. Interestingly, the lower eye in males supports optomotor responses involved in course control, whilst the upper eye does not, adding weight to the view that the task of detecting females is quite distinct from "ordinary getting about". Dietrich (1909) gives many other examples of divided eyes in Diptera. In other insect orders, the best-known double eyes are in male mayflies (Ephemeroptera), where the upper eyes stick up like turrets, and are often described as "turbanate". In spite of several anatomical studies (Horridge 1976; Wolburg-Buchholz 1976; Nilsson this Vol.) the optical system is not fully understood, having features in common with both apposition and superposition types. The function of the dorsal eye, however, is the same as in swarming dipterans. A rather similar division of function is found in the superposition eyes of the neuropteran owl-flies (*Ascalaphus* and related genera), but here both sexes have divided eyes, and the upper eye is used for prey-capture (Schneider et al. 1978). The mode of life is rather like a dragonfly's, but the upper eyes, at least, are superposition not apposition.

In the sea, daylight comes only from above, and many animals feed by detecting prey objects against this background of downwelling light. The situation is very similar to that of an insect in a swarm, and divided eyes remarkably like those of male dipterans occur in a number of crustacean groups, notably the hyperiid amphipods, the euphausiids and a small number of mysids. Most hyperiids have apposition eyes with an unusual lens-cylinder construction (Land 1981b; Nilsson 1982) and nearly all have smaller $\Delta\Phi$'s in the upward-pointing part (Fig. 10). There may be a continuous gradation between the two parts (*Hyperia*), the halves may be distinct but touching (*Platyscelus, Phrosina*; Fig. 10), or quite separate (*Phronima*). The most extreme, and deepest-living, is *Cystisoma*, which only has dorsal eyes, presumably because there is no light in any other direction. The differences in $\Delta\Phi$ can be very large. In *Hyperia* Nilsson (1982) estimates 0.8° dorsally and 10.5° ventrally, in *Phronima* the range is similar, 0.44° to 12°, and it is similar also in *Phrosina* (Fig. 10) where $\Delta\Phi$ varies between 0.6° in the dorsal eye and 8° in the ventral. These $\Delta\Phi$ values are much smaller than the acceptance angles, and Land (1981b) argues that this situation arises partly from the nature of the detection task and partly because there is neural pooling. In the lower eyes the p values are high, 8 μm in *Hyperia*, 17 μm in *Phronima*; in the upper eyes, however, they are closer to the dipteran values, 0.8 and 1.4 μm respectively. The euphausiids and mysids both have superposition eyes, and some genera in each order have divided eyes. Typically, $\Delta\Phi$ in the upper eyes is about half that in the lower (1.2° compared with 2.6° in *Stylocheiron maximum;* Land et al. 1979), and the upper eyes have fields of view ranging from 45–90°. They have an interesting nonconcentric structure, suggesting they have better-corrected optics than the spherically symmetrical ventral eyes (Land et al. 1979). There is a particularly interesting series of *Stylocheiron* species, where the effective aperture of the upper eye increases with the depth that the animals inhabit by the addition of extra crystalline cones. In *S.suhmi* there are only three cones per row, in *S.affine* there are four to eight in *S.longicorne* 7–19, and *S.elongatum* 13–16. The corresponding daytime depths are 0–50, 40–140, 100–380, and 180–420 m. This seems to be a crude but effective way of ensuring a roughly similar level of illumination on the retinae of the different species.

Returning from the sea to the air, amongst the most striking dorsal acute zones are those of dragonflies (Fig. 11). In fact, there is great variety in the distribution of resolution across the eyes of different dragonflies, and this appears to be closely related to the life-styles of the different groups (Sherk 1978). Primitive zygopterans have only a weakly developed frontal acute zone, whereas in the faster-flying corduliids this is more pronounced, and there is also a second nearly vertical zone.

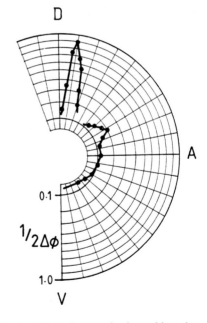

Fig. 10. Double eye of the deep-sea amphipod *Phrosina.* The photograph shows binocular pseudopupils in the lower eyes, and the enlarged facets of the upper eyes. The plot on the *right* shows the resolution (represented by the sampling frequency $1/2\Delta\Phi$) plotted along a sagittal meridian. The upper eye has a dorsally directed high-resolution field that is only 10° wide; the lower eye covers the rest of the field with much lower resolution

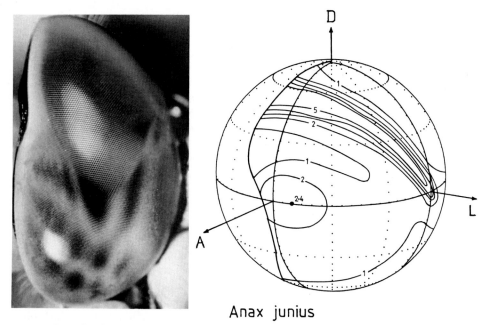

Anax junius

Fig. 11. Eyes of aeshnid dragonflies. *Left Aeshna multicolor,* right eye showing the dorsal wedge of enlarged facets. *Right* axis density plot (axes per square degree) of the field of view of the left eye of *Anax junius.* There is a strip of high resolution across the whole of the dorso-frontal quadrant corresponding to the "wedge" in the photograph. There is also a weaker acute zone pointing forewards, along the line of flight. (Photograph and data from Sherk 1978)

In perching libellulids, the frontal acute zone is minimal, but there is a high-resolution region about 40° across in the fronto-dorsal region. The most impressive eyes are those of the migratory fast-flying aeshnids; they have the greatest numbers of ommatidia — Sherk (1978) counted 28,672 in *Anax junius* — amongst the largest facets (62 μm) and certainly the smallest interommatidial angles (0.24° in the dorsal acute zone of *A.junius*). The dorsal acute zone takes the form of a relatively narrow band of high resolution extending right across the upper eye along a great circle intersecting the midline about 30° in front of the dorsal pole (Fig. 11). It is clearly visible on the eye as a wedge of enlarged facets, whereas the frontal acute zone is not obviously different from the rest of the eye. The value of p in the dorsal zone is 0.48 μm, and 0.55 in the frontal zone. Interestingly, in the Australian dragonfly, *Zyxomma obtusum,* whose habits are crepuscular (Horridge 1978), p in the dorsal acute zone is higher (0.93 μm) than would be expected, although in the frontal zone it is 0.63, hardly different from its value in *Anax.* From the point of view of the present study, the interesting thing about dragonfly eyes is that they combine acute zones with different functions; the frontal zone is presumably concerned with forward flight, like its counterpart in bees and butterflies, whereas the dorsal zone is for finding prey. One imagines the great stripe in *Anax* trawling through the air, picking out small insects against the sky rather in the manner of the scan line in a radar set. Interestingly, the predatory larvae of aeshnids also have an acute zone with vertical interommatidial angles as small as 0.2° (Sherk 1977). Here, however, the acute zone is frontal, much as in mantids. Most of the eye in the adult, including both acute zones, is new growth, only a small region of facets at the back being retained from the larval eye.

3.3 Horizontal Acute Zones

Some animals inhabit a world where nearly all the structures of interest lie in a narrow band around the horizon. These include crabs that live on flat sand or mud, and insects that hunt either just above or just below the water surface. The ocypodid crabs come into the first category, and both Horridge (1978) and Zeil et al. (1986) found that in *Ocypode ceratophthalmus* there was a horizontal band of high vertical resolution, only about 30° high, corresponding to the horizon and the region just below it (Fig. 12). Vertical $\Delta\Phi$'s here were as low as 0.5°, and horizontal $\Delta\Phi$'s nearly four times as great. The facet diameters are largest in this band, 45 μm, and fall to 25 μm at the upper and lower periphery. Zeil et al. found that this extreme concentration on the horizon only occurred in crabs with closely-set elongated eyes with long stalks, and these always inhabited mud- or sand flats. Crabs from further up the beach and from rocky shores had more spherical eyes without the band (Fig. 12). One of their conclusions was that the long stalks and high vertical acuity enable the "flatland" crabs to judge distance by measuring the angle between the horizontal and the base of an object, whereas normal crabs with eyes set wide apart use binocular clues.

    The other flat world is the surface of water, and several insects that hunt there have horizontal acute zones. That of the backswimmer, *Notonecta*, is perhaps the most elaborate (Schwind 1980). These animals hang suspended from the surface

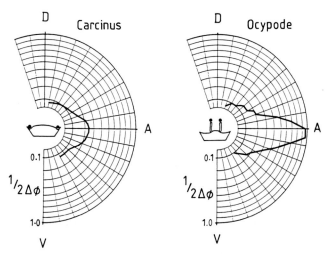

Fig. 12. Vertical resolution in the eyes of a crab from the upper shore (*left*) and a sand-flat (*right*); taken along sagittal meridians. Both crabs show higher resolution around the horizon, but the effect is five times greater in *Ocypode* compared with *Carcinus*. (Data from Zeil et al. 1986)

and their eyes have two separate bands of high resolution corresponding to the images of the surface seen from above (looking out into air just above the critical angle) and below (through the water). Pond skaters (*Velia, Gerris*), which operate just above the surface, also have a horizontal band of greatly increase vertical resolution (Dr H-J Dahmen pers commun). Interestingly in none of these bugs are there noticeable differences in facet diameter corresponding to these bands, which have to be located by pseudopupil measurements. This is not the case, however, in certain empid flies. These small dipterans cruise a few mm above the water surface, and some species, at least, have a clearly defined band of enlarged facets corresponding to the projection of the water surface (Fig. 9). A section of the eye of such a fly (*Hilara maura*) is figured by Dietrich (1909).

3.4 Oddities

I can think of few regional specializations of compound eyes that cannot be squeezed into one of the above categories, but there are two that deserve attention. The first was mentioned by the pioneer in the field, Mallock. In his 1922 paper on "divided composite eyes" he mentions that whitefly (*Aleyrodes proletella*; Hemiptera) have well-divided eyes with the larger lenses *below*. It would be convenient if these animals flew upside down, but it seems they do not! Recently, another example of a ventral acute zone has come to light, and in this case its function is clear. Certain predatory flies (Phoridae) follow ants or termites by flying above them, prior to laying eggs on them. The most extreme example is in *Apocephalus laceyi* where the lower facets are 35 μm across and the upper ones 26 μm (Disney and Schroth 1988).

4 Conclusions

Many arthropods overcome the tyranny of the diffraction limit by squeezing into their compound eyes zones of higher resolution which cover regions of the surrounding environment that are particularly important. The capture of prey and mates has led to the development of the most prominent acute zones, but other factors, such as the way the image flows across the eye during flight, have also affected the distribution of acuity.

To insert an acute zone into an eye requires anatomical modifications of various kinds, and these have been discussed by Stavenga (1979). It also requires modifications of the optical design of the ommatidia, and these modifications seem to vary considerably between species. Perhaps the simplest to understand are the "constant eye-parameter" eyes, like those of male *Calliphora*, where D and $\Delta\Phi$ vary inversely to give a constant value of p, which will mean that the eye is equally sensitive all over, provided the acceptance angle $\Delta\rho$ varies in the same way as $\Delta\Phi$. At the other extreme, the frontal acute zone in butterflies is achieved with almost no variation either facet diameter D or $\Delta\rho$ in spite of changes in $\Delta\Phi$ of up to three-fold. Clearly p is not constant, but since both D and $\Delta\rho$ are unchanging, the sensitivity does not alter. What does change is the way the image is sampled. Praying mantids seem to be intermediate; there is a constant relationship between $\Delta\rho$ and $\Delta\Phi$ across the eye, but the change in facet diameter does not keep pace with the change in $\Delta\Phi$, which means that p is not constant and the sensitivity in the center of the acute zone is lower than elsewhere. It seems that there is no one "optimal" design strategy for compound eyes. Having said that, however, the general predictions about p-values based on optimization theory (Snyder et al. 1977) have stood up well; diurnal insects have p around 0.5 μm, with a few bright-light hoverers having slightly smaller values, but never below 0.3 μm. Recently, Howard and Snyder (1983) have argued that even at the highest light levels p should not quite reach the diffraction-limited value of 0.25 μm, because of noise in the photoreceptors associated with transduction. This noise prevents the exploitation of the highest spatial frequencies in the image, as their contrast is too low to be detected. At light intensities lower than the daylight range, measured p-values rise very much as Fig. 3 predicts. In crepuscular insects p is around 1, indicating a substantial sacrifice of resolution to sensitivity; and in some deep-sea and nocturnal crustaceans p can be much very much higher.

The question that still will not go away is: why did most arthropods retain compound eyes, when single lens eyes offer the potential of much better resolution? Exner believed that compound eyes were especially good at movement detection, but there is little modern evidence to support this. Maybe insects are simply unable, evolutionarily, to change from a design that works well at low resolution; but even this is difficult to accept because many larvae have perfectly good simple eyes, and many adults have dorsal ocelli. Whatever the reason, we are left with a wonderful series of complicated compromises to contemplate.

References

Autrum H, Wiedemann I (1962) Versuche über den Strahlengang im Insektenauge. Z Naturforsch 17b:480–482

Barlow HB (1952) The size of ommatidia in apposition eyes. J Exp Biol 29:667–674

Barlow HB (1964) The physical limits of visual discrimination. Photophysiology 2:163–202

Barrós-Pita JC, Maldonado H (1970) A fovea in the praying mantis eye. II. Some morphological characteristics. Z Vergl Physiol 67:79–92

Baumgärtner H (1928) Der Formensinn und die Sehschärfe der Bienen. Z Vergl Physiol 7:56–143

Collett TS, Land MF (1975) Visual control of flight behaviour in the hoverfly *Syritta pipiens* L. J Comp Physiol A 99:1–66

del Portillo J (1936) Beziehungen zwischen den Öffnungswinkeln der Ommatidien, Krümmung und Gestalt der Insekten-Augen und ihrer funktionellen Aufgabe. Z Vergl Physiol 23:100–145

De Vries H (1956) Physical aspects of the sense organs. Prog Biophys Chem 6:208–264

Dietrich W (1909) Die Facettenaugen der Dipteren. Z Wiss Zool 92:465–539

Disney RHL, Schroth M (1988) Observations on *Megaselia persecutrix* Schmitz (Diptera: Phoridae). Entomol Mon Mag (in press)

Exner S (1891) Die Physiologie der facettirten Augen von Krebsen und Insecten. Deuticke, Leipzig

Hardie RC, Franceschini N, Ribi W, Kirschfeld K (1981) Distribution and properties of sex-specific photoreceptors in the fly *Musca domestica*. J Comp Physiol A 145:139–152

Hateren JH van (1984) Waveguide theory applied to optically measured angular sensitivities of fly photoreceptors. J Comp Physiol A 154:761–771

Hausen K, Strausfeld N (1980) Sexually dimorphic interneuron arrangements in the fly visual system. Proc R Soc London Ser B 208:51–71

Horridge GA (1976) The ommatidium of the dorsal eye of *Chloeon* as a specialization for photoreisomerization. Proc R Soc London Ser B 193:17–29

Horridge GA (1978) The separation of visual axes in apposition compound eyes. Philos Trans R Soc London Ser B 285:1–59

Howard J, Snyder AW (1983) Transduction as a limitation on compound eye function and design. Proc R Soc London Ser B 217:287–307

Hughes A (1977) The topography of vision in mammals of contrasting life style: comparative optics and retinal organisation. In: Crescitelli F (ed) Handbook of sensory physiology, vol VII/5. Springer, Berlin Heidelberg New York pp 613–756

Kirschfeld K (1974) The absolute sensitivity of lens and compound eyes. Z Naturforsch 29c:592–596

Kirschfeld K (1976) The resolution of lens and compound eyes. In: Zettler F, Weiler R (eds) Neural principles in vision Springer, Berlin Heidelberg New York, pp 354–370

Kirschfeld K (1979) The visual system of the fly: physiological optics and functional anatomy as related to behaviour. In: Schmitt FO, Worden FG (eds) The neurosciences 4th study program. MIT, Cambridge, Mass, pp 297–310

Kirschfeld K, Wenk P (1976) The dorsal compound eye of simuliid flies: an eye specialized for the detection of small rapidly moving objects. Z Naturforsch 31c:764–765

Land MF (1981a) Optics and vision in invertebrates. In: Autrum H (ed) Handbook of Sensory Physiology, vol VII/6B. Springer, Berlin Heidelberg New York, pp 472–592

Land MF (1981b) Optics of the eyes of *Phronima* and other deep-sea amphipods. J Comp Physiol A 145:209–226

Land MF (1984) The resolving power of diurnal superposition eyes measured with an ophthalmoscope. J Comp Physiol A 154:515–533

Land MF, Collett TS (1974) Chasing behaviour of houseflies (*Fannia canicularis*). J Comp Physiol 89:331–357

Land MF, Eckert H (1985) Maps of the acute zones of fly eyes. J Comp Physiol A 156:525–538

Land MF, Burton FA, Meyer-Rochow VB (1979) The optical geometry of euphausiid eyes. J Comp Physiol A 130:49–62

Mallock A (1894) Insect sight and the defining power of composite eyes. Proc R Soc London Ser B 55:85–90

Mallock A (1922) Divided composite eyes. Nature (London) 110:770–771

Nilsson D-E (1982) The transparent compound eye of *Hyperia* (Crustacea): examination with a new method for analysis of refractive index gradients. J Comp Physiol A 147:339–349

Praagh JP van, Ribi W, Wehrhahn C, Wittmann D (1980) Drone bees fixate the queen with the dorsal frontal part of their compound eyes. J Comp Physiol A 136:263–266

Rossel S (1979) Regional differences in photoreceptor performance in the eye of the praying mantis. J Comp Physiol A 131:95–112

Rossel S (1980) Foveal fixation and tracking in the praying mantis. J Comp Physiol A 139:307–331

Rossel S (1983) Binocular stereopsis in an insect. Nature (London) 302:821–822

Schiff H, Manning RB, Abbott BC (1986) Structure and optics of ommatidia from eyes of stomatopod crustaceans from different luminous habitats. Biol Bull 170:461–480

Schneider L, Draslar K, Langer H, Gogala M, Schlecht P (1978) Feinstruktur und Schirmpigment-Eigenschaften der Ommatidien des Doppelauges von *Ascalaphus* (Insecta, Neuroptera). Cytobiologie 16:274–307

Schwind R (1980) Geometrical optics of the *Notonecta* eye: Adaptations to optical environment and way of life. J Comp Physiol A 140:59–68

Seidl R (1982) Die Sehfelder und Ommatidien-Divergenzwinkel von Arbeiterin, Königin und Drohne der Honigbiene (*Apis mellifica*) D Thesis, Tech Hochsch Darmstadt

Sherk TE (1977) Development of the compound eyes of dragonflies (Odonata). I. Larval compound eyes. J Exp Zool 201:391–416

Sherk TE (1978) Development of the compound eyes of dragonflies (Odonata). III. Adult compound eyes. J Exp Zool 203:61–80

Snyder AW (1979) Physics of vision in compound eyes. In: Autrum H (ed) Handbook of sensory
    physiology, vol VII/6A. Springer, Berlin Heidelberg New York, pp 225–313
Snyder AW, Stavenga DG, Laughlin SB (1977) Spatial information capacity of compound eyes. J Comp
    Physiol 116:183–207
Stavenga DG (1979) Pseudopupils of compound eyes. In: Autrum H (ed) Handbook of sensory
    physiology vol VII/6A. Springer, Berlin Heidelberg New York, pp 357–439
Wagner H (1986) Flight performance and visual control of flight of the free-flying housefly (*Musca
    domestica* L.) II. Pursuit of targets. Philos Trans R Soc London Ser B 312:553–579
Walls GL (1942) The Vertebrate Eye and its Adaptive Radiation. Hafner, New York
Wehrhahn C (1979) Sex-specific differences in the chasing behaviour of houseflies (*Musca*). Biol
    Cybernet 32:239–241
Wolburg-Buchholz K (1976) The dorsal eye of *Chloëon dipterum* (Ephemeroptera). A light- and
    electron microscopical study. Z Naturforsch 31c:335–336
Zeil J (1983a) Sexual dimorphism in the visual system of flies: the compound eyes and neural
    superposition in Bibionidae (Diptera). J Comp Physiol A 150:379–393
Zeil J (1983b) Sexual dimorphism in the visual system of flies: the free flight behaviour of male
    Bibionidae (Diptera). J Comp Physiol A 150:395–412
Zeil J, Nalbach G, Nalbach H-O (1986) Eyes, eyestalks and the visual world of semi-terrestrial crabs.
    J Comp Physiol A 159:801–811

Chapter 6

Visual Pigments of Compound Eyes — Structure, Photochemistry, and Regeneration

JOACHIM SCHWEMER, Bochum, FRG

1 Introduction

Visual pigments comprise a family of chromoproteins which are incorporated in highly specialized membrane areas of photoreceptor cells. The function of these molecules is to absorb light quanta, which are often directed onto these membrane areas by special optical means. Absorbed light quanta cause the 11-cis chromophore of the visual pigment to isomerize to the all-trans form, the only reaction of the visual process that requires light. This in turn leads to conformational changes of the protein. These changes initiate a complex sequence of biochemical and biophysical events that eventually lead to the excitation of the photoreceptor cell. Following transmission of these electrical signals to higher-order neurons and their processing, the information gathered through the absorption of a few light quanta may finally lead to a comprehensive behavioral response. Thus, the knowledge of how visual pigments are constructed, transformed by light, and regenerated is fundamental to the understanding of photoreceptor function. Although this brief survey will concentrate on the visual pigments of compound eyes, data from other pigment systems will be quoted for comparison whenever appropriate.

2 Structure of Visual Pigments

All visual pigments that have been investigated so far are membrane-bound and constitute the major fraction of the total membrane protein. Covalently bound to the protein opsin is the chromophoric group, the 11-cis isomer of a vitamin A derivative. In photoreceptors of compound eyes, these molecules are located within the membrane of an array of highly ordered microvilli which together build the light-sensitive rhabdomere of the cell.

2.1 Visual Pigment Protein

Opsins have been isolated and partially characterized from only a few compound eyes. The molecular weights range between 32 kDa and 39 kDa and are quite similar to that of vertebrate opsin, whereas cephalopod opsins have a somewhat higher molecular weight (43 k Da-51 k Da; for review see Stavenga and Schwemer

Stavenga/Hardie (Eds.) Facets of Vision
© Springer-Verlag Berlin Heidelberg 1989

1984). Enormous progress with respect to the molecular structure of opsin has recently been made through the amino acid sequencing of vertebrate opsins (Ovchinnikov 1982; Hargrave et al. 1983; Pappin et al. 1984). Moreover, the application of recombinant DNA technology has resulted in a burst of additional information on the structure of visual pigments. Besides the isolation and sequencing of several vertebrate opsin genes including those of man (for review see Baehr and Applebury 1986, Applebury and Hargrave 1986, Ovchinnikov 1987), the genes coding for the visual pigment of photoreceptors R1–6 (O'Tousa et al. 1985, Zuker et al. 1985) and the central photoreceptors R7 and R8 (Cowman et al. 1986, Zuker et al. 1987, Montell et al. 1987, Fryxell and Meyerowitz 1987) have been isolated in *Drosophila*. The deduced amino acid sequence for the visual pigment in R1–6 contains 373 residues (compared to 348 residues in bovine opsin) and shows regions that are highly conserved (see Goldsmith this Vol.).

On the basis of experimental data and theoretical considerations, structural models of opsin have been constructed which in general show the polypeptide chain being folded in seven hydrophobic segments of 24–28 predominantly nonpolar amino acids each. Several lines of evidence point to a helical arrangement of these membrane-spanning regions, which are connected by hydrophilic polypeptide loops on both sides of the membrane (Fig. 1). The three loops exposed to the cytoplasm most probably provide sites of interaction with cytoplasmic proteins which are involved in the biochemical amplification cascade (e.g., Applebury and Hargrave 1986). The carboxyl-terminal region, which is also exposed to the cytoplasmic surface, carries serines and threonines which are sites of the light-dependent phosphorylation of the visual pigment (Paulsen and Bentrop 1986). The functional significance of the extracellular loops is still unclear as is the question of whether or not the N-terminus of *Drosophila* opsin contains oligosaccharides as vertebrate opsins do. The hydrophobic domain of the protein forms some kind of a pocket in which the chromophore is covalently bound. For a detailed presentation and an extensive discussion of visual pigment structure and function, the reader is referred to recent review articles by Applebury and Hargrave (1986), Findlay (1986) and Ovchinnikov (1987).

2.2 Visual Pigment Chromophores

Until recently it has generally been assumed that the chromophoric group of invertebrate visual pigments is the widely spread retinaldehyde (Fig. 1b), although direct evidence for this was given in only one insect (Paulsen and Schwemer 1972). Whilst this assumption may still hold for the majority of invertebrate visual pigments, the most interesting findings by Vogt (1983) and Suzuki et al. (1984) demonstrate that the identity of the chromophores of invertebrate visual pigments deserves a closer look. Suzuki et al. found that the eyes of a crayfish (*Procambarus clarkii*) contained not only retinal, but also 3,4-dehydro-retinal (Fig. 1b), which so far had been found only in freshwater fishes and amphibians (Bridges 1972). Interestingly, the content of 3,4-dehydro-retinal in crayfish eyes was shown to vary with the season: approximately 40% of the total chromophore content was present in the 3,4-dehydro form during the winter, whereas a negligible amount was found

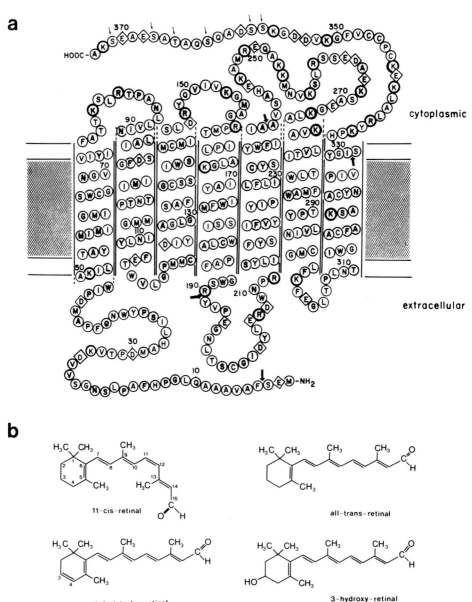

a

cytoplasmic

extracellular

b

11-cis-retinal

all-trans-retinal

3,4-dehydro-retinal
(all-trans)

3-hydroxy-retinal
(all-trans)

Fig. 1a,b. Components of the visual pigment molecule. *a* A model for the structure of *Drosophila* opsin in R1-6. The seven transmembrane segments (*I* to *VII from left to right*) are drawn as α-helices and are connected by hydrophilic loops. About half of the protein mass is immersed in the lipid bilayer. Lysine K 319 in segment VII is proposed to be the site at which the chromophore is covalently bound. Amino acids are represented by *one letter symbols. Diamonds* residues with negative charge. *Heavy circles* residues with positive charge. *Light arrows* indicate possible phosphorylation sites at the C-terminal end (O'Tousa et al. 1985). *b* Structural formulas of chromophores which have been isolated from compound eyes. The chain of the carbon atoms C_1-C_{15} building the backbone of the molecule is linked by alternating single and double bonds which allows formation of a series of different stereo-isomers. However, only two of them, the 11-cis (visual pigment) and the all-trans isomer (photoproducts), are involved in the visual cycle

during the summer. Although it has not yet been unequivocally demonstrated that 3,4-dehydro-retinal is actually used as a visual pigment chromophore, this finding, which has been confirmed for the American crayfish (see Goldsmith and Bernard 1985), is certainly of interest, and the possible pigment(s) formed as well as the physiological significance need clarification.

Another exception from what was thought to be the rule was found by Vogt (1983) and Vogt and Kirschfeld (1984), who showed that the chromophore of fly visual pigment was not retinal, but 3-hydroxy-retinal (Fig. 1b). This novel chromophore was later shown to occur in other insect species as well (see Vogt this Vol.). A functional role of the hydroxyl-group at C3 is not known so far.

Besides the unusual chromophore, flies exhibit another interesting phenomenon: in photoreceptors R1–6, R7y and probably R8y, a second chromophore, which is close or even attached to opsin, absorbs UV quanta and transfers the energy onto the visual pigment, i.e., the activation of the visual pigment occurs via a sensitizing pigment, most probably 3-hydroxy-retinol (reviews: Kirschfeld 1986; Vogt this Vol.). Most recently, retinol is suggested to have a sensitizing function in photoreceptors of simuliids (Kirschfeld and Vogt 1986).

The chromophores are derived from their respective alcohols and C_{40}-carotenoids. Interconversion of retinol and 3,4-dehydro-retinol is known to occur in the pigment epithelium of freshwater fishes and amphibians (Bridges 1972). Similarly, the hydroxylation of exogenous retinal at C3 was found to be carried out in the eye of the blowfly (Schwemer 1986), i.e., dietary uptake of zeaxanthin or lutein is not a basic requirement for using 3-hydroxy-retinal as chromophore.

2.3 Chromophore Binding Site

In vertebrate rod visual pigments, 11-cis retinal is bound to the ε-amino group of lysine 296 located in helix VII by a protonated Schiff base linkage (Bownds 1967; Oseroff and Callender 1974; Ovchinnikov 1982; Applebury and Hargrave 1986). Thereby, parts of the hydrophobic domains of the seven helices must somehow provide an appropriate pocket for binding of the chromophore.

The binding site of the chromophore, as well as the binding itself, is less well established in invertebrate visual pigments and, except for a few experimental data, our knowledge is primarily based on indirect evidence and conclusions drawn by analogy with vertebrate pigments. But recent studies on *Drosophila* opsins show that helix VII also contains a single, approximately centrally located lysine [R1–6: lysine(K)319, O'Tousa et al. 1985; Zuker et al. 1985; R7: lysine 328, Fryxell and Meyerowitz 1987] which, by analogy with bovine opsin, is proposed to be the chromophore attachment site.

As far as the binding is concerned, a protonated Schiff base (Fig. 2a) between chromophore and opsin has been proposed since the thermostable photoproduct in the visual pigment cycle of invertebrate pigments can be reversibly shifted between two states, acid and alkaline metarhodopsin, by changing the pH (see Fig. 3), a property which is characteristic of a Schiff base linkage. Since transamination of the chromophore during the visual pigment cycle has never been observed, it is most likely that the chromophore in rhodopsin is also bound via a Schiff base. Moreover, protonated model Schiff bases of retinal show a bathochromic shift and

a

$$C_{19} H_{27} - \underset{\underset{H}{|} \; \underset{H}{|}}{C} = \overset{+}{N} - \text{opsin} \quad \underset{+ \; H^+}{\overset{- \; H^+}{\rightleftharpoons}} \quad C_{19} H_{27} - \underset{\underset{H}{|}}{\overset{\cdot}{C}} = N - \text{opsin}$$

protonated form   unprotonated form

(rhodopsins, acid metarhod-   (alkaline metarhodopsins:
opsins: absorbance maxima in   absorbance maxima in the
the visible spectral range)   ultraviolet at ~375 nm)

b

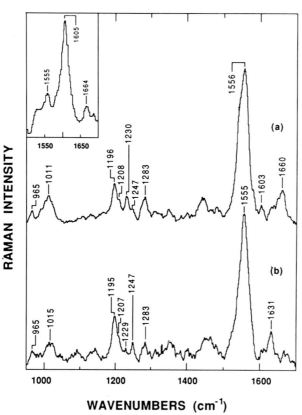

Fig. 2a,b. Binding of the chromophore to opsin. *a* Protonated and unprotonated form of Schiff base formed between retinal and the visual pigment protein opsin. *b* Resonance Raman spectra of *Ascalaphus* acid metarhodopsin in detergent (trace *a*) and following deuteration of the sample (trace *b*). The C=N stretching mode at 1660 cm^{-1} in H$_2$O (*a*) which is shifted upon deuteration to 1631 cm^{-1} (*b*) indicates a protonated Schiff base linkage between all-trans retinal and opsin. The structure sensitive fingerprint region (1100–1350 cm^{-1}) shows similarities to both the all-trans protonated Schiff base of model retinal chromophores as well as to that of octopus acid metarhodopsin and bovine meta-rhodopsin I. *Inset* shows the Schiff base region in the UV-rhodopsin measured at 170 K. (Pande et al. 1987b)

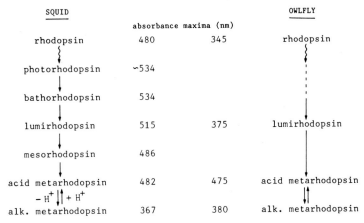

Fig. 3. Photochemical cycle of visual pigment from squid and owlfly. Squid after Yoshizawa and Shichida 1982, and Shichida et al. 1984; owlfly after Hamdorf et al. 1973

absorb maximally at about 440 nm. In most visual pigments, the wavelength of peak absorbance is shifted to longer wavelengths, most probably due to a modulation by the protein environment. Recent resonance Raman spectra obtained from the UV-rhodopsin of the owlfly *Ascalaphus* and its acid metarhodopsin are of special interest, since the absorbance maximum of the rhodopsin at 345 nm might be expected to result from an unprotonated Schiff base. However, the resonance Raman data, the only ones which are available for a visual pigment of compound eyes so far, support the view of a protonated Schiff base. In the acid metarhodopsin (λ_{max} 475nm) a C $=$ N stretching mode is seen at 1660 cm^{-1} in H$_2$O, and upon deuteration of the sample this is shifted to 1631 cm^{-1} (Fig. 2b). These data clearly demonstrate that the Schiff base in acid metarhodopsin is protonated. The inset of Fig. 2b shows the resonance Raman spectrum of the UV-rhodopsin at -100°C: The observed Schiff base mode at 1664 cm^{-1} strongly suggests that the rhodopsin also contains a protonated Schiff base (Pande et al. 1987b). Evidence for a protonated Schiff base has also been obtained by resonance Raman spectroscopy for another invertebrate visual pigment system (octopus: Pande et al. 1987a): The Schiff base vibrational mode appears at 1660 cm^{-1} in rhodopsin and its photoproducts batho- and metarhodopsin, which agrees with that reported for bovine rhodopsin (Oseroff and Callender 1974). The protonation of the Schiff base linkage in the UV-rhodopsin, as well as in vertebrate and octopus rhodopsin, suggests that a charged chromophore is essential for visual transduction.

3 Photochemistry of Visual Pigments

Invertebrate visual pigments have been studied by a variety of different techniques (for review see Stavenga and Schwemer 1984). Besides their biochemical and biophysical attributes, the pigments are mainly characterized by their photochemical properties. In cephalopods and crustaceans, the absorbance maxima lie within the wavelength range 470 to 535 nm, whereas those of insects span a much broader part of the spectrum ranging from 345 to about 590 nm. Multiple visual pigment systems, the basis for color vision, have been identified in compound eyes of several insect species. The first trichromatic visual pigment system that has been isolated and analyzed to some extent is that of a moth, *Deilephila elpenor* (Hamdorf et al. 1973; Schwemer and Paulsen 1973), with pigments absorbing maximally at 345, 440, and 520 nm. The chromophore of these pigments is 3-hydroxy-retinal. The bathochromic shift in the absorbance maximum of retinal from about 380 nm towards the visible part of the spectrum can, as mentioned above, be partly explained by the protonated Schiff base linkage. A further spectral shift of the maxima is most probably due to the protein environment of the chromophore within the hydrophobic pocket, possibly through charged groups, as has been proposed by Honig et al. (1979).

Absorption of a light quantum by the bound chromophore initiates a sequence of reactions which leads to various pigment states. Since one of these states triggers a chain of biochemical events that ultimately causes the excitation of the photoreceptor cell (see Fein and Payne this Vol.), investigation of the photochemical sequence is of fundamental interest.

The most extensively analyzed cycle in invertebrates is that of cephalopod rhodopsin (Fig. 3). Following the absorption of a photon by visual pigment, the highly unstable photoproduct photorhodopsin is formed within about 20 ps (Shichida et al. 1984). The difference between the absorbance characteristics of rhodopsin and photorhodopsin is most probably caused by the 11-cis → all-trans isomerization of the chromophore. Photorhodopsin decays within several hundred ps to bathorhodopsin which is then followed by lumi- and mesorhodopsin from which a final state is formed, a fairly thermostable acid metarhodopsin. By raising the pH this pigment state can be reversibly shifted to the alkaline form, which, in contrast to the preceding intermediates, absorbs in the near ultraviolet. The different spectralproperties of all intermediates following photorhodopsin are due to conformational changes of opsin which lead to different interactions between opsin and the chromophore. The transition from acid to alkaline metarhodopsin involves in addition the deprotonation of the Schiff base (Fig. 2a). This reversible reaction is thought to be the main reason for the large spectral shifts of the absorbance maxima which, in the blowfly for example, may be as much as 200 nm (Schwemer 1979).

In contrast to cephalopods, the visual pigment systems of compound eyes have not been studied in such detail, mainly because the quantity of pigment needed for spectral studies of the intermediates at low temperatures is difficult to obtain from the small eyes. A lumirhodopsin has been characterized in the UV-sensitive rhodopsin system of *Ascalaphus* at −50°C (Fig. 3; Hamdorf et al. 1973), and evidence for short-lived intermediates has been presented for the blowfly visual pigment system (Kruizinga et al. 1983). All other pigment systems from compound eyes have been characterized only by the spectral absorbance properties of rhodopsin and acid metarhodopsin (for review see Schwemer and Langer 1982; Stavenga and Schwemer 1984; White 1985). Although the intermediates are not known, it is assumed that probably all visual pigments decay through a cycle similar to that found for vertebrate and cephalopod rhodopsin. However, it seems questionable whether the intermediates of invertebrates, which have been designated in accordance with the terminology of the vertebrate rhodopsin cycle, also exhibit the corresponding biochemical and biophysical characteristics.

Since rhodopsin triggers the electrical response of a photoreceptor, the absorbance spectrum determines the spectral sensitivity of the receptor, although this can be modulated by several factors (see Stavenga and Schwemer 1984). The cascade of biochemical events which link the visual pigment cycle to the generation of the receptor potential is presently not well understood. However, the available data indicate that rather than one of the short-lived intermediates, the acid metarhodopsin represents the activated state which starts the biochemical amplification cascade by interacting, probably at the cytoplasmic loop which connects helices V and VI (Fig. 1; see, e.g., Applebury and Hargrave 1986) with a G-protein (Paulsen and Bentrop 1986, for review see Tsuda 1987). In order to prevent a continuous activation of G-protein, the enzymatically active meta-state must be deactivated. Amongst other mechanisms discussed, multiple phosphorylation of opsin at the sites indicated in Fig. 1 has been proposed for inactivating the meta-state. Light-induced phosphorylation of opsin has been shown to occur in fly rhabdoms (Paulsen and Bentrop 1986).

4 Pathways of Visual Pigment Regeneration

After having triggered the excitation of the photoreceptor cell (Fein and Payne this Vol.), the visual pigment molecules are inactivated and, in invertebrates, remain in the microvillar membrane in the acid metarhodopsin state (most probably phosphorylated). Since continuous light activation would gradually reduce the rhodopsin content of the rhabdomere, and hence the sensitivity of the photoreceptor, the inactive state has to be regenerated to rhodopsin in a reaction sufficiently fast to maintain photoreceptor function. Besides this fast regeneration, damaged or inactivated molecules are recognized and removed by an additional, much slower mechanism.

4.1 Photoregeneration

Most probably all invertebrates are provided with a unique pathway for regenerating rhodopsin, first described for cephalopod visual pigment systems (Hubbard and St. George 1958; Brown and Brown 1958; Kropf et al. 1959; Hamdorf et al. 1968; Schwemer 1969) and later extensively studied in compound eyes (Hamdorf et al. 1973, for review see Hamdorf and Schwemer 1975; Hamdorf 1979; Stavenga and Schwemer 1984). This pathway is based on the property that the all-trans chromophore is not released from opsin, as is the case in vertebrate metarhodopsin, but remains covalently bound to the protein. Consequently, the thermostable metarhodopsin has the chance to absorb light. Absorption of light by metarhodopsin reisomerizes the all-trans chromophore to the 11-cis form, which in turn causes the opsin to refold to yield rhodopsin. This photoregeneration may occur through several intermediates, one of which (11-cis metarhodopsin) has been found in the UV-rhodopsin of *Ascalaphus* (Hamdorf et al. 1973). The photochemistry of a visual pigment system which consists basically of two photointerconvertible states has been extensively studied by using monochromatic light. It was demonstrated that the photoequilibria between rhodopsin (P) and metarhodopsin (M) which are formed upon saturating irradiation are determined by the absorbance coefficients α_P and α_M at the wavelength of irradiation (λ), the quantum yield for both photoreactions (P $\rightarrow$ M and M $\rightarrow$ P) being almost equal and independent of the wavelength of irradiation (Schwemer 1969, for review see Stavenga and Schwemer 1984). This finding has recently been confirmed by Dixon and Cooper (1987), who determined the absolute quantum efficiencies in a cephalopod visual pigment system. Under these conditions, the fraction f of P and M in a rhabdomere at photoequilibrium (eq) are given by

$$f_{Peq}(\lambda = \frac{\alpha_M(\lambda)}{\alpha_M(\lambda) + \alpha_P(\lambda)} \; ; \text{and } f_{Meq}(\lambda) = 1 - f_{Peq}(\lambda).$$

According to these relations $f_{(P)}$ in a rhabdomere is reduced by irradiation with wavelengths at which the ratio $\alpha_M/(\alpha_M + \alpha_P)$ is small; irradiation by the wavelength at the isosbestic point ($\alpha_P = \alpha_M$) leads to $f_{(P)} = f_{(M)} = 0.5$; on the other hand $f_{(P)}$ increases and even approaches 1 at wavelengths of irradiation which will primarily be absorbed by M as exemplified in Fig. 4, which shows the visual pigment system P490 and M570 in blowfly. The light intensity does not influence the photoequilibrium, but determines rather the rate at which the photoequilibrium is reached. For a more detailed presentation of the photochemistry, the reader is referred to reviews by Hamdorf and Schwemer (1975), Hamdorf (1979) and Stavenga and Schwemer (1984). Besides the restitution of the chromophore, photoregeneration also leads to a dephosphorylation of the protein (Paulsen and Bentrop 1986). The physiological significance of this photoregeneration was first demonstrated in the compound eye of the owlfly (Hamdorf et al. 1971) as an increase in sensitivity of the UV-receptor during adaptation to blue light which converts M to P.

Under physiological conditions, $f_{(P)}$ will be determined by the spectral composition of the ambient light and the absorbance characteristics of P and M which, often in combination with screening pigment effects, favor photoregeneration and thus lead to a high P content of the rhabdomere.

4.2 Renewal of Photoreceptor Membrane and Visual Pigment

In photoreceptor cells of both vertebrates and invertebrates, a continuous renewal of the light-sensitive membrane is known to occur, which includes the breakdown of membranes and the addition of newly synthesized visual pigment and membrane to the existing light-sensitive structure. Although still a matter of debate, the rationale for this turnover is generally seen in a replacement of "old" membranes to prevent inefficient function of the receptor due to aged or light-damaged constituents. Moreover, the drastic diurnal changes in rhabdom size observed in

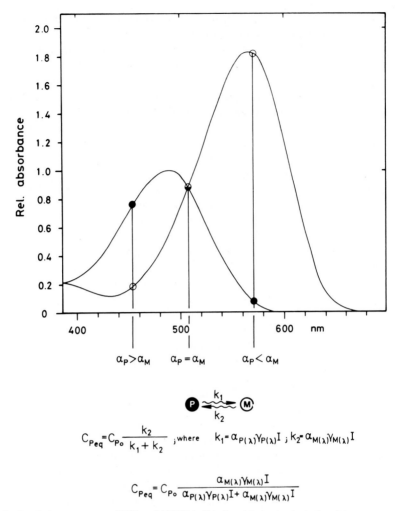

$$\alpha_P > \alpha_M \qquad \alpha_P = \alpha_M \qquad \alpha_P < \alpha_M$$

$$P \underset{k_2}{\overset{k_1}{\rightleftharpoons}} M$$

$$C_{Peq} = C_{Po}\frac{k_2}{k_1+k_2} \quad ;where \quad k_1 = \alpha_{P(\lambda)}\gamma_{P(\lambda)}I \; ; \; k_2 = \alpha_{M(\lambda)}\gamma_{M(\lambda)}I$$

$$C_{Peq} = C_{Po}\frac{\alpha_{M(\lambda)}\gamma_{M(\lambda)}I}{\alpha_{P(\lambda)}\gamma_{P(\lambda)}I + \alpha_{M(\lambda)}\gamma_{M(\lambda)}I}$$

Fig. 4. Blowfly visual pigment system (P490 and M570) in R1–6 and its basic photochemistry

several compound eyes (for review see Autrum 1981) suggest that the turnover of photoreceptor membrane also serves as part of an adaptation process to ambient light conditions.

4.2.1 Membrane Turnover

The pioneering work of Eguchi and Waterman (1967) and White (1968) demonstrating a light-dependent turnover of photoreceptor membrane in compound eyes presaged a great many studies which paid special attention to changes of the rhabdom, rhabdomeres, and microvilli, as well as to cytological changes within the photoreceptor during light and dark adaptation. From the wide diversity of phenomena observed (for reviews see Waterman 1982; Blest et al. 1984; Schwemer 1986) only the most salient results can be summarized here.

The general picture that evolved from studies on membrane breakdown shows that the photoreceptor membrane is internalized by endocytosis: at the base of the microvilli, small pinocytotic vesicles are formed which pinch off from the cell membrane and build so-called multivesicular bodies by secondary endocytosis. Subsequently, primary lysosomes, small vesicles which most likely originate from Golgi bodies and contain lysosomal enzymes, fuse with the multivesicular bodies and cause a sequential degradation of the internalized membranes.

Besides pinocytosis, a variety of different mechanisms of extracellular membrane degradation has been reported which start with the shedding of groups of microvilli (e.g., Itaya 1976) or tips of microvilli (Williams and Blest 1980) into the extracellular medium. These microvillar debris may then be phagocytosed by either photoreceptors (e.g., Stowe 1983), glial cells (Blest 1980), or by granular hemocytes which invade the retina (Waterman 1982). In addition, a massive disruption of rhabdoms has been reported in some cases, but how these membranes are internalized remains unclear.

The counterpart of this degradation, membrane biosynthesis and assembly of microvilli, is even less well understood that the breakdown. It is very difficult to correlate morphological features with such complex processes, and the mechanisms which must occur in order to compensate for the membrane breakdown remain for the most part obscure. In principle it is assumed that membrane-bound visual pigment, after it has been synthesized in the rough endoplasmic reticulum, is transported by some sort of vesicle towards the rhabdomere, where it is thought to fuse with the existing microvillar membrane (see Waterman 1982; Blest et al. 1984; Schwemer 1986).

Turnover has been shown to be influenced by the light regime. In general, light causes a breakdown of membrane and a decrease in rhabdom size, whereas darkness favors membrane synthesis and leads to an increase in rhabdom size. Moderate or drastic changes in rhabdom size have been reported in most of the studies (e.g. Nässel and Waterman 1979), and in only a few of the large variety of animals studied does the size of the rhabdomeres remain fairly constant under changing light regimes (e.g. in the fly, Williams 1982), despite a continuous renewal of the membrane. However, little is known about the possible mechanisms whereby membrane turnover is triggered by light or by darkness (Stowe 1983; Chamberlain and Barlow 1984; Barlow et al. this Vol.).

4.2.2 Visual Pigment Turnover

Until recently the few data available on visual pigment regeneration other than photoregeneration were contradictory and gave no insight into the mechanisms involved (Schwemer 1969; Stavenga et al. 1973; Goldman et al. 1975; Bruno et al. 1977). The first evidence for a turnover of visual pigment in invertebrates came from studies on insect compound eyes (Stein et al. 1978; Schwemer 1979). Visual pigment synthesis was reported by Stein et al., who demostrated a light-dependent incorporation of labeled amino acids into opsin of mosquito visual pigment. The amount of labeled opsin was found to be larger in dark- than in light-adapted eyes, a result that was interpreted, by analogy with the morphological data, as there being an increased breakdown of visual pigment in the light.

A more detailed and comprehensive view into the complex process of visual pigment turnover came from studies of photoreceptors R1–6 in blowfly using spectroscopic, ultrastructural and biochemical methods (Schwemer 1979, 1983, 1984; Paulsen and Schwemer 1983; Schwemer and Henning 1984). Important insight into the biosynthesis of visual pigment was gained from experiments on vitamin A-deficient blowflies (P⁻) which had only 2–4% of the visual pigment of "normal" flies (P⁺). Besides the loss of visual pigment, vitamin A deficiency manifests itself in a reduction in the density of intramembrane particles which can be observed in freeze-fracture replicas of rhabdomeral membranes (Boschek and Hamdorf 1976; Harris et al. 1977; Schwemer 1979). This lack of intramembrane particles could be demonstrated to be due to the loss of opsin in these membranes (Paulsen and Schwemer 1979), a result which is quite surprising, since opsin constitutes about 65% of the total membrane protein in P⁺-flies (Paulsen and Schwemer 1983). All these changes could be reversed by, e.g., injection of 11-cis retinal into the eyes of P⁻-flies, which caused an exponential increase of visual pigment in the dark reaching the value of P⁺-flies about 40 hrs after injection. However, when all-trans retinal was used, no increase of visual pigment content was observed in P⁻flies (Schwemer 1983). These results implied two important facts: first, synthesis of visual pigment protein requires the presence of an 11-cis retinoid, and second, the fly's eye does not contain an enzyme capable of isomerizing the all-trans to the essential 11-cis form in the dark.

Further experiments demonstrated that the visual pigment P490 is rather stable in darkness and undergoes only a slow degradation with a halftime of about 130 h, whereas the meta-state M570 decreases about 60 times faster, leading to the conclusion that M570 is selectively degraded (Fig. 5). Since the visual pigment content of the photoreceptors could be restored, not only by injection of 11-cis retinal, but also by simply returning the animals to room light (Fig. 5b), it was concluded that light plays an important role in that it is somehow involved in isomerizing the all-trans chromophore which is released from opsin as a result of the degradation process. Moreover, these results clearly indicate that the released chromophore is stored somewhere within in the eye until further use.

Degradation of visual pigment has also been analyzed during light adaptation and, as in the dark, it is the meta-state which is preferentially degraded. However, since the meta-state is continuously formed in the light from P490, the time-course of the decrease of the total visual pigment in R1–6 becomes a function of the amount of meta-state present in the photoreceptor membrane. Besides the amount of the meta-state, the light intensity was found to have a profound effect on the rate of degradation: counterintuitively, the rate of breakdown is inversely proportional to the light intensity, i.e., degradation of the meta-state increases with decreasing intensity and vice versa. In order to explain this surprising result, the hypothesis was put forward that M570 is converted to a less stable state, M_u, which has the same absorbance characteristics as M570 and can, like M570, be reconverted to P490 by light. As far as the M_u-state is concerned it has been speculated that it may be the fully phosphorylated meta-state (Schwemer 1984), since phosphorylation, which

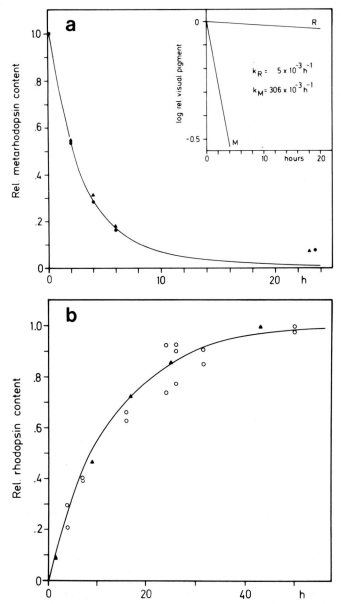

Fig. 5a,b. Degradation and biosynthesis of visual pigment in fly photoreceptors 1–6. *a* Degradation of the meta-state in the dark. *Inset* shows a comparison of degradation of P490 and M570. *b* Increase of visual pigment content in R1–6 during exposure to room light. Before the animals were transferred to polychromatic light, they had been kept in green light in order to degrade the visual pigment (P^+_g-flies *open circles*). For comparison, the time-course of synthesis of P490 in vitamin A-deficient (P^-) flies after application of exogenous 11-cis retinal is shown (*triangles*, animals were kept in the dark after injection). (Schwemer 1984)

does not alter the spectral properties of M and which can be reversed by light (Paulsen and Bentrop 1986), is known as one means of marking intracellular proteins for breakdown. Thus, phosphorylation of opsin may serve two functions, to deactivate the active meta-state (see above), and at the same time to "label" the molecule for breakdown.

The light-induced isomerization of the all-trans chromophore released from opsin during degradation plays a key role in the visual pigment cycle of the fly because it links the two processes, degradation and biosynthesis. This isomerization is caused by light of the blue/violet spectral range. Consequently, flies which are kept in light lacking this spectral component will lose their visual pigment because, due to the increasing deficiency of 11-cis chromophore, the continuous breakdown cannot be compensated for by visual pigment synthesis. Based on the spectral quality of light and on the stereospecific isomerization to the 11-cis form, the hypothesis was put forward that the all-trans chromophore is bound to a specific protein by a protonated Schiff base which would not only shift the absorbance spectrum of the chromophore from about 380 nm to longer wavelengths, but also provide a highly stereospecific isomerization. Following isomerization, the complex is proposed to hydrolyze: the resulting 11-cis chromophore then regulates opsin synthesis, whereas the postulated protein binds another all-trans chromophore. This hypothesis is supported by recent investigations of a soluble protein which has been isolated from the honeybee retina and which meets all these requirements. This protein binds all-trans retinal by a Schiff base and absorbs maximally at 440 nm. Irradiation of this pigment isomerizes the chromophore to its 11-cis form. When the protein is incubated with an excess of all-trans retinal, subsequent irradiation isomerizes the total amount of chromophore almost exclusively to 11-cis retinal, which indicates a cycling of the chromophore (Pepe et al. 1982, 1987; Schwemer et al. 1984). In this respect, the pigment resembles the extensively studied retinochromes of molluscs (Hara and Hara 1980, 1982; Ozaki et al. 1986; Uematsu et al. 1986) which, in addition to the isomerization of all-trans retinal, are also postulated to be involved in the reconstitution of rhodopsin in squid (Seki et al. 1980).

The results obtained from the studies of R1–6 in the fly's ommatidium are summarized in the visual pigment cycle of Fig. 6. This cycle provides functioning visual pigment molecules at any time during the life-span of the animal and thus prevents malfunction of the photoreceptor due to accumulation of "aged" membrane constituents.

An exponential decrease of metarhodopsin and an increase of rhodopsin in the dark has also been verified in photoreceptors of several butterflies (Bernard 1983). However, both processes (which depend strongly on temperature) occur at much higher rates than those found in the fly. The underlying mechanisms of these fast processes have not yet been analyzed, but it has been suggested that the all-trans chromophore of metarhodopsin is replaced by the 11-cis form, while the opsin remains in the membrane (Bernard 1984; Goldsmith and Bernard 1985). A recovery of rhodopsin in the dark has also been reported for the photoreceptors of the crayfish *Procambarus* (Cronin and Goldsmith 1984). This recovery is very slow and takes about 10 days, but is speeded up and needs less than 2 days when the

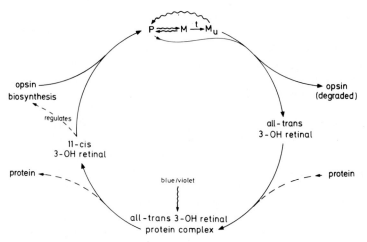

Fig. 6. Visual pigment cycle in photoreceptors R1–6 of the blowfly. (*Wavy lines* photoreactions; *solid lines* thermal reactions; *dashed lines* hypothetical reactions.) (After Schwemer 1984)

animals are exposed to blue light prior to transferring them to darkness, indicating that a retinochrome-like pigment might also be involved in providing the 11-cis chromophore. The authors suggest that the recovery of visual pigment is due to a replacement of metarhodopsin by newly synthesized rhodopsin. A metabolic regeneration of rhodopsin has also been described in the crayfish *Astacus* (Hamacher 1981), but the mechanism by which rhodopsin is restored remains to be elucidated, as is the case for butterflies and *Procambarus*.

4.2.3 Chromophore Cycle

The fate of all-trans 3-hydroxy-retinal released from opsin during visual pigment degradation in the fly has recently been investigated in more detail (Schwemer in prep.). Retinoids were extracted and analyzed by high performance liquid chromatography from flies with "normal" visual pigment content (P^+, controls), from "normal" flies after they had been kept in green light in order to reduce the visual pigment content of the receptors to about 5–8% of the original value (P_g^+-flies), and finally from P_g^+-flies after they had been returned to "white" light to restore the original visual pigment content (P_{gw}^+). The visual pigment content of the animals was determined spectrophotometrically.

Figure 7 shows that the retina of P^+-flies contains about equal amounts of 3-hydroxy-retinal and 3-hydroxy-retinol. About 85% of the 3-hydroxy-retinal was found to be covalently bound and this value was in agreement with the visual pigment content of about 2.2 pmol/retina. The finding of a comparable amount of alcohol points to its possible function as sensitizing pigment (see Vogt this Vol.). A surprising result has been obtained from the isolated retinae of P_g^+-flies which show a drastic decrease in their total content of 3-hydroxy-retinal by approximately 85% (Fig. 7), whereas the alcohol content is reduced by 50% as compared to the controls.

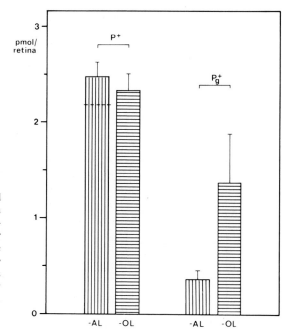

Fig. 7. Content of 3-hydroxy-retinal (*-AL*) and 3-hydroxy-retinol (*-OL*) in the fly retina with "normal" high visual pigment content (*P*+) and after degradation of the pigment in the green light (P_g^+). Amount *above the dashed line* (*left column*) corresponds to aldehyde which is not covalently bound. *Bars* indicate standard deviation

These results demonstrate that visual pigment degradation leads to a decrease of both aldehyde and alcohol within the retina, i.e., both are obviously stored neither within the receptors nor in the surrounding secondary pigment cells. Since, however, the visual pigment is completely restored when isomerization is allowed to occur, the retinoids are most probably stored in some other part of the eye. Therefore, the most distal part of the eye, which is left over after the isolation of the retinae, was analyzed. This part (referred to as the cornea preparation) consists mainly of cornea and crystalline cones, as well as Semper and primary pigment cells. In the cornea preparation of P+-flies, approximately 0.8 pmol/cornea preparation 3-hydroxy-retinal were found, which amounts to about 25% of the total content/eye (Fig. 8a). The amount of 3-hydroxy-retinol was found to be 2.6 pmol, which corresponds to about 53% of the total per eye, i.e., almost half of the alcohol resides in the most distal part of the eye. In P_g^+-flies, however, the ratio of 3-hydroxy-retinal between retina and cornea preparation is reversed: 86% of the total 3-hydroxy-retinal of the eye is now located in the cornea preparation. When compared to P+-flies, the total aldehyde content in P_g^+-flies seems to be reduced. These results indicate that the aldehyde which is released from opsin is indeed transported from the photoreceptor cell to some other, as yet unidentified, compartment which is located close to the cornea. The relative distribution of the alcohol within the eye of P_g^+-flies is almost identical with that in P+-flies, although the absolute concentration seems to be reduced in the latter.

These changes in retinoid distribution in P_g^+-flies caused by the degradation of the visual pigment were reversed by returning these animals to room light for isomerization and visual pigment synthesis. Figure 8b shows the results of a series

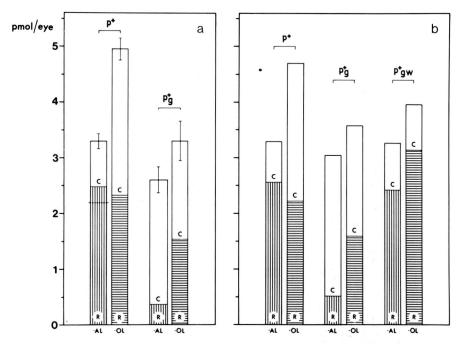

Fig. 8. Amount of 3-hydroxy-retinal (*-AL*) and 3-hydroxy-retinol (*-OL*) in the isolated retina (*R*) and in the remaining cornea preparation (*C;* see text) of P^+, P_g^+ and P_{gw}^+ flies. The latter were returned to polychromatic light after visual pigment degradation in green light (cf. Fig. 5). For explanation see text

of experiments carried out with a single culture of flies. The data obtained for P^+- and P_g^+-flies agree well with those of Fig. 8a. The distribution of 3-hydroxy-retinal in the eyes of P_{gw}^+-flies which had resynthesized about 93% of the visual pigment content during the 40 h exposure to room light is almost identical to that in the beginning of the experiment (Fig. 8b, first column). Similarly, the 3-hydroxy-retinol increased again in the retina.

Analysis of the C_{40}-carotenoids showed two major components, which were identified as lutein and zeaxanthin (see also Vogt and Kirschfeld 1983). For both carotenoids, a specific distribution between cornea preparation and retina was found which did not change during visual pigment breakdown or biosynthesis. Although it cannot be ruled out that a small amount of the carotenoids may serve as precursor for the visual pigment chromophore, they are most probably not involved in the chromophore cycle as long as the chromophore requirements can be covered from the store. The results are summarized in the flow chart of Fig. 9.

These preliminary results show that the chromophore which is set free during visual pigment degradation is transported from the photoreceptor cell to a store which is localized close to the cornea. While incorporated in this store, the chromophore is isomerized by light to the 11-cis form which is then transported back to the receptor cell as either aldehyde or alcohol. A similar shuttle exists for 3-hydroxy-retinol. Besides isomerization of the chromophore, exogenous retinal which is injected into the eye becomes hydroxylated and reduced within this store.

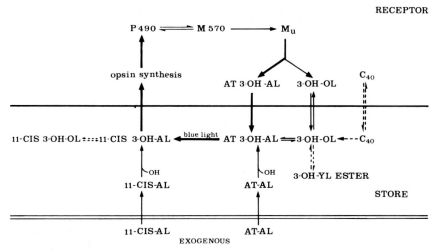

Fig. 9. Flow diagram of 3-hydroxy-retinoids and C_{40}-carotenoids in the eye of the blowfly. 11-cis 3-OH- AL and AT 3-OH -AL: isomers of 3-hydroxy-retinal 3-OH -OL: 3-hydroxy-retinol; C_{40}: C_{40}-carotenoids; 11-cis -AL and AT -AL: isomers of retinal which were injected into the eye. *Heavy arrows* indicate the presumptive main route of the chromophore

Although not yet unequivocally identified, it seems that the primary pigment cells which surround the crystalline cones provide the cellular compartment of this store. The shuttle of the retinoids is most probably achieved by specific retinoid-binding proteins. From recent experiments, it seems quite possible that the pigment isolated from the honeybee retina is involved not only in the isomerization but also in the transport of these retinoids since it also binds retinol (Pepe et al. 1987).

The basic cycle of the visual pigment chromophore in the fly's eye shows surprising similarities to that known from the vertebrate eye in that the all-trans chromophore also leaves the photoreceptor cell and is transported to another ocular tissue which also serves as retinoid store, the pigment epithelium. After arriving in this tissue, the all-trans chromophore is isomerized to the 11-cis form (most probably the alcohol, Bernstein et al. 1987) and shuttled back to the photoreceptor cell by well-characterized retinoid-binding proteins to form rhodopsin.

Biosynthesis, assembly, and degradation of visual pigment and photoreceptor membrane involve a highly regulated interplay of complex molecular mechanisms. Although the few data which are presently available give only a faint insight into these complex cellular mechanisms, they nevertheless provide a solid basis for further research which, through the application of the rapidly developing techniques in molecular biology, promises a deeper understanding of the basic mechanisms involved and, in addition, may provide solutions to the most intriguing puzzle of how all these processes are linked to each other, how the different rates of synthesis and degradation are regulated and their limits determined in order to maintain continuous photoreceptor function.

Acknowledgments. This work was supported by the Deutsche Forschungsgemeinschaft through a Heisenberg Stipendium, the SFB 114 and Schw 269/3-1.

References

Applebury ML, Hargrave P (1986) Molecular biology of the visual pigments. Vision Res 26:1881–1895

Autrum H (ed) (1981) Light and dark adaptation in invertebrates. In: Handbook of sensory physiology, vol VII/6C. Springer, Berlin Heidelberg New York, pp 1–91

Baehr W, Applebury ML (1986) Exploring visual transduction with recombinant DNA techniques. TINS 9:198–203

Bernard GD (1983) Bleaching of rhabdoms in eyes of intact butterflies. Science 219:69–71

Bernard GD (1984) Dark-regeneration of butterfly rhodopsin at physiological temperatures. Invest Ophthalmol 25 (Suppl): 60

Bernstein PS, Law WC, Rando RR (1987) Isomerization of all-trans retinoids to 11-cis retinoids in vitro. Proc Natl Acad Sci USA 84:1849–1853

Blest AD (1980) Photoreceptor membrane turnover in arthropods: Comparative studies of breakdown processes and their implications. In: Williams TP, Baker BN (eds) The effect of constant light in visual processes. Plenum, New York, pp 217–245

Blest AD, Stowe S, deCouet HG (1984) Turnover of photoreceptor membranes in arthropods. Sci Progr 69:83–100

Boschek CB, Hamdorf K (1976) Rhodopsin particles in the photoreceptor membrane of an insect. Z Naturforsch 39c:762

Bownds D (1967) Site of attachment of retinal in rhodopsin. Nature (London) 184:1178–1181

Bridges CDB (1972) The rhodopsin-porphyropsin visual system. In: Dartnall HJA (ed) Handbook of sensory physiology, vol VII/1. Springer, Berlin Heidelberg New York, pp 417–480

Brown PK, Brown PS (1958) Visual pigments of the octopus and cuttle fish. Nature (London) 182:1288–1290

Bruno MS, Barnes SN, Goldsmith TH (1977) The visual pigment and visual cycle of the lobster, Homarus. J Comp Physiol A 120:123–142

Chamberlain SC, Barlow RB (1984) Transient membrane shedding in Limulus photoreceptors: Control mechanisms under natural lighting. J Neurosci 4:2792–2810

Cowman AF, Zuker CS, Rubin GM (1986) An opsin gene expressed in only one photoreceptor cell type of Drosophila eye. Cell 44:705–710

Cronin TW, Goldsmith TH (1984) Dark regeneration of rhodopsin in crayfish photoreceptors. J Gen Physiol 84:63–81

Dixon SF, Cooper A (1987) Quantum efficiencies of the reversible photo-reaction in octopus rhodopsin. Photochem Photobiol 46:115–119

Eguchi E, Waterman TH (1967) Changes in the retinal fine structure induced in the crab Libinia by light and dark adaptation. Z Zellforsch 79:209–229

Findlay JBC (1986) The biosynthetic, functional and evolutionary implications of the structure of rhodopsin. In: Stieve H (ed) The molecular mechanism of photoreception. Dahlem Konf. Springer, Berlin Heidelberg New York, pp 11–30

Fryxell KJ, Meyerowitz EM (1987) An opsin gene is expressed only in the R7 photoreceptor cell of Drosophila. EMBO J 6:443–451

Goldman LJ, Barnes SN, Goldsmith TH (1975) Microspectrophotometry of rhodopsin and meta-rhodopsin in the moth Galleria. J Gen Physiol 66:383–404

Goldsmith TH, Bernard GD (1985) Visual pigments of invertebrates. Photochem Photobiol 42:805–809

Hamacher K (1981) Absorptionsspektroskopische Analyse des Astacus Rhodopsinsystems und Nachweis einer metabolischen Regeneration des Rhodopsins nach Helladaptation. Thesis, Ber Kfa Jülich 1718, Jülich, FRG

Hamdorf K (1979) The physiology of invertebrate visual pigments. In: Autrum H (ed) Handbook of sensory physiology, vol VII/6A. Springer, Berlin Heidelberg New York, pp 145–224

Hamdorf K, Schwemer J (1975) Photoregeneration and the adaptation process in insect photoreceptors. In: Snyder AW, Menzel R (eds) Photoreceptor optics. Springer, Berlin Heidelberg New York, pp 263–289

Hamdorf K, Schwemer J, Täuber U (1968) Der Sehfarbstoff, die Absorption der Rezeptoren und die spektrale Empfindlichkeit der Retina von *Eledone moschata.* Z vergl Physiol 60:375–415

Hamdorf K, Gogala M, Schwemer J (1971) Beschleunigung der Dunkeladaptation eines UV-Rezeptors durch sichtbare Strahlung. Z Vergl Physiol 75:189–199

Hamdorf K, Paulsen R, Schwemer J (1973) Photoregeneration and sensitivity control of photoreceptors of invertebrates. In: Langer H (ed) Biochemistry and physiology of visual pigments. Springer, Berlin Heidelberg New York, pp 155–166

Hara T, Hara R (1980) Retinochrome and rhodopsin in the extraocular photoreceptor of the squid, *Todarodes.* J Gen Physiol 75:1–19

Hara T, Hara R (1982) Cephalopod retinochrome. In: Packer L (ed) Methods in enzymology. Pt H: Visual pigments and purple membranes I, vol 81. Academic Press, New York London, pp 827–833

Hargrave PA, McDowell JH, Curtis DR, Wang JK, Juszczak E, Fong S-L, Rao JKM Argos P (1983) The structure of bovine opsin. Biophys Struct Mech 9: 235–244

Harris AW, Ready DF, Lipson ED, Hudspeth AJ, Stark WS (1977) Vitamin A deprivation and *Drosophila* photopigments. Nature (London) 266:648–650

Honig B, Dinur U, Nakanishi K, Balogh-Nair V, Gawinowicz MA, Arnaboldi M, Motto MG (1979) An external point-charge model for wavelength regulation in visual pigments. J Am Chem Soc 101:7084–7086

Hubbard R, St. George RCC (1958) The rhodopsin system of the squid. J Gen Physiol 41:501–528

Itaya SK (1976) Rhabdom changes in the shrimp, *Palaemonetes.* Cell Tissue Res 166:265–273

Kirschfeld K (1986) Activation of visual pigment: Chromophore structure and function. In: Stieve H (ed) The molecular mechanism of photoreception. Dahlem Konf. Springer, Berlin Heidelberg New York, pp 31–49

Kirschfeld K, Vogt K (1986) Does retinol serve a sensitizing function in insect photoreceptors? Vision Res 26:1771–1777

Kropf A, Brown PK, Hubbard R (1959) Lumi-and metarhodopsin of squid and octopus. Nature (London) 183:446–448

Kruizinga B, Kamman RL, Stavenga DG (1983) Laser-induced visual pigment conversions in fly photoreceptors measured in vivo. Biophys Struct Mech 9:299–307

Montell C, Jones K, Zuker CS, Rubin G (1987) A second opsin gene expressed in the ultraviolet sensitive R7 photoreceptor cell of *Drosophila melanogaster.* J Neurosci 7:1558–1566

Nässel DR, Waterman TH (1979) Massive diurnally modulated photoreceptor membrane turnover in crab light and dark adaptation. J Comp Physiol A 131:205–216

Oseroff AR, Callender RH (1974) Resonance Raman spectroscopy of rhodopsin in retinal disk membranes. Biochemistry 13:4243–4248

O'Tousa JE, Baehr W, Martin RL, Hirsh I, Pak WL, Applebury ML (1985) The *Drosophila* nina E gene encodes an opsin. Cell 40:839–850

Ovchinnikov YA (1982) Rhodopsin and bacterio-rhodopsin: structure function relationships. FEBS Lett 148:179–191

Ovchinnikov YA (1987) Structure of rhodopsin and bacteriorhodopsin. Photochem Photobiol 45:909–914

Ozaki K, Terakita A, Hara R, Hara T (1986) Rhodopsin and retinochrome in the retina of a marine gastropod, *Conomulex luhuanus.* Vision Res 26:691–705

Pande C, Pande A, Yue KT, Callender R, Ebrey TG, Tsuda M (1987a) Resonance Raman spectroscopy of octopus rhodopsin and its photoproducts. Biochemistry 26:4941–4947

Pande C, Deng H, Rath P, Callender R, Schwemer J (1987b) Resonance Raman spectroscopy of an UV-sensitive insect rhodopsin. Biochemistry 26:7426–7430

Pappin PJC, Eliopoulos E, Brett M, Findlay JBC (1984) A structural model for bovine rhodopsin. Int J Biol Macromol 6:73–76

Paulsen R, Bentrop J (1986) Light-modulated biochemical events in fly photoreceptors. In: Lüttgau HCh (ed) Membrane control. Fortschr Zool 33:299–319

Paulsen R, Schwemer J (1972) Studies on the insect visual pigment sensitive to ultraviolet light: retinal as the chromophoric group. Biochim Biophys Acta 283:520–529

Paulsen R, Schwemer J (1979) Vitamin-A deficiency reduces the concentration of visual pigment protein within blowfly photoreceptor membranes. Biochem Biophys Acta 557:385–390

Paulsen R, Schwemer J (1983) Biogenesis of blowfly photoreceptor membranes is regulated by 11-cis retinal. Eur J Biochem 137:609–614

Pepe IM, Schwemer, J Paulsen R (1982) Characteristics of retinal-binding protein from the honeybee retina. Vision Res 22:775–781

Pepe IM, Cugnoli C, Peluso M, Vergani L, Boero A (1987) Structure of a protein catalyzing the formation of 11-cis retinal in the visual cycle of invertebrate eyes. Cell Biophys 10:15–22

Schwemer J (1969) Der Sehfarbstoff von *Eledone moschata* und seine Umsetzungen in der lebenden Netzhaut. Z Vergl Physiol 62:121–152

Schwemer J (1979) Molekulare Grundlagen der Photorezeption bei der Schmeißfliege *Calliphora erythrocephala* Meig. Habil Schr, Fak Biol, Ruhr-Univ Bochum, FRG

Schwemer J (1983) Pathways of visual pigment regeneration in fly photoreceptors. Biophys Struct Mech 9:287–298

Schwemer J (1984) Renewal of visual pigment in photoreceptors of the blowfly. J Comp Physiol A 154:535–547

Schwemer J (1986) Turnover of photoreceptor membrane and visual pigment in invertebrates. In: Stieve H (ed) The molecular mechanism of photoreception. Dahlem Konf. Springer, Berlin Heidelberg New York, pp 303–326

Schwemer J, Henning U (1984) Morphological correlates of visual pigment turnover in photoreceptors of the fly. Cell Tissue Res 236:293–303

Schwemer J, Langer H (1982) Insect visual pigments. In: Packer L (ed) Methods in enzymology. Pt H: Visual pigments and purple membranes I, vol 81. Academic Press, New York London, pp 182–190

Schwemer J, Paulsen R (1973) Three visual pigments in *Deilephila elpenor* (Lepidoptera, Sphingidae). J Comp Physiol 86:215–229

Schwemer J, Pepe IM, Paulsen R, Cugnoli C (1984) Light-activated trans-cis isomerization of retinal by a protein from honeybee retina. J Comp Physiol A 154:549–554

Seki T, Hara R, Hara T (1980) Reconstitution of squid rhodopsin in rhabdomal membranes. Photochem Photobiol 32:469–479

Shichida Y, Matuoka S, Yoshizawa T (1984) Formation of photorhodopsin, a precursor of bathorhodopsin, detected by picosecond laser photolysis at room temperature. Photobiochem Photobiophys 7:221–228

Stavenga DG, Schwemer J (1984) Visual pigments of invertebrates. In: Ali MA (ed) Photoreception and vision in invertebrates. Plenum, Oxford New York, pp 11–61

Stavenga DG, Zantema A, Kuiper JW (1973) Rhodopsin processes and the function of the pupil mechanism in flies. In: Langer H (ed) Biochemistry and physiology of visual pigments. Springer, Berlin Heidelberg New York, pp 175–180

Stein PJ, Brammer JO, Ostroy SE (1978) Renewal of opsin in the photoreceptor cells of the mosquito. J Gen Physiol 74:565–582

Stowe S (1983) Light-induced and spontaneous breakdown of the rhabdoms in a crab at dawn; depolarization versus calcium levels. J Comp Physiol A 153:365–375

Suzuki T, Makino-Tasaka M, Eguchi E (1984) 3-dehydroretinal (vitamin A2 aldehyde) in crayfish eye. Vision Res. 24:783–789

Tsuda M (1987) Photoreception and phototransduction in invertebrate photoreceptors. Photochem Photobiol 45:915–931

Uematsu J, Hara R, Nishimura I, Wada K, Matsubara H, Hara T (1986) Amino-terminal sequence of squid retinochrome. Photobiochem Photobiophys 13:197–201

Vogt K (1983) Is the fly visual pigment a rhodopsin? Z Naturforsch 38c:329–333

Vogt K, Kirschfeld K (1983) C 40-Carotinoide in Fliegenaugen. Verh Dtsch Zool Ges 76:330

Vogt K, Kirschfeld K (1984) Chemical identity of the chromophores of fly visual pigment. Naturwissenschaften 71:211–213

Waterman TH (1982) Fine structure and turnover of photoreceptor membranes. In: Westfall JA (ed) Visual cells and evolution. Raven, New York, pp 23–41

White RH (1968) The effect of light and light deprivation upon the structure of the larval mosquito eye. III. Multivesicular bodies and protein uptake. J Exp Zool 169:261–278

White RH (1985) Insect visual pigments and color vision. In: Kerkut GA, Gilbert Ll (eds) Comprehensive insect physiology, biochemistry and pharmacology, vol 6. Pergamon, New York, pp 431–493

Williams DS (1982) Rhabdom size and photoreceptor membrane turnover in a muscoid fly. Cell Tissue Res 226:629–639

Williams DS, Blest AD (1980) Extracellular shedding of photoreceptor membrane in the open rhabdom of a tipulid fly. Cell Tissue Res 205:423–438

Yoshizawa T, Shichida Y (1982) Low-temperature spectrophotometry of intermediates of rhodopsin. In: Packer L (ed) Methods in enzymology. Pt H: Visual pigments and purple membranes I, vol 81. Academic Press, New York London, pp 333–353

Zuker CS, Cowman AF, Rubin GM (1985) Isolation and structure of a rhodopsin gene from *D. melanogaster*. Cell 40:851–858

Zuker CS, Montell C, Jones K, Raverty T, Rubin GM (1987) A rhodopsin gene is expressed in photoreceptor cell R7 of the *Drosophila* eye: homology with other signal transducing molecules. J Neurosci 7:1550–1557

Chapter 7

Distribution of Insect Visual Chromophores: Functional and Phylogenetic Aspects

KLAUS VOGT, Freiburg, FRG

1 Introduction

The study of visual pigments started in earnest with the work of Boll and Kühne, both of whom were contemporaries of Exner. Whilst Boll (1876) discovered that the color of a frog retina fades when exposed to light, Kühne (1878) was the first to extract the visual pigment, finding its spectral absorptivity similar to the human spectral sensitivity under scotopic conditions.

No more major advances were made until the 1930's with the isolation of retinal from photo-bleached frog rhodopsin by Wald (1934, 1936), followed by the characterization of the structure of retinal by Morton (1944) and the discovery of the cis-trans isomerization of retinal as the trigger of visual transduction (rev. Wald 1968). Wald (1939) also discovered that vitamin A_2 aldehyde was used as a chromophore by some fishes and amphibians, the resulting visual pigments being known as porphyropsins (cf. Suzuki et al. 1984).

We are now experiencing a third major period of discovery notable for the success in sequencing several opsin molecules, including those of *Drosophila* (Ovchinnikov 1982; Hargrave 1982; O'Tousa et al. 1985; Zuker et al. 1985; Cowman et al. 1986; Goldsmith this Vol.; Schwemer this Vol.) and by impressive new insights into the transduction process (rev. Stryer 1986; Fein and Payne this Vol.).

Generally speaking, the chromophores of the insect visual pigments have not been a major subject of investigation in recent years, since the question as to their identity seemed to be settled. With the exception of the vitamin A_2 aldehyde in the porphyropsins, retinal has been regarded as the universal chromophore and it has been widely believed and asserted that insects also have a retinal-based visual pigment. However, in Diptera, Lepidoptera, and several other insect orders a visual pigment chromophore has been found that is drastically more polar than both known chromophores and which was identified as 3-hydroxyretinal (Vogt 1983, 1984a,b; Vogt and Kirschfeld 1984). For visual pigments based on this chromophore the common name xanthopsin has been proposed (Fig. 1, see Vogt 1987).

In this chapter I first consider the recent literature on the identity and function of the novel chromophore. Most of the detailed analysis has been performed upon the visual pigments of the higher dipterans, which in addition to the novel chromophore also harbor an accessory chromophore (sensitizing pigment), which extends the spectral bandwidth of the photoreceptors. Secondly, I consider the

Stavenga/Hardie (Eds.) Facets of Vision
© Springer-Verlag Berlin Heidelberg 1989

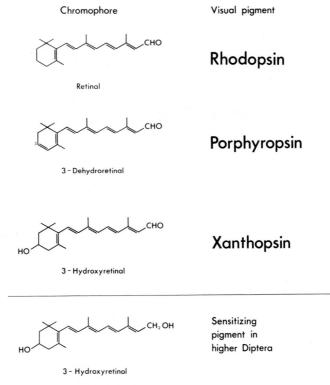

Fig. 1. Structural formulas of the known visual pigment chromophores

distribution of the novel chromophore amongst the different insect orders. This in turn leads to speculation on both insect phylogeny and also the evolutionary pressures which may be involved in determining the choice of visual pigment chromophore.

2 Visual Pigment Chromophores in Flies

2.1 The Normal Chromophore

The initial discovery of the novel chromophore was made using high performance thin layer chromatography (HPTLC) of retina extracts from blowflies (*Calliphora erythrocephala*). Quite unexpectedly, an aldehyde was isolated which was much more polar than any isomer of retinal or 3-dehydroretinal. In order to show that this compound was indeed the visual pigment chromophore, use was made of the known bi-stability of fly visual pigment, whereby blue illumination converts ca. 80% of the pigment to a stable meta-state which in turn is completely reisomerized

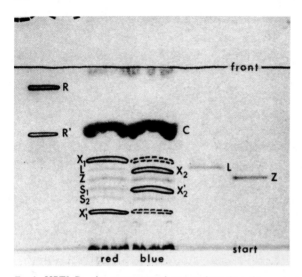

Fig. 2. HPTLC – chromatogram of extracts from fly retinae, previously adapted with red or blue light. Retina extracts: *C* cholesterol; *L* lutein; *Z* zeaxanthin; S_1, S_2 3-hydroxyretinol isomers; S_1 all-trans; X_1, X_1 syn, anti of 11-cis 3-hydroxyretinal; X_2, X_2 syn, anti of all-trans 3-hydroxyretinal. References: *R,R* syn, anti of all-trans retinal; *L* lutein; *Z* zeaxanthin. Retinoids are actually present as oximes due to pre-treatment with NH_2OH (Vogt 1983)

by red light (rev. Stavenga and Schwemer 1984; Schwemer this Vol.). As would be predicted from this behavior, extracts from flies which had been illuminated with red light yielded a single isomer (X1, Fig. 2), whereas blue light reduces this isomer to some 20%, the rest being changed into a new isomer (X2, Fig. 2) (Vogt 1983). Furthermore, the compound was found to be absent from flies raised on a carotenoid free diet which induces blindness.

Several lines of evidence were used to identify the novel chromophore as 3-hydroxyretinal: (1) the chromophore is spectrally indistinguishable from retinal, suggesting the same conjugated system; (2) C_{40} carotenoids with 3-hydroxy ß-end-rings serve as chromophore precursors. (3) the presence of a single hydroxyl group could be demonstrated by acetylation of the fly chromophore, and its position narrowed down by methylation experiments (Vogt 1983; Vogt and Kirschfeld 1984). Finally, the identity of the chromophore has now been conclusively confirmed by co-chromatography of synthetic 3-hydroxyretinal, along with 4-hydroxyretinal and 2-hydroxy compounds for comparison (Vogt 1987). In the meantime, several laboratories have also confirmed both the presence of the novel chromophore in flies and butterflies (Goldsmith et al. 1985), and also its identity as 3-hydroxyretinal (Seki et al. 1986; Goldsmith et al. 1986; Tanimura et al. 1986).

As would be expected by analogy with retinal-containing visual pigments, the normal form of the novel chromophore (in *Drosophila*) has been shown to be the 11-cis isomer (Seki et al. 1986), and this is converted to the all-trans form in the meta-state of the pigment (Tanimura et al. 1986).

2.2 Carotenoid Metabolism

Like most other animals, insects are apparently unable to synthesize carotenoids, and thus depend on green plants, fungi, or bacteria for their supply. However, the basic carotenoid skeleton may subsequently be modified, most commonly by insertions of hydroxyl- and/or oxo-groups at the various positions of the ß-end-ring (Kayser 1982). *Calliphora* can obviously hydroxylate carotenes since, when reared on a ß-carotene diet, the eyes contained only zeaxanthin and the inter-mediate monohydroxy-compound ß,ß-cryptoxanthin. This pathway is apparently not reversible, since no traces of ß,ß-cryptoxanthin or ß,ß-carotene could be found in animals reared on a zeaxanthin diet. Whereas there was no indication that retinal could be formed by splitting ß,ß-carotene, the presence of a dioxygenase-like enzyme cleaving zeaxanthin to 3-hydroxyretinal was clearly shown (Vogt 1983a; Vogt and Kirschfeld 1983). C_{20} compounds can also be hydroxylated, as has been shown in the tsetse fly, which has a high content of 3-hydroxyretinal when reared on swine blood which contains retinol but virtually no C_{40}-carotenoids (Hardie and Vogt, in prep.). Figure 3 summarizes our present rather limited knowledge of carotenoid metabolism in flies.

A rough measure of th ability of an animal to hydroxylate a carotene is given by the amount of the intermediate monohydroxy compound (cryptoxanthin or "hidden pigment") it contains compared to the amount in its food. Using this kind of data (rather than tedious dietary experiments), it seems that many, though not all, Lepidoptera are unable to hydroxylate ß,ß-carotene and they depend on

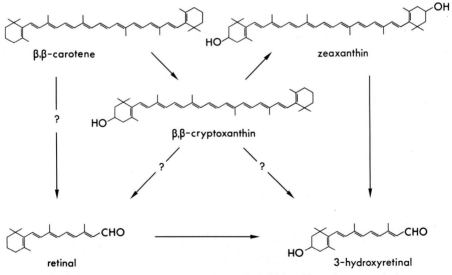

Fig. 3. Carotenoid metabolism in higher flies. Pathways for which evidence is not conclusive are indicated by a question mark. Retinal and 3-hydroxyretinal are interchangeable with the corresponding alcohols (Vogt 1987)

xanthophylls in their diet to form their chromophore (Vogt 1984a). Interestingly, in the gut of many Lepidoptera xanthophylls such as lutein are even absorbed in preference to ß,ß-carotene (Kayser 1982).

2.3 The Sensitizing Pigment

The chromatographic analysis of fly retina extracts was originally undertaken in an attempt to find chemical evidence for the existence of a sensitizing pigment. Such a pigment had been postulated by Kirschfeld et al. (1977) in order to explain the long-standing enigma of the dual-peaked spectral sensitivity of fly photoreceptors. Thus, as first shown by Burkhardt (1962), the action spectrum of the most common receptor type (R1–6) has two maxima: one close to 500 nm, and one (usually even higher) in the near-ultraviolet at 350 nm (Fig. 4). Such a dual peak sensitivity cannot be explained by the absorbance spectra of normal visual pigments, which only have a small secondary UV peak (ß-peak). The explanations which had previously been proposed (e.g., two visual pigments in one and the same cell, or waveguide effects that enhance the ß-peak absorption) all turned out to be inconsistent with the experimental evidence.

The essence of the sensitizing pigment hypothesis (Kirschfeld et al. 1977) is that a UV-absorbing photostable pigment transfers the energy of absorbed quanta to the visual pigment by means of Förster's (1951) radiationless dipole-dipole mechanism. The visual pigment is thereby excited just as if it had directly absorbed a light quantum. Apart from absorption in the appropriate UV range such a mechanism requires the donor (sensitizing) molecule to exhibit high fluorescence spectrally overlapping with the acceptor's (visual pigment) absorption, and also a

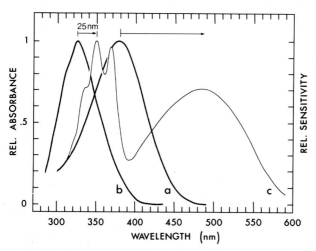

Fig. 4. Absorbance spectra of the fly chromophore aldehyde (*a*) and the alcohol (*b*). *Thin line* the spectral sensitivity of the photoreceptors R1–6. *Arrows* indicate the spectral shift between the two chromophores (measured in ethanol) and the corresponding sensitivity peaks. (Vogt and Kirschfeld 1984)

close proximity of donor and acceptor molecules. The hypothesis was supported by the demonstration that the rhabdomeres did indeed contain a photostable substance absorbing at the UV peak of the action spectrum and that UV light was able to isomerize the visual pigment whilst the UV extinction remained unaltered.

This hypothesis is also consistent with two older observations: firstly that a carotenoid-deficient diet leads to a relatively greater loss of sensitivity in the UV than in the green (Goldsmith et al. 1964) — a result which is difficult to reconcile with the notion of a single chromophore exhibiting two absorption maxima — and secondly the fact that it is impossible to modify the relative heights of the two sensitivity peaks by selective chromatic adaptation (Burkhardt 1962), which suggests that there is only one visual pigment.

Further evidence comes from the finding that, in contrast to the visible range, fly photoreceptors show no polarization sensitivity (PS) in the UV (Fig. 5) (Hardie 1978; Guo 1981; Vogt and Kirschfeld 1983b). In carotenoid-deprived flies, however, where UV sensitivity is presumed to derive from ß-peak absorption, PS *is* found in the UV with the same phase and similar magnitude as in the visible (Vogt and Kirschfeld 1983b). This is consistent with an additional UV chromophore with an absorbing dipole oriented differently from that of the principal chromophore. The lack of dichroism in the UV can then be explained if the dipole of the sensitizing chromophore is, on average, tilted at a specific angle (ca. 35°) to the plane of the membrane.

Since the UV sensitivity and extinction depend upon a carotenoid diet (Goldsmith et al. 1964; Stark et al. 1977; Kirschfeld et al. 1983), it had already been inferred that the UV chromophore is a carotenoid derivative. Gemperlein et al.'s (1980) demonstration of a pronounced fine structure in the UV sensitivity peak (see Fig. 4), initiated a flurry of speculation about the chemical identity of the sensitizing pigment. Several carotene derivatives exhibiting a structured UV extinction, e.g., phytofluene, retroretinal, and retinol bound to a protein as in the retinol- ß-lactoglobulin complex (Fugate and Song 1980) were discussed as possible candidates (Paul 1981; Kirschfeld et al. 1983; Franceschini 1983). However, only one

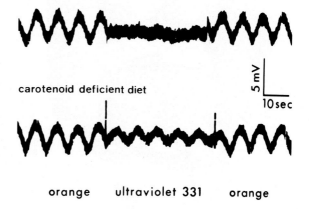

carotenoid rich diet

Fig. 5. Responses of photoreceptors R 1–6 in *Calliphora* to a steady light delivered through a continuously rotating polaroid filter. Following orange light, the color was switched to UV of equivalent effective intensity, and then back to orange. In the fly raised on bovine liver (*upper trace*) there is no modulation during the UV, whilst a polarization-sensitive response is obvious in the carotenoid deprived fly, raised on horse skeletal muscle. (Vogt and Kirschfeld 1983b)

carotenoid deficient diet

5 mV

10 sec

orange ultraviolet 331 orange

substance could be isolated from fly eyes that fulfilled the prerequisites of a sensitizing pigment (see above). This substance was even more polar than the chromophore aldehyde (isomers S1, S2 in Fig. 2) and was identified as 3-hydroxyretinol on spectral, chromatographic and biochemical evidence (Vogt 1983; Vogt and Kirschfeld 1984).

The dominant isomer in extracts from fly retina appears to be the all-trans form (S1) with only small quantities of a cis-isomer (S2), which after isolation and UV-irradiation is mainly converted to the all-trans form.

The absorbance spectrum of 3-hydroxyretinol does not exhibit the characteristic fine structure of the UV sensitivity and also peaks at shorter wavelengths (Fig. 4). However, this discrepancy may be explained by analogy with the binding of retinol to certain proteins: a bathochromic shift and a fine structure result if retinol is bound in an approximately planar conformation which allows fuller conjugation of the ring double bond (Fugate and Song 1980; Chytil and Ong 1984). In retinol-binding protein this has been attributed to hydrophobic interactions within a cleft in the protein (Fugate and Song 1980). However, the structure of 3-hydroxyretinol, with a hydroxy group at each end of the molecule suggests an alternative or additional mechanism. Thus the coplanar conformation could be maintained by hydrogen bonds between the two hydroxy groups and appropriate polar groups on the opsin (Fig. 6, Vogt and Kirschfeld 1984).

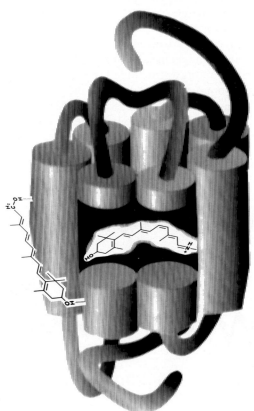

Fig. 6. Tentative model of the fly visual pigment with two chromophores: the covalently bound isomerizable chromophore 11-cis 3-hydroxyretinal, and the non-isomerizable, sensitizing chromophore all-trans 3-hydroxyretinol bound by hydrogen bonds to the opsin (cf. Vogt and Kirschfeld 1984). The drawing of the opsin is adapted from the structure proposed in *Drosophila* (O'Tousa et al. 1985; Zuker et al. 1985), which in turn follows the model for bovine opsin (Ovchinnikov 1982; Hargrave 1982). (Vogt 1987)

An intimate association between the sensitizing chromophore and the opsin has in fact already been indicated by measurements of the quantum efficiency of energy transfer between sensitizing and ordinary chromophore. This is surprisingly high, and can be conservatively estimated at ≥ 0.8 (Vogt and Kirschfeld 1982; cf. Smakman and Stavenga 1986). Using the Förster equation, this implies a maximal distance between the chromophores of 2.5 nm, which is less than the diameter of the opsin. A fixed distance between the two chromophores is also suggested by the fact that the transfer efficiency remains constant if the density of visual pigment in the rhabdomeric membrane is considerably reduced (Vogt and Kirschfeld 1983b). Besides the proximity to the ordinary chromophore, a binding of the sensitizing chromophore to the opsin could be functionally important because of an enhancement of the fluorescence efficiency, e.g., by a factor of 7–8 in the case of retinol when bound to a cellular binding protein (Chytil and Ong 1984).

The molar ratio between ordinary and sensitizing chromophores is not known with certainty, but a 1:1 relationship seems to be just sufficient to explain the maximal ratios of UV/visible sensitivity. If ß-peak absorption and waveguide effects are taken into account, the ratio of sensitivity due to sensitizing pigment and peak sensitivity in the visible range is 1.5 at the most (Hardie 1985; cf. Smakman and Stavenga 1986). This is about the same as the ratio between the molar extinction coefficients of all-trans 3-hydroxyretinol (assumed to be the same as that of all-trans retinol) and of fly visual pigment (Stavenga and Schwemer 1984). Recently Schwemer (this Vol.) has quantitatively extracted 3-hydroxyretinol and 3-hydroxyretinal from blowfly eyes, and finds them present in approximately equal quantities.

Figure 6 summarizes the concepts developed above, showing a tentative model of the fly visual pigment with two chromophores (cf. Vogt and Kirschfeld 1984). However, the evidence for a binding of the sensitizing pigment to the opsin is by no means conclusive, and the possibility cannot be excluded, for example, that the sensitizing pigment binds to another protein, in turn closely attached to the opsin.

A visual pigment complex with two chromophores has obvious selective advantages. Although occupying virtually no more space in the membrane than a simple visual pigment, the complex would have a considerably greater overall absorption when integrated over the spectrum (cf. Kirschfeld 1983).

3 Visual Pigment Chromophores Amongst the Insect Orders

The following comparative study had two principal aims: firstly, by correlating different parameters, including retinoids, C_{40}-carotenoids and UV sensitivity, it may be possiblee to gain new insights and in particular to address the question of whether the new chromophore 3-hydroxyretinal and the corresponding alcohol are prerequisites for a sensitized UV vision. Secondly, if the new chromophore proves to be a stable chemotaxonomic marker, we may be able to address such questions as when the new chromophore was acquired and "why" this happened. The study is based on a sample of 55 species embracing 17 different insect orders. To improve on random sampling, whenever possible the species were chosen from different main branches of the order in question. As well as the retinoids and carotenoids in both eyes and body, we also examined the spectral sensitivity, searching for the characteristic UV fine structure which can be conveniently detected in the electroretinogram (Vogt 1983, 1984a; Vogt and Kirschfeld in prep.)

3.1 Retinoids, UV-Sensitivity, and Carotenoids

Not only was the distribution pattern of the retinoids remarkably straightforward, but it also showed striking correlations with the other parameters determined. Thus all species which exhibited the characteristic UV fine structure were also found to have the novel 3-hydroxyretinal chromophore. This thus appears to be a necessary, though not sufficient prerequisite for this sort of sensitized UV vision. Looking in more detail, it turns out that the predominant form of the chromophore alcohol in extracts from these species (higher flies) is free all-trans 3-hydroxy-retinol. By contrast, in species with xanthopsin, but no UV fine structure (lepidopterans and tabanoid flies), the alcohol is found in either an esterified form, or as a cis-isomer. This correlation is thus a strong additional argument for the identity of all-trans 3-hydroxy-retinol as the sensitizing pigment.

In irradiated solutions the all-trans isomer of 3-hydroxy-retinol is the dominant form, so that the presence of the cis-isomer in lepidopterans and horseflies may indicate the presence of a 3-hydroxyretinol-isomerase or a specific storage protein. It is tempting to relate the above finding to Bernard's (1983a,b) results on the dark-regeneration of the butterfly visual pigment, which suggest a visual pigment cycle involving bleaching of the metapigment and replacement of the 11-cis chromophore following dark reisomerization. In this respect it would be interesting to see whether horsefly metaxanthopsins show the same relative instability as those of butterflies.

C_{40} carotenoids extracted from insect bodies showed no clear correlation with the type of visual pigment chromophore, but those extracted from the eye did. In fact, the major carotenoid type in the eye is virtually diagnostic for the chromophore and in eyes with 3-hydroxyretinal as chromophore, xanthophylls (zeaxanthin, lutein or ß, ß-cryptoxanthin) dominate, whilst ß, ß-carotene is found in eyes with a retinal chromophore.

3.2 Correlation with Phylogeny

3.2.1 The Overall Situation

Superimposing the chromophore distribution on a family tree of the insect orders instantly reveals a clustering of xanthopsin in the more recent holometabolic orders, namely flies (Diptera), scorpionflies (Mecoptera), butterflies and moths (Lepidoptera), caddisflies (Trichoptera) and lacewings (Neuroptera). In the remaining insect orders — from the mayflies (Ephemeroptera) to the alderflies (Megaloptera) — rhodopsin seems to be the visual pigment (Fig. 7). The four exceptions found so far, the order Odonata, a lacewing, a single fly family and a cicada, will be discussed separately. These exceptions apart, the transition from

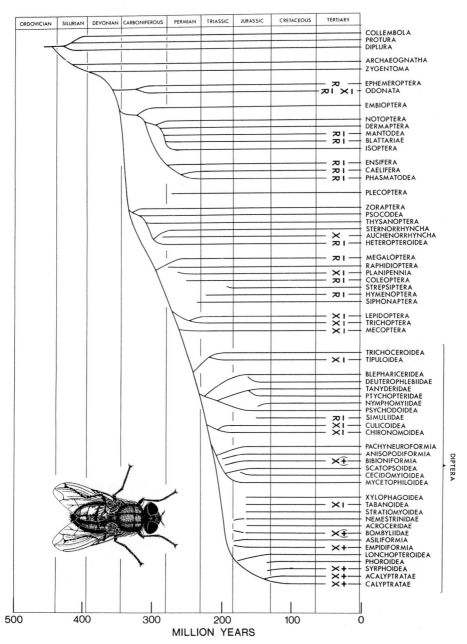

Fig. 7. Phylogenetic tree of the insect orders with geological periods (*left*) and the corresponding time scale (*right*); compiled from several figures of Hennig (1969, 1973). *R* denotes the occurrence of retinal (rhodopsin) and X, the occurrence of 3-hydroxyretinal (xanthopsin). + denotes a fine structure in the UV sensitivity, indicate of 3-hydroxyretinol as sensitizing pigment; − denotes its absence. The *stars* mark the approximate time of acquisition of 3-hydroxyretinal and the sensitizing pigment respectively. (Vogt and Kirschfeld in prep.)

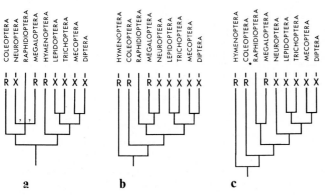

Fig. 8a-c. Three hypotheses for the relation of the holometabolic orders. *R* denotes the occurrence of rhodopsin; *X* the occurrence xanthopsin. *a* the view favored nowadays (Kristensen 1975); *b* derived from older workers (e.g., Weber 1954; Ross 1965); *c* proposed scheme. All hypotheses assume that the Lepidoptera (butterflies and moths), Trichoptera (caddisflies), Mecoptera (scorpionflies) and Diptera (flies), are descended from a common ancestor. The occurrence of xanthopsin in the Neuroptera (lacewings), and rhodopsin in the Coleoptera (beetles), Hymenoptera (bees etc.) and Megaloptera (Alderflies) supports hypothesis *c* which, in contrast to *b*, implies a non-monophyletic origin of the neuropteroid orders

rhodopsin to xanthopsin seems to have occurred somewhere amongst the holometabolic orders. However, the evidence for the phylogenetic grouping of these orders is actually rather poor, and at least three alternative hypotheses can be considered (Fig. 8). Even though all assume that the Lepidoptera, Trichoptera, Mecoptera and Diptera (collectively known as the Mecopteroidea) are descended from a common ancestor, the evidence for this has been somewhat limited (Hennig 1969) and the occurrence of xanthopsin in all these orders is important corroborative evidence of this relationship. The currently favoured hypothesis (Fig. 8a, Kristensen 1975) in which the Hymenoptera[1] are regarded as a sister group of the Mecopteroidea, clashes with the occurrence of xanthopsin in the Neuroptera (Fig. 8a). An alternative view, however, deriving mainly from older workers (e.g., Weber 1954; Ross 1965), considers the Neuropteroidea as the sister group of the Mecopteroidea (Fig. 8b) and this tallies somewhat better with the distribution of xanthopsin. Even so, within the Neuropteroidea rhodopsin is found in the Megaloptera and xanthopsin in the Neuroptera. This means that, if we accept the monophyletic origin of these orders, then one or other of the pigments must have been acquired convergently. However, since the monophyly of the Neuropteroidea is rather weakly founded (Kristensen 1975) a third, alternative hypothesis can be proposed (Fig. 8c) in which only the order Neuroptera forms a sister group of the Mecopteroidea (Fig. 8c).

Whatever the correct phylogeny, the straightforward pattern of chromophore distribution testifies to the stability of the chromophore as a chemotaxonomic marker. This in turn suggests that several proteins had to be adapted in parallel in

---

[1] In the eyes of bees Goldsmith (1958) was the first to demonstrate the occurrence of retinoids in insects.

order to use 3-hydroxyretinal as a visual pigment chromophore, e.g., the dioxygenase cleaving the C_{40} precursors, the opsin protein itself, a monoxygenase in muscoid flies, and possibly a 3-hydroxyretinal/ol isomerase. For opsin at least, the specificity for 3-hydroxyretinal is in principle testable, and comparison of recent sequence data may point to the crucial parts of the protein.

The straightforward pattern amongst insects contrasts markedly with the situation with rhodopsins and porphyropsins in fishes. Firstly, the same fish can actually change its visual pigment chromophore (retinal or 3-dehydro-retinal) during its lifetime, although the opsin itself does not change. Furthermore, the distribution pattern of porphyropsin and rhodopsin shows no obvious correlation with the fish phylogenetic tree, implying that the step from retinal to 3-dehydroretinal has probably occurred several times in fish evolution (Bridges 1972).

If we accept that xanthopsin only evolved once amongst the holometabolic orders, we may return to the time scale (Fig. 7) in order to estimate the approximate time of its acquisition. This turns out to be in the Upper Carboniferous, i.e., about 300 million years ago.

What properties of the new chromophore may have been crucial in determining its acquisition?

Perhaps two of the most obvious attributes of xanthopsins, as exemplified by the well-studied dipteran pigments, are: (1) the possibility of photoregeneration of the visual pigment (in contrast to vertebrate visual pigments the meta-state is thermostable and can be reconverted to xanthopsin without a costly metabolic cycle (revs. Stavenga and Schwemer 1984; Schwemer this Vol.); and (2) the additional sensitizing pigment, whose selective advantage has already been emphasized. However, neither of these factors can have had anything to do with the initial acquisition of the 3-hydroxyretinal chromophore. Thus firstly, photoregeneration is of general occurrence in most invertebrates, including many which have a rhodopsin, and in fact the only insects in which the meta-pigment has been reported to be relatively unstable are the butterflies, which actually have xanthopsin (Bernard 1983a,b). Secondly, many orders have xanthopsin as their visual pigment without possessing sensitized UV vision (Fig. 7). In fact the exclusive occurrence of the sensitizing chromophore amongst the higher Diptera also suggests a monophyletic origin of this trait and if we again refer to the time scale (Fig. 7), we can estimate that about a hundred million years elapsed from the acquisition of xanthopsin to the invention of the sensitizing pigment – a long time to realize the full potential of the new chemistry. However, once established, its high adaptational value should make sensitized UV vision a very persistent trait and no higher flies have yet been found which lack a sensitizing pigment.

Having dismissed the obvious functional attributes of xanthopsins, let us consider the proposition that the new chromophore may have been a dietary imperative. For example, the common ancestor of the Mecopteroidea may have been specialized, like many present-day caterpillars, to feed on a single plant species. With this possibility in mind, I determined the carotenoid content of a number of the surviving representatives of the Carboniferous flora, such as mosses, ferns, and conifers, to see if any of them showed a significant lack of carotenes at the expense of xanthophylls. But this was not the case.

What, however, if the ancestral food source was rotting vegetation? Thus according to some zoologists, one of the initial advantages of holometabolic development may have been that the absence of wing buds allowed the larvae to burrow into more solid decomposing plant material and thus exploit hitherto untapped food resources. I thus also extracted some samples of mulch and old compost and indeed found the xanthophylls, lutein and zeaxanthin, to be the dominant carotenoids, with only very faint traces of ß, ß-carotene. Thus, this larval way of life, which is still found, for instance, in the scorpionflies (Tillyard 1926) suggests a plausible scenario.

A final consideration relates to the fact that retinol is in fact very toxic. Thus its amphipathic character (polar head and hydrophobic tail) means that free retinol can act as a detergent. Mammals avoid the toxic action by transporting retinol attached to plasma and cellular retinol-binding proteins (Shichi 1983; Goodman 1984). However, an even simpler way of removing the toxicity — without losing the functional group — would be to put another polar group on the opposite end of the molecule, i.e., to synthesize 3-hydroxyretinol. Possibly, the acquisition of this molecule allowed insects to save the metabolic costs of specific transport proteins (thereby also reducing the risk of blindness which can be caused by protein deficiency, as in the cgse of kwashiorkor which is a very common blinding disease in the third world — Goodman 1984).

3.2.2 The Exceptions

Odonata (Dragonflies and Damselflies). The eyes of both damselflies (*Calopteryx*) and dragonflies (*Aeshna*) contain both 3-hydroxyretinal and retinal. The Odonata are thus the only animals where both chromophores have been found to be present in one and the same eye. Interestingly, in *Aeshna cyanea* the dorsal part of the eye (which has mainly short wavelength receptors) contains virtually only retinal whilst both chromophores are present in the ventral eye (Vogt unpubl.).

As far as phylogeny is concerned, the Odonata can obviously not be regarded as the "missing link" between insects with rhodopsin and those with xanthopsin, since this would mean complete revision of all accepted insect systematics. The presence of 3-hydroxyretinal in the Odonata must thus be considered as an independent acquisition.

Ascalaphus (Owlfly). As shown by Paulsen and Schwemer (1972), and recently confirmed (Vogt unpubl.), the eyes of *Ascalaphus* contain retinal. However, in two other neuropteran species (the closely related *Euroleon nostras* and the more distantly related *Chrysopa carnea*), only 3-hydroxyretinal was present (Vogt et al. in prep.). Therefore we assume that the Neuroptera originally had a 3-hydroxy chromophore, and we again regard the retinal in *Ascalaphus* as a secondary acquisition.

Simulium (Blackfly). Of the 19 dipteran species investigated, only the primitive blackfly has retinal rather than 3-hydroxyretinal as its chromophore. In a detailed study of the dorsal eye of the male *Simulium* (Kirschfeld and Vogt 1986), the

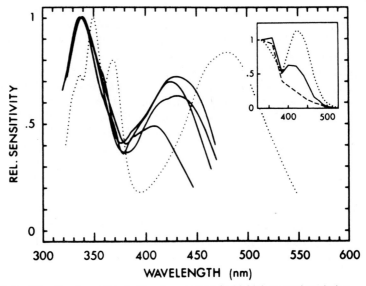

Fig. 9. Spectral sensitivity of the dorsal eye of male *Simulium* measured with high spectral resoluting using the electroretinogram (*continuous curves*). For comparison spectral sensitivity of *Drosophila* (female, white eye) is also shown (*dotted curve*) *Inset* spectral sensitivity measured intracellularly at discrete wavelengths with axial illumination. *Continuous curve* mean of 14 cells (7 animals); *dashed* and *dotted curves* are extreme cases to illustrate the variability in the relative heights between the UV and 430 nm peaks. (Kirschfeld and Vogt 1986)

photoreceptors were found to be basically UV receptors (Fig. 9), with a peak at 340 nm and a variable secondary shoulder or peak at about 430 nm. Microspectro-photometric and electrophysiological evidence indicate that these photoreceptors contain both a blue absorbing rhodopsin (λ_{max} 430 nm) and a UV-sensitizing pigment. In general these photoreceptors are remarkably similar to the so-called 7y receptors in the higher flies (cf. Hardie 1985), however, the UV sensitivity peak (340 nm) of *Simulium* is shifted to shorter wavelengths by about 10 nm and furthermore the characteristic fine structure is absent. Significantly, the only candidate for a sensitizing pigment extractable from these eyes was retinol. We have already suggested that the fine structure and bathochromic shift characteristic of the 3-hydroxy-sensitizing pigment may be due in part to hydrogen bonds with the two hydroxy groups at either end of the molecule (Sect 2.3). Since retinol, the suspected sensitizing pigment of simuliids, lacks the C3-hydroxygroup, the above hypothesis also offers an explanation for the simuliid case: i.e., the smooth UV peak and the smaller bathochromic shift may be attributed to greater torsional freedom of the molecule.

It seems clear that the ancestors of the simuliids had a 3-hydroxyretinal based visual pigment (Fig. 7). This is further supported by the presence of an atavistic trait in their carotenoid metabolism: simuliids are able to hydroxylate ß,ß-carotenes. Curiously, they insert the hydroxygroup at C2 rather than at C3 and can only hydroxylate one ß-ring. Thus their heads contains plenty of ß,ß-carotene-2-ol, but not the corresponding diol (Vogt unpubl.). Possibly the retinal of these animals is produced from the ß,ß-carotene-2-ol using a C_{40} cleaving enzyme which (still) requires a ß-end-ring hydroxygroup on the substrate.

Since the dorsal parts of the eyes of the exceptions discussed so far — *Ascalaphus, Aeshna,* and the male *Simulium* — all have a similar optical task, i.e., detecting small objects against the sky, it is tempting to look for a general adaptive advantage of a retinal chromophore in this direction. The swarming *Simulium* males have to detect their tiny mates against the sky, a task which could obviously be facilitated by improving the acuity of the photoreceptors. Optical resolution in a diffraction-limited lens receptor system is proportional to D/λ, where D is the lens diameter and λ the wavelength of light. That the *Simulium* males are indeed trying to optimize their resolution is already indicated by the very large lens diameters in the dorsal eye, and we should expect λ also to be as small as possible. In this respect, retinol would be a better choice than 3-hydroxyretinol as the sensitizing pigment. The 10-nm difference in peak sensitivity (340 vs. 350 nm) would correspond to a ca. 5% improvement in the probability of detecting a female (Kirschfeld and Vogt 1986), which should be sufficient evolutionary incentive to favor the selection of the trait.

Cicada (Tettigia) orni. The final example which does not fit readily into the overall phylogenetic scheme concerns a cicada (order Auchenorrhynca, Fig. 7) which contains only 3-hydroxyretinal in its eyes, and only xanthophylls as C_{40} carotenoids (Vogt unpubl.). Consideration of the animal's lifestyle suggests that here again a "dietary imperative" may have been the major factor in the acquisition of the novel chromophore. Thus, with the exception of the mycetoma, both adult and larval cicadas feed exclusively on phloem or xylem sap which they extract with specialized stylets. Cicadas can thus be expected to experience problems in obtaining sufficient carotenoids since phloem cells contain no chloroplasts, the main source of plant carotenoids. If carotenoids are present at all in phloem sap, then traces of xanthophylls may be expected in the leucoplasts and plastid fragments (H. Mohr pers. commun.). If this interpretation is correct we can predict that the 3-hydroxy chromophore may yet be discovered in other insect species facing similar extreme nutritional dilemmas.

4 Concluding Remarks

With 3-hydroxyretinal there are now three visual pigment chromophores known to occur in nature; is this likely to be the end of the story or are there still some more waiting to be discovered? I am prepared to venture that the list is now complete. Although this may appear to be a brash statement, it is in fact based on a number of considerations.

To use an analogy from the world of economics, we have to take into account both supply and demand. With respect to demand, the consumer (an opsin or photoreceptor cell), imposes a number of strict functional constraints on the chromophore. Firstly, the chromophore should have a high extinction coefficient; secondly, it should have a very high quantum efficiency of isomerization within the visual pigment. This requirement is particularly important. Whereas low extinction could in principle be compensated for by a higher pigment concentration or a longer receptor, low quantum efficiency inevitably results in a loss of sensitivity due to self-screening, since, for reasons of spatial acuity, the visual pigment is packaged into long narrow rods. Finally, in order to minimize noise, the chromophore should have a high thermostability with respect to spontaneous isomerization.

Although a large number of synthetic retinal analogs have been found which can combine with cattle opsin, several of these artificial visual pigments are in fact either unstable (e.g., Liu 1982) or else have a low quantum efficiency of isomerization (e.g., Nelson et al., 1970). The number of suitable compounds is thus, in fact probably rather low.

With respect to supply, it would seem that only abundant and ubiquitous carotenoids come into question as possible precursors of the chromophore. In fact, the annual production of carotenoids by plants (ca. 10^8 tons) is restricted to four major compounds: fucoxanthin and its derivatives (the pigments of brown algae), lutein and zeaxanthin (including their epoxy derivatives) and ß-carotene. Since animals have only a very limited ability to modify carotenoids, a further constraint is placed on the number of possible chromophores. The discovery of 3-hydroxyretinal, however, must always have been a possibility when one bears in mind the abundance of its precursors, lutein and zeaxanthin.

References

Bernard GD (1983a) Bleaching of rhabdoms in eyes of intact butterflies. Science 219:69–71

Bernard GD (1983b) Dark-processes following photoconversion of butterfly rhodopsins. Biophys Struct Mech 9:227–286

Boll F (1876) Zur Anatomie und Physiologie der Retina. Mber Berlin Akad 41:783–787

Bridges CDB (1972) The rhodopsin-porphyropsin visual system. In: Dartnall HJA (ed) Handbook of sensory physiology, vol VII/1. Springer, Berlin Heidelberg New York, pp 417–480

Burkhardt D (1962) Spectral sensitivity and other response characteristics of single visual cells in the arthropod eye. Symp Soc Exp Biol 16:86–109

Chytil F, Ong DE (1984) Cellular retinoid-binding proteins. In: Sporn MB, Roberts AB, Goodman DS (eds) The retinoids, vol 2. Academic Press, New York London, pp 89–123

Cowman AF, Zuker CS, Rubin GM (1986) An opsin gene expressed in only one photoreceptor cell type of *Drosophila* eye. Cell 44:705–710

Förster T (1951) Fluoreszenz organischer Verbindungen. Vandenhoeck & Ruprecht, Göttingen

Franceschini N (1983) In-vivo microspectrofluorimetry of visual pigments. In: Cosens DJ, Vince-Price D (eds) The biology of photoreception. Soc Exp Biol Symp 36:53–85

Fugate RD, Song PS (1980) Spectroscopic characterization of ß-lactoglobulin-retinol complex. Biochim Biophys Acta 625:28–42

Gemperlein R, Paul R, Lindauer E, Steiner A (1980) UV fine structure of the spectral sensitivity of flies visual cells. Revealed by FIS (Fourier Interfermotoric Stimulation). Naturwissenschaften 67:565–566

Goldsmith TH (1958) The visual system of the honeybee. Proc Natl Acad Sci 44:123–126

Goldsmith TH, Barker RJ, Cohen CF (1964) Sensitivity of visual receptors of carotenoid-depleted flies: a vitamin A deficiency in an invertebrate. Science 146:65–67

Goldsmith TH, Bernard GD, Marks BC, Tongoren I (1985) HPLC of unusual retinoids from arthropods. Invest Ophthalmol Vision Sci 26:293

Goldsmith TH, Marks BC, Bernard GD (1986) Separation and identification of geometric isomers of 3-hydroxyretinoids and occurrence in the eyes of insects. Vision Res 26:1763–1769

Goodman DS (1984) Plasma retinol-binding protein. In: Sporn MB, Roberts AB, Goodman DS (eds) The retinoids, vol 2. Academic Press, New York London, pp 41–88

Guo AK (1981) Elektrophysiologische Untersuchungen zur Spektral- und Polarisations-Empfindlichkeit der Sehzellen von *Calliphora erythrocephala* III. Sci Sin 24:272–286

Hardie RC (1978) Peripheral visual function in the fly. PhD Thesis, Aust University, Canberra

Hardie RC (1985) Functional organization of the fly retina. In: Ottoson D (ed) Progress in sensory physiology, vol 5. Springer, Berlin Heidelberg New York, pp 1–79

Hargrave PA (1982) Rhodopsin chemistry, structure and topography. In: Osborne N, Chader G (eds) Progress in retinal research. Oxford Univ Press, pp 1–51

Hennig W (1969) Die Stammesgeschichte der Insekten. Kramer, Frankfurt

Hennig W (1973) Handbuch der Zoologie, 2. edn, IV 2–2/31. De Gruyter, Berlin

Kayser H (1982) Carotenoids in insects. In: Britton G, Goodwin TW (eds) Carotenoid chemistry and biochemistry. Pergamon, Oxford New York, pp 195–210

Kirschfeld K (1983) Are photoreceptors optimal? TINS 6:97–101

Kirschfeld K, Vogt K (1986) Does retinol serve a sensitizing function in insect photoreceptors? Vision Res 26:1771–1777

Kirschfeld K, Franceschini N, Minke B (1977) Evidence for a sensitising pigment in fly photoreceptors. Nature (London) 269:386–390

Kirschfeld K, Feiler R, Hardie R, Vogt K, Franceschini N (1983) The sensitizing pigment in fly photoreceptors. Properties and candidates. Biophys Struct Mech 10:81–92

Kristensen NP (1975) The phylogeny of hexapod "orders". A critical review of recent accounts. Z Zool Syst Evolutionsforsch 13:1–44

Kühne W (1878) On the photochemistry of the retina and on visual purple. Macmillan, London

Liu RSH (1982) Synthetic and structural studies of visual pigments. In: Britton G, Goodwin TW (eds) Carotenoid chemistry and biochemistry. Pergamon, Oxford New York, pp 253–264

Morton RA (1944) Chemical aspects of the visual process. Nature (London) 153:69–71

Nelson R, DeRiel JK, Kropf A (1970) 13-desmethyl rhodopsin and 13-desmethyl isorhodopsin visual pigment analogues. Proc Natl Acad Sci USA 66:531–538

O'Tousa JE, Baehr W, Martin RL, Hirsh J, Pak WL, Applebury ML (1985) The *Drosophila* ninaE gene encodes an opsin. Cell 40:839–850

Ovchinnikov YA (1982) Rhodopsin and bacteriorhodopsin: structure-function relationships. FEBS Lett 148:179–189

Paul R (1981) Neue Aspekte der spektralen Empfindlichkeit von *Calliphora erythrocephala*, gewonnen durch Fourier-interferometrische Stimulation (FIS). Diss, Univ München

Paulsen R, Schwemer J (1972) Studies on the insect visual pigment sensitive to ultraviolet light: retinal as the chromophoric group. Biochim Biophys Acta 283:520–529

Ross HH (1965) A textbook of entomology, 3rd edn. Sidney, New York London

Seki T, Fujishita S, Ito M, Matsuoka N, Kobayashi C, Tsukida K (1986) A fly, *Drosophila melanogaster*, forms 11-cis 3-hydroxyretinal in the dark. Vision Res 26:255–258

Shichi H (1983) Biochemistry of vision. Academic Press Orlando

Smakman JGJ, Stavenga DG (1986) Spectral sensitivity of blowfly photoreceptors: Dependence on waveguide effects and pigment concentration. Vision Res 26:1019–1025

Stark WS, Ivanyshyn AM, Greenberg RM (1977) Sensitivity and photopigments of R1–6, a two peaked photoreceptor in *Drosophila, Calliphora* and *Musca*. J Comp Physiol A 121:289–305

Stavenga DG, Schwemer J (1984) Visual pigments of invertebrates. In: Ali MA (ed) Photoreception and vision in invertebrates. Plenum, Oxford New York, pp 11–61

Stryer L (1986) The cyclic GMP cascade of vision. Ann Rev Neurosci 9:87–119

Suzuki T, Makino-Tusaka M, Eguchi E (1984) 3-dehydroretinal (Vitamin A_2 aldehyde) in crayfish eye. Vision Res 24:783–787

Tanimura T, Isono K, Tsukahara Y (1986) 3-Hydroxyretinal as a chromophore of *Drosophila melanogaster* visual pigment analyzed by high pressure liquid chromatography. Photochem Photobiol 43:225–228

Tillyard RJ (1926) Kansas Permian insects. VII. The order Mecoptera. Am J Sci 11:133–164

Vogt K (1983) Is the fly visual pigment a rhodopsin? Z Naturforsch 38c:329–333

Vogt K (1984a) The chromophore of the visual pigment in some insect orders. Z Naturforsch 39c:196-197

Vogt K (1984b) Zur Verteilung von Rhodopsin und Xanthopsin bei Insekten. Verh Dtsch Zool Ges 77:258

Vogt K (1987) Chromophores of insect visual pigments. Photobiochem Photobiophys Suppl:273-296

Vogt K, Kirschfeld K (1982) Die Quantenausbeute der Energieübertragung in Photorezeptoren von Fliegen. Verh Dtsch Zool Ges 75:337

Vogt K, Kirschfeld K (1983a) C_{40}-Carotinoide in Fliegenaugen. Verh Dtsch Zool Ges 1983, 330

Vogt K, Kirschfeld K (1983b) Sensitizing pigment in the fly. Biophys Struct Mech 9:319-328

Vogt K, Kirschfeld K (1984) Chemical identity of the chromophores of fly visual pigment. Naturwissenschaften 71:211-213

Wald G (1934) Carotenoids and the vitamin A cycle in vision. Nature (London) 134:65

Wald G (1936) Carotenoids and the visual cycle. J Gen Physiol 19:351-357

Wald G (1939) On the distribution of vitamins A_1 and A_2. J Gen Physiol 22:391-415

Wald G (1968) Molecular Basis of visual excitation. Science 162:230-239

Weber H (1954) Grundriss der Insektenkunde. Fischer, Stuttgart

Zuker CS, Cowman AF, Rubin GM (1985) Isolation and structure of a rhodopsin gene from *Drosophila melanogaster*. Cell 40:851-858

Chapter 8

Pigments in Compound Eyes

DOEKELE G. STAVENGA, Groningen, The Netherlands

1 Introduction

Vision has evolved because of the universal presence of pigments, i.e., substances which absorb light in the visible and thus provide objects with their color. The visual instruments themselves, the eyes, can also only properly execute their function by virtue of the pigments they contain. Three classes of eye pigments can readily be distinguished. Primarily, of course, the photolabile visual pigments, which absorb the light entering the eye and then trigger the phototransduction chain, and the related retinoid-binding proteins, which together maintain visual pigment concentration and thus light sensitivity. Secondly, the photostable screening pigments, whose main function is to block out unwanted stray light. A third class comprises the pigments in the mitochondrial respiratory chain, universally present in eukaryotic cells; the aptly called cytochromes, together with the functionally related flavoproteins and NADH.

Exner (1891) was intrigued by the variety in pigmentation of compound eyes and made extensive studies of the interrelation of the optics of compound eyes and their pigment layers. His most significant findings related to the function of the displacement of the screening pigments upon light and dark adaptation. Whereas the presence of the screening pigments was quite evident because of their high optical density, the visual pigments could only be inferred to exist from the work on vertebrates. Exner actually attempted to demonstrate their presence by bleaching but was unsuccessful; we now understand why this was so (see Schwemer this Vol.; Vogt this Vol.). The mitochondrial pigments were only discovered considerably after Exner's studies when Keilin (1925) performed his revolutionary studies on the respiratory pigments (see Keilin 1966).

In this chapter I will survey our present knowledge of pigments in compound eyes and in doing so will pursue Exner's quest to understand their interrelationships. I will make particular reference to the experimental avenues recently opened by microspectrofluorometry.

Stavenga/Hardie (Eds.) Facets of Vision
© Springer-Verlag Berlin Heidelberg 1989

2 Visual Pigments

2.1 Distribution of Absorption Spectra

The visual pigments of compound eyes are embedded in the rhabdomeric membrane of the visual sense cells. Their molecular and photochemical characteristics have already been extensively reviewed in this volume by Schwemer (this Vol.) and Vogt (this Vol.). Because of the theme of this chapter I add some spectral properties. As shown in Fig. 1, the native visual pigment state, be it a rhodopsin, porphyropsin, or xanthopsin (see Schwemer this Vol. and Vogt this Vol.), can have an absorption peak anywhere from the ultraviolet to the yellow. (Probably there exist visual pigments peaking in the red, but they have not been properly characterized; see, e.g., Bernard 1979.) In particular, the visual pigments of insects cover a very broad spectral range. The thermostable photoproduct, i.e., the meta-state, however, has its absorption peak in a much narrower range, as a rule in the blue. A notable exception are the flies, whose main visual pigment has a meta-state absorbing in the orange. The functional significance of this special property is returned to in Section 5.5.

The absorption spectrum of a visual pigment can be determined by a number of different methods (reviewed by Stavenga and Schwemer 1984). A simple method for studying a visual pigment in the completely intact eye of a living

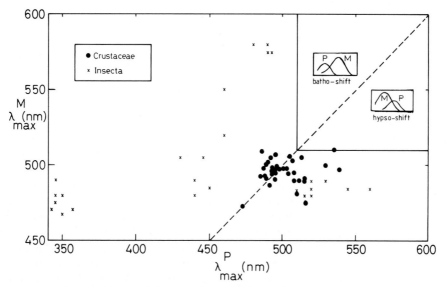

Fig. 1. Peak wavelengths of the visual pigments of compound eyes. The peak wavelengths of the native visual pigment state (λ_{max}^P) ranges from the ultraviolet to the yellow, those of the thermostable photoproduct, the metastate M (λ_{max}^M), from the blue to the orange; cf. Stavenga and Schwemer (1984, Fig. 14), extended with data from Cronin and Forward (1988) and Bennett and Brown (1985). The visual pigment P has a bathochromic meta-state when $\lambda_{max}^M > \lambda_{max}^P$, and hypsochromic when $\lambda_{max}^M < \lambda_{max}^P$. The latter appears to be the case for all visual pigments with $\lambda_{max}^P > 510$ nm

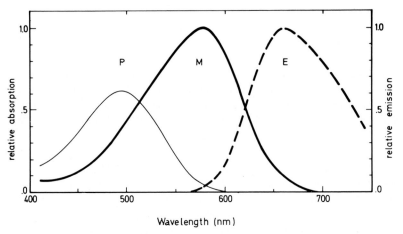

Fig. 2. Spectra of the main visual pigment of blowflies located in the photoreceptor cells R1–6. The xanthopsin P has an absorption band in the blue-green and the metaxanthopsin M in the orange. Only the meta-state fluoresces, in the red (E). (Stavenga 1983)

animal has recently been developed by Tinbergen and Stavenga (1986). The transmission of the eye of a blowfly (white eyed mutant chalky) can be measured directly by making use of the roughly hemispherical shape of the compound eye, thus allowing light to be shone through the eye, across a large number of om-matidia. After illuminating the eye alternately with red and blue light, the measured absorbance spectra differ distinctly. The derived difference spectrum clearly reveals the characteristics of the main visual pigment, i.e. the blue-green absorbing xanthopsin of the photoreceptor cells R1–6 and its orange-red absorbing metax-xanthopsin (Figs. 2,3a,b).

2.2 Fluorescence of Photoproducts

An alternative method for studying visual pigments in completely intact, living animals is by fluorescence. In general it appears that the native visual pigment, i.e., with the chromophore in the 11-cis configuration, fluoresces negligibly, whilst the meta-state, with the all-trans chromophore, exhibits substantial fluorescence (crayfish: Cronin and Goldsmith 1981, 1982a,b; mantis shrimp: Cronin 1985; fruitfly: Stark et al. 1977; housefly: Franceschini 1977; Franceschini et al. 1981; Stavenga et al. 1984; see also Stavenga 1983). The emission spectra determined so far all peak in the red around 650 nm. In flies, two different fluorescent meta-states have been identified, probably with an all-trans (M, Fig. 2) and a 13-cis chromophore (M'), respectively (Stavenga et al. 1984; Kruizinga and Stavenga, in preparation).

The fluorescence of visual pigments can be exploited experimentally for several purposes, e.g., the mapping of the distribution of different spectral classes of photoreceptors (Franceschini et al. 1981) and also for studying the dynamics of the pupil mechanism of the photoreceptor cells in the living animal (see Sect. 5.3).

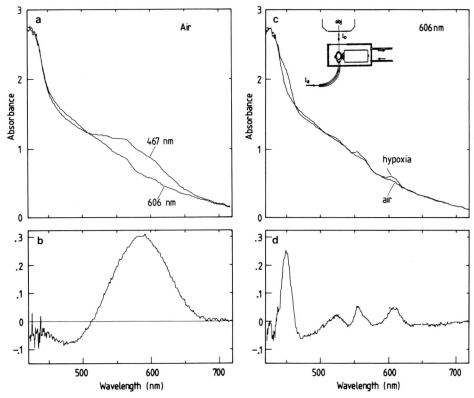

Fig. 3a-d. Absorbance spectra of the photoreceptor layer of the blowfly eye (mutant chalky) and their absorbance difference spectra. *a* Absorbance spectra of the eye adapted to blue (467 nm) and red (606 nm) light, respectively, with the fly in air. *b* Absorbance difference between the two spectra in *a*, revealing the characteristic xanthopsin-metaxanthopsin difference spectrum of the visual pigment in fly photoreceptors R1–6. *c* As *a*, but with the visual pigment purely in the xanthopsin state, due to 606 nm preadaptation, and with the fly in air and in a nitrogen atmosphere, respectively. *d* The absorbance difference between the spectra of *c*, revealing the characteristic redox difference spectrum for mitochondrial pigments (from Tinbergen and Stavenga 1986). The absorbance spectra were calculated from measurements (with a microspectrophotometer, obj) of light transmitted through the photoreceptor layer (*c*, inset)

3 Mitochondrial Pigments

3.1 Cytochromes, Flavoproteins, and NADH

In aerobic organisms, ATP is produced in the mitochondria by the process of oxidative phosphorylation. The number of mitochondria in a cell is therefore a direct indicator of its energy requirements. Most photoreceptors score highly in this respect. The numerous mitochondria are usually found close to the cell membrane, and are sometimes also concentrated in the distal part of the cell (Trujillo-Cenóz 1972). The main pigments of the mitochondria, the cytochromes, play a crucial role: they generate the energy necessary to recycle ADP into ATP by transferring electrons to oxygen. In this process the cytochromes undergo redox changes which are measurable as a change in absorption. Low oxygen pressure drives the cytochromes into the reduced state, as was first demonstrated in vivo in insect wing muscle by an optical method (Keilin 1925).

Measurement of respiration by optical means has only recently been explored in compound eyes (Tinbergen and Stavenga 1986). Although the absorbance curves of the blowfly eye (chalky mutant) determined alternately in air and under hypoxia differ little, they do so significantly (Fig. 3c). The difference spectrum shows peaks near 550 and 605 nm (Fig. 3d), characteristic for the redox changes of cytochromes a_1, a_3, c and c_1, respectively. The trough in the blue at 480 nm is the hallmark of another type of pigment in the mitochondrial respiratory chain, the flavoproteins.

In addition to the cytochromes, which absorb throughout the visible range, and the flavoproteins, absorbing mainly in the blue, the mitochondria still contain a third important pigment, namely NADH, which absorbs mainly in the ultraviolet. During the respiratory process all pigments undergo redox changes, reflected in substantial changes in absorption, which can be used as a diagnostic for measuring mitochondrial activity (Chance and Williams 1956; Lehninger 1970).

Both the flavoproteins and NADH exhibit a substantial fluorescence, whilst the cytochromes do not fluoresce. Since the fluorescence properties of flavoproteins and NADH depend strongly on the redox state, it is possible to monitor mitochondrial activity in vivo by measuring either the ultraviolet-excited blue emission of NADH or the blue-excited green emission of the flavoproteins (Fig. 4; Tinbergen and Stavenga 1986; see Scholz et al. 1969).

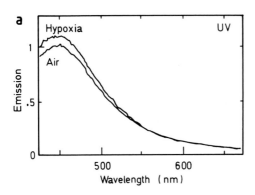

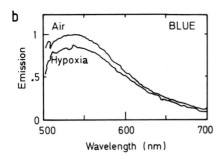

Fig. 4a,b. Emission spectra of the blowfly compound eye (mutant chalky), measured in air and under hypoxia, respectively, and induced by ultraviolet excitation (*a*) and blue excitation (*b*) by epi-illumination; see inset Fig. 3c. (Tinbergen and Stavenga 1986)

3.2 Light-Induced Mitochondrial Activity

Illumination of a dark-adapted blowfly eye (chalky mutant) causes a transient increase in blue-excited green fluorescence of the mitochondrial flavoproteins, indicating a transient increase in oxygen consumption. The spectral sensitivity of this phenomenon is identical to that of the photoreceptor cells R1–6 (Tinbergen and Stavenga 1987). This finding confirmed previous measurements of light-dependent oxygen consumption by isolated blowfly retinae, with a microrespirometer (Hamdorf and Langer 1966). Jones and Tsacopoulos (1987) measured the light-induced oxygen consumption by the photoreceptors of the honeybee drone in eye slices, with oxygen-sensitive electrodes, and also found that the spectral sensitivity was identical to the spectral sensitivity of the receptor potential. Evidently, in fly and bee both phototransduction and mitochondrial activity are ultimately triggered by visual pigment conversion.

The simplest explanation for the light-induced mitochondrial activity is that visual pigment conversion induces an increase in the intracellular calcium concentration which then stimulates mitochondrial respiratory processes (Tsacopoulos et al. 1986; Tinbergen and Stavenga 1987; Fein and Tsacopoulos 1988). Hamdorf et al. (1988), however, question this hypothesis and argue that the light-induced oxygen consumption is mainly caused by an increase in ADP, this being due to a light-enhanced sodium pump activity and, at high light levels, also due to the energy-requiring processes of biochemical amplification and adaptation. More work is necessary here, but these studies unequivocally demonstrate that mitochondrial activity is intimately linked with the visual function of the photoreceptors in the compound eye. Furthermore, studying mitochondrial activity in photoreceptors can have direct implications for current insights into the universal cellular machinery of energy supply (Fein and Tsacopoulos 1988).

4 Fluorophores

In addition to visual pigments, mitochondrial flavoproteins and NADH, yet other fluorescing pigments exist in blowfly eyes, i.e., in the rhabdomeres, Semper cells and cornea (Schlecht et al. 1987). The fluorophore in the rhabdomeres and Semper cells is a photochromic substance with an ultraviolet and a blue-absorbing state. It is speculated to be part of an extramitochondrial redox system. The fluorescing substance in the cornea is also subject to photoconversion by short wavelength light. Although the function of these fluorophores has still to be clarified, Schlecht et al.'s (1987) study clearly demonstrates the value of fluorescence in revealing pigments which hitherto have completely escaped attention.

5 Screening Pigments

5.1 Location and Chemical Nature

Members of several classes of photostable pigments are found in compound eyes and their functions are manifold (Sect. 5.5). Prevalent, however, are the universally encountered screening pigments. The most prominent location of these pigments are the pigment cells in the distal part of the eye. They determine an eye's coloration and, because of the similarity to the pigment in the iris of the vertebrate eye, Exner (1891) called them the iris pigment cells.

In the more recent literature, alternative nomenclatures have been introduced. The classification scheme which is commonly used for insects (Goldsmith and Bernard 1974) distinguishes the primary pigment cells, enveloping the crystalline cone, and the secondary pigment cells, which traverse the retina from cornea to basement membrane (Chi and Carlson 1976; Hardie 1985). In the eyes of crustaceans, the pigment layers are generally separated into distal and proximal pigments, but this distinction is unsatisfactory because of the multitude of pigments and their occurrence in both photoreceptors and a variety of accessory pigment cells. For instance, in the crayfish, *Cherax*, the distal pigment consists of a mixture of black and reflecting pigment, contained in two types of distal pigment cells, whilst the proximal pigment consists of red pigment in the photoreceptive, retinula cells and reflecting pigment in the proximal pigment cells (e.g., Bryceson and McIntyre 1983); in the mysid *Praunus*, Hallberg (1977) distinguishes distal pigment cells with dark pigment granules and distal pigment cells with little or lightly colored pigment granules; near or below the basement membrane are the basal red pigment cells and the proximal reflecting pigment cells, together with the pigment in the retinula cells. Because each species has its own characteristic eye pigmentation, with its associated photomechanical effects and/or circadian rhythms, it is virtually impossible to practice a generally applicable, clear-cut nomenclature (see Exner 1891; Autrum 1981; Rao 1984; Nilsson this Vol.).

The pigments of insect eyes include the ommochromes (the name indicating where they were first found), and also the pterins and carotenoids (Langer 1975; Bouthier 1981; Kayser 1985). The crustaceans derive their eye pigments from the same classes, and have in addition melanins and purines (Kleinholz 1957; Linzen 1967; Shaw and Stowe 1982). Chemical studies of the eye pigments have benefited greatly from the existence of eye color mutants of a number of insects (the flies *Drosophila* and *Lucilia*, the bee *Apis* and the moth *Ephestia*; for references see Dustmann 1975; Summer et al. 1982; Fuzeau-Braesch 1985; Kayser 1985).

5.2 Pigment Migration in Accessory Cells

Extensive changes in the position of the pigment cells and/or the pigment granules they contain are a common phenomenon in compound eyes. These effects generally occur under the influence of changes in the ambient light intensity, as was already well known a century ago (e.g., Exner 1891; see Autrum 1981, for an indepth review). The pigment migrations can also be under neurohormonal control via a circadian rhythm system (Walcott 1975; Stavenga 1979; Autrum 1981; Barlow et al. this vol.), or may depend exclusively on temperature (Veron 1974a). The pigment distribution is often determined by a combined effect of some or all the agents mentioned (e.g., crayfish, Frixione and Aréchiga 1981).

The changes in eye pigmentation can be readily recognized with the naked eye in a number of insects (e.g., damsel fly, Veron 1973; praying mantis, Stavenga 1979; planthopper, Howard 1981) and crustaceans (e.g., prawn, Sandeen and Brown 1952). A very helpful tool in observing the pigment migrations is the optical phenomenon of the pseudopupil, first described and exploited in the study of compound eyes by Exner (1891) and more recently elaborated by Franceschini (1975) and Stavenga (1979). Pseudopupil studies are most rewarding when the eye under investigation has a tapetum, i.e., a reflecting layer lying proximally in the eye. Outstanding in this respect are the lepidopterans, which have evolved a special, reflective function for the air sacs in their head (Miller 1979); in many crustacean eyes the lightly colored or whitish pteridine pigment granules furnish the tapetum (e.g., shrimp, Doughtie and Rao 1984). In such eyes, incident light microscopy reveals a so-called luminous pseudopupil (Exner 1891; Stavenga 1979), which changes in appearance and intensity when pigment migrations are elicited in the eye. The classical case is that of the moth. Figure 5 presents the optical changes in the eye of *Ephestia*, occurring during the process of light adaptation induced by continuous illumination of the eye through only one facet. Because the moth eye is of the optical superposition type (Nilsson this Vol.) and pigment migration is induced already at low light levels, it is obvious to assume that light capture by the retinula cells triggers the distal pigment migration, as is the case in the pupillary system of the human eye. Spectral sensitivities of the pupil mechanism in the eye of the moths *Amyelois* (Bernard et al. 1984), *Ephestia* (Weyrauther 1986) and *Manduca* (White et al. 1983) appeared to be perfectly compatible with the above interpretation, but *Deilephila* did not conform (Hamdorf and Höglund 1981). Furthermore, illumination of pigment cells isolated from the retina still induced pigment migration, and thus founded the hypothesis that the pigment cells have

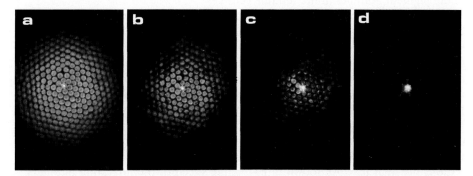

Fig. 5a-d. The moth *Ephestia kuehniella* observed with an epi-illumination microscope. The broadband white illumination was focused at one facet. In the dark-adapted state light is reflected by the tapetum through a few hundred facets (*a*), but upon light adaptation this number is reduced, due to migration of the screening pigment into the light path. Photographs were taken after 20 s illumination (*a*), 3 min (*b*), 10 min (*c*), and 20 min (*d*). (Courtesy Dr E. Weyrauther)

their own light-sensitive machinery (Hamdorf and Höglund 1981; Hamdorf et al. 1986). Land (1986), in an elegant optical study on one moth, directly demonstrated that the effector resided distally in the eye. Nilsson (in prep.) now has evidence that pigment migration in pigment cells can be induced by light absorption in both the retinula cells and in the pigment cells proper. (Interestingly, the pupillary iris sphincter of the frog *Rana* contains visual pigment, in the smooth muscle cells, and is fully operational even when detached from the retina; see Kargacin and Detwiler 1985.) From the consideration that the pigment cells are essentially epi(hypo)dermal cells (Friza 1929; Veron 1974b; Stavenga 1979), like the chromatophores in the cuticle of crustaceans, which have their own light sensory apparatus (Kleinholz 1957), it may be presumed that the mechanisms of pigment migration in both types of pigmented cells have much in common (see further Section 5.4).

5.3 Pupillary Pigment in Photoreceptor Cells

The light-induced migration of the pigment granules in the photoreceptor or retinula cells of compound eyes was first described in *Limulus* by Miller (1958) and has since been demonstrated in numerous insect and crustacean species, with both histological and in vivo optical methods (revs Stavenga 1979; Autrum 1981). Upon illumination with bright light, the granules migrate towards the lightguiding rhabdom(ere), where they regulate the light flux by a combination of absorption and scattering. This type of pupil mechanism has been investigated in most detail in flies. In particular, Franceschini contributed important knowledge by measuring the optical changes occurring in the deep pseudopupil (Franceschini 1975). Franceschini and Kirschfeld (1976) investigated various *Drosophila* mutants with eye pigment defects and found no pupillary pigment migration in those mutants where ommochrome synthesis was selectively blocked, whilst in a mutant with no pterin synthesis pupillary pigment migration was similar to that in the normal, wild-type fly. In the fruitfly, ommochromes thus appear to be the sole constituent of the pigment granules in the retinula cells.

In flies and some hymenopterans, the photoreceptor pigment migration and its effect on the light flux in the rhabdom(ere)s can be studied fairly easily in living, intact animals by means of transmission and/or reflection measurements (Franceschini 1975; Stavenga 1979), but this becomes difficult in species with darkly and heavily pigmented eyes. Fluorescence then provides an attractive alternative (Fig. 6). The red emission of the fluorescing meta-states of the visual pigments (section 2.2) can be observed in most compound eyes with an epi-illumination fluorescence microscope using blue and/or green excitation light, subsequent to a sufficiently long dark adaptation time. Usually then, this emission fades because the illumination activates the retinula cell pupil mechanism which reduces the light flux. As shown in Fig. 6, the speed of the migrating pupillary granules varies widely between species. The pupillary speed in flies (time constant ca. 2 s), hymenopterans (ca. 20 s) and orthopterans (ca. 100 s) seems to be correlated with the strongly species dependent dynamics of the receptor potential (cf. Pinter

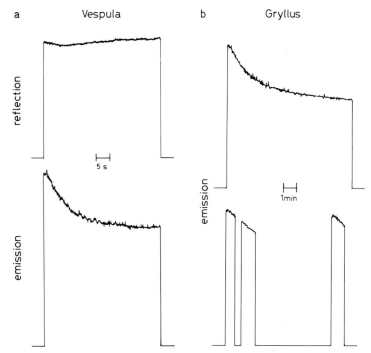

Fig. 6a,b. Pigment migration in photoreceptor cells. *a* The reflection of a 467 nm illuminating beam measured from the deep pseudopupil in the eye of a dark-adapted wasp *Vespula germanica* (*upper curve*) shows only a minor change in signal (compare Stavenga 1979, Fig. 41), but the red (> 665 nm) fluorescence excited by the same light clearly reveals the closing pupil (*lower curve*). *b* In the dark-adapted eye of the cricket *Gryllus bimaculata* broad-band blue illumination evokes a much slower closure of the pupil (*upper curve*). In the dark the pupil reopens (*lower curve*), but the reversal of the pigment migration process takes longer than 0.5 min, as appears after the first dark interval

1972; Laughlin 1981). The pupil time constants in the mantid shrimps, the diurnal *Gonodactylus* (1–5 s) and the crepuscular *Squilla empusa* (100 s) are also correlated with life style (Cronin 1988).

5.4 Mechanism of Pigment Migration

Although several studies have focused on the mechanism of pigment migration, little definitive can be stated, except that the cytoskeleton is involved and also that the action of a number of agents has been substantiated. With respect to compound eyes, Miller and Cawthon (1974) first demonstrated in *Limulus* that colchicine, a drug which disrupts microtubuli, can block retinula cell pigment migration. Subsequently, this approach has been extended to the retinula cells of the crayfish where pigment migrations are extensive (Exner 1891; Walcott 1975; Frixione 1983a,b; Bryceson and McIntyre 1983). Here, a massive column of microtubules, along which pigment granules and other organelles are aligned, extends longitudinally throughout each retinula cell and its axon (Tsutsumi et al. 1981; Frixione 1983a,b). Frixione and Aréchiga (1981) concluded from their ionic studies that sodium influx induced by light results in calcium release from internal stores, which in turn triggers dispersion of the pigment granules within the retinular cells. However, recent phototransduction studies have shown that sodium influx rather follows an increase in calcium (see Fein and Payne this Vol.). The importance of $Ca^{2)}$ for the migration of retinula cell pigment granules was also demonstrated in fly photoreceptor cells by Kirschfeld and Vogt (1980) and Howard (1984).

Wilcox and Franceschini (1984a) injected colchicine into the retina of the housefly *Musca* and found that light-induced pigment migration was affected in a restricted number of retinula cells which had taken up the drug (see Sect. 6). However, the receptor potential also diminished gradually in a manner symptomatic of anoxia (cf. Payne 1981). The pigment granules finally stayed put in an intermediate position as if the metabolic energy which the process required was depleted. Indeed, colchicine not only interferes with microtubules, but can also block membrane processes like the sodium-potassium pump (see Lambert and Fingerman 1976, 1978). Experimental results using colchicine must therefore be interpreted with caution. Nevertheless, most studies strongly suggest that the cytoskeletal microtubules are essential for pigment migration. Note that no filamentous network was resolved in retinula cells of tipulid flies (Blest et al. 1984) and that pigment granules in these cells do not migrate in response to light and dark adaptation (Williams 1980); in the cave shrimp neither microtubules nor pigment migration could be demonstrated (Meyer-Rochow and Juberthie-Jupeau 1983).

Pigment migration in retinula cells can be driven by factors other than light. For instance, in the crayfish a diurnal rhythm drives the red proximal screening pigment to its fully dark-adapted position below the basement membrane at night, and in the day time towards its light-adapted position where it screens the rhabdom (Bryceson and McIntyre 1983). Such effects are presumably mediated via second messenger systems activated by neurotransmitters or hormones via specific membrane receptor molecules (for the exemplary case of *Limulus*, see Barlow et al. 1985, this Vol.). Chromatophores are directly comparable to the crayfish retinula cells and the knowledge gained on pigment movements in the different cells may be mutually beneficial. In crustacean chromatophores Ca^{2+} presumably acts directly on the cytoskeletal elements and also indirectly by controlling cyclic nucleotide levels (Rao and Fingerman 1983). (Generally, Ca^{2+}, cAMP and ATP play key roles in pigment migration; chromatophores, Luby-Phelps and Schliwa 1982, melanophores, e.g., Rozdzial and Haimo 1986). Pigment movements in crustacean chromatophores are regulated by pigment-concentrating and pigment-dispersing neurosecretory hormones (Rao and Fingerman 1983). The chemical structure of a number of these hormones has been determined. The light-adapting hormone (LAH), acting on the distal retinal pigment of the prawn *Pandalus* (Fernlund and Josefsson 1972), and the pigment-dispersing hormone (PDH), acting on melanophores of the fiddler crab *Uca* (Rao et al. 1985), are closely related octadecapeptides; LAH and PDH are released from the sinus glands in the eye stalks (review Kleinholz 1957; prawn, Aoto and Hisano 1985). In the sphingid moth *Deilephila* light-induced pigment migration in the screening pigment cells is reversed by noradrenaline, suggesting the involvement of catecholamines in pigment regulation (Juse et al. 1987; see also Hardie this Vol.). Both pigment aggregation and dispersal appear to be energy-requiring processes (Banister and White 1987). To make things even more complicated, two different mechanisms of pigment granule translocation operating in two separate regions exist in the retinula cells of the crayfish (Frixione et al. 1979).

5.5 Functions of Screening Pigments

Several and various functions of the screening pigments in compound eyes can be listed.

— To screen out stray light; i.e., to protect photoreceptors and their visual pigment from activation by off-axial light, so that only light incident within a narrow spatial angle can contribute to the visual signal. This is the most general function of pigments, especially in apposition eyes (Exner 1891; Goldsmith and Bernard 1974). The photoreceptor cells in compound eyes without screening pigment, e.g., in white-eyed mutants, have distinctly broadened visual fields (blowfly, Streck 1972; bee, Gribakin 1988). Consequently, contrast detection by white-eyed insects is reduced (Hengstenberg and Götz 1967; Gribakin 1988). Moreover, the visual sense cells lose the ability to process contrast at high photon absorption rates (Howard et al. 1987). This occurs in photoreceptors directly radiated by the sun, even in a well-pigmented eye. However, because the sun has a subtense of approximately 0.5 degree, i.e., well within interommatidial angles of compound eyes, only b few ommatidia are thus inactivated at any one time. But, in pigment-less eyes sunlight will result in completely blinding of most or all photoreceptor cells.

— To enhance light sensitivity by reflecting light which has already passed once through the receptor layer thus providing another chance for light absorption (the tapetum; Exner 1891; Doughtie and Rao 1984).

— To provide eye coloration for camouflage, display or other purposes. This holds for the light-coloured distal pigments which overlay black pigment (see Stavenga 1979).

— To control light absorption for temperature regulation (Veron 1974b).

— To control the light flux incident at the photoreceptors and thus to extend the working range of the visual sense cells. This is most dramatically exemplified by the screening pigments in the secondary pigment cells of moth superposition eyes, which can effectively reduce the light flux by 2–3 log units (Höglund and Struwe 1970); but a quite impressive light control (ca. 2 log units) is also executed by the pupil mechanism in the retinula cells of flies (Leutscher-Hazelhoff and Van Barneveld 1983; Howard et al. 1987).

— To control the angular sensitivity of the receptors and thus optimize the visual acuity for the ambient light intensity. This can be achieved by pigment migration in the retinula cells (cockroach, Butler and Horridge 1973; blowfly Hardie 1979; Smakman et al. 1984), by moving the pigment of the pigment cells (praying mantis,

Rossel 1979; Stavenga 1979; crab, Leggett and Stavenga 1981), or by a combined action of pigment migration in all pigment-containing cells, eventually together with other structural changes, e.g., in rhabdom size and shape of crystalline cone (crayfish, Walcott 1974; Bryceson and McIntyre 1983; *Limulus*, Barlow et al. 1980, this Vol.; ant, Menzi 1987; see Autrum 1981; Nilsson this Vol.).

— To photoregenerate visual pigment photoproducts back into their native state. This is the case with visual pigments with appreciably bathochromic meta-states (see Fig. 1). The classic case is the fly where the red leaky screening pigment serves to maintain a high visual pigment level (Stavenga et al. 1973, revs. Langer 1975; Hardie 1986). Visual pigments absorbing in the green, which have a hypsochromic meta-state (Fig. 1), cannot rely on photoregeneration by stray light and require a black screen (Stavenga 1979).

— To shift the spectral sensitivity and thus improve the animal's colour and/or contrast discrimination capacity. Spectral changes by retinula cell pigment have been documented for the crayfish (Goldsmith 1978; Bryceson 1986), crab (Stowe 1980), fly (Hardie 1979), cabbage butterfly (Steiner et al. 1987) and firefly (Lall et al. 1988). In all these cases, the spectral shifts are brought about by pigment granules located distally and/or adjacent to the rhabdom(ere)s. A most special case is that of the fly photoreceptor cell R7y, which has carotenoid pigment molecules embedded in the rhabdomeric membrane. These act as a dense blue filter greatly affecting the spectral sensitivity not only of the R7 cell but also of the underlying R8 (Kirschfeld et al. 1978; for a detailed review, see Hardie 1985). R7y furthermore has a sensitizing pigment, like the receptors R1-6, presumably 3-hydroxyretinol (see Hardie 1985, Vogt this Vol.)

— To protect against photodestruction by intense ultraviolet light. This is an additional function of the carotenoids in fly R7y receptors (Zhu and Kirschfeld 1984). According to Dontsov et al. (1984) ommochromes take part in the anti-oxidative protection system in invertebrate photoreceptors.

— To act as a Ca^{2+} store. This is the case for ommochrome pigment granules in some insects; no calcium was detected in the pigment granules of crayfish retinula cells (White and Michaud 1980; see Frixione and Ruiz 1988 for discussion).

Often, of course, several of these functions are combined, but the spectral characteristics of the pigments and their locations will presumably always be geared to the optimal functioning of the complete visual system. It is the task of future research to understand better the precise optimization strategies, and how they are related. One challenging case, for instance, is the complete absence of screening pigments found in the dorsal marginal area of the cricket compound eye (Burghause 1979). In this region the facet lenses are flat, which is thought to be correlated with the strong polarization sensitivity of the photoreceptors (Labhart et al. 1986). That the dorsal ommatidia are devoid of pigment can be readily observed in vivo with an epi-illumination fluorescence microscope applying blue excitation light. Intriguingly, it appears that not all the pigmentless ommatidia are associated with flat facet lenses, but that a quite substantial proportion of these ommatidia also have normal, convex facetlenses (Fig. 7).

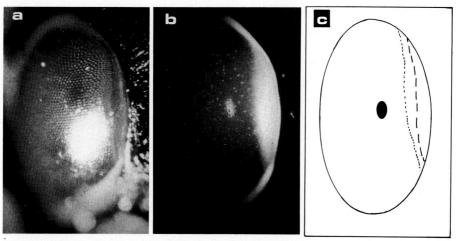

Fig. 7a-c. The dorsal area of the cricket *Gryllus bimaculata* in reflected light (*a*) and under epi-fluorescence (*b*). The area of the eye with a smooth corneal surface (*c* between interrupted line and eye margin) is distinctly smaller than the area where blue-induced green fluorescence is high, due to absence of screening pigment (*c* between dotted line and eye margin). Note the pseudopupil which is dark in reflection, due to the strongly light-absorbing screening pigments surrounding the crystalline cone, and bright in fluorescence, due to the fluorescing visual pigment in the rhabdoms

6 Dyes in Compound Eyes

The use of dyes for marking cells and for coloring cellular structures is a standard technique in histological research. In the last decades newly developed, highly fluorescent dyes have become a major research tool. In compound eyes as well, lucifer yellow is now most widely used (e.g., Hardie et al. 1981; Strausfeld 1984). Wilcox and Franceschini (1984b), in a study of retinal transport, injected lucifer yellow through a small hole in the cornea into the retina of an otherwise completely intact, living housefly, and subsequently illuminated a restricted number of ommatidia via the objective of a fluorescence microscope. They discovered that the dye was selectively taken up by the illuminated cells (see also Shaw this Vol.). Although the mechanism of the light-induced dye uptake still has to be clarified, the occurrence of dye-filled cytoplasmic vesicles suggests that endocytosis of the photoreceptor membrane is part of the process.

Actually, Wilcox and Franceschini (1984a) made their discovery while studying the effect of colchicine on the pupil mechanism (Sect. 5.3). Like lucifer yellow, colchicine also appears to enter brightly illuminated cells only, and there it blocks pigment migration. Apparently it binds to microtubules, as witnessed by the blue fluorescence induced by ultraviolet excitation light. The fluorescence waned, however, in parallel with the blockage of the pigment migration, and after 1 h in the dark the photoreceptor cells were back to normal. Furthermore, light-induced uptake of lucifer yellow or colchicine by the photoreceptors was only possible within a limited time after injection into the retina (Wilcox and Franceschini 1984a,b). Evidently the substances were subject to an active clearing process, located in or near the retina.

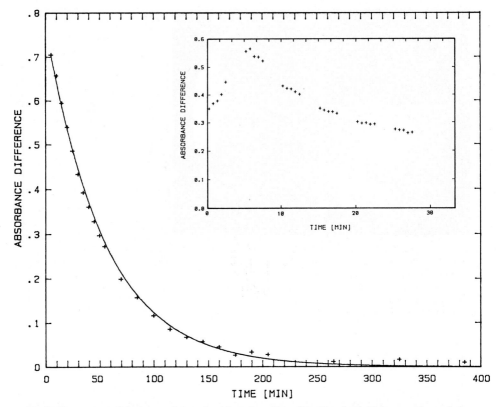

Fig. 8. Time course of clearance of phenol red from the retina of the blowfly *Calliphora erythrocephala* (mutant chalky). The absorbance change at 560 nm was calculated from measurements of the transmission across the photoreceptor layer (see inset Fig. 3c). The phenol red was injected laterally into the retina. Upon diffusion through the retina the dye enters the path of the measurement beam, as is seen from the absorbance upswing in the first minutes (*inset*; experiment from another fly). The last phase of diffusion overlaps with the early phase of the clearing process. (Weyrauther et al. 1988)

This mechanism has been studied in more detail by Weyrauther et al. (1988). The clearance of the absorbing dye phenol red (Fig. 8) has an exponential time course ($t^{1/2} = 38$ min). Owing to the high fluorescence of lucifer yellow, which behaves similarly, it could be shown that the dyes are not degraded in the retina, but that they are transported to other parts of the body, i.e., the thorax and the abdomen. Part of the necessary transport system must be located near the basement membrane, as follows from injections of lucifer yellow into the thorax. The reverse effect now occurs; namely the dye readily reaches the head and the cell layers below the basement membrane, but is then only gradually pumped into the retina (Fig. 9). Presumably the transport system is located in the basal glial cells (Shaw 1977). Its most obvious function is to supply the eye with metabolic substrates and to remove superfluous substances.

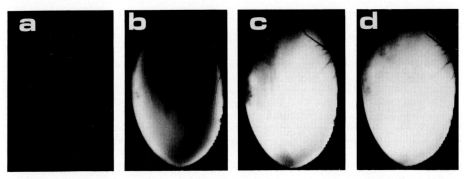

Fig. 9a-d. Fluorescence of the eye of the blowfly (mutant chalky). Blue-induced green-yellow emission, *a* before injection, and *b* 5 min, *c* 1 h, *d* 2 h after injection of lucifer yellow in the thorax. The dye appears to be pumped gradually into the retina. (Weyrauther et al. 1988)

7 Conclusion

Compound eyes are intricate systems consisting of the principal, photosensitive, retinula cells and the accessory, supporting, screening pigment cells. The spectral properties of the visual pigments and the screening pigments are tuned to each other, and the spatial properties of the sense cells are, among other factors, greatly improved by the presence of the pigment cells. The mitochondrial pigments have no optical function whatsoever, but their presence can be experimentally exploited to reveal the role of the energy supply by the mitochondria in photoreceptor function.

We may presume that Exner would have been pleased to see how the field of compound eye physiology has developed since his pioneering studies. It is certainly exciting to speculate on the state of compound eye research after yet another century. Phototransduction will doubtlessly have been clarified in exhaustive molecular detail, and cellular mechanical processes will have left few secrets. However, the enormous variety of compound eyes, which was what so fascinated Exner, is likely still to pose countless fascinating questions of functional and ecological significance.

References

Aoto T, Hisano S (1985) Ultrastructural evidence for the existence of the distal retinal pigment light-adapting hormone in the sinus gland of the prawn *Palaemon paucidens*. Gen Comp Endocrinol 60:468–474

Autrum H (ed) (1981) Light and dark adaptation in invertebrates. In: Handbook of sensory physiology, vol VII/6C. Springer, Berlin Heidelberg New York, pp 1–91

Banister MJ, White RH (1987) Pigment migration in the compound eye of *Manduca sexta*: effects of light, nitrogen and carbon dioxide. J Insect Physiol 33:733–743

Barlow RB, Chamberlain SC, Levinson JZ (1980) *Limulus* brain modulates the structure and function of the lateral eyes. Science 210:1037–1039

Barlow RB, Kaplan E, Renninger GH, Saito T (1985) Efferent control of circadian rhythms in the *Limulus* lateral eye. Neurosci Res (Suppl) 2:S65–S78

Bennett RR, Brown PK (1985) Properties of the visual pigments of the moth *Manduca sexta* and the effects of the two detergents, digitonin and chaps. Vision Res 25:1771–1781

Bernard GD (1979) Red-absorbing visual pigment of butterflies. Science 203:1125–1127

Bernard GD, Owens ED, Hurley AV (1984) Intracellular optical physiology of the eye of the pyralid moth *Amyelois*. J Exp Zool 229:173–187

Blest AD, deCouet HG, Howard J, Wilcox M, Sigmund C (1984) The extrarhabdomeral cytoskeleton in photoreceptors of Diptera. I. Labile components in the cytoplasm. Proc R Soc London Ser B220:339–352

Bouthier A (1981) Les ommochromes, pigments absorbantes des yeux des Arthropodes. Arch Zool Exp Gen 122:237–252

Bryceson K (1986) The effect of screening pigment migration on spectral sensitivity in a crayfish reflecting superposition eye. J Exp Biol 125:401–404

Bryceson K, McIntyre P (1983) Image quality and acceptance angle in a reflecting superposition eye. J Comp Physiol A 151:367–380

Burghause FMHR (1979) Die stukturelle Spezialisierung des dorsalen Augenteils der Grillen (Orthoptera, Grylloidea). Zool Jahrb Physiol 83:502–525

Butler R, Horridge GA (1973) The electrophysiology of the retina of *Periplaneta americana* L. 1. Changes in receptor acuity upon light/dark adaptation. J Comp Physiol 83:263–278

Chance B, Williams GR (1956) The respiratory chain and oxidative phosphorylation. In: Nord FF (ed) Advances in enzymology, vol 17. Interscience, New York, pp 65–134

Chi C, Carlson SD (1976) The large pigment cell of the compound eye of the house fly *Musca domestica*. Fine structure and cytoarchitectural associations. Cell Tissue Res 170:77–88

Cronin TW (1985) The visual pigment of a stomatopod crustacean, *Squilla empusa*. J Comp Physiol A 156:679–687

Cronin TW (1988) Visual pigments and spectral sensitivity in the stomatopods. Boll Zool (in press)

Cronin TW, Forward RB, Jr. (1988) The visual pigments of crabs. I. Spectral characteristics. J Comp Physiol A 162:463–478

Cronin TW, Goldsmith TH (1981) Fluorescence of crayfish metarhodopsin studied in single rhabdoms. Biophys J 35:653–664

Cronin TW, Goldsmith TH (1982a) Photosensitivity spectrum of crayfish rhodopsin measured using fluorescence of metarhodopsin. J Gen Physiol 79:313–332

Cronin TW, Goldsmith TH (1982b) Quantum efficiency and photosensitivity of the rhodopsin ⇌ metarhodopsin conversion in crayfish photoreceptors. Photochem Photobiol 36:447–454

Dontsov AE, Lapina VA, Ostrovsky MA (1984) O^-_2 photogeneration by ommochromes and their role in the system of antioxidative protection of invertebrate cells. Biofyzika 29:878–882

Doughtie DG, Rao KR (1984) Ultrastructure of the eyes of the grass shrimp, *Palaemonetes pugio*. General morphology, and light and dark adaptation at noon. Cell Tissue Res 238:271–288

Dustmann JH (1975) The pigment granules in the compound eye of the honey bee *Apis mellifica* in wild type and different eye-color mutants. Cytobiology 11:133–152

Exner S (1891) Die Physiologie der facettirten Augen von Krebsen und Insecten. Deuticke, Leipzig

Fein A, Tsacopoulos M (1988) Activation of mitochondrial oxidative metabolism by calcium ions in *Limulus* ventral photoreceptors. Nature (London) 331:437–440

Fernlund P, Josefsson L (1972) Crustacean color-change hormone: Amino acid sequence and chemical synthesis. Science 177:173–175

Franceschini N (1975) Sampling of the visual environment by the compound eye of the fly: Fundamentals and applications. In: Snyder AW, Menzel R (eds) Photoreceptor optics. Springer, Berlin Heidelberg New York, pp 98–125

Franceschini N (1977) In vivo fluorescence of the rhabdomeres in an insect eye. Proc Int Union Physiol Sci XII, 237. XXVIIth Int Congr, Paris

Franceschini N, Kirschfeld K (1976) Le contrôle automatique du flux lumineux dans l'oeil composé des Diptères. Propriétés spectrales, statiques et dynamiques du mécanisme. Biol Cybernet 21:181–203

Franceschini N, Kirschfeld K, Minke B (1981) Fluorescence of photoreceptor cells observed in vivo. Science 213:1264–1267

Frixione E (1983a) The microtubular system of crayfish retinula cells and its changes in relation to screening-pigment migration. Cell Tissue Res 232:335–348

Frixione E (1983b) Firm structural associations between migratory pigment granules and microtubules in crayfish retinula cells. J Cell Biol 96:1258–1265

Frixione E, Aréchiga H (1981) Ionic dependence of screening pigment migrations in crayfish retinal photoreceptors. J Comp Physiol A 144:35–43

Frixione E, Ruiz L (1988) Calcium uptake by smooth endoplasmic reticulum of peeled retinal photoreceptors of the crayfish. J Comp Physiol A 162:91–100

Frixione E, Aréchiga H, Tsutsumi V (1979) Photomechanical migrations of pigment granules along the retinula cells of the crayfish. J Neurobiol 10:573–590

Friza F (1929) Zur Frage der Färbung und Zeichnung des facettierten Insektenauges. Z Vergl Physiol 8:289–336

Fuzeau-Braesch S (1985) Colour changes. In: Kerkut GA, Gilbert LI (eds) Comprehensive insect physiology, biochemistry and pharmacology, vol 10. Biochemistry. Pergamon, Oxford New York, pp 549–589

Goldsmith TH (1978) The effects of screening pigments on the spectral sensitivity of some crustacea with scotopic (superposition) eyes. Vision Res 18:475–482

Goldsmith TH, Bernard GD (1974) The visual system of insects. In: Rockstein M (ed) The physiology of Insecta, vol 2. Academic Press, New York San Francisco, pp 165–272

Gribakin FG (1988) Photoreceptor optics of the honeybee and its eye colour mutants: The effect of screening pigments on the long wave subsystem of colour vision. J Comp Physiol A 164:123–140

Hallberg E (1977) The fine structure of the compound eye of mysids (Crustacea: Mysidacea). Cell Tissue Res 184:45–65

Hamdorf K, Höglund G (1981) Light induced retinal screening pigment migration independent of visual cell activity. J Comp Physiol A 143:305–309

Hamdorf K, Langer H (1966) Der Sauerstoffverbrauch des Facettenauges von Calliphora erythrocephala in Abhängigkeit von der Temperatur und dem Ionenmilieu. Z Vergl Physiol 52:386–400

Hamdorf K, Höglund G, Juse A (1986) Ultra-violet and blue induced migration of screening pigment in the retina of the moth Deilephila elpenor. J Comp Physiol A 159:353–362

Hamdorf K, Hochstrate P, Höglund G, Burbach B, Wiegand U (1988) Light activation of the sodium pump in blowfly photoreceptors. J Comp Physiol A 162:285–300

Hardie RC (1979) Electrophysiological analysis of the fly retina. I. Comparative properties of R1–6 and R7 and R8. J Comp Physiol A 129:19–33

Hardie RC (1985) Functional organization of the fly retina. In: Ottoson D (ed) Progress in sensory physiology, vol 5. Springer, Berlin Heidelberg New York, pp 1–79

Hardie RC (1986) The photoreceptor array of the dipteran retina. Trends Neurosci 9:419–423

Hardie RC, Franceschini N, Ribi W, Kirschfeld K (1981) Distribution and properties of sex-specific photoreceptors in the fly Musca domestica. J Comp Physiol A 145:139–15

Hengstenberg R, Götz KG (1967) Der Einfluß des Schirmpigmentgehalts auf die Helligkeits – und Kontrastwahrnehmung bei Drosophila Augenmutanten. Kybernetik 3:276–285

Höglund G, Struwe G (1970) Pigment migration and spectral sensitivity in the compound eye of moths. Z Vergl Physiol 67:229–237

Howard FW (1981) Pigment migration in the eye of Myndus crudus (Homoptera: Cixiidae) and its relationship to day and night activity. Insect Sci Appl 2:129–133

Howard J (1984) Calcium enables photoreceptor pigment migration in a mutant fly. J Exp Biol 113:471–475

Howard J, Blakeslee B, Laughlin SB (1987) The intracellular pupil mechanism and photoreceptor signal: noise ratios in the fly Lucilia cuprina. Proc R Soc London Ser B 231:415–435

Jones GJ, Tsacopoulos M (1987) The response to monochromatic light flashes of the oxygen consumption of honeybee drone photoreceptors. J Gen Physiol 89:791–813

Juse A, Höglund G, Hamdorf K (1987) Reversed light reaction of the screening pigment in a compound eye induced by noradrenaline. Z Naturforsch 42c:973–976

Kargacin GJ, Detwiler PB (1985) Light-evoked contraction of the photosensitive iris of the frog. J Neurosci 5:3081–3087

Kayser H (1985) Pigments. In: Kerkut GA, Gilbert LI (eds) Comprehensive insect physiology, biochemistry and pharmacology, vol 10. Biochemistry. Pergamon, Oxford New York, pp 367–415

Keilin D (1925) On cytochrome, a respiratory pigment, common to animals, yeast, and higher plants. Proc R Soc London Ser B 98:312–339

Keilin D (1966) The history of cell respiration and cytochrome. Univ Press, Cambridge

Kirschfeld K, Vogt K (1980) Calcium ions and pigment migration in fly photoreceptors. Naturwissenschaften 67:516–517

Kirschfeld K, Feiler R, Franceschini N (1978) A photostable pigment within the rhabdomere of fly photoreceptors no. 7. J Comp Physiol 125:275–284

Kleinholz LH (1957) Endocrinology of invertebrates, particularly of crustaceans. In: Scheer BT, Bullock TH, Kleinholz LH (eds) Recent advances in invertebrate physiology. Univ Press Oregon, Eugene, pp 173–196

Kruizinga B, Stavenga DG (1989) Fluorescence spectra of blowfly metaxanthopsins (submitted)

Labhart T, Hodel B, Valenzuela I (1986) The physiology of the cricket's compound eye with particular reference to the anatomically specialized dorsal rim area. J Comp Physiol A 155:289–286

Lall AB, Strother GK, Cronin TW, Seliger HH (1988) Modification of spectral sensitivities by screening pigments in the compound eyes of twilight-active fireflies (Coleoptera: Lampyridae). J Comp Physiol A 162:23–33

Lambert DT, Fingerman M (1976) Evidence for a non-microtubular colchicine effect in pigment granule aggregation in melanophores of the fiddler crab, *Uca pugilator*. Comp Biochem Physiol 53C:25–28

Lambert DT, Fingerman M (1978) Colchicine and cytochalasin B: A further characterization of their actions on crustacean chromatophores using the ionophore A23187 and thiol reagents. Biol Bull 155:563–575

Land MF (1986) Screening pigment migration in a sphingid moth is triggered by light near the cornea. J Comp Physiol A 160:355–357

Langer H (1975) Properties and functions of screening pigments in insect eyes. In: Snyder AW, Menzel R (eds) Photoreceptor optics. Springer, Berlin Heidelberg New York, pp 429–455

Laughlin SB (1981) Neural principles in the peripheral visual systems of invertebrates. In: Autrum H (ed) Handbook of sensory physiology, vol VII/6B. Springer, Berlin Heidelberg New York, pp 133–280

Leggett LMW, Stavenga DG (1981) Diurnal changes in angular sensitivity of crab photoreceptors. J Comp Physiol A 144:99–109

Lehninger AL (1970) Biochemistry. Worth, New York

Leutscher-Hazelhoff JT, Barneveld HH van (1983) The *Calliphora* pupil and phototransduction modelling. Rev Can Biol Exp 3:263–270

Linzen B (1967) Zur Biochemie der Ommochrome. Unterteilung, Vorkommen, Biosynthese und physiologische Zusammenhänge. Naturwissenschaften 21b:259–267

Luby-Phelps KJ, Schliwa M (1982) Pigment migration in chromatophores: A model system for intracellular particle transport. In: Weiss DG (ed) Axoplasmic transport. Springer, Berlin Heidelberg New York, pp 17–26

Menzi U (1987) Visual adaptation in nocturnal and diurnal ants. J Comp Physiol A 160:11–21

Meyer-Rochow VB, Juberthie-Jupeau L (1983) An open rhabdom in a decapod crustacean: the eye of *Typhlatya garcia* (Atyidae) and its possible function. Biol Cell 49:273–282

Miller WH (1958) Fine structure of some invertebrate photoreceptors. Ann NY Acad Sci 74:204–209

Miller WH (1979) Ocular optical filtering. In: Autrum H (ed) Handbook of sensory physiology, vol VII/6A. Springer, Berlin Heidelberg New York, pp 69–143

Miller WH, Cawthon DF (1974) Pigment granule movement in *Limulus* photoreceptors. Invest Ophthalmol 13:401–405

Payne R (1981) Suppression of noise in a photoreceptor by oxidative metabolism. J Comp Physiol A 142:181–188

Pinter RB (1972) Frequency and time domain properties of retinula cells of the desert locust (*Schistocerca gregaria*) and the house cricket (*Acheta domestica*). J Comp Physiol 77:383–397

Rao KR (1984) Pigmentary effectors. In: Bliss DE, Mantel H (eds) The biology of Crustacea, vol 9. Academic Press, New York London, pp 395–462

Rao KR, Fingerman M (1983) Regulation of release and mode of action of crustacean chromatophorotropins. Am Zool 23:517–527

Rao KR, Riehm JR, Zahnow CA, Kleinholz LH, Tarr GE, Johnson L, Norton S, Landau M, Semmes OJ, Sattelberg RM, Jorenby WH, Hintz MF (1985) Characterization of a pigment-dispersing hormone in eyestalks of the fiddler crab *Uca pugilator*. Proc Natl Acad Sci USA 82:5319-5322

Rossel S (1979) Regional differences in photoreceptor performance in the eye of the praying mantis. J Comp Physiol A 131:95-112

Rozdzial MM, Haimo LT (1986) Bidirectional pigment granule movements of melanophores are regulated by protein phosphorylation and dephosphorylation. Cell 47:1061-1070

Sandeen MI, Brown FA, Jr. (1952) Responses of the distal retinal pigment of *Palaemonetes* to illumination. Physiol Zool 25:222-230

Schlecht P, Juse A Höglund G, Hamdorf K (1987) Photoreconvertible fluorophore systems in rhab-domeres, Semper cells and corneal lenses in the compound eye of the blowfly. J Comp Physiol A 161:227-243

Scholz R, Thurman RG, Williamson JR, Chance B, Bucher T (1969) Flavin and pyridine nucleotide oxidation-reduction changes in perfused rat liver. I. Anoxia and subcellular localization of fluorescent flavoproteins. J Biol Chem 9:2317-2324

Shaw SR (1977) Restricted diffusion and extracellular space in the insect retina. J Comp Physiol 113:257-282

Shaw SR, Stowe S (1982) Photoreception. In: Sandeman DC, Atwood HL (eds) The biology of Crustacea, vol 3. Academic Press, New York London, pp 291-367

Smakman JGJ, Hateren JH van, Stavenga DG (1984) Angular sensitivity of blowfly photoreceptors: Intracellular measurements and wave-optical predictions. J Comp Physiol A 155:239-247

Stark WS, Ivanyshyn AM, Greenberg RM (1977) Sensitivity and photopigments of R1-6, a two-peaked photoreceptor, in *Drosophila*, *Calliphora* and *Musca*. J Comp Physiol 121:289-305

Stavenga DG (1979) Pseudopupils of compound eyes. In: Autrum H (ed) Handbook of sensory physiology, vol VII/6A. Springer, Berlin Heidelberg New York, pp 357-439

Stavenga DG (1983) Fluorescence of blowfly metarhodopsin. Biophys Struct Mech 9:309-317

Stavenga DG, Schwemer J (1984) Visual pigments of invertebrates. In: Ali MA (ed) Photoreception and vision in invertebrates. Plenum, New York, pp 11-61

Stavenga DG, Zantema A, Kuiper JW (1973) Rhodopsin processes and the function of the pupil mechanism in flies. In: Langer H (ed) Biochemistry and physiology of visual pigments. Springer, Berlin Heidelberg New York, pp 175-180

Stavenga DG, Franceschini N, Kirschfeld K (1984) Fluorescence of housefly visual pigment. Photo-chem Photobiol 40:653-659

Steiner A, Paul R, Gemperlein R (1987) Retinal receptor types in *Aglais urticae* and *Pieris brassicae* (Lepidoptera), revealed by analysis of the electroretinogram obtained with Fourier interferometric stimulation (FIS). J Comp Physiol A 160:247-258

Stowe S (1980) Spectral sensitivity and retinal pigment movement in the crab *Leptograpsus variegatus* (Fabricius). J Exp Biol 87:73-98

Strausfeld NJ (ed) (1984) Functional neuroanatomy. Springer, Berlin Heidelberg New York

Streck P (1972) Der Einfluss des Schirmpigmentes auf das Sehfeld einzelner Sehzellen der Fliege *Calliphora erythrocephala* Meig. Z Vergl Physiol 76:372-402

Summer KM, Howells AJ, Pyliotes NA (1982) Biology of eye pigmentation in insects. Adv Insect Physiol 16:119-166

Tinbergen J, Stavenga DG (1986) Photoreceptor redox state monitored *in vivo* by transmission and fluorescence microspectrophotometry in blowfly compound eyes. Vision Res 26:239-243

Tinbergen J, Stavenga DG (1987) Spectral sensitivity of light induced respiratory activity of photo-receptor mitochondria in the intact fly. J Comp Physiol A 160:195-203

Tsacopoulos M, Fein A, Poitry S (1986) Stimulus-induced increase of mitochondrial respiration in a single neuron. Experientia 42:642

Trujillo-Cenóz O (1972) The structural organization of the compound eye in insects. In: Fuortes MGF (ed) Handbook of sensory physiology, vol VII/2. Springer, Berlin Heidelberg New York, pp 5-61

Tsutsumi V, Frixione E, Aréchiga H (1981) Transformations in the cytoplasmic structure of crayfish retinula cells during light- and dark-adaptation. J Comp Physiol A 145:179-189

Veron JEN (1973) Physiological control of the chromatophores of *Austrolestes annulosus* (Odonata). J Insect Physiol 19:1689-1703

Veron JEN (1974a) Physiological colour changes in Odonata eyes. A comparison between eye and epidermal chromatophore pigment migrations. J Insect Physiol 20:1491-1505

Veron JEN (1974b) The role of physiological colour change in the thermoregulation of *Austrolestes annulosus* (Selys) (Odonata) Aust J Zool 22:457–469

Walcott B (1974) Unit studies on light-adaptation in the retina of the crayfish, *Cherax destructor*. J Comp Physiol 94:207–218

Walcott B (1975) Anatomical changes during light-adaptation in insect compound eyes. In: Horridge GA (ed) The compound eye and vision of insects. Oxford, Clarendon, pp 20–33

Weyrauther E (1986) Do retinula cells trigger the screening pigment migration in the eye of the moth *Ephestia kuehniella*? J Comp Physiol A 159:55–60

Weyrauther E, Roebroek JGH, Stavenga DG (1988) Dye transport across the retinal basement membrane of the blowfly *Calliphora erythrocephala*. J Exp Biol (in press)

White RH, Michaud NA (1980) Calcium is a component of ommochrome pigment granules in insect eyes. Comp Biochem Physiol 65A:239–242

White RH, Banister MJ, Bennett RR (1983) Spectral sensitivity of pigment migration in the compound eye of *Manduca sexta*. J Comp Physiol A 153:59–66

Wilcox M, Franceschini N (1984a) Stimulated drug uptake in a photoreceptor cell. Neurosci Lett 50:187–192

Wilcox M, Franceschini N (1984b) Illumination induces dye incorporation in photoreceptor cells. Science 225:851–854

Williams DS (1980) Organisation of the compound eye of a tipulid fly during the day and night. Zoomorphology 95:85–104

Zhu H, Kirschfeld K (1984) Protection against photodestruction in fly photoreceptors by carotenoid pigments. J Comp Physiol A 154:153–156

Chapter 9

Phototransduction in *Limulus* Ventral Photoreceptors: Roles of Calcium and Inositol Trisphosphate

ALAN FEIN and RICHARD PAYNE, Woods Hole, Massachusetts, USA

1 Introduction

The rhabdomere of invertebrate photoreceptors has long been considered to be the site of visual transduction. Exner wrote in 1891 (p. 96) "This arrangement suggests that the most fundamental visual process — the transduction of light energy into nervous excitation — be it by photochemical or other means takes place, or is at least initiated during the passage of light along the strongly refracting rods." Modern electrophysiological experiments (Hagins et al. 1962; Lasansky and Fuortes 1969) conclusively demonstrated that the microvillar membrane of the rhabdomere, which contains the visual pigment, is the site where electrical current flows into the cell upon illumination. This photocurrent underlies the receptor potential, which is a depolarization of the receptor cell membrane (Millecchia and Mauro 1969).

The purpose of this paper is to briefly review our knowledge of the intermediate biochemical events that occur between the absorption of light by visual pigment molecules and the photocurrent. We will endeavor to relate these intermediate events to structural elements of the photoreceptor by considering the spatial localization of these events within a single cell, the *Limulus* ventral photoreceptor. Specifically, we will attempt to link the biochemical events occurring in the rhabdomeral membrane to the release of stored calcium from submicrovillar cisternae of smooth endoplasmic reticulum (SMC), which closely appose the rhabdom. These cisternae, or "palisades", have been described in many microvillar photoreceptors, beginning with the study of beetle compound eyes by Kirchhoffer [1908; reviewed by Whittle (1976)]. A variety of functions have been ascribed to them, including improving the optical performance of the rhabdom as a light guide (Horridge and Barnard 1965; Snyder and Horridge 1972; Nilsson this Vol.) and as a site for membrane storage (Whittle 1976). Perrelet and Bader (1978) first suggested that the SMC might be a store for calcium in a study of bee photoreceptors. Subsequent studies confirmed the ability of the SMC in fly, leech, and *Limulus* ventral photoreceptors to actively sequester calcium using ATP as an energy source (Walz 1982a,b; Walz and Fein 1983).

2 Anatomy of *Limulus* Ventral Photoreceptors

The cell bodies of *Limulus* ventral photoreceptors are divided into two segments, only one of which contains microvilli. The microvillar, photoreceptive segment is

Stavenga/Hardie (Eds.) Facets of Vision
© Springer-Verlag Berlin Heidelberg 1989

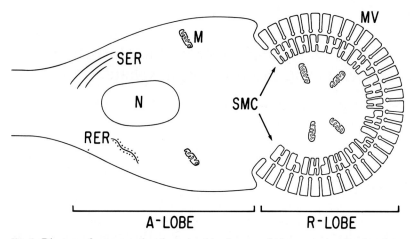

Fig. 1. Diagram of a cross-section through a *Limulus* ventral photoreceptor, showing the morphology of the A- and R-lobes of the cell. The A-lobe contains the nucleus (*N*), rough endoplasmic reticulum (*RER*), smooth endoplasmic reticulum (*SER*) and mitochondria (*M*). The R-lobe contains microvilli (*MV*), subrhabdomeral cisternae of smooth ER (*SMC*) and mitochondria

referred to as the rhabdomeral (R) lobe and the nonphotoreceptive segment as the arhabdomeral (A) lobe (Stern et al. 1982; Calman and Chamberlain 1982; Chamberlain and Barlow 1984). Figure 1 is a highly schematized representation of a *Limulus* ventral photoreceptor, emphasizing the segmentation of the cell body into the two lobes (Calman and Chamberlain 1982; Stern et al. 1982). Because of their large size and obvious segmentation, it has been possible to characterize the anatomical and functional properties of the R- and A-lobes. In particular, the R-lobe contains the microvilli and the closely apposed SMC. As we shall see in the following, the SMC appear to be intimately involved in phototransduction.

3 Localization of Excitation

Microvillar photoreceptors respond to light with a depolarization that is caused by current flowing into the cell through ionic channels (Fuortes and O'Bryan 1972; Hagins 1972; Bacigalupo and Lisman 1983). Excitation begins with the absorption of a photon by a single visual pigment molecule and ends with the opening of many ionic channels (Cone 1973). This process of amplification must involve some spatial spread of intermediates, but excitation is nonetheless confined, in *Limulus* ventral photoreceptors, to the rhabdomeral membrane. The extent of the rhabdomeral membrane can be mapped by plotting the profile of the cell's local sensitivity to a small spot of light (Fig. 2). Only those areas which contain microvilli will absorb the light and respond. We also determined the local density of current flowing into the cell after a flash of light that uniformly illuminated the entire photoreceptor by using a vibrating current probe. The close match of the inward current density with the local sensitivity to light (Fig. 2c) indicates that current only

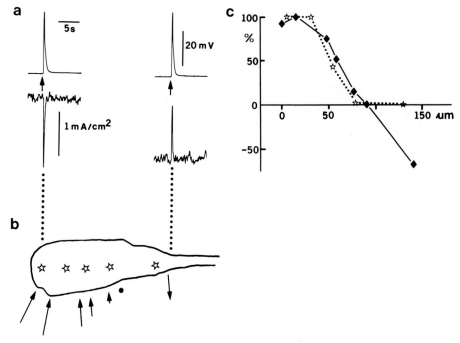

Fig. 2. a Intracellular voltage (*upper*) and extracellular current (*lower*) recorded in response to a diffuse light flash that uniformly illuminated the entire photoreceptor. The extracellular currents were recorded at the extreme left and right of the cell body *b* as indicated by the *dotted lines. b* Sketch of a ventral photoreceptor, showing positions (*stars*) at which a 5 μm diameter spot of light was placed to determine the local sensitivity to light. Length and direction of *arrows* give the magnitude and direction of the extracellular current resulting from a diffuse light flash that uniformly illuminated the entire photoreceptor. *c* The peak amplitude of the extracellular current (*diamonds*) and the local sensitivity to light (*stars*) are plotted as a function of position along the cell. Both are plotted as a percentage of their respective maxima (see Payne and Fein 1986, for further details)

flows into those areas containing microvilli, i.e., the R-lobe at the distal end of the cell. The ionic channels opened by light are therefore found primarily in the R-lobe (Payne and Fein 1986).

Localization of excitation within the rhabdomeral membrane itself was originally demonstrated by Hagins et al. (1962), using cephalopod retinae. They found that the current flowing into the cell, as the result of illumination, was localized to an area within a few microns of the site of illumination.

4 Localization of Adaptation

In addition to exciting the photoreceptor by opening ionic channels, light also initiates a slower process, called adaptation, that reduces the cell's sensitivity to light. Thereby, the range of light intensities over which the photoreceptor can respond to light is greatly expanded (see review by Autrum 1981). When they first

a

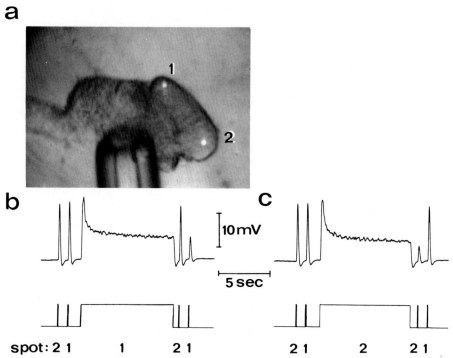

b

c

10 mV

5 sec

spot: 2 1   1   2 1   2 1   2   2 1

Fig. 3a-c. Localized adaptation in the R-lobe of a ventral photoreceptor. *a* Photomicrograph of a photoreceptor after it has been stripped of its glia (Stern et al. 1982). The photoreceptor is being held in place by suction applied to a glass electrode, the tip of which is visible in the micrograph. A portion of the cell can be seen protruding into the suction electrode. The intracellular microelectrode used for recording from the cell is not visible in the plane of focus. Two spots of light (*1,2*) can be seen illuminating separate regions of the R-lobe. (*b*) Localized desensitization of region 1 illuminated by the upper spot in *a. c* Localized desensitization of region 2 illuminated by the lower spot in *a.* In *b* and *c* the lower trace indicates the occurrence of the light stimuli at regions *1* and *2* in *a.* The two spots had the same relative intensity. The adapting stimulus in a given region had the same intensity as the test flash in that region (see Payne and Fein 1983)

reported their discovery that ventral photoreceptors were segmented into two lobes, Stern et al. (1982) pointed out that adaptation, as well as excitation, is initiated in the R-lobe. We subsequently showed that adaptation is localized to a region surrounding the site of photon absorption that is significantly smaller than the dimensions of an R-lobe (Fig. 3; Payne and Fein 1983). This report confirmed previous studies in other retinae that showed similar localization of adaptation within the rhabdomeral membrane (Hagins et al. 1962; Hamdorf 1970).

Any molecular mechanism thought to be involved in excitation and adaptation of ventral photoreceptors must be able to account for the spatial localization of these two processes within the rhabdomeral membrane.

5 Adaptation of Ventral Photoreceptors by Calcium

The suggestion that a light-induced rise in the intracellular cytoplasmic concentration of calcium ions (Ca_i) is involved in the adaptation of ventral photoreceptors originated with Lisman and Brown (1972). Their suggestion was based on the observation that the intracellular ionophoretic injection of calcium reversibly reduced the response to light, which is the essential effect of light adaptation. In support of calcium's role, injection of calcium chelators opposes light adaptation of the cell (Lisman and Brown 1975). Additionally, if calcium is involved in adaptation, then Ca_i should increase upon illumination. Indeed, a light-induced rise in Ca_i has been demonstrated in the ventral photoreceptor (Brown and Blinks 1974) and, like adaptation, it is also initiated in the R-lobe (Levy and Fein 1985). The light-induced rise in Ca_i is diminished by at most 50% when ventral photoreceptors are bathed in calcium-free sea water. Therefore, a large fraction of the rise in Ca_i is apparently due to release of calcium from internal stores and not to influx from the external medium (Brown and Blinks 1974; Levy and Fein 1985).

If calcium is the messenger of adaptation, then localized adaptation within the R-lobe (Fig. 3) requires that illumination of a subregion of the R-lobe should cause a rise in Ca_i that is localized to that subregion. As illustrated in Fig. 4, the results confirm this prediction. The cell shown in Fig. 4a and b was stripped of its glia and injected with aequorin, (a water-soluble protein that luminesces in the presence of calcium, Shimomura et al. 1962). As shown in Fig. 4c, when the entire photoreceptor was uniformly illuminated by a diffuse flash of light there was a subsequent rise in Ca_i throughout the cell's R-lobe but not in the A-lobe (see also Levy and Fein 1985). When only a subregion of the R-lobe was illuminated by a spot of light (Figure 4b) there was a subsequent rise in Ca_i that was confined to the vicinity of the region illuminated (Figure 4d).

We can pursue the localization of adaptation further along the chain of events initiated by light. If localized light adaptation of ventral photoreceptors results from a localized rise in Ca_i, then injection of calcium into the cell should also cause a localized desensitization. Fein and Lisman (1975) found that ionophoretic injection of calcium into ventral photoreceptors did indeed lead to a desensitization that was localized to the region of the cell around the injection site.

In their description of the R- and A-lobe of ventral photoreceptors, Calman and Chamberlain (1982) noted that a small number of ventral photoreceptors possessed two R-lobes. These cells should have two regions in which light can initiate a depolarization and in these same regions there should be a concomitant light-induced rise in Ca_i. As illustrated in Fig. 5, ventral photoreceptors exhibiting such properties are sometimes observed.

6 Excitation of Ventral Photoreceptors by Calcium

Pulsed pressure injection of calcium into ventral photoreceptors allows the introduction of large, rapid increases of Ca_i to localized regions of the cell. As illustrated in Figure 6b, pressure injection of calcium into the R-lobe of a ventral

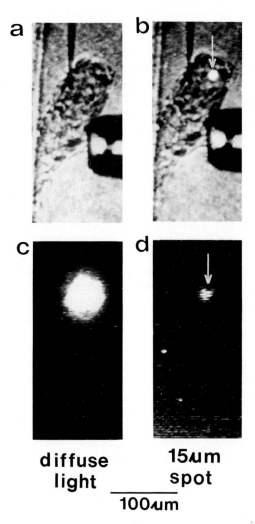

Fig. 4a-d. Localized rise in Ca_i in the R-lobe of a ventral photoreceptor. *a* and *b* Photomicrograph of a ventral photoreceptor after it had been stripped of its glia. The photoreceptor is being held in place by suction applied to a fire polished glass electrode, the tip of which is visible at the right. The intracellular electrode which was used to pressure inject aequorin into the photoreceptor can be seen at the top. *c* Aequorin luminescence recorded from the cell following a diffuse flash that uniformly illuminated the cell. The luminescence is confined to the distal region of the cell, the region containing the R-lobe. *d* Aequorin luminescence recorded from the cell after flash illumination of the cell with the spot of light shown in *b*. The aequorin luminescence was detected using a microscope objective to focus light coming from the cell onto the photocathode of an image intensifier (see Payne and Fein 1987a)

photoreceptor causes a depolarization of the cell (Payne et al. 1986a) as well as a subsequent temporary desensitization to light, which was previously observed using ionophoretic injection of calcium (Lisman and Brown 1972). The depolarization caused by calcium injection results from an inward current having the same reversal potential as that of the light induced current. Thus both light and calcium appear to open the same ionic channels in the cell's plasma membrane. Pressure injection of calcium into the A-lobe (Fig. 6a) is much less effective in depolarizing and desensitizing the cell. This is reminiscent of the action of light (see Sect. 4). Illumination of the R-lobe is much more effective in depolarizing and desensitizing the cell than is illumination of the A-lobe (Stern et al. 1982).

The findings reviewed so far could be simply explained if the light-, and calcium-sensitive mechanism that activates ionic channels were primarily located in the cell's R-lobe along with the visual pigment. This would explain the R-lobe's high sensitivity to light, the light-induced current flowing into the R-lobe, and the

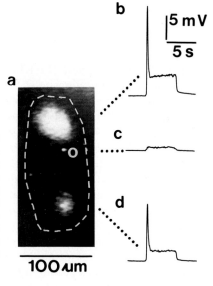

Fig. 5a-d. Ventral photoreceptor with two R-lobes. *a* A ventral photoreceptor, outlined by *broken line*, injected with aequorin was uniformly illuminated with a diffuse flash of light, which gave rise to luminescence from two regions of the cell. Illumination of either region that exhibited luminescence (*b* and *d*) gave rise to a large membrane depolarization. While illumination of the region *c* of the cell in between the two regions that exhibited luminescence was much less effective in depolarizing the cell. The *O* indicates the site of injection of aequorin into the cell

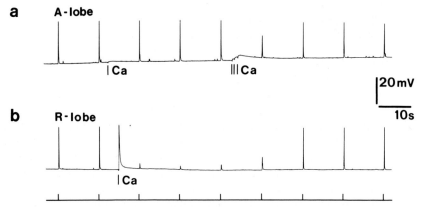

Fig. 6a,b. Excitation and adaptation of a ventral photoreceptor by pressure injection of calcium into the R-lobe. Pulse injection of calcium into the R-lobe depolarized the cell and desensitized the responsivess to light (*b*) while a similar injection into the the A-lobe was almost without effect (*a*). Calcium injections are labeled below traces in *a* and *b*, and the *lower trace* in *b* indicates the occurrence of dim light flashes (see Payne et al. 1986a)

depolarization of the cell by injection of calcium into the R-lobe. But why is the light-induced rise in Ca_i confined to the R-lobe? If the rise in Ca_i comes about by release of calcium from an intracellular compartment (Brown and Blinks 1974; Levy and Fein 1985), then how does absorption of light by visual pigment in the microvilli which are part of the cell's plasma membrane cause this release of calcium? Lisman and Strong (1979) proposed the need for an intracellular messenger which, after release from the microvillar membrane by light, travels to the SMC where it releases stored calcium. It is to the identity of this messenger that we now turn.

7 Inositol Trisphosphate: An Intracellular Messenger for Calcium Release

Inositol 1,4,5-trisphosphate (Ins1,4,5P$_3$) is the water-soluble product of the hydrolysis of phosphatidylinositol(4,5)bisphosphate (PtdIns4,5P$_2$), a minor component of plasma membrane phospholipid (Grado and Ballou 1961; Tomlinson and Ballou 1961). Berridge (1983) proposed that, in cells responding to hormones, hydrolysis of PtdIns4,5P$_2$ could be stimulated by a hormone-receptor complex, releasing Ins1,4,5P$_3$ that, in turn, releases calcium from internal stores. Thus it was of interest to investigate the action of Ins1,4,5P$_3$ in ventral photoreceptors, where photoactivated visual pigment might take the place of the hormone-receptor complex. Pressure injection of Ins1,4,5P$_3$ into ventral photoreceptors caused both a depolarization and a desensitization of the cell (Brown et al. 1984; Fein et al. 1984). Both of these effects were found to result from an Ins1,4,5P$_3$-induced rise in Ca$_i$ (Payne et al. 1986b), due to the release of calcium from internal stores (Brown and Rubin 1984; Payne et al. 1986b). Thus Ins1,4,5P$_3$ releases calcium from internal stores in sufficient amounts to mimic the effects of light. Furthermore, there is a light-induced rise in Ins P$_3$ in ventral photoreceptors (Brown et al. 1984).

Having identified a possible messenger of calcium release, we can return to the question of why the rise in Ca$_i$ is primarily confined to the R-lobe? The findings presented in Fig. 7 provide a reasonable explanation for this observation (Payne and Fein 1987a). When Ins1,4,5P$_3$ was injected into the R-lobe there was an immediate depolarization of the cell and a rise in Ca$_i$ within a limited region of the R-lobe. In contrast, when Ins1,4,5P$_3$ was injected into the A-lobe there was no detectable rise in Ca$_i$ or depolarization of the cell. This would suggest that the compartment from which calcium is released is primarily found in the R-lobe and not the A-lobe. An accumulating body of evidence indicates that Ins1,4,5P$_3$ releases calcium from the endoplasmic reticulum (for example see Streb et al. 1984). Thus, the compartments from which calcium is released are most probably the SMC, which is primarily found in the R-lobe (Calman and Chamberlain 1982).

With regards to the localization of adaptation within the R-lobe (Fig. 3), injection of Ins1,4,5P$_3$ into the R-lobe only desensitizes the subregion of the R-lobe surrounding the injection site (Fein et al. 1984). The finding (Fig. 7) that Ins1,4,5P$_3$ raises Ca$_i$ only around the injection site can provide an explanation for this localized adaptation. Illumination of a subregion of the R-lobe causes a rise in Ins1,4,5P$_3$ in that region of the cell, causing a local rise in calcium which locally desensitizes the cell's responsiveness to light.

We can now suggest a plausible model for the mechanism of calcium release within the R-lobe of ventral photoreceptors. Photoactivation of visual pigment leads to the release of Ins1,4,5P$_3$ from the plasma membrane; Ins1,4,5P$_3$ then diffuses through the cytoplasm to the nearby SMC where it causes the release of calcium. The released calcium can then lead to excitation and adaptation of the cell. Interestingly, all the pieces of this cascade from visual pigment, to the SMC (the calcium store) and the ion channels opened by light may be localized to the R-lobe. We suggest that the complex of the microvilli and the adjacent cisternae of smooth

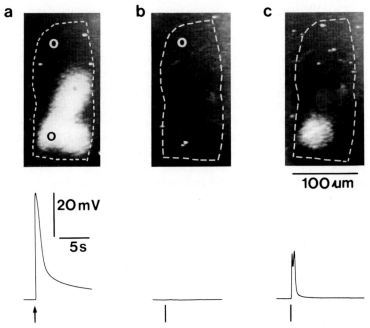

Fig. 7a-c. Injection of Ins1,4,5P$_3$ into the R-lobe of a ventral photoreceptor causes a rise in Ca$_i$ localized to a subregion of the R-lobe. In the *upper panel* of *a, b,* and *c* the aequorin luminescence coming from a ventral photoreceptor was detected using a microscope objective to focus light coming from the cell onto the photocathode of an image intensifier. The *lower panel* shows the corresponding effect of the stimulus on the cell's transmembrane potential. *a* Stimulation of the cell with a diffuse light flash (*arrow*) results in a depolarization and rise in Ca$_i$ throughout the R-lobe. *b* Injection of Ins1,4,5P$_3$ (*bar*) into the A-lobe (*white circle*) is without a detectable effect. *c* Injection of Ins1,4,5P$_3$ into the R-lobe (*black circle* in *a*) depolarizes the cell and causes a rise in Ca$_i$ in a subregion of the R-lobe centered on the injection site (see Payne and Fein 1987a)

endoplasmic reticulum are an important structure for phototransduction, just as the triad junctions of sarcoplasmic reticulum and T-tubule are important for calcium release in skeletal muscle.

The model of the R-lobe just described provides a reasonable explanation for the release of calcium in ventral photoreceptors without any major discrepancies between the model and experimental observation. The release of calcium by light is sufficient to explain *adaptation* of the cell. However, this model may not be sufficient to explain *excitation* in ventral photoreceptors. The difficulty is that the calcium buffer EGTA blocks excitation by injected Ins1,4,5P$_3$ (Payne et al. 1986b) but does not block excitation by light (Lisman and Brown 1975). This may mean that another, as yet unidentified, messenger may be required for excitation. Johnson et al. (1986) have proposed that guanosine 3':5'-cyclic monophosphate (cGMP) may be a suitable candidate for a messenger released by light that does not act via a rise Ca$_i$. Their observation of critical areas within the R-lobe at which injection of cGMP opens light-sensitive ionic channels presents a new and challenging problem in the localization of the biochemical pathway of visual transduction.

8 The Phosphoinositide Cascade in Phototransduction

A light-induced rise in Ca_i has been reported in all microvillar photoreceptors where measurements of Ca_i have been possible (see reviews by Payne 1986; Tsuda 1987). However, the relative contributions of calcium influx from the surrounding extracellular fluid and calcium release from internal stores seem to vary according to species. In those photoreceptors where the light-induced rise in calcium occurs by release from the SMC, we think it probable that the release is initiated by $Insl,4,5P_3$. The requisite light-induced production of $Insl,4,5P_3$ or an isomer, $Ins1,3,4P_3$, has now been demonstrated in the eyes of three species: *Limulus* ventral photoreceptors (Brown et al. 1984), squid retinae (Szuts et al. 1986) and a preparation of housefly photoreceptor membranes (Devary et al. 1987). Recent experiments have indicated that $Insl,4,5P_3$ is also effective in other microvillar photoreceptors. We have pressure-injected $Insl,4,5P_3$ into photoreceptors of the lateral eye of *Limulus* and observed excitation and adaptation of the light response, similar to the effects of injection into ventral photoreceptors (Payne and Fein 1987b). Also, introduction of $Insl,4,5P_3$ into photoreceptors of the housefly causes membrane voltage-noise similar to that elicited by light (Devary et al. 1987).

However, it is possible that the release of $Insl,4,5P_3$ is not the only biochemical cascade initiated by light in invertebrate photoreceptors. Analysis of the amino acid sequence of *Drosophila* visual pigment (O'Tousa et al. 1985; Zuker et al. 1985) reveals the opsin to be a member of a family of receptor proteins, each of which initiates a different enzyme cascade via the activation of specific GTP-binding proteins. Other members include the muscarinic acetylcholine receptor, the ß-adrenergic receptor and, of course, vertebrate opsin (Hall 1987). Activated vertebrate rod rhodopsin initiates a reduction in levels of cGMP in the rod, causing cGMP-activated channels to close (see review by Stryer 1986). The ß-adrenergic receptor initiates the activation of an adenylate cyclase to increase intracellular levels of cAMP (Levitzki 1986) and the muscarinic acetylcholine receptor cloned by Kubo et al. (1986) probably activates a phospholipase-C to increase intracellular levels of $InsP_3$. Whether activation of a single kind of receptor can initiate the simultaneous production of more than one internal messenger in a given cell, however, is not known. Certainly, receptors can cross-react with different G-proteins under artificial conditions (Ebrey et al. 1980; Bitensky et al. 1982; Vandenberg and Montal 1984; Saibil and Michel-Villaz 1984). Thus it is possible that $Insl,4,5P_3$ is not the only messenger released upon photoactivation of invertebrate visual pigment and that cyclic nucleotide levels are also modulated. cGMP levels in squid retinae and photoreceptor membrane preparations increase upon illumination (Saibil 1984; Johnson et al. 1986). Therefore, cGMP may be the additional messenger that accounts for the persistence of excitation by light in the presence of intracellular calcium buffers that block excitation by $Insl,4,5P_3$ (see Sect. 7). If cGMP is an additional messenger, the apparent evolutionary gap between microvillar and vertebrate rod photoreceptors would be considerably reduced.

Acknowledgments. We thank Ms. Susan Wood and Dr. Marco Tsacopoulos for their constructive criticisms of this manuscript. This work was supported by N.I.H. grant EY03793.

References

Autrum H (ed) (1981) Light and dark adaptation in invertebrates. In: Handbook of sensory physiology, vol VII/6c. Springer, Berlin Heidelberg New York, pp 1–92

Bacigalupo J, Lisman J (1983) Single channel currents activated by light in *Limulus* ventral photoreceptors. Nature (London) 304:268–270

Berridge MJ (1983) Rapid accumulation of inositol trisphosphate reveals that agonists hydrolyse polyphosphoinositides instead of phosphatidylinositol. Biochem J 212:849–858

Bitensky MW, Wheeler MA, Rasenick MM, Yamazaki A, Stein P, Halliday KR, Wheeler GL (1982) Functional exchange of components between light-activated photoreceptor phosphodiesterase and hormone-activated adenylate cyclase systems. Proc Natl Acad Sci USA 179:3408–3412

Brown JE, Blinks JR (1974) Changes in intracellular free calcium during illumination of invertebrate photoreceptors. Detection with aequorin. J Gen Physiol 64:643–665

Brown JE, Rubin LJ (1984) A direct demonstration that inositol-trisphosphate induces an increase in intracellular calcium in *Limulus* photoreceptor. Biochem Biophys Res Commun 125:1137–1142

Brown JE, Rubin LJ, Ghalayini AJ, Tarver AP, Irvine RF, Berridge MJ, Anderson RE (1984) Myo-inositol polyphosphate may be a messenger for visual excitation in *Limulus* photoreceptors. Nature (London) 311:160–163

Calman BG, Chamberlain SC (1982) Distinct lobes of *Limulus* ventral photoreceptors II. Structure and ultrastructure. J Gen Physiol 80:839–862

Chamberlain SC, Barlow RB, Jr (1984) Transient membrane shedding in *Limulus* photoreceptors: control mechanisms under natural lighting. J Neurosci 4:2792–2810

Cone RA (1973) The internal transmitter model for visual excitation. Some quantitative implications. In: Langer H (ed) Biochemistry and physiology of visual pigments. Springer, Berlin Heidelberg New York, pp 275–282

Devary O, Heichal O, Blumenfeld A, Cassel A, Suss A, Barash A, Rubinstein T, Minke B, Selinger Z (1987) Coupling of photoexcited rhodopsin to phospholipid hydrolysis in fly photoreceptors. Proc Natl Acad Sci USA 84:6939–6943

Ebrey TG, Tsuda M, Sassanrath G, West JL, Waddell WH (1980) Light-activation of bovine rod phosphodiesterase by non-physiological visual pigments. FEBS Lett 116:217–219

Exner S (1891) Die Physiologie der facettirten Augen von Krebsen und Insecten. Deuticke, Leipzig

Fein A, Lisman JE (1975) Localized desensitization of *Limulus* photoreceptors produced by light or intracellular calcium ion injection. Science 187:1094–1096

Fein A, Payne R, Corson DW, Berridge MJ, Irvine RF (1984) Photoreceptor excitation and adaptation by inositol 1,4,5 trisphosphate. Nature (London) 311:157–160

Fuortes MGF, O'Bryan PM (1972) Physiology of Photoreceptor Organs. In: Fuortes MGF (ed) Handbook of sensory physiology, vol II/2. Springer, Berlin Heidelberg New York, pp 279–319

Grado C, Ballou CE (1961) Myo-inositol phosphates obtained by alkaline hydrolysis of beef brain polyphosphoinositide. J Biol Chem 236:54–60

Hagins WA (1972) The visual process: Excitatory mechanisms in the primary receptor cells. Annu Rev Biophys Bioeng 1:131–158

Hagins WA, Zonana HV, Adams RG (1962) Local membrane current in the outer segments of squid photoreceptors. Nature (London) 194:843–844

Hall ZW (1987) Three of a kind: the ß-adrenergic receptor, the muscarinic acetylcholine receptor, and rhodopsin. Trends Neurosci 10:99–101

Hamdorf K (1970) Korrelation zwischen Sehfarbstoff und Empfindlichkeit bei Photorezeptoren. Verh Dtsch Zool Ges 64:148–157

Horridge GA, Barnard PBT (1965) Movement of palisade in locust retinula cells when illuminated. QJ Microsc Sci 106:131–135

Johnson EC, Robinson PR, Lisman JE (1986) Cyclic GMP is involved in the excitation of invertebrate photoreceptors. Nature (London) 324:468–470

Kirchhoffer O (1908) Untersuchungen über die Augen Käfer pentamerer Käfer. Arch Biontol 2:237–287

Kubo T, Fukuda K, Mikami A, Maeda A, Takahashi H, Mishina M, Haga T, Haga K, Ichiyama A, Kangawa K, Kojima M, Matsuo H, Hirose T, Numa S (1986) Cloning, sequencing and expression of complementary DNA encoding the muscarinic acetylcholine receptor. Nature (London) 323:411–416

Lasansky A, Fuortes MGF (1969) The site of origin of electrical responses in visual cells of the leech *Hirudo medicinalis*. J Cell Biol 42:241–252

Levitski A (1986) β-Adrenergic receptors and their mode of coupling to adenylate cyclase. Physiol Rev 66:819–854

Levy S, Fein A (1985) Relationship between light sensitivity and intracellular free Ca concentration in *Limulus* ventral photoreceptors. A quantitative study using Ca-selective microelectrodes. J Gen Physiol 85:805–841.

Lisman JE, Brown JE (1972) The effects of intracellular iontophoretic injection of calcium and sodium ions on the light response of *Limulus* ventral photoreceptors. J Gen Physiol 59:701–719

Lisman JE, Brown JE (1975) Effects of intracellular injection of calcium buffers on light adaptation in *Limulus* ventral photoreceptors. J Gen Physiol 66:489–506

Lisman JE, Strong JA (1979) The initiation of excitation and light adaptation in *Limulus* ventral photoreceptors. J Gen Physiol 73:219–243

Millecchia R, Mauro A (1969) The ventral photoreceptor cells of *Limulus*. III. A voltage-clamp study. J Gen Physiol 54:331–351

O'Tousa JE, Baehr W, Martin RL, Hirsch W, Pak WL, Applebury ML (1985) The *Drosophila* ninaE gene encodes an opsin. Cell 40:839–850

Payne R (1986) Phototransduction by microvillar photoreceptors of invertebrates: mediation of a visual cascade by inositol 1,4,5 trisphosphate. Photobiochem Photobiophys 13:373–397

Payne R, Fein A (1983) Localized adaptation within the rhabdomeral lobe of *Limulus* ventral photoreceptors. J Gen Physiol 81:767–769

Payne R, Fein A (1986) Localization of the photocurrent of *Limulus* ventral photoreceptors using a vibrating probe. Biophys J 50:193–196

Payne R, Fein A (1987a) Inositol 1,4,5 trisphosphate releases calcium from specialized sites within *Limulus* photoreceptors. J Cell Biol 104:933–938

Payne R, Fein A (1987b) Rapid desensitization terminates the response of *Limulus* photoreceptors to brief injections of inositol trisphosphate. Biol Bull (Abstr) 173:447–448

Payne R, Corson DW, Fein A (1986a) Pressure injection of calcium both excites and adapts *Limulus* ventral photoreceptors. J Gen Physiol 88:107–126

Payne R, Corson DW, Fein A, Berridge MJ (1986b) Excitation and adaptation of *Limulus* ventral photoreceptors by inositol 1,4,5 trisphosphate result from a rise in intracellular calcium. J Gen Physiol 88:127–142

Perrelet A, Bader ChR (1978) Morphological evidence for calcium stores in the photoreceptors of the honeybee drone retina. J Ultrastruct. Res 63:237–243

Saibil HR (1984) A light-stimulated increase of cyclic GMP in squid photoreceptors. FEBS Lett 168:213–216

Saibil HR, Michel-Villaz M (1984) Squid rhodopsin and GTP-binding protein cross-react with vertebrate photoreceptor systems. Proc Natl Acad Sci USA 81:5111–5115

Shimomura O, Johnson FH, Saiga Y (1962) Extraction, purification and properties of aequorin, a bioluminescent protein from the luminous hydromedusan *Aequorea*. J Cell Comp Physiol 59:223–240

Snyder AW, Horridge GA (1972) The optical function of changes in the medium surrounding the cockroach rhabdom. J Comp Physiol A 81:1–8

Stern J, Chinn K, Bacigalupo J, Lisman JE (1982) Distinct lobes of *Limulus* ventral photoreceptors. I. Functional and anatomical properties of lobes revealed by removal of glial cells. J Gen Physiol 80:825–837

Streb H, Bayerdorffer E, Haase W, Irvine RF, Schulz I (1984) Effect of inositol 1,4,5 trisphosphate on isolated subcellular fractions of rat pancreas. J Membrane Biol 81:241–253

Stryer L (1986) The cGMP cascade of vision. Annu Rev Neurosci 9:87–119

Szuts EZ, Wood SF, Reid MA, Fein A (1986) Light stimulates the rapid formation of inositol trisphosphate in squid retinae. Biochem J 240:929–932

Tomlinson RV, Ballou CE (1961) Complete characterization of myo-inositol polyphosphates obtained from beef brain polyphosphoinositides. J Biol Chem 236:1902–1906

Tsuda M (1987) Photoreception and phototransduction in invertebrate photoreceptors. Photochem Photobiol 45:915–931

Vandenberg CA, Montal M (1984) Light-regulated biochemical events in invertebrate photoreceptors. I. Light-activated guanosine-triphosphatase, guanine nucleotide binding, and cholera toxin labeling of squid photoreceptor membranes. Biochemistry 23:2339–2347

Walz B (1982a) Calcium-sequestering smooth endoplasmic reticulum in retinular cells of the blowfly. J Ultrastruct Res 81:240–248

Walz B (1982b) Ca^{2+}-sequestering smooth endoplasmic reticulum in an invertebrate photoreceptor. II. Its properties as revealed by microphotometric methods. J Cell Biol 93:849–859

Walz B, Fein A (1983) Evidence for calcium-sequestering smooth ER in *Limulus* ventral photo-receptors. Invest. Opthalmol. Visual Sci 24 (Suppl):281

Whittle AC (1976) Reticular specializations in photoreceptors: A review. Zool Scr 5:191–206

Zuker CS, Cowman AF, Rubin GM (1985) Isolation and structure of a rhodopsin gene from *D. melanogaster*. Cell 40:851–858

Chapter 10

The Retina-Lamina Pathway in Insects, Particularly Diptera, Viewed from an Evolutionary Perspective

STEPHEN R. SHAW, Halifax, Canada

1 Introduction. Insect Neural Systems: The Evolutionary Perspective

The optical novelties peculiar to arthropod compound eyes would be of little use, were not the associated neural machinery tailored usefully to suit the optical design. We know very little about neural solutions to such matching problems. Neurobiologists are turning increasingly to thinking in evolutionary terms about changes in form and function in eyes, including alterations in the photopigments (Vogt this Vol.) and in the optics (Land and Nilsson this Vol.), but we are fundamentally ignorant about the ways that the *neural* apparatus might have changed during phylogeny. One progenitor of this volume, Exner, is remembered mostly for having exercised his talents upon optical aspects of compound eye design. By contrast, until recently, very little has been written about the basis for neural design and its evolution in insects or any other group of animals. Simple but powerful comparative methods are still of use in initial assaults upon such problems, an approach associated with the other mentor of the volume, Professor Autrum.

What form might evolutionary changes take, in nervous systems? In the existing literature, this avenue has been explored empirically mostly for the vertebrate nervous system, and usually at the level of volumetric changes in entire brains (review: Jerison 1985). Whilst this is interesting as far as it goes, and is the only level at which the important fossil evidence can be approached (indirectly, via cranial endocasts), it falls short by many orders of magnitude of the levels of explanation that have proved most fruitful recently in helping to understand how the nervous system functions: molecular, cellular, and small systems of neurons. This chapter addresses the problem of interpreting neural evolution at the cellular level, on which we have practically no useful information in the literature. The study introduced here, made on a wide range of Diptera (the true flies), actually says little about the optical problems or even the matching problem outlined at the beginning, but the results are some of the first that serve to define what has changed and what has not, during the neural evolution of the eye system of this insect group.

The fossil record for the dipteran lineage is fragmentary but relatively good, and establishes that the ancestral lines of several families were already extant by at least $\sim$ 205 Ma/ago (upper Triassic), with more doubtful specimens from

Stavenga/Hardie (Eds.) Facets of Vision
© Springer-Verlag Berlin Heidelberg 1989

ORDER DIPTERA

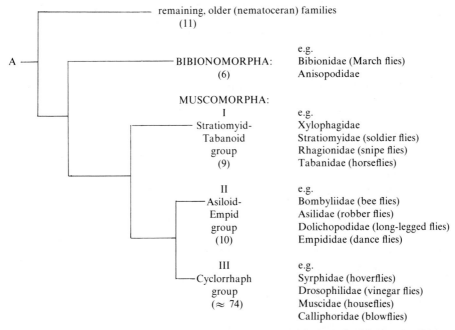

Fig. 1. Partial summary of the general taxonomic relationships within the Order Diptera. The division of the most recently derived subgroup Muscomorpha into three basic subsections I-III follows Hackman and Väisänen (1982), commentators on the analyses of Hennig (1973) and Griffiths (1972). Figures in parentheses give the number of extant families recognized by Steyskal (1974), of which only a few examples are named; some of the common names included are not in universal use. There is substantial agreement about the phylogenetic relationships of most families in I and II, but more disagreement about the affinities of some of the many families within III (cf. Steyskal 1974 and Colless and McAlpine 1970). Mecoptera-like insects are believed to be the ancestors (*A*) of the Order.

Permian even earlier (Hennig 1973, 1981). The Triassic fauna include the Bibionomorpha (Anisopodidae), the probable sister group to the 'recent' subsection Muscomorpha, that contains most of the flies of interest here (see Fig. 1). In Hennigian cladistics, branch points are always dichotomous, so that finding one sister group implies that the other, the Muscomorpha, also must already have been founded by the late Triassic, though actual fossils of the older subgroups of these muscomorphs appeared in the record a little later, in the lower and upper Jurassic (e.g., rhagionids; Fig. 1). The most ancient nematocerous families of the Order, represented already in the Triassic, most likely originated even earlier, from a pool of mecopteroid ancestors that also founded the present sister order Mecoptera (scorpion flies). Whilst thought to be one of the more recently formed insect Orders, the Diptera thus must have existed for approaching half of multicellular life's total span on Earth, now estimated at around $\approx$ 680 Ma. The Diptera currently fall about fourth in the league of most successful groups in the animal kingdom, in terms of species diversity, surpassed only by other insect Orders. Discoveries about neural evolution in such a cosmopolitan, successful and long-standing group of animals are unlikely to be totally idiosyncratic, and might be applicable more generally to other taxa.

2 Taxonomy, Phylogeny and Neuronal Selection

In order to try to follow the evolutionary progression in any structure, it is necessary to use a group whose ancestry both has been charted is agreed upon as monophyletic at its source, and the progression of which is not too strongly confounded by evolutionary convergences. Nature may help in the latter regard, since, as Dumont and Robertson (1986, 1987) comment, in an important expansion of a parallel argument used by embryologists, evolution must have selected for the overt phenotype produced by the nervous system, behavior, and not for the underlying, protected phenotype of the neurons' structure that produce the behavior. With little cost per se invested in having one set of neurons and connections rather than another, neurons could in practice be cobbled together in any makeshift manner into functional circuits, provided that these could produce usefully adaptive modifications of sensory input or behavioral output. Dumont and Robertson (1986) give examples of apparent designer idiocy that seem to support this view, in which the implied evolutionary changes represent a design by accretion upon pre-existing circuits.

A corollary, if the neural microstructure itself has little selective value, is that the neurons themselves may be an unrecognized repository of the evolutionary history of a group. This would be especially useful in unraveling the origin of neural structure, if it turns out that neuronal changes come about sufficiently slowly that some remain in place even in the modern representatives of the groups that pioneered them. This will be argued below.

In the Diptera, the considerable disagreement about the proper taxonomic affinities of several groups (e.g., Steyskal 1974; Hennig 1981) is most severe within the bounds of the more ancient group of families often collectively termed the Nematocera, that we have studied little. The higher flies, where most attention has been focused, have been ranked traditionally as two equal subdivisions (sometimes suborders), Brachycera and Cyclorrhapha. Hackman and Väisänen (1982), following Hennig (1973), find little but historical precedence to commend the division into equally weighted branches. They consider that all of the advanced (non-nematoceran) Diptera — their Muscomorpha — form the sister group to the Bibionomorpha, a smaller collection of 'nematoceran' families including the Bibionidae and in particular the small family Anisopodidae, which several authors, including Hackman and Väisänen, implicate as the group possibly closest to the ancestor line that led to the Muscomorpha (Fig. 1).

Most modern authors believe that the Muscomorpha are a natural group with a monophyletic origin[1] from an anisopodid-like ancestor, although some of the smaller subunits (families) within the Muscomorpha, such as the dance flies

(Empididae), may be paraphyletic assemblages (incompletely formulated, by unnaturally excluding some of the descendants of the most recent common ancestor, sensu Hennig 1981). Hackman and Väisänen (1982) recognize three major subdivisions of the Muscomorpha. A smaller group containing a few families (e.g., asilids, empids), forms the sister group to the largest assembly containing all the remaining "higher" families (originally the Cyclorrhapha). This excludes only the remaining most ancient taxon, the tabanid-stratiomyid group of flies, which are assigned sister group status to the two preceding (Fig. 1). Some phylogenetic affinities within the cyclorrhaph group are obscure, but except for a few families (e.g., Syrphidae, Phoridae), they are not groups that we have had to deal with yet.

   It happens that our dendrogram of neuronal characters, described later, matches the lower branches of the most recent taxonomic schemes fairly well, and so itself serves doubly as an approximate, much condensed guide to the presumed phylogenetic relationships (see Fig. 6).

3 The Rhabdomere Arrays and Axonal Projections to the Lamina

3.1 Comparative Neurology of Rhabdomere Patterning in Diptera

One of the most accessible features of the neural organization is the pattern of light-absorbing rhabdomeres of the photoreceptors under each lens in the compound eye. Such patterns can be inspected with simple optical methods in vivo from without the fly's eye, using a microscope (review: Franceschini 1975). It is known already, from studies confined to a few higher muscomorphs (*Musca*, *Drosophila*), that the seven rhabdomeres are separated laterally from each other by small optically isolating gaps, and are arranged in a characteristic asymmetric

---

[1]Disney (1986a,b) doubts this monophyly, based on an inability to homologize parts of the genitalia of a presumed primitive cyclorrhaph family, Phoridae, with anything in the empid group that earlier taxonomists conjectured stands closest to the cyclorrhaph ancestor (but cf. Hennig 1973). One problem with this view is that it rests narrowly on comparing only two groups, for structures that are complex in flies, presumably potent in promoting species isolation, and therefore likely to evolve rapidly in bizarre directions, perhaps obliterating traces of homology. Moreover, the Phoridae are not usually treated by taxonomists as a central founder-like group upon whose taxonomic fate mainstream cyclorrhaphan affinities stand or fall, but as the group with the most tenuous cyclorrhaphan affinities (Hennig 1973). Possession of the trapezoidal rhabdomere array by phorids (Table 1) and the implied possession of the corresponding axonal divergence pattern by all muscomorphs including phorids (Sect. 3.2) stand with other characters (Hennig 1973) as formidable obstacles to regarding the muscomorphan lineage as polyphyletic. It is unlikely that evolutionary convergence could account for the uniform optical design. Whilst neural superposition is optically advantageous enough to have warranted discovery more than once, its implementation could follow many different routes, and does not depend at all upon adopting the standard trapezoidal pattern. A different pattern would have been expected upon re-invention, as indeed is found in the nematoceran family Bibionidae (Sect. 3.3). Hennig's verdict on the muscomorph half of the Diptera was likely correct (1973, p. 35): "mit Sicherheit als monophyletische Gruppe..."

Table 1. Occurrence of symmetrical and asymmetrical rhabdomere patterns in the retinas of 38 families of Diptera

Family	No. of species		Reference	Family	No. of species		Reference
				Cyclorrhaph groups:			
Nematocera				*Aschiza*			
Tipulidae	(1)	*H*	1, 4	Phoridae	(1)	■	2
Culicidae	(1)	*H*	5	Lonchopteridae	(2)	■	2
Psychodidae	(1)	*H*	6	Pipunculidae	(2)	■	1, 2
Simuliidae	(2)	*H*	1, 3, 7	Syrphidae	(2)	■	1, 2, 3
Bibionidae	(1)	*H*	3, 8				
Anisopodidae	(2)	*H*	1	*Schizophora* — Acalyptrates			
				Conopidae	(3)	■	1, 2
Muscomorpha				Micropezidae	(1)	■	2
Orthorrhaph groups:				Otitidae	(1)	■	1
Rhagionidae	(4)	■	1, 2	Platystomatidae	(1)	■	1
Tabanidae	(8)	■	1, 2, 3	Tephritidae	(1)	■	2
Xylophagidae	(1)	■	1	Sepsidae	(3)	■	1, 2
Stratiomyidae	(5)	■	1, 2, 3*	Lauxaniidae	(3)	■	1
Therevidae	(3)	■	2	Heliomyzidae	(2)	■	1
Asilidae	(6)	■	1, 2, 3	Sphaeroceridae	(2)	■	1, 2
				Drosophilidae	(1)	■	1, 2, 10
Bombyliidae	(6)	■	1, 2, 3	Ephydridae	(2)	■	1, 2
Acroceridae	(1)	■	2	Chloropidae	(1)	■	1
Empididae	(5)	■	1, 2, 3				
Dolichopodidae	(7)	■	1, 2, 9	*Schizophora* — Calyptrates			
				Anthomyiidae	(1)	■	1
				Muscidae	(3)	■	1, 2, 11
				Hippoboscidae	(2)	■	1, 2
				Calliphoridae	(4)	■	1, 2, 3
				Sarcophagidae	(2)	■	1, 2
				Tachinidae	(2)	■	1, 2

1. S.R. Shaw, unpublished observations; 2. Wada (1975); 3. Dietrich (1909); 4. Williams (1980); 5. Brammer (1970); 6. Seifert and Smola (1984); 7. Boschek (1971); 8. Altner and Burkhardt (1982), Zeil (1983a); 9. Trujillo-Cenóz and Bernard (1972); 10. Waddington and Perry (1960); 11. Trujillo-Cenóz (1965). *shown symmetrical by Dietrich (1909).
‡□ having a circularly or bilaterally symmetrical rhabdomere pattern
■ having the asymmetrical, trapezoidal pattern of rhabdomeres

(trapezoidal) pattern (see Fig. 3c). The seven rhabdomeres of one ommatidium image adjacent points in visual space, through their tiny common lens. Because the eye is curved to the degree that the interommatidial angle matches the angle subtended by adjacent rhabdomeres, particular different rhabdomeres under seven adjacent lenses in fact point at the same direction in space (Kirschfeld 1967; Nilsson this Vol.; in fact the visual axes converge slightly upon a point about 4 mm from the eye: Pick 1977). It is precisely the axons of these seven cells, and no others, that early in the eye's development are programmed to weave between those of their neighbors, to come to lie within one and the same cartridge of the lamina. The regular array of lamina cartridges in higher Diptera, not the ommatidial array, therefore holds the eye's first properly ordered map of external visual space. Despite the seeming complexity of the developmental problem, hardly any errors of connection exist in the final axonal array, which remains one of the supreme examples of naturally ordered patterning known (Horridge and Meinertzhagen 1970).

How widespread is the trapezoidal rhabdomere pattern? Dietrich (1909), Wada (1975) and others have demonstrated the rhabdomere pattern in a number of species. We have added extra examples, using either optical inspection (deep pseudopupil, or antidromic illumination; see Stavenga 1979), or anatomical methods. The collected list, summarized in Table 1, demonstrates impressively that all of the 38 families from across the Muscomorpha surveyed so far, without exception, have the advanced trapezoidal pattern, even the ectoparasitic hipposcids with much reduced eyes. None of the lower groups has this pattern. The Bibionidae and Anisopodidae, closest to the stem line (Fig. 1), possess circularly or bilaterally symmetrical patterns (Fig. 2a).

It is rare to find a single descriptor that defines an entire large taxon (Sneath and Sokal 1973), but the trapezoidal array of rhabdomeres appears to fulfil the requirement for the entire Muscomorpha having eyes (i.e., excepting a handful of eyeless parasitic species), and is apparently the single adult character known with this distinction. Antennal segmentation, for instance, becomes much reduced in the Muscomorpha, but the transition is far from clean, several of the more ancient muscomorph families having multi-segmented antennae. The stability of the rhabdomere patterns, maintained throughout the emergence of the Muscomorpha, supports the hope expressed earlier, that phylogenetic history can persist, uneroded, to be read from the neuronal arrays. It strongly supports the proposed monophyletic origin of the muscomorphan lineage.

3.2 Did the Asymmetric Axonal Divergence Pattern Develop in its Final Form at the Same Time as the Asymmetrical Rhabdomere Pattern?

The advantage of the fly's optical-neural system can be understood simply, since it allows this type of eye to gather and sum photons through seven facets instead of one, funneling the resulting neural excitation into a single cartridge (the principle of neural superposition, Kirschfeld 1967; see Nilsson this Vol.). The arrangement markedly increases the light-gathering aperture of a cartridge without degrading its visual resolution, since high acuity depends upon retaining thin, unfused rhabdomeres. The muscomorph subsection of the Diptera comprises diurnal species almost exclusively, which should benefit positively from retaining maximum visual resolution.

Neural superposition is optimized in practice if the pattern of ommatidial origin of axons converging on to rows of lamina cartridges precisely replicates the pattern of the rhabdomeres at the other end of the same ommatidia, with the axon bundles first undergoing an obligatory 180° twist, to compensate for the 180° image rotation imposed by the lens. This condition is fulfilled in the housefly *Musca* (Kirschfeld 1967), and in the bluebottle *Calliphora* from the adjacent family (Horridge and Meinertzhagen 1970). In more ancient flies (Bibionidae, March

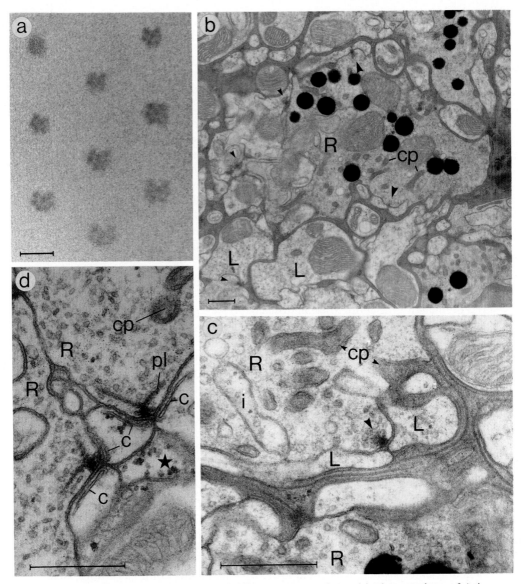

Fig. 2. a "Primitive" horseshoe-shaped rhabdomere patterns observed in the ventral eye of Anisopodidae (*Sylvicola dubius* ♂) using antidromic illumination and corneal neutralization. *b* Electron
micrograph of a photoreceptor terminal (*R*) in *S. fenestralis* ♀, showing pigment granules and two
dyadic afferent synaptic sites (*large arrowheads*). *Small arrows* other synapses, engaging monopolar (*L*)
cells or unidentified profiles. *c, d* Synapse in primitive Anisopodidae contrasted with the homologous
site in the more advanced Dolichopodidae (*D. cuprinus* ♀). The two dolichopodid synapses in *d* both
show the advanced characteristic of a T-shaped presynaptic bar, comprising a pedestal surmounted by
an extended platform (pl). In this section, three of the postsynaptic elements of the synapse, a tetrad,
appear, of which the two central L cell processes contain large, flattened subsynaptic cisternae, *c*, while
the third (amacrine) element (*) does not. Capitate projections (*cp*) from glia into R terminals have
well-differentiated "heads" in dolichopodids, as in more advanced flies (cf. Fig. 7). *b, c* By contrast, the
homologous synapse in the anisopodid is dyadic, both the presynaptic platform and the subsynaptic
cisternae are absent, and the capitate projections are undifferentiated. The L fibres in this family make
extensive invaginations (*i*) into the R terminals. *b–d* from a study by Shaw and Meinertzhagen (1986).
Scale bar 10 μm in *a*, 0.5 μm in *b–d*

flies), where the alignment of the axes of the cartridges appears not to match the optical axes of the rhabdomeres precisely, Zeil (1983a) demonstrates that even a less than complete optimization of the (different) bibionid solution to neural superposition would be better than having no vestige of superposition at all. It thus becomes an interesting empirical question to ask whether the "advanced" axonal projection of the muscomorphs co-evolved with the trapezoidal rhabdomere pattern, or whether it developed fully only at some later stage in dipteran phylogeny. How the axonal projection at one end of the ommatidium comes to match exactly the positioning of the rhabdomere tips at the other, as needed for optimization, is an unsolved developmental riddle in all these systems. The two processes would seem to be unrelated morphogenetic events, suggesting again that they could have arisen independently in the ancestry of the Muscomorpha.

It is much more effort to trace out a complete axonal projection reliably than it is to simply observe the rhabdomere pattern, and there are no reports in the literature on Muscomorpha outside the original muscid-calliphorid group. We have recently traced out the projection for three ommatidia in serial light micrographs from one dolichopodid specimen, confirming that the axons diverge from each ommatidium as predicted from the pattern known from higher flies. The Dolichopodidae form one of the major families in the immediate sister group to the largest, cyclorrhaphid group of flies (Fig. 1), importantly extending the generality of the projection.

Of greater interest are the families standing at the base of the muscomorph lineage, in the stratiomyid-tabanid complex (Fig. 1). There is little information in the literature on the likely identity of the closest living relatives of the muscomorph founder group, although a branch of the stratiomyids seems to be favored by some taxonomists. I have therefore attempted to demonstrate the axonal projection in a stratiomyid by extending the elegant discovery of Wilcox and Franceschini (1984) with *Musca*, that prolonged illumination of a restricted group of ommatidia will lead to uptake of a marker dye, Lucifer Yellow, previously infused into the retina. By refining their method to stimulate single facets, it ought to be possible to trace the projection relatively painlessly, by looking for patterns of dye-filled terminals in adjacent lamina cartridges. Stimulated dye uptake is observed quite often when many ommatidia are illuminated, but the refinement to single ommatidial level is much more capricious in my hands and seldom works, and all cases recovered are suggestive of a pathological cellular reaction. The few preparations recovered successfully (either as dye-filled terminals, or holes where these have degenerated) have, however, served to demonstrate the exact projection predicted from muscids, and confirmed above for a dolichopodid (Fig. 3).

This result therefore shows that in at least one modern stratiomyid and presumably in that family's ancestral stock, perhaps the closest extant group to the ancestral muscomorph, the "advanced" trapezoidal axonal projection had already been developed along with the trapezoidal rhabdomere pattern. Presumably the projection, like the rhabdomere pattern (Table 1), is common to all muscomorphs. This co-evolution of structures makes good sense in terms of optical optimization, but deepens the enigma of what developmental process could inseparably link two such disparate and differently scaled morphogenetic events, occurring at opposite ends of the cellular array.

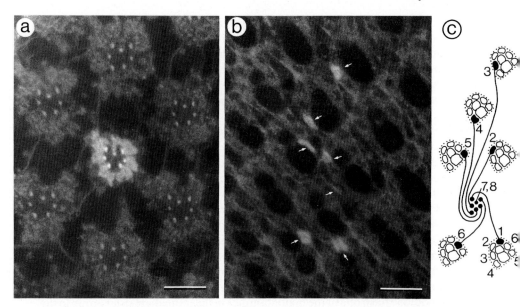

Fig. 3a-c. The axonal projection pattern from an ommatidium in a primitive stratiomyid fly, demonstrated by light-stimulated dye uptake. *a* One facet was strongly illuminated through a fiber optic, after pressure injection of the eye with 54 nl of 0.5% Lucifer Yellow in 330 mOsm sucrose, from a microelectrode. Dye was taken up by all photoreceptors R1-R7 in that ommatidium (although not into R8 which lies under R7 — not shown). The site is in the ventral half of the eye a few ommatidial rows below the equator. The asymmetric pattern of rhabdomeres R1-6 surrounding central cells R7, 8 has been found in all higher Diptera, shown in *c* as the pattern observed at distal end of the ommatidium. *b* The axons of R1-6 from this ommatidium diverge to terminate in lamina cartridges in exactly that pattern already determined for higher flies, shown for comparison in *c*. R7 forms a thin, nearly invisible, flattened process alongside its cartridge; R8 cannot be traced by this method. The terminals in a cartridge form a horseshoe-shaped array in some of the more primitive families, in contrast to the circularly symmetrical ring found in muscids. The dark areas in *b* are the nonfluorescent profiles of monopolar cells L1 and L2, occupying the axis of each cartridge. Scale bars: 10 μm

3.3 A Comment on Axonal Projections in the Bibionomorpha

This probable sister group to the Muscomorpha (Fig. 1) contains the only other family that has been subjected to the same kind of optical investigation as the higher muscid-calliphorid group. Zeil (1979, 1983a, b) and Altner and Burkhardt (1981) discovered a hexagonally symmetrical rhabdomere pattern in the dorsal eye of male Bibionidae. Zeil found that the interrhabdomeric angles are about twice as large as the interommatidial angles, so that optimum information collection only obtains if axons converge upon cartridges from subadjacent rather than directly adjacent ommatidia; he found evidence for this in Golgi preparations. A puzzling feature is that photoreceptor axons seem to associate with two cartridges, thus potentially degrading the resolution offered by the optimum projection to single cartridges (Zeil 1983a, Fig. 7). Zeil sensibly suggests that the first association is informal, and only the second, after axonal bifurcation, is actually synaptic.

Our only pertinent observations relating to this come from analysis of a series of EM sections from *Bibio* sp., made to determine the structure of the synapses (Shaw and Meinertzhagen 1986). We noticed that the terminals in a cartridge arise as branches from nonsynapsing bundles of photoreceptor axons — pseudocartridges — that actually run between cartridges rather than associating with them. This confirms Zeil's latter suspicion that the axons terminate in single cartridges, as required for optimal resolution, but we have not attempted to check whether the destinations of the axons is that predicted by Zeil's optical analysis.

We have also examined cartridges in the Anisopodidae, the group variously favoured as closest to the ancestor line of the muscomorphs. There is a bilaterally symmetrical rhabdomere pattern (Fig. 2a), similar to that in the eye of female and the ventral eye of male Bibionids. The lamina neurones unfortunately have convoluted processes, difficult to follow, and we are not sure whether the projection is *Bibio-* or *Musca*-like.

4 Lamina Neurons and Their Synaptology in Recent Diptera (Muscomorpha)

This area has been reviewed at greater length elsewhere for Diptera (e.g., Strausfeld and Campos-Ortega 1977; Shaw 1981, 1984), and there is only a little new information.

In recent flies like *Musca* or *Drosophila*, about 14 different types of neuron inhabit the lamina (Fig. 4). The qualification "about 14" signifies my developing suspicion based on Golgi studies, that there may be only one type of medium-sized efferent tangential (TAN) neurone, and that either TAN 1 or TAN 2 may have been mistaken for the distal network from the more recently discovered TAN 3, perhaps because as frequently happens, it co-stains along with amacrine cell processes. TAN 3 ramifies throughout the entire eye field with its main arbors just distal to the lamina synaptic zone proper (Nässel et al. 1983), and has recently been renamed LBO5HT, to reflect its origin in the brain and its 5-HT-like immunoreactivity. There are two mirror-symmetrical LBO5HT cells in the protocerebrum, each ramifying bilaterally into all neuropils of each optic lobe (review: Nässel 1988; Hardie this Vol.). Thus 14 neurons (including TAN 3/LBO5HT) have well established identities.

About 28 classes of connection are established by the 14 neurons. The qualification this time encompasses such categories as the few types of connection that were found very infrequently, or those that are doubtful because they do not get mentioned subsequently by the original investigator, or those that may have been identified erroneously (review: Shaw 1981). Undoubtedly, some connections remain to be discovered when a detailed investigation of the inadequately studied distal cartridge region is undertaken.

The axons of the photoreceptors R1-R6 dominate the structure of each cartridge of a fly like *Musca*, forming a circlet of six easily identified terminals around a core of some of the other neurons. Of the 14 well-validated neurons

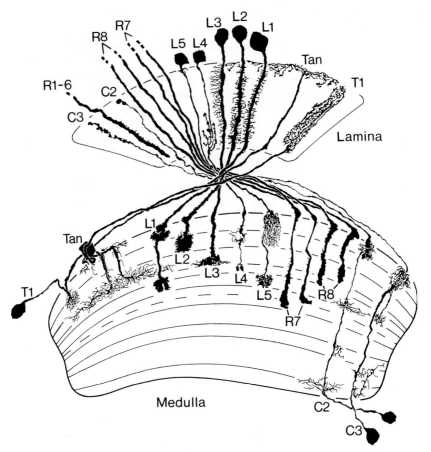

Fig. 4. The neuronal types of the *Drosophila* lamina, based originally upon a Golgi study by Fischbach (1983). Most of the cells are immediately recognizable from the descriptions of similar neurons by Strausfeld (1976), from calyptrate flies like *Musca*. Differences occur in the branching of L4 (see Fig. 5), in the more uniform distribution of L3's neurites, and in the radically different medulla arborization of C3, but most of the neurons are remarkably similar to those of *Musca*, given the difference in size and the rather large phylogenetic distance. Two subtypes of axon terminals of photoreceptors R7 and R8 are now recognized, but amacrine neurons, known to exist from Hauser-Holschuh (1975), fail to impregnate. Only one type of tangential cell is known (see text). Figure courtesy Dr. K. F. Fischbach, from Fischbach and Dittrich (in preparation)

(counting R1–6 as one class), 10 are rigidly periodic, appearing once only at a well-determined location in each cartridge. The 10 are the axons of the remaining photoreceptors R7 and R8, en route through the lamina, where they make no synapses; the three second-order monopolar cells, L1-L3, all postsynaptic to R1–6; three neurons, L4, L5, and T1, possessing connections that are third-order or higher; and two efferent (centrifugal) neurons C2 and C3 originating in the medulla (the nomenclature follows Strausfeld and Campos-Ortega (1977) who, to prevent confusion, retained the names C2 and C3 from a time when existence of another efferent, C1, was suspected).

From the point of view of studying the microanatomy, one of the convenient things about the lamina in higher Diptera is that each cartridge is completely cut off from its neighbors by an encircling sheath of three epithelial glial cells (Boschek 1971). There is no possibility for interconnections between cartridges along their length, except at the proximal and distal borders of the lamina. One can therefore assert with unusual certainty that the connections of all the above ten cell types are made within the borders of their own cartridge, as delimited by the glia. This needs to be qualified only for L4, which does have distal inputs within its own cartridge, but makes output connections in two additional, adjacent cartridges, via a pair of proximal processes, in muscids (see Fig. 5a).

The three to four remaining cells are aperiodic, the efferents TAN 1/2 and TAN 3, and the lamina amacrine, Am. All cartridges are visited by six alpha-processes stemming from more than one amacrine (Fig. 7), the cell which has the greatest variety of synaptic connection of any lamina nuerons. Presumably every cartridge is visited by TAN 1/2 also, but there is no easy way to recognize the processes in EM sections. Along with all other connections made in the distal cartridge, those of the tangentials are only perfunctorily described in the literature, and are in need of re-evaluation. The final cell type mentioned already, TAN 3 or LBO5HT, is not a normal synaptic participant and probably releases transmitter extra-synaptically, just distal to the lamina (Nässel et al. 1983, Nässel 1988). Most likely it has some sort of neuromodulatory role there, either on lamina or retina targets. We have speculated that it may be involved in long-term light adaptational changes that we recorded extracellularly from the lamina, but the evidence is circumstantial and preliminary (Donovan et al. 1986).

There has been welcome recent progress in the neglected area of defining candidate neurotransmitters associated with lamina neurons. In addition to the serotonin-like immunoreactivity of TAN 3, immunocytochemistry implicates efferent C2 as a probable GABAergic fiber (Meyer et al. 1986; Datum et al. 1986). Recent evidence points to histamine as the most likely candidate for a transmitter in photoreceptors R1–6 (Hardie 1987 and this Vol.). A degree of GABA-like immunoreactivity has been reported for axons of one of the central photoreceptors, R7 (Datum et al. 1986).

5 Glia, Local Diffusion Barriers and Possible Glia-Neuronal Interactions

The glia of the lamina are unusual in that they not only structurally invest a cartridge, but also electrically isolate it from its neighbors, forming the substrate for large extracellular voltage gradients following local light activation, which are expected to affect synaptic activity. It appears that photocurrent can spread between cartridges, interfering with normal synaptic transmission (review: Shaw 1984). Saint Marie and Carlson (1983) found extensive runs of tight junctions suggestively disposed on the glial membranes so as to interrupt the extracellular route between cartridges. A. Fröhlich and I (unpublished) have confirmed that cell-cell junctions are indeed present on some of these faces, and not just at autocellular contacts that these glia make so frequently, but which would not contribute to local barrier formation. An interpretational problem is that even given such epithelial glial obstructions to lateral current flow, there are other more circuitous extracellular routes by which photocurrent could leak from one cartridge to another, at the distal border, for example. In fact all extracellular routes outwards appear to be blocked. We recently found that the fluorescent tracer Lucifer Yellow ejected from a micropipette into lamina slices observed in vitro remains there, isolated in compartments of a size similar to cartridges, without diffusing away. When ejected elsewhere in the optic lobe or in the retina, dye clears visibly by diffusion within a few seconds (Järvilehto et al. 1985; Stavenga, this Vol.).

Like glia everywhere, the glial cells of the lamina are connected by gap junctions and are presumed to be electrically coupled. I suggested earlier that the glia may be the conduit for the lateral movement of photocurrent in the lamina, difficult to explain otherwise as an extracellular current flux as I originally had thought, because of the extracellular barriers (Shaw 1984). An exotic but perhaps related finding is that these "glia" are frequent recipients of synaptic contacts, from R1-6, T1, and amacrine neurons, falling second only to the main afferent synapse in connection frequency, both in Boschek's (1971) partial attempt to quantify the overall types and numbers of synapses, and in mine (Shaw 1984). At present these are the only two accounts from which some quantitative idea of the types and numbers of connections is available. Complete reconstruction of cartridges of representative species by serial EM is technically possible, and badly needed. The synaptic counts available so far are compatible with the view that the "epithelial glia" might fulfil a quasi-neuronal function, in forming a laterally conducting neural network, transmission through which can be modulated by light, analogous to the role played by horizontal cell networks in vertebrate retinae.

6 Evolutionary Change in the Dipteran Peripheral Visual System

6.1 Options for Evolutionary Change at the Cellular Level

Using the lamina as a model system, we have recently pursued a question of general interest, asking what it is that changes during the evolution of nervous systems, that could account for the known or assumed changes in behavior or neural processing. These neuronal changes must necessarily accompany, for instance, the evolving changes in external morphology, for the animal to be able to make appropriate use of its mechanical innovations (Shaw 1984; Shaw and Meinertzhagen 1985, 1986). Most of the existing literature on neural evolution focuses disappointingly at the macroscopic level of changes in brain size or on commissural homologies, not at the much finer level at which questions about neural evolution may be addressed more profitably (Sect. 1).

Considering in principle what means are available at the cellular level to a nervous system to modify its treatment of its own input, or alter its output, it seems unlikely that even large changes in the number of neurons in a nucleus alone would be qualitatively effective. A change in numbers could refine the precision of a movement or percept, as is well established for the control of mammalian ocular muscles, or the cortical representation of the fovea, but this would not alter the neural information flow in a qualitative sense. Changes in the time-scale of synaptic action or even in its polarity hold out more promise, through substitution of different neurotransmitters or postsynaptic channels, but the former is likely to be of secondary importance, whilst the latter is limited (only two options) and potentially too drastic.

By default, two broad, obvious strategies remain that an intelligent planner might exploit, were the means available. First, new neurons might be generated by modification of the ancestral developmental plan. This must have happened at some points in phylogeny, since despite very broad homologies known in other domains (in respiratory biochemistry for instance, or visual pigment structure), there is no recognizable cellular homology between the neuronal ensembles of major phyla. It would then still remain a requirement to connect these newly generated cells into the existing neural circuitry in a novel way, in order to alter the performance. The first "option" — if it is one, in the evolutionary short term — really involves a double jeopardy, the successful genesis both of new populations of cells, and, in addition, their effective synaptic connection to each other and to the pre-existing neuronal population.

A second option seemed to us more likely from the outset: the creation of novel connections between pre-existing categories of neuron. We were struck by isolated bits of evidence from comparative studies, that identified neurons must have persisted during phylogeny for long periods, even when measured against the geological time-scale (Shaw and Meinertzhagen 1986). In particular, the commonalities of early neural development amongst widely different orders of insects, involving apparently homologous cells (Thomas et al. 1984), and the cellular homologies between adult motor neurons in crickets and grasshoppers (Wilson et al. 1982) suggest this. Even vertebrates show obvious homologies between the main cell morphs, in the retina and in the cerebellum from the fishes onwards, although these are actually cell classes each subsuming, probably, not one but a number of unique cell types (e.g., Cajal 1973). One way to interpret this fragmentary evidence favoring long-term conservation of cell types is that the developmental programmes for neurogenesis are too elaborately interlocking to permit frequent successful modification. To alter development substantially might introduce profound lethal disruptions, so that major alterations would survive to perpetuate themselves only infrequently.

Dumont and Robertson (1986) provide persuasive examples that nervous system design more resembles the compass of a blind helmsman than the ambit of an intelligent architect, and at the outset of our study (Shaw and Meinertzhagen 1985, 1986), it remained an empirical question which, if either, of the two "strategies" above actually resembled the evolutionary progression observed in the nervous system. We could find nothing in the literature to indicate that synaptic connections between homologous cells can evolve. The alternative case for the introduction of new modifier neurons is unusually hard to prosecute, because in a large neuropile it is almost impossible to rule out the alternative that a neuron of some importance to a circuit was present ab initio, but remained undiscovered. It is here that the almost unique virtues of the lamina synaptic neuropile of Diptera for this kind of comparative study come to the fore. The lamina is made up of precisely constructed, reduplicated, self-contained cartridges of modest (sectionable) size, each possessing the same small handful of individually identifiable cells, the connections of which have been described extensively already in muscids and calliphorids.

This groundwork on recent flies like *Musca* provides a valuable reference set of neurons to which to relate departures from that plan in other species, in a practical model system for exploration of the evolutionary alternatives above.

6.2 Problems of Homology and Convergent Evolution

Homology between neurons in different taxa is appealingly simple idea — that the "same" neurons occur in related taxa because of their recent, common genetic heritage — but difficult to define rigorously and in practice usually not defined, because of the current lack of useful insight into developmental mechanisms. Possession of a common developmental lineage for two purportedly homologous neurons should normally be a strong indicator of homology, but is not practical for lamina cells in insects, because the lineages are not known. The idea is confounded anyway in the nematode *Caenorhabditis,* where some "homologous" cells that occupy bilaterally and functionally equivalent positions do have differing lineages (Sulston et al. 1983). One could categorize these as nonhomologous, convergent neurons, but alternatively, they could be homologous in a deeper sense of representing the common expression of a set of instructions in the genome, perhaps read out in a different temporal sequence. Until the underlying developmental mechanisms are known, it is fruitless to speculate whether cases like this represent homology or convergence in some sense or not, but, ultimately, the definitions need to be formulated at the level of molecular expression.

Two main practical problems remain in pursuing homology at the neuronal level in the dipteran lamina. The first is to make sure that the homologies are appropriate, if, as argued below, two hypothetically homologous cells in different species turn out not to be precisely identical. It becomes possible to make incisive comparisons only if the cells in one shape category (such as monopolar cells) can all be described, so that the most likely homology can be assigned; this is possible in the lamina. An operational definition of homology in which the assignment of homology identities is fitted to produce the maximum correspondence between members (largest similarity, fewest misfits) is one commonly used by numerical taxonomists (Sneath and Sokal 1973).

The second problem, the spectre of convergent evolution which could lead to the false assignment of homology, is one that troubles commentators regularly. One might plausibly argue for a small probability that two neurons of different developmental origin might have converged in shape to resemble one another, for some functional reason over millions of years, if these are two otherwise unidentified individuals set against a background of a myriad unknown others, in similar ganglia of different species. To argue that the cast of 14 neurons in a lamina cartridge of one species has independently evolved to resemble all 14 in another requires an astonishingly high probability for convergence to have occurred in the average neuron over the course of its evolution. Nothing like this degree of convergence is argued for the rest of the body plan even in mimetic insects, where selection *is* presumed to be acting upon the structure. Thus there is strength in numbers in this case, where the larger the replicate set of identifiable neurons being compared, the much more remote the possibility of convergent evolution of the entire set becomes. The similar sets of neurons described in the following sections are accordingly assumed to be actual homologs, and not at all mimetic impostors.

6.3 Axon Counts Give an Index of Neuronal Conservation During Evolution

As a simple step towards assessing whether there has been any change in the neuronal complement in the lamina, relative to the situation known from more recently evolved flies like muscids and drosophilids, we counted the axons in the bundles connecting lamina and medulla cartridges in species from the two other (orthorrhaph) subsections of the Muscomorpha (Fig. 1), the Dolichopodidae from the nearer subsection, and the Rhagionidae from the more distant stratiomyid-tabanoid complex. In each case the smallest number of axons observed in a bundle was ten, with some bundles having one or two more, as expected from the aperiodic intrusions of tangential neurons into the basic clusters (Shaw and Meinertzhagen 1986). Ten is indeed the number expected from studies on higher flies, comprising neurons R7-8, L1-5, T1, and C2-3, and suggests that these cells may have been conserved throughout the entire series of the Muscomorpha. This important conclusion is largely borne out by my more recent Golgi studies.

6.4 Conservation of the Neuronal Set of Lamina Cells in Muscomorph Diptera, Assessed Comparatively from Golgi Impregnations

Golgi studies are necessarily less dependable for revealing all the cell types, since some neurons may fail to impregnate regularly with silver salts, but unlike axon counting, Golgi methods identify the cell types actually staining. Results with two stratiomyids and one rhagionid species from the oldest branch, and two dolichopodids from the middle branch of the muscomorphan series indicate that most of the lamina cells can be recognized according to the *Musca* holotypes, but that some have radically altered arborizations. The main departures from the muscid plan observed so far are summarized in Table 2, which also includes a few observations based on electron microscopy. The following examples give some details of the variety observed.

R7, R8. At least one long visual fibre has been recognized running through the lamina in all five species examined, terminating in a rough, somewhat expanded ending just above the main serpentine layer in the medulla, as in higher flies. Two slightly different types of ending can be recognized in *Rhagio*, presumably corresponding to R7,8; so far, the other species have not been examined sufficiently to be sure whether there is more than one. A major departure from the advanced plan is that one through-running axon puts out several fine medially directed spines within the lamina, in "stratiomyid 2", a large but as yet unidentified species. This points to some undefined synaptic involvement of this fibre in the lamina in this

Table 2. Some anatomical nonconformities in lamina neurons from more ancient dipterans, compared to neurons of more recent muscids or calliphorids

Family in which nonconformity was observed	Observation	Method used[a]	Implied functional change, compared to a muscid or calliphorid
Stratiomyid 2	R8 makes local spines in lamina[b], en route to medulla	G	Extra input or output, to or from R8
Dolichopodid 3 Rhagionid[c]	R7 ends as a complete terminal in lamina; normal cartridge has 7 terminals	EM LM	R7 synapses in lamina, like R1–6
Stratiomyid 2	Efferent C2 has enormous horns matching similar horns on L1	G	Suggests enhanced C2 > L1 connection
Drosophila[d] Dolichopodid 2	Monopolar neuron L4 has the usual basal processes, but lacks any distal neurities	G	Distal input from amacrine abolished or confined to trunk of L4
Dolichopodid 1	L4 has 2 basal processes, no distal arborization, but with ladder-like lateral processes throughout cartridge	G	As above, but with extra input or output on ladder-like spines
Rhagionid	'L4' stout, aberrant, with luxuriant distal arbor, no basal processes, and paired ladder-like neurites throughout cartridge	G	Enhanced (amacrine?) input distally, presumed output to neighbouring L4s at ladder-like neurites
Rhagionid	Amacrine-like arborizations present, but no soma in chiasm; axon extends to medulla	G	'Amacrine' precursor is medulla-lamina projection neurone, not an amacrine
Rhagionid, Tabanid[e]	Axon of T1 becomes a parallel fiber in distal medulla, instead of a discrete terminal bush	G	T1 output in medulla may be distributed more widely
Rhagionid Asilid Stratiomyid 2	One or both L1, L2 lie at edge of the cartridge; neurites of L1, 2 directionally oriented	G EM LM	L1, L2 may selectively contact some of R1–6
Most groups examined	'L3' does not have exclusively distal neurites, in the lamina	G	Unclear, but suggests synaptic rearrangement
All groups	Usual trilaminar C3 arborization in medulla, but not T-branched	G	Unclear, perhaps none (see text)

[a] G, Golgi method; EM, transmission electron-microscopy; LM, light microscopy. [b] also illustrated for some axons in syrphids (Strausfeld 1970); [c] ventral eye, dolichopodid, dorsal 3/4 of male eye, rhagionid. [d] KF Fischbach (see Fig. 4 here); [e] Cajal and Sánchéz (1915)

species, in contrast to muscids and calliphorids. Spines are not observed in the smaller "stratiomyid 1", *Microchrysa* sp. A different distribution of spines has been illustrated for R7,8 in a family near the base of the cyclorrhaph Muscomorpha, the hoverflies (Syrphidae, Strausfeld 1970), and spines with synaptic involvement occur commonly on the long visual fibres of other orders of insects, in dragonflies for instance.

Efferents C2, C3. These are the two most frequently stained fibres in Golgi preparations. C3 looks basically similar in all the species examined, originating from an axon with a three-tiered arbor in the medulla and ending in an inflorescence of variable prominence throughout the length of one cartridge, as illustrated for *Musca* (Strausfeld 1976) and *Drosophila* (Fig. 4). C2 also arborizes at three similar levels in the medulla, and ends uniformly in a clasp-like terminal in the extreme distal lamina, in four of my species and in *Drosophila* (Fig. 4).

One variation has been found involving C2. In a large stratiomyid, the fiber is much thicker and the lamina termination enlarges into two gargantuan prongs like a pitchfork, in the fenestrated zone above the usual limit of the neuropil. The prongs match those on one of the monopolar cells, L1, identified from its medulla termination and its position in the lamina cartridge, matching L1 in muscids (Braitenberg and Hauser-Holschuh 1972). The medulla arborization of the pitchfork cell appears conventional and points to either C2 or C3, because no other medulla-lamina cells have remotely similar medulla branching patterns. Since C3 can be accounted for, C2 is indicated. The implied synaptic connection C2 → L1 is similar but more specific to that proposed for higher flies, C2 → L1, L2 (Strausfeld and Campos-Ortega 1977), and, judged by the size of C2's ending, may be functionally stronger. The result implies that a functional disparity exists between L1 and L2 in terms of their efferent connectivity, that has not been recognized in muscids.

A more radical, perplexing aberration occurs in an advanced group, calliphorid flies, for the other efferent, C3. Cajal and Sanchéz (1915, Figs. 37m, 38c) illustrate, from *Calliphora*, an arciform medullary ending belonging to an efferent, the lamina termination of which they were unable to trace, but which from its medullary pattern must be either C2 or C3. It is identified as C3 by subsequent authors, and traced to its lamina terminus (e.g., Strausfeld 1970). The fiber arborizes at the three levels expected, but the two proximal arbors spring from a descending ramus that arises from a T-branch in the distal medulla. In all five species from older groups that I examined, in the Tabanidae (Cajal and Sanchéz 1915, their Fig. 40c) and in the more recent family Drosophilidae (Fischbach 1983; Fig. 4), the triply-tiered arbor arises instead directly from the straight trans-medullary axon, which runs on to enter the lamina. The surprising implication is that the early growth trajectory of C3 is radically transformed in calliphorids from the earlier condition, but that the neuron could be making similar functional connections, despite its modified gross branching pattern.

L4. L4 is one of the most readily identified lamina neurons in higher flies (Fig. 5a), because it has distinctive twin proximal processes that allow it to interconnect three cartridges (Strausfeld and Braitenberg 1970; Strausfeld and Campos-Ortega 1973), plus a distal tuft of branches within its own cartridge, where it receives input from descending amacrine rami (Campos-Ortega and Strausfeld 1973). It is one of the most variable neurons in the series of flies examined (Fig. 5). In *Drosophila* (Fischbach 1983), the proximal branches are present but the distal tuft is missing, and I find a similar condition in a small sun-loving dolichopodid *Condostylus* (Fig. 5b,d). In a larger woodland genus *Dolichopus*, additional short spines are found on the axon trunk (Fig. 4c). These variations suggest that in some of the middle and

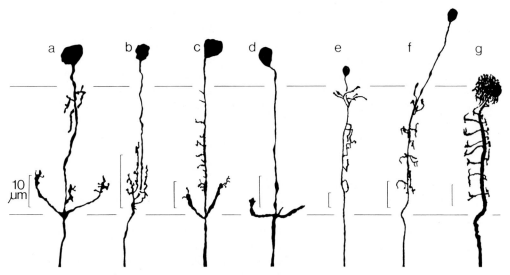

Fig. 5a-g. Lamina monopolar neurone L4, from Golgi impregnations. *a* The "type" L4 originally described from *Musca*, redrawn from Strausfeld (1976); *b* L4 in *Drosophila*, redrawn from Fischbach (1983); *c-g* are representative L4 homologs in more primitive Diptera, from an unpublished study by the author: *c, d* Dolichopodids *Dolichopus* ♀ and *Condostylus* ♀, *e, f* Stratiomyids, unidentified (♀) and *Microchrysa* ♀. *g* The most extreme divergent form equated with L4, from the rhagionid *Rhagio* ♂. Scale bars: 10 μm

most recent muscomorph groups, the input to L4 may be modified or reduced, relative to that described originally for *Musca*. In the putatively oldest group examined (Fig. 5e-g), the situation is reversed. Long neurites, often arranged ladder-like in pairs, are present throughout the depth of the cartridge, but the proximal processes are missing. By contrast, a distal plexus is present and spreads more widely, possibly beyond one cartridge in stratiomyids. The most extreme form is found frequently in the genus I have examined so far most thoroughly, *Rhagio,* where the L4 homolog has a mop of descending neurites that spring from the soma of the cell, reaching elaborate luxuriance in the male's dorsal eye (Fig. 5g).

What evidence indicates that such an extremely variant form as this is really homologous to L4 in, say, muscids? First, only five types of monopolar can be identified, and the cartridge bundles contain the predicted minimum of ten axons. As argued earlier, this implies that the five monopolars probably are homologs of those in higher flies. In all the species I examined, the forms corresponding to L1 and L2 can be recognized readily by their larger calibre, and their two forms of medulla termination, very similar to muscids. In *Rhagio*, two additional thinner monopolars with much shorter nondescript neurites can also be identified, having different termini that again conform closely to those of their putative holotypes in *Musca*, L3 and L5. Of the alternatives, the bushy 'L4' most closely fits the pattern of L4 seen in related groups (Fig. 5e,f), and it too has a slender, simple terminal in the distal medulla that is the most similar of the five types to that of L4 in muscids. The conservative appearance of the medulla terminations of L1-L5 in different species contrasts with the variations between their lamina arbors, a feature which holds for some other neurons, but not for all. For T1, for example, the pattern is reversed.

The results with L4 illustrate the strength of the comparative approach using Golgi methods — easy breadth of coverage — but also its weakness in terms of functional implications. A dedicated sceptic could probably argue that all the cell types shown in Fig. 5 operate in essentially the same manner, taking input at the distal dendrites, or, where these are missing, on the naked shaft. The proximal output to neighbouring cartridges known for type a could be maintained in types b-d, or subsumed by the paired branches of cells f, g, where the different spatial emphasis need not be critical for cable-like cells over distances of 20–60 μm. Thus even for this rather favorably exposed class of identifiable neuron, L4, the Golgi method is a blunt instrument for approaching the question of whether all this variety of form is really expressing a similarly rich variety of function in cell homologs (as one might suppose), simply because the method does not reveal synaptic connectivity directly. The Golgi studies, however, contribute a clear picture that the set of neurons available to make these connections has not changed detectably in the course of over 200 Ma of evolution of the Muscomorpha. To see if the actual connections have changed during evolution, as our earlier hypothesis suggested, has required the use of electron microscopy on identified connections between known, selected neurons.

6.5 Comparative Electron-Microscope Evidence for Radical Changes in Synaptic Connectivity in the Dipteran Lamina

Most of our attention has been directed to the chief afferent "tetrad" synapse, which in higher flies connects single axons of photoreceptors R1–6 to four post-synaptic elements in a cartridge (Burkhardt and Braitenberg 1976). Each terminal in a cartridge makes about 200 of the tiny, stereotyped tetrads, at which the larger central postsynaptic recipients are always processes of monopolar cells L1 and L2, whilst the two outlying 'polar' postsynaptic sites are usually occupied by alpha neurites from an amacrine cell (Nicol and Meinertzhagen 1982a, b). L3 or glia can replace the amacrine in the polar positions, in the distal cartridge. The presynaptic photoreceptor terminal is always easily recognizable in all the species we examined, and the postsynaptic spines can be traced at many of the tetrads, without ambiguity of cell identity. An example of a sectioned tetrad from the middle group of muscomorphs (Dolichopodidae) is given in Fig. 2d, and could represent any of the three dolichopodid species that we examined.

Our major finding (Shaw and Meinertzhagen 1985, 1986) is that the tetradic configuration is confined to the upper reaches of the dendrogram of dipteran evolution, from about the level of the Bombyliidae (bee flies) onwards (Fig. 6). In modern representatives of the families that diverged prior to this point, the synapse has remained dyadic, with the two postsynaptic participants coming again from L1 and L2 (Fig. 2c). It appears that the synaptic connection has changed during evolution of the flies by the attraction of amacrine processes and L3 into the

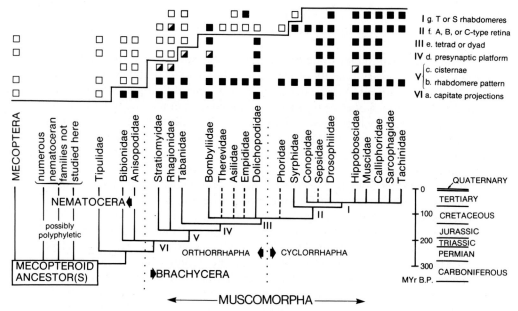

Fig. 6. Dipteran phylogeny assessed by acquisition of recently derived characteristics (■) associated with the photoreceptor afferent synapse, or with eye design, listed at the top. □, ancestral condition; ◪, intermediate state. The characteristics have been arranged in an ascending order that would reflect the temporal order in which they were acquired, if the dendrogram given accurately reflects orthodox views of the phylogenetic sequence, which in fact it largely does (cf. steps I - VI, taken from Hackman and Väisänen 1982, and Steyskal 1974). The derived characteristics signify the possession of: (a) glial capitate projections; (b) a trapezoidal rhabdomere pattern; (c) postsynaptic cisternae in L1, L2; (d) a distinct platform surmounting the presynaptic ribbon; (e) a tetradic synaptic configuration; (f) a C-type retina, defined according to the position of R8 in the ommatidium, by Wada (1975); (g) a tiered (T) as opposed to the older segmented (S) configuration of the central rhabdomeres at the dorsal margin of the eye (Wada 1974). The geological sequence at the right gives a rough idea of the time scale. From Shaw and Meinertzhagen (1986), courtesy of the National Academy of Sciences, USA

postsynaptic cluster during its development, presumably serving to generalize the photoreceptor output to a wider set of neurons. Once this modification was acquired, it did not reverse itself: all members of the higher groups that we examined had tetradic synapses.

The devout functionalist might argue that all advanced flies "need" the tetradic configuration in order to see "better" in advanced flight maneuvres, perhaps through faster visual responses resulting from direct amacrine feedback loops to R1–6, which form a prominent synaptic class in calliphorids (Shaw 1984). The assumption is that the transient response of the tetrad synapse would be enhanced by negative feedback, accounting for its adoption by the higher groups. Some of the higher flies (calliphorids) have evolved a modified wing stop mechanism that may allow them to make tighter pursuit turns (Ennos 1987). Such groups would benefit from faster visual responsiveness, since the visual reaction times are remarkably short, as little as 13 ms overall for course corrections during flight pursuit, in the closely related *Musca* (Wehrhahn 1985).

Amongst the more recently evolved groups of flies, many of the numerous small acalyptrate families in particular inhabit ground vegetation and are neither strong nor maneuverable fliers, and some fly little or not at all. However, the tetradic connection has not shown itself to be a labile one, that reverts at whim to a dyad when the visual demands subside. All the advanced species examined had tetradic synapses, including a small flightness lauxaniid (*Steganopsis*), which has curiously bent, unusable wings, and hops about low grasses. A particularly decisive case of resistance to reversion is presented by the Hippoboscidae, a parasitic family of flies probably related to the recent family Muscidae. The sluggish flightless species we examined has almost vestigial eyes with only a few rows of ommatidia, appropriate to its natural dark habitat, deep in the fleece of sheep. Despite the reduced need for vision and the disorderly appearance of the lamina, the synapses are clearly advanced and tetradic in their structure (Shaw and Meinertzhagen 1986). As in all muscomorphs, the ommatidia have retained the trapezoidal rhabdomere pattern (Table 1).

6.6 Progressive Evolutionary Changes in Synaptic Ultrastructure in the Lamina

Several ultrastructural innovations change the appearance of the same afferent synapse and its associated organelles, during progression from the presumed precursor state to the advanced, derived condition. A curious, perhaps coincidental observation is that almost all the changes seem to involve an increased elaboration of the structure, as if evolution led unidirectionally towards increased complexity (anamorphosis). A second finding is that the individual alterations in structure apparently arose independently in different families (Fig. 6), indicating that they are not directly causally interlinked.

The earliest change in the series occurred in the capitate projections, striking club-like invaginations by glia of photoreceptor terminals in higher flies (Trujillo-Cenóz 1965), but still of unknown functional significance. We could not find these in an older nematoceran group, but they are obvious from at least the level of the Anisopodidae onwards, where they are rod-like and undifferentiated (cp, Fig. 2b, c). A club-like head emerges and enlarges throughout the series, becoming very prominent only in the phylogenetically recent families.

The two central elements L1 and L2 at the recent tetrad synapse possess elaborate subsynaptic cisternae, connected to the cell's SER (Burkhardt and Braitenberg 1976; Fig. 2d). No cisternae could be detected in nematocerans including the Anisopodidae (Fig. 2c). Stratiomyids and rhagionids have small tubular cisternae, which change to the much larger sac-like variety in the bombyliids and beyond (Fig. 6).

The most obvious character associated with the synapses of the Diptera examined prior to our study is a large presynaptic platform disposed orthogonally above the presynaptic ribbon, a feature producing a characteristic 'T' appearance in sections, found in other parts of the CNS and also at neuromuscular junctions, but apparently unique to Diptera. Of the older families examined so far, only the Dolichopodidae possess this feature, along with the higher groups. In fact the 'T' reaches its most exaggerated form in all three dolichopodids examined (Fig. 2d) compared even to the most recent groups, our only reversal so far countering the case for evolutionary anamorphosis. In all the older families, the presynaptic structure is represented only by a simple unadorned presynaptic ribbon (Fig. 2b, c), as in other insects, though in a few cases we detected a slight increase of density above the ribbon, that might suggest an intermediate condition (Fig. 6).

The ultrastructural changes have been tabulated in the implied order of their acquisition in phylogeny in Fig. 6, along with two structural features of the retina, studied comparatively by Wada (1974, 1975). The dendrogram that results essentially replicates the phylogenetic scheme proposed independently, based on many more characters, by Hackman and Väisänen (1982), with later steps from Steyskal (1974). This provides support for proposals by Sulston et al. (1983) and Dumont and Robertson (1986, 1987), that the nervous system may be a reliable repository of information about the evolutionary history of a species. The most severe discrepancy in the scheme of Fig. 6 arises from displacement of the Tabanidae away from the Rhagionidae, on account of the intermediate development of the presynaptic table in our tabanid. This is in fact our weakest character, since it is probably susceptible to the vagaries of EM staining, which can vary in intensity. The development of the table needs to be examined more critically in groups at this level in future, to help resolve the discrepancy.

6.7 Postscript: Other Changes, Other Groups

Did other connections change during evolution? This question has not been pursued in detail yet, but if the main contrast-coding channel of the compound eye can change radically, it seems a foregone conclusion that other, arguably less immediate, accessory channels also may have been altered. In the older families where the amacrine or L3 neuron exist but have not gained access at the (dyadic) synapse, it is reasonably sure that these cells must have some other, differing connections, to justify their existences. To cite a concrete example, the reciprocal (feedback) synapse from amacrines back to R1–6 that is so prominent in Calliphoridae (Trujillo-Cenóz 1965; Shaw 1984; Fig. 7) is totally absent in *Musca domestica* from the closely related family Muscidae (e.g., Campos-Ortega and Strausfeld 1973; A. Fröhlich, I.A. Meinertzhagen, S.R. Shaw, collective observations). This example incidentally illustrates nicely the importance of current mindset for scientific interpretation, since thinking earlier from a more uniformitarian standpoint, I argued that one or other of these connections must have been incorrectly identified (Shaw 1981). This particular synapse is most usually a dyad, Am → R, T1 in calliphorids (Shaw 1984). In a recent survey of several families across the spectrum of the Muscomorpha, similar dyads could be found in nearly all the groups, involving what appears to be an amacrine presynaptic element, with a photoreceptor axon at one postsynaptic locus, and an element of variable identity at the other (Shaw unpublished). The absence of this synaptic class in *Musca* thus qualifies as a case of selective synaptic deletion, either of this entire synaptic category, or, more selectively, of its photoreceptor component. If the latter were the case, the appearance of the synaptic site in micrographs would be so radically altered as to render it not immediately recognizable. Am → T1 synapses with no

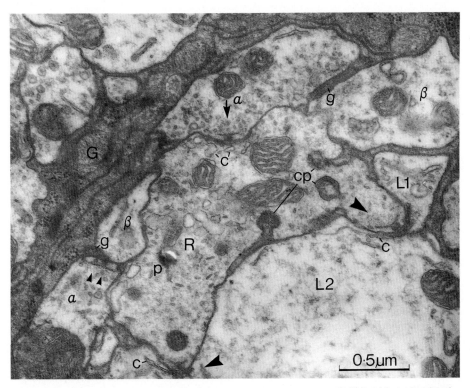

Fig. 7. The edge of a lamina cartridge in the blowfly *Lucilia cuprina* ♂ (Calliphoridae), showing the frequently occurring feedback synapse (*arrow*) at which an amacrine process ($a = \alpha$) is presynaptic to a photoreceptor terminal (*R*). This connection is characterized by extensive, irregular subsynaptic cisternae (*c*), opposite the small T-shaped presynaptic bar. The synapse is missing in the closely related fly *Musca domestica* (Muscidae). The amacrine is itself postsynaptic in both families to the same set of R terminals, forming the polar elements at the larger afferent tetrad synapses, two of which appear (*large arrowheads*), but which have been cut in the middle where only the two central L processes are evident, identified by their flat cisternae (*c*), and thus appear dyadic. The *double arrowhead* indicates the second most common type of synapse in the *Lucilia* cartridge, in which the amacrine is presynaptic to invaginations (*g*) from the surrounding glial cell (*G*). *p* pigment granule; *cp* capitate projections from glia into R terminals; *β* beta process of basket cell T1

R1–6 component have in fact been reported from *Musca* (Boschek 1971). Deletion of components stands in contrast to the addition of processes, described above for the afferent dyad to tetrad transition.

At this early stage of investigation, it is anyone's guess as to just how labile the lamina connections in the end will turn out to have been in the evolution of this neuropil. There is no reason to believe that the changes found are isolated examples, or that lamina neurons are unusually promiscuous. Judged from the variety of form seen even in my small Golgi sample, and in the emerging variety of synaptic connections, the few synaptic changes detected so far could represent just the tip of the iceberg, in just one dipteran ganglion. Whilst it remains to try to extend the list of changes, to generalize these to other ganglia and to other animal groups, it seems very unlikely that the findings synopsized here will turn out to be just a dipteran idiosyncracy. More likely, synaptic reaffiliation of conserved, identified neurons will emerge as an important avenue of neural evolution in general. Moreover, documenting the range of reaffiliations occurring naturally through the experiments of Nature may give some clue to the operating rules of the cell recognition processes active during neural development, by which the sets of connections peculiar to each species establish themselves during each cycle of generation.

Acknowledgments. It is a pleasure to thank Ian Meinertzhagen and Ami Fröhlich for much valued discussions and collaboration on ultrastructural analyses; Karl Fischbach for information about *Drosophila* and for the original of Fig. 4; and Patricia Dickson, Maryse Mallet and Gary Chernenko for their excellent technical help; supported by NSERC, Canada.

References

Altner I, Burkhardt D (1982) Fine structure of the ommatidia and the occurrence of rhabdomeric twist in the dorsal eye of male *Bibio marci* (Diptera, Nematocera, Bibionidae). Cell Tissue Res 215:607-623
Boschek CB (1971) On the fine structure of the peripheral retina and lamina ganglionaris of the fly, *Musca domestica*. Z Zellforsch 118:369-409
Braitenberg V, Hauser-Holschuh H (1972) Patterns of projection in the visual system of the fly II Quantitative aspects of second-order neurons in relation to models of movement perception. Exp Brain Res 16:184-209
Brammer JD (1970) The ultrastructure of the compound eye of a mosquito *Aedes aegypti* L. J Exp Zool 175:181-196
Burkhardt W, Braitenberg V (1976) Some peculiar synaptic complexes in the first visual ganglion of the fly *Musca domestica*. Cell Tissue Res 173:287-308
Cajal SR y (1973) The vertebrate retina. Engl Transl in: Rodieck RW, The vertebrate retina. Freeman, San Francisco, pp 772-904 [Orig Publ Cellule 9:17-257 (1893)]
Cajal SR y, Sanchéz D (1915) Contributions to the knowledge of the nerve centers of insects. Trab Lab Invest Biol Univ Madrid 13:1-164 [Orig Span; Engl translation by Power ME, Truscott BL (1942) Yale Univ Libr]
Campos-Ortega JA, Strausfeld NJ (1973) Synaptic connections of intrinsic cells and basket arborizations in the external plexiform layer of the fly's eye. Brain Res 59:119-136
Colless DH, McAlpine JF (1970) Diptera. In: Waterhouse DF (ed) The insects of Australia. Melbourne Univ Press, pp 656-740
Datum KH, Weiler R, Zettler F (1986) Immunocytochemical demonstration of gamma-amino butyric acid and glutamic acid decarboxylase in R7 photoreceptors and C2 centrifugal fibres in the blowfly visual system. J Comp Physiol A 159:241-249
Dietrich W (1909) Die Facettenaugen der Dipteren. Z Wiss Zool 92:465-539
Disney RHL (1986a) Morphological and other observations on *Chonocephalus* (Phoridae) and phylogenetic implications for the Cyclorrhapha (Diptera). J Zool London (A) 210:77-87
Disney RHL (1986b) Two remarkable new species of scuttle-fly (Diptera: Phoridae) that parasitize termites (Isoptera) in Sulawesi. Syst Entomol 11:413-422
Donovan LA, Thompson PM, Shaw SR (1986) Light adaptation, serotonin, and efferent neurons all may affect visual responsiveness in a circadian-insensitive insect retina. Soc Neurosci Abstr 12:856
Dumont JPC, Robertson RM (1986) Neuronal circuits: an evolutionary perspective. Science 233:849-853
Dumont JPC, Robertson RM (1987) Neuronal circuits and evolution. Science 236:1681-2
Ennos AR (1987) A comparative study of the flight mechanism of Diptera. J Exp Biol 127:355-372

Fischbach KF (1983) Neurogenetik am Beispiel des visuellen Systems von *Drosophila melanogaster*. Habil Thesis, Univ Würzburg, FRG

Franceschini N (1975) Sampling of the visual environment by the compound eye of the fly: fundamentals and applications. In: Snyder AW, Menzel R (eds) Photoreceptor optics. Springer, Berlin Heidelberg New York, pp 98–125

Griffiths GCD (1972) The phylogenetic classification of Diptera Cyclorrhapha with special reference to the structure of the male postabdomen. Junk, The Hague

Hackman W, Väisänen R (1982) Different classification systems in the Diptera. Ann Zool Fenn 19:209–219

Hardie RC (1987) Is histamine a neurotransmitter in insect photoreceptors? J Comp Physiol A 161:201–214

Hauser-Holschuh H (1975) Vergleichende quantitative Untersuchungen an den Sehganglien der Fliegen *Musca domestica* und *Drosphila melanogaster*. Diss, Univ Tübingen, FRG

Hennig W (1973) Ordnung Diptera (Zweiflügler). In: Beier M (ed) Kukenthal's Handbuch der Zoologie, 2nd edn, vol IV/2 (2), 31, pp 1–337 (Engl Transl Entomol Res Inst Libr, Agric Can, Ottawa)

Hennig W (1981) Insect phylogeny. (Engl Transl by Pont AC) Wiley & Sons, New York

Horridge GA, Meinertzhagen IA (1970) The accuracy of the patterns of connexions of the first- and second-order neurons of the visual system of *Calliphora*. Proc R Soc London Ser B 175:69–82

Järvilehto M, Meinertzhagen IA, Shaw SR (1985) Diffusional restriction and dye coupling in insect brain slices. Soc Neurosci Abstr 11:240

Jerison HJ (1985) Issues in brain evolution. Oxford Surveys Evol Biol 2:102–134

Kirschfeld K (1967) Die Projektion der optischen Umwelt auf das Raster der Rhabdomeren im Komplexauge von *Musca*. Exp Brain Res 3:248–270

Meyer EP, Matute C, Streit P, Nässel DR (1986) Insect optic lobe neurons identifiable with monoclonal antibodies to GABA. Histochemistry 84:207–216

Nässel DR (1988) Serotonin and serotonin-immunoreactive neurons in the nervous system of insects. Progr Neurobiol 30:1–86

Nässel DR, Hagberg M, Seyan HS (1983) A new, possibly serotonergic neuron in the lamina of the blowfly optic lobe: an immunocytochemical and Golgi-EM study. Brain Res 280:361–367

Nicol D, Meinertzhagen IA (1982a) An analysis of the number and composition of the synaptic populations formed by photoreceptors of the fly. J Comp Neurol 207:29–44

Nicol D, Meinertzhagen IA (1982b) Regulation of the fly photoreceptor synapses: the effects of alterations in the number of pre-synaptic cells. J Comp Neurol 207:45–60

Pick B (1977) Specific misalignments of rhabdomere visual axes in the neural superposition eye of dipteran flies. Biol Cybernet 26:215–224

Saint Marie RL, Carlson SD (1983) Glial membrane specializations and the compartmentalization of the lamina ganglionaris of the housefly compound eye. J Neurocytol 12:243–275

Seifert P, Smola U (1984) Morphological evidence for interaction between retinula cells of different ommatidia in the eye of the moth-fly *Psychoda cinerea* Banks (Diptera, Psychodidae). J Ultrastruct Res 86:176–185

Shaw SR (1981) Anatomy and physiology of identified non-spiking cells in the photoreceptor-lamina complex of the compound eye of insects, especially Diptera. In: Roberts A, Bush BMH (eds) Neurons without impulses. Cambridge Univ Press, pp 61–116

Shaw SR (1984) Early visual processing in insects. J Exp Biol 112:225–251

Shaw SR, Meinertzhagen IA (1985) Evolutionary progression in an identified synaptic contact. Soc Neurosci Abstr 11:626

Shaw SR, Meinertzhagen IA (1986) Evolutionary progression at synaptic connections made by identified homologous neurons. Proc Natl Acad Sci USA 83:7961–7965

Sneath PHA, Sokal RR (1973) Numerical taxonomy. Freeman, San Francisco

Stavenga DG (1979) Pseudopupils of compound eyes. In: Autrum H (ed) Handbook of sensory physiology, vol VII/6A. Springer, Berlin Heidelberg New York, pp 357–439

Steyskal GC (1974) Recent advances in the primary classification of the Diptera. Ann Entomol Soc 67:513–517

Strausfeld NJ (1970) Golgi studies on insects Part II. The optic lobes of Diptera. Philos Trans R Soc London Ser B 258:135–223

Strausfeld NJ (1976) Atlas of an insect brain. Springer, Berlin Heidelberg New York

Strausfeld NJ, Braitenberg V (1970) The compound eye of the fly (*Musca domestica*): connections between the cartridges of the lamina ganglionaris. Z Vergl Physiol 70:95–104

Strausfeld NJ, Campos-Ortega JA (1973) The L4 monopolar neurone: a substrate for lateral interaction in the visual system of the fly *Musca domestica* (L). Brain Res 59:97–117

Strausfeld NJ, Campos-Ortega JA (1977) Vision in insects: pathways possibly underlying neural adaptation and lateral inhibition. Science 195:894–897

Sulston JE, Schierenberg E, White JG, Thompson JN (1983) The embryonic cell lineage of the nematode *Caenorhabditis elegans*. Dev Biol 100:64–119

Thomas JB, Bastiani MJ, Bate M, Goodman CS (1984) From grasshopper to *Drosophila*: a common plan for neuronal development. Nature (London) 310:203–207

Trujillo-Cenóz O (1965) Some aspects of the structural organization of the intermediate retina of dipterans. J Ultrastruct Rest 13:1–33

Trujillo-Cenóz O, Bernard GD (1972) Some aspects of the retinal organization of *Sympycnus lineatus* Loew (Diptera, Dolichopodidae). J Ultrastruct Res 38:149–160

Wada S (1974) Spezielle randzonale Ommatidien der Fliegen (Diptera: Brachycera): Architektur und Verteilung in den Komplexaugen. Z Morphol Tiere 77:87–125

Wada S (1975) Morphological duality of the retinal pattern in flies. Experientia 31:921–923

Waddington CH, Perry MM (1960) The ultrastructure of the developing eye of *Drosophila*. Proc R Soc London Ser B 153:155–187

Wehrhahn C (1985) Visual guidance of flies during flight. In: Kerkut GA. Gilbert LI (eds) Comprehensive insect physiology, biochemistry and pharmacology, vol 6. Pergamon, Oxford New York, pp 673–684

Wilcox M, Franceschini N (1984) Illumination induces dye incorporation in photoreceptor cells. Science 225:851–854

Williams D (1980) Organisation of the compound eye of a tipulid fly during the day and night. Zoomorphologie 95:85–104

Wilson JA, Phillips CE, Adams ME, Huber F (1982) Structural comparison of an homologous neuron in gryllid and acridid insects. J Neurobiol 13:459–467

Zeil J (1979) A new kind of neural superposition eye: the compound eye of male Bibionidae. Nature (London) 278:249–250

Zeil J (1983a) Sexual dimorphism in the visual system of flies: the compound eyes and neural superposition in Bibionidae (Diptera). J Comp Physiol A 150:379–393

Zeil J (1983b) Sexual dimorphism in the visual system of flies: the divided brain of male Bibionidae (Diptera). Cell Tissue Res 229:591–610

Note added in proof. We can now supply a limited and (to us at least) rather tantalizing answer and corollary to the major question posed in the Postscript, Section 6.7: have synaptic connectivity changes been widespread during evolution? In a recent TEM and Golgi-TEM study of several cell types in the lamina of Stratiomyidae compared to Muscidae, we find that more than half of the approximately 20 connections traced so far differ between the families, with only a minority shared in common (S.R. Shaw and D. Moore in preparation). This embarrassment of riches must mean that transformation of information by the lamina is shown to differ radically between Stratiomyidae and Muscidae when examined in finest detail. Contrarily, at the higher systems level, common design requirements for information transfer must be met at the periphery of any visual system (Srinivasan et al. 1982). The emerging discrepancy between the circuit and systems descriptions could be resolved plausibly if more than one connectional scheme had evolved to implement the same visual transfer function. The need to examine this new idea further through comparative exploration of lamina function (Moore and Shaw 1988) is evident.

Moore D, Shaw SR (1988) Fast oscillatory responses in peripheral visual neurones of Diptera. Soc Neurosci Abstr 14:375

Srinivasan MV, Laughlin SB, Dubs A (1982) Predictive coding: a fresh view of inhibition in the retina. Proc R Soc B 216:427–459

Chapter 11

Coding Efficiency and Design in Visual Processing

SIMON B. LAUGHLIN, Cambridge, England

1 Introduction

In this article I will discuss some of the design principles that underlie the coding of pictures in compound eyes. I will start by putting this work in its historical context and by establishing the credentials of an approach that uses theoretical models to elaborate design principles. The role of models and analytical techniques in understanding design is illustrated by the study of physiological optics. Here, concepts of design, such as the matching of components to each other and the tailoring of structure to optimize performance are established. I will then review the applications of this approach to a well-characterised neural circuit, the retina and lamina of the fly. The article concludes by discussing the role played by coding efficiency in promoting the evolution of neural circuits.

2 Lessons from Physiological Optics

Eyes are favourable material for an analysis of structure and function and have yielded a number of new optical principles, as illustrated by the work of Exner. This is primarily because the optics of image formation provides an appropriate theoretical framework for assessing function. Few would deny that the purpose of an eye's optics is to cast an image onto the receptor mosaic, and that image formation and image quality are constrained by the Laws of Optics. Clearly, Exner already had both a relevant definition of function, and an appropriate theoretical framework. By exploiting the comparative ease with which orderly structures such as compound eyes can be described, and by using his knowledge of optics to generate critical theories and tests of function, Exner was able to establish the optical principles governing apposition and superposition eyes, and demonstrate the way in which superposition is implemented by an array of lens cylinders, one in each ommatidium. In fact, he discovered the lens cylinder and, in collaboration with his brother and at about the same time as Matthiesen, laid the foundations of the field of graded refractive index optics — a field that had not, at that stage, progressed much beyond James Clerk Maxwell's breakfast (rev. Pumfrey 1961).

   Since Exner, the design principles governing retinal image formation in compound eyes have been analyzed using a powerful combination of appropriate experimental description and theory (revs. Kunze 1979; Snyder 1979; Land 1981,

Stavenga/Hardie (Eds.) Facets of Vision
© Springer-Verlag Berlin Heidelberg 1989

this Vol.; Nilsson this Vol.). For example, Exner showed that the critical parameter in the design of a lens cylinder is a parabolic refractive index gradient. This gradient has been confirmed by measurement, ray tracing shows that it is sufficient for superposition, and superposition has been observed directly (Horridge et al. 1972; Hausen 1973). More recently, the quality of superposition images was measured ophthalmoscopically, using techniques from Optical Systems Analysis (Land 1984). In addition, comparative studies have been initiated, relating changes in the design of superposition eyes to differences in function. Again this study depends upon optical measurements, precise anatomy and ray tracing (Caveney and McIntyre 1981). For apposition eyes, the limitations imposed upon image quality by quantum catch (photon noise), lens diffraction and receptor spacing have been investigated theoretically and this theory has provided the foundation required for the experimental analysis of eye design (Snyder et al. 1977; Snyder 1979; Land 1981).

In summary, optics has established a sound basis for analyzing the design of eyes. The goal of a design is an image of sufficient brightness and clarity, the constraints are physical size (determining photon catch and diffraction limits) and the optical properties of biological materials. The relevant parameters to measure can be derived from optical theory and the performance of the system analyzed by combining theory and measurement. Can a similar theoretical framework be found for studying information processing by neurons?

3 Coding Principles in Compound Eyes

Having worked on the optical constraints governing visual acuity (Snyder et al. 1977), it occurred to me that a similar approach could be taken to analyzing the design of neural circuits. Neurons transmit and process information. The efficiency with which they do this can be assessed using Information Theory and by simulation. In addition, just as optical properties define constraints, neural processing is limited by the biophysical and biochemical properties of nerve cells. These neural limitations have not been fully explored, but we know enough to interpret some aspects of design. In particular, intrinsic noise added during signal generation and transmission can limit the smallest resolvable amplitude of signal, just as photon noise limits optical resolution. The finite time constants of the processes that generate neural signals will smooth signals. This smoothing obliterates fine temporal details (e.g. phototransduction limits the flicker fusion frequency) in a manner that is analogous to the blurring of retinal image by the optical point spread function (Srinivasan and Bernard 1975). In addition, neurons can produce only a narrow range of signal amplitudes, set by reversal potentials and spike coding mechanisms. Clearly these elementary cellular properties have the potential to limit the range of signals produced, and will add noise. Fortunately, data transmission in noisy channels with a limited repertoire of responses is a classic problem of communications engineering. Shannon developed Information Theory to handle many of these problems (Shannon and Weaver 1949). Thus there is some hope of using the concept of coding within defined constraints to analyze the design of neural circuits.

We have developed and applied this approach to the first stage in visual processing: the transfer of information across chemical synapses from photoreceptors to second-order cells. The fly visual system is ideal for this type of work because its first optic ganglion, the lamina, is one of the best-described neuropiles in the Animal Kingdom. A number of studies, spanning more than 70 years, have defined the structure of the circuits and key aspects of their function, so providing a firm foundation for erecting and testing models. Thus a brief historical introduction to coding in the fly retina and lamina is appropriate.

4 An Overview of the Structure of the Lamina

We will restrict our interest to those aspects of structure and function that are directly related to our later discussion of coding. To appreciate the complexity of the lamina readers are referred to several comprehensive reviews (Laughlin 1981b; Strausfeld and Nässel 1981; Shaw 1981, 1984, this Vol.; Meinertzhagen and Fröhlich 1983; Strausfeld this Vol.). At the beginning of this century, Vigier (rev. Braitenberg and Strausfeld 1973) and Cajal and Sanchez (1916) drew attention to the orderly structure of the lamina. The lamina is an array of repeated subunits, the lamina cartridges, and there is one cartridge for each ommatidium. Each cartridge receives axons from the set of photoreceptors that look at the same point in space. In fly these happen to be in different ommatidia, an organisation that enhances the quantum catch of the system at least six fold (Nilsson this Vol.; Shaw this Vol.).

Every cartridge contains a well-defined set of interneurons. The most up to date account of the connections within and between cartridges is given by Shaw (this Vol.). The interneurons that concern us are the large monopolar cells, designated L1 and L2, and conveniently referred to as LMC's. These two cells form the central elements in every cartridge and project retinotopically to the medulla. Not only are the LMC's the largest cells in the lamina, they receive the majority of the photoreceptor synapses; approximately 200 from each of the six receptor axons that look at the same point in space and terminate in the lamina. Inputs to LMC's from other cell types are much less frequent. This means that an LMC codes information about the intensity received at a single sampling station on the retina and is a major conduit for the transmission of information from photoreceptors to the medulla.

5 The Responses of the LMC's, L1 and L2

Early electrophysiological work on coding in insect eyes studied the electroretinogram (ERG): the massed extracellular response of the retina and optic lobes. In fly, the ERG is dominated by two components, a slow tonic receptor wave and sharp "on" and "off" transients generated by neural elements in the lamina (Autrum 1958; Heisenberg 1971). The "on" and "off" transients dominate the ERG under many conditions and are responsible for the high flicker fusion frequency of the response. This electrical flicker fusion frequency correlates with an excellent behavioural performance (Autrum and Stöcker 1952). Autrum suggested that good resolution in the time domain compensated for the poor spatial acuity of the eye; a suggestion that is certainly correct in the sense that good temporal resolution boosts the information capacity of the eye (Autrum 1984). Later intracellular analyses of the responses of photoreceptors and LMCs provide a ready explanation for the properties of the ERG (e.g. Burkhardt 1962; Autrum et al. 1970; Järvilehto and Zettler 1971). Both photoreceptors and LMC's respond with graded potentials which generate much of the ERG. In particular, the LMC signal is an inverted, amplified and transient version of the photoreceptor response, and these three properties account for the large neural transients in the ERG; an observation that has recently been elegantly confirmed by genetic dissection (Coombe 1986) and by correlation of the intracellular and extracellular responses (McCook and Laughlin 1986).

The three transformations, inversion, amplification and high pass filtering are illustrated in Fig. 1, which compares the responses of a fly photoreceptor and a fly LMC to the same set of stimuli. Consider first the dark-adapted retina (upper records). A fly sits in the dark and a small light is presented for 100 ms in the centre

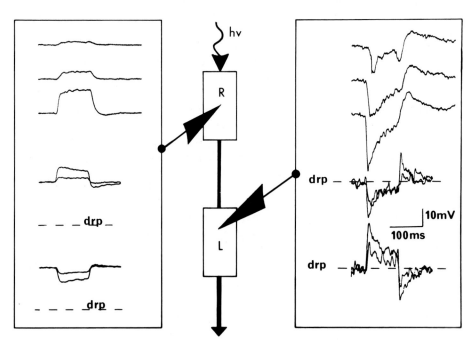

Fig. 1. The graded membrane potential responses of a photoreceptor (*R*) and an LMC (*L*) to identical stimuli. For each cell type the upper three responses are to flashes of light presented to the dark adapted animal (stimulus intensity increases from top to bottom). The middle responses are to 100 ms intensity increments of contrast 0.2 and 1.0, applied in the presence of a background that fully light adapts the LMC. The lower responses are recorded at the same background intensity but the stimulus is dimmed for 100 ms, giving contrasts of –0.2 and –0.8. The *dashed line, drp,* shows the level of the resting potential recorded in darkness. Note that the background continuously depolarises the receptor but does not polarize the LMC. (Laughlin 1982)

of the cell's receptive field, (i.e. at the point in space the cell samples). The receptor responds with a graded depolarization, its amplitude increasing with stimulus intensity. The LMC responds with a graded hyperpolarization which also increases with intensity but exhibits a hyperpolarizing on transient and a depolarizing off transient. These transients are not seen in the receptor and must be generated during transmission from the receptor cell body to the LMC. In addition, the amplitude of the LMC response is approximately six times greater than the receptors (Laughlin et al. 1987). From the point of view of coding and transmitting visual information, the inversion is not of direct significance, (e.g. vertebrate rods and cones hyperpolarize in response to light and invertebrate retinula cells depolarize, yet both types of animal distinguish light from dark). However, the generation of transients and amplification has a profound effect on coding.

Both photoreceptors and LMC's code intensity as graded potentials within a restricted range (ca. 70 mV in receptors and ca. 60 mV in LMC's). Because signals are amplified, the LMC response saturates at lower light levels than the receptor (Järvilehto and Zettler 1971). Thus amplification restricts the intensity range coded by dark-adapted LMC's to about 2 log units. To cope with a useful intensity range, extending over 5 log units to full sunlight, a neural adaptation mechanism regulates the LMC response (Laughlin and Hardie 1978). It is this mechanism that generates the response transients. The effects of this mechanism on intensity coding can be seen by comparing the responses of photoreceptors and LMC's to a steady bright "background" light (Fig. 1). The photoreceptor is continuously depolarized by the background, up to a level of 30 mV . By comparison, the background produces little sustained response in the LMC. During constant illumination, the LMC response decays within a few hundred milliseconds; the membrane potential returning close to the dark level. Clearly LMC's do not code the background, but they continue to respond transiently to deviations from the background level (Fig. 2).

In essence, the neural adaptation mechanism has subtracted away the sustained background component from the response, leaving the LMC's free to signal increments and decrements about the background (Laughlin and Hardie 1978). The responses of LMC's to such increments and decrements are seen in Fig. 1, where they can be compared with the responses of a photoreceptors to identical stimuli. Increments generate hyperpolarizations, and decrements depolarizations. Note that the LMC response to a given small increment or decrement is larger than the receptor's. The amplification remains constant, at approximately x 6, from the dark-adapted state to full sunlight (Laughlin et al. 1987). Amplification and subtraction are essential elements of a simple strategy for retinal gain control. This strategy enables the retina to cope with signals over a wide range of intensities.

6 A Simple Overview of Coding and Gain Control

From sunrise to midday on a cloudless day, the intensity of light in a given location changes one hundredfold. Variations in cloud conditions and shading by foliage extends the full diurnal range of intensity to over five orders of magnitude (e.g. Martin 1983). A photographer will tell you that it is no easy matter to take pictures rapidly and accurately over this intensity range. The majority of eyes have evolved a number of mechanisms to cope with large variations in background intensity. Many of these mechanisms involve the area summation of signals. As intensity falls, signals are pooled from more photoreceptors to produce a statistically significant photon count in a reasonably short time. The operations of fly photo-receptors and LMC's typify a different coding strategy that operates at high intensities, where acuity is limited more by the density of photoreceptors than the density of photons. This strategy allows a single set of receptors and neurons, R 1–6 and the LMC's, to code intensity over the entire diurnal range. The five orders of magnitude in intensity are accommodated within an LMC's 60 mV response range without sacrificing high sensitivity to small increments and decrements at any one mean intensity level (background intensity).

This operational problem is solved by coding contrast. Contrast is a measure of relative intensity in which all values of absolute intensity are expressed as a proportion of the mean light level. The advantage of coding contrast is that it is a natural measure (rev. Laughlin 1981a; Shapley and Enroth-Cugell 1984). Few natural objects are self-luminous. Their visibility stems from the degree to which they reflect or transmit incident light, so that an object transmits to the eye a fraction of the light falling upon it. This fraction remains constant, irrespective of the overall level of illumination. As observed by von Helmholtz (1924), when the visual system scales the intensity of an object by expressing it as a proportion of the mean light level, the object will have the same scaled value at all background levels. By dividing through by the background intensity, i.e. by coding contrast, the five orders of magnitude of diurnal intensities are reduced to a smaller range of scaled values. This reduced set of values is more easily accommodated within a neuron's limited response range. In our studies of non-periodic stimuli we use the general definition, contrast $= \Delta I/I$, where I is the mean level and ΔI the difference between the intensity at a particular point and the mean.

If one presents stimuli of constant contrast at a number of background intensities, one sees that receptors and LMC's co-operate to code contrast extremely effectively (Fig. 2). The stimulus is a light whose intensity is modulated with a contrast of 0.4 to produce responses similar to those shown in the lower half of Fig. 1. It is presented at a number of intensity levels, the highest of which are equivalent to natural sunlight. The receptor response consists of two components, the contrast response (i.e. the change in potential generated by the change in intensity) and a sustained response to the background component. The contrast response increases with background intensity over the intensity range $10^2–10^5$ effective photons/receptor/second. At higher intensities, the modulation generated by a constant contrast is of constant amplitude to within $\pm 4\%$. Contrast is being coded virtually independent of background intensity. This behavior approximates to a logarithmic transformation of stimulus intensity (Zettler 1969). However, the receptor contrast response suffers from two deficiencies. First, the contrast signal is a fraction of the total range of receptor response amplitudes. Second, this small contrast signal is superimposed on a larger background component. The processes of neural adaptation and amplification now come into play. As the signal is transferred from the site of phototransduction to the LMC, neural

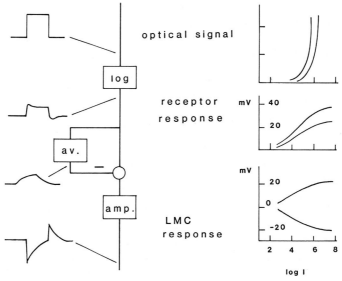

Fig. 2. A diagrammatic summary of the log-substract-amplify strategy for retinal adaptation in the fly. The block diagram on the *left* shows the sequence of transformations implementing the strategy, and the waveforms of the signals generated by a pulse of light. An approximately logarithmic transformation (*log*) is performed on the optical signal by the receptors. Neural mechanisms acting in the lamina take a spatio-temporal average (*av.*) of receptor signals, which is subtracted from the receptor input. The residue is amplified (*amp.*) by a constant factor (approx. x 6). The effect of this transformation on stimuli of constant contrast is shown by the curves on the *right hand side.* The amplitude of an optical signal of contrast 0.4 (the mean environmental level, Laughlin, 1981b) increases in proportion to the background intensity (*upper graph*). The receptors compress this signal to small but workable proportions (*middle graph*). The steady background component is then removed, and the modulating component amplified to fill the LMC dynamic range (*lower graph*). Figure adapted from Laughlin and Hardie (1978) using more recent measurements of photoreceptor and LMC responses to signals of fixed contrast

adaptation removes the background component, a process equivalent to subtraction. The remaining contrast signal is then amplified to fill the LMC's response range. Note that amplification is increasing the amount of information available to the brain by making it easier for higher-order neurons to resolve fine pictorial detail of low contrast.

This simple log-subtract-amplify procedure for contrast coding (Laughlin and Hardie 1978) forms the basis of the analysis of design. The object is to code contrast efficiently within the limited dynamic response range in LMC's. The question of design is approached by asking, "If I were an engineer how would I design this system to achieve its goal most effectively?". The first question to ask is, "Why amplify the signals in LMC's? Why not generate a larger response to contrast in the photoreceptors?". A possible answer is that the log transformation performed by receptors is itself an optimization of a different set of constraints, the non-linear summation of conductance events on the photoreceptor membrane (Laughlin 1981a). Optimization with respect to this constraint leaves little room for improvement of the receptor signal in the retina. The deficiencies must be corrected by later processing in the lamina. Thus the questions of design are restricted to the subtraction of the mean receptor signal and the amplification of the residue. In particular, how does one derive an appropriate mean value to be subtracted away at each lamina catridge and precisely what level of amplification should be used?

7 Computing the Mean — Predictive Coding

The computation of the mean signal appears to be a trivial problem. One simply sums receptor outputs over the entire retina and divides by the number of receptors. Unfortunately, this global mean is usually inappropriate at most retinal regions. The world is generally unevenly illuminated, with patches of highlight and shade producing retinal areas of different mean intensity. A local measure of the mean intensity would be better, but just how local should this mean be? If it is computed over too small an area the local mean begins to resemble the patterns of intensity one wishes to see, and its subtraction would remove valuable pictorial detail. Is there an optimum measure of the mean somewhere between the two extremes of the global and the too local?

Computer image processing suggested an attractive solution — predictive coding. This technique was developed to compress pictorial information for efficient transmission and storage (rev. Gonzalez and Wintz 1977). Raw pictorial information is simply a matrix of numbers, each value representing the intensity at a spatial sampling point. Predictive coding reduces the size of the number coded at each point by subtracting from it the local mean. Smaller numbers require fewer bits to code them and the quantity of data transmitted and stored is reduced. No information is lost by this subtraction because the computation of the local mean is based upon statistical correlations in images. Two neighbouring points in a picture are more likely to have the same intensity than two distant points. Using these correlations one can predict the value expected at a certain point from the values in its neighbourhood. This prediction is derived by taking a weighted mean of surrounding points. The weighting function takes into account the expected pattern of correlation. The prediction is the value of the local mean that is subtracted, and by definition it contains no information. One is removing redundant signal components that can be inferred from their context. Thus predictive coding is a reversible process. Pictures can be completely restored following coding because no information is removed.

Predictive coding resembles the centre-surround organization of retinal neurons. Furthermore, because it uses an estimate of the local mean that contains no pictorial information, it is eminently suitable for contrast coding in the lamina. It is quite straightforward to see whether predictive coding is implemented by fly LMC's. The theory of predictive coding describes the weighting functions that must be used for different degrees of correlation and these are compared with the patterns of antagonism found in LMC's (Srinivasan et al. 1982). Take first the more obvious case of predictive coding in space where, as outlined above, a local spatial mean is computed using a spatial weighting function (Fig. 3). There is lateral

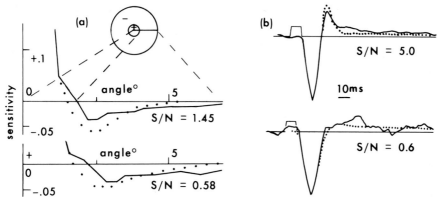

Fig. 3a,b. Intensity dependent inhibition and the implementation of predictive coding in space (*a*) and time (*b*). *a* The angular sensitivity profile through the receptive field of an LMC showing the relatively narrow centre (marked + ve) and the antagonistic (–ve) surround. To implement predictive coding the surround narrows as the signal-to-noise ratio (S/N) increases (compare lower and upper curves). The *dotted lines* show the theoretical receptive fields required for predictive coding at the given S/N's. *b* Impulse responses of LMC's (continuous curves), showing that the antagonistic depolarising component increases in amplitude and decreases in time constant as the signal to noise ratio increases, as required for predictive coding (*dotted curve*). The duration of the brief intensity increment that generated these responses is indicated on the *thin baseline trace*. (Laughlin 1982, after Srinivasan et al. 1982)

antagonism within LMC receptive fields (Zettler and Järvilehto 1972; Dubs 1982), giving rise to a centre-surroundorganisation. The antagonistic surround corresponds to the mechanism computing the local mean because it is subtracted from the signal at the centre. Theory shows that the shape of the surround is not influenced much by the precise pattern of correlation in a given scene. This insensitivity is rather convenient for the animal. One does not have to reorganize one's lamina as one moves from a meadow to a wood. However, the shape of the surround should be critically dependent on the amount of noise contaminating the photoreceptor signals. At high intensities, receptor signals are reliable because they are generated by large numbers of photons. Under these conditions, satisfactory predictions are generated by signals in the nearest neighbours. The reliability of one's prediction is not increased by looking further afield. At low intensities relatively large amounts of noise are generated because photons dribble in at random. The random fluctuations in photon count render individual photoreceptor signals unreliable. One must extend the surround to average many more inputs in order to obtain a reliable prediction. In physiological terms this means that lateral antagonism should become weaker and wider spread at low intensities, as is observed (Dubs 1982). The relationship between signal-to-noise ratio and surround size provides a test for predictive coding.

   LMC receptive fields and signal to noise ratios were measured at two intensities. The signal-to-noise ratio was then used to compute the receptive field required for predictive coding, and this was compared with the measurements (Fig. 3). There is reasonable agreement. The antagonistic surround narrows with intensity, but it is still too broad to be a perfect predictive encoder. We argued that a broader surround is advantageous for coding moving images (Srinivasan et al. 1982), but this possibility has not been analyzed theoretically. Moreover, the signal-to-noise ratios attained in these experiments were limited by the low intensity of the oscilloscope screen used to generate the stimuli. There is a need to examine spatial coding by LMC's at higher intensities.

If we transfer the predictive coding argument from the spatial domain to the time domain, the experimental findings are more compelling. To perform predictive coding in time, one subtracts a weighted mean of previous signals from the signal being received. To see how an LMC undertakes this procedure, consider its response to a sudden brief increase in intensity — the impulse response (Fig. 3). The initial response of the LMC must code the present signal and the cell hyperpolarizes, signifying an intensity increase here and now. A short time later the brief flash is history and it must now contribute to the weighted mean of previous stimulation. The contribution to the mean is equivalent to a response of opposite polarity because it is to be subtracted from any new signal that might be received. Thus predictive coding is implemented by a biphasic impulse response. The brief initial phase coding the present and the longer-delayed phase of opposite polarity representing the weighted mean of previous stimulation. Predictive coding in time also requires a pattern of antagonism that becomes weaker and extends over longer time intervals as the signal to noise ratio decreases (Fig. 3). This intensity dependence provides a straightforward test for predictive coding. There is good agreement between theory and experiment, demonstrating that LMC's perform predictive coding in time. In doing so they utilise a time dependent antagonism that is assumed to result from the signal received at a single sampling point, i.e. it corresponds to the self-inhibition described in *Limulus* lateral eye (e.g. Hartline 1969). It should be noted that the major source of temporal correlation in the visual signal derives from the relatively slow time course of the photoreceptor response. No matter how the image moves, it is coded by the receptors and this helps explain the excellent correspondence between theory and experiment.

In summary we see that neural adaptation in the lamina executes predictive coding in the LMC's. Predictive coding removes a local estimate of the mean signal without removing pictorial information and is, therefore, well designed for removing the background signal. The design involves using intensity-dependent adaptation mechanisms that take into account the quality of the incoming signal. The mechanisms mediating this antagonism have not been elucidated, although electric field effects and synaptic feed-back mechanisms have been implicated (e.g. Shaw 1981, 1984). In addition, a number of points concerning the optimization of coding remain to be resolved. The first is the precise nature of spatial coding at higher intensities. The second is the spatial origin of the antagonistic processes seen in the impulse response — is it entirely self-inhibition? The third is intellectually more challenging. In general, eyes see only moving images. Movement relates spatial and temporal correlations. In addition, spatial predictions are made in time. The preliminary evidence suggests that the lateral antagonism has a slower time constant than the processes seen in the impulse response (Dubs 1982). Is there a predictive coding strategy that operates in both space and time to code moving images? Will it account for the temporal properties of lateral inhibition and the spatial properties of self-inhibition?

8 A Strategy for Amplification — Matched Coding

Having removed the large standing potential generated by the mean intensity, the contrast signal is amplified. Assuming that it is desirable to maximize amplification in order to make fine detail more visible at higher levels of neural processing, what sets the level? An upper limit is imposed by the 60 mV response range of the LMC. Signals that are too large to be accommodated in this range will saturate the response and be lost. To prevent saturation, amplification should be related to the amplitudes of signals to be coded. Information theory suggests a simple solution — matched coding.

A simple analogy can be drawn between the letters of an alphabet and the graded responses of an LMC. The number of response levels that can be discriminated in the output of an LMC is ultimately limited by noise. If the change in LMC membrane potential falls below the noise level, then it cannot be reliably assigned to a change in contrast in the retina. Thus, even in the presence of a completely reliable input signal, the noise added by neural processing divides the LMC response range into a finite number of levels, much as a ruler of given length is divided into a number of intervals by gradations. Each discriminable level of LMC response is equivalent to a letter in an alphabet, because it is equivalent to a symbol denoting a particular state. Shannon and Weaver (1949) demonstrated that an alphabet would carry information most efficiently if all symbols were used equally often. By analogy, the limited response range of an LMC will be used most effectively when all response levels are used equally often. This condition is realised when the function relating response amplitude to input level, the intensity-response curve, is matched to the statistical distribution of intensity levels — hence the term matched coding.

The matching that ensures that all outputs occur equally often is shown in Fig. 4. The levels of input are distributed according to a particular probability distribution. The levels of LMC response are related to the input levels by an intensity-response curve. The probability of observing responses within a given range of amplitudes depends upon the probability of encountering the corresponding range of input intensities, given by the area under the probability distribution within this intensity range. Equal levels of response will occur equally often when equal increments in response correspond to equal areas under the distribution. Under this condition the matched intensity-response function is equivalent to the cumulative probability function for the input. To confirm this equivalence, note that the cumulative probability function has intensity on the x-axis and has a y-axis on which equal increments are, by definition, equal probabilities.

To test for matched coding, a model fly eye was used to measure the probability distribution of contrast in natural scenes. This distribution was then compared with the relationship between response amplitude and contrast, measured in LMC's under light-adapted conditions (Laughlin 1981b). The contrast-response function follows the cumulative distribution of contrast (Fig. 4). Recently, we have analyzed the dependence of coding on light level. At low light levels significant amounts of photon noise are added to the contrast signal. Does the shape of the contrast coding

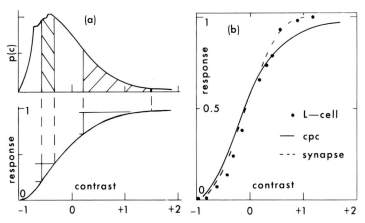

Fig. 4a,b. The matched coding strategy in theory (*a*) and in practice (*b*). *a* Theory — a given probability distribution of contrasts (*upper curve*) must be coded by a contrast-response curve (*lower curve*) to ensure that all levels of response are utilized equally often. This condition is met when the contrast-response curve is constructed so that equal areas under the contrast distribution (e.g. *hatched areas* upper curve) correspond to equal intervals in response amplitude (projections to lower curve). In this case the contrast-response curve is equivalent to the cumulative probability distribution for contrasts. *b* Implementation by LMC's — the contrast-response curve for a light-adapted LMC (·), follows the measured cumulative probability distribution for contrast in natural scenes (*cpc*), as viewed by a model fly eye (Laughlin 1981b). The *measured curve* also follows the transfer characteristics for transmission across a chemical synapse. (Laughlin 1982)

curve change to accommodate this noise? Our analysis (Laughlin et al. 1987) shows that as intensity falls, the slope of the contrast-response curve decreases. This decrease is far greater than is necessary to accommodate the measured levels of photon noise power within the LMC response range. It turns out that the decrease in the slope of the LMC contrast-response curve originates in the photoreceptors. They are also less sensitive to contrast at low intensities (Laughlin and Hardie 1978) and the amplification is approximately constant at all background light levels. It is probably unnecessary for the visual system to compensate for the decrease in receptor signal by boosting amplification at low intensities. At low intensities the photoreceptor inputs are so noisy that little information will be gained by amplifying them. In addition, at low background intensities, inputs from many LMC's may be summed at a higher level and, in this case, a wider linear range of response is advantageous.

What mechanisms are responsible for matching the LMC contrast-response curve to environmental contrast levels? Matching requires two factors. The first is a sigmoidal shape of the curve and the second is the correct slope in the mid-region of the curve. Voltage-sensitive conductances in the LMC membrane are an obvious candidate for producing a sigmoidal contrast response curve by reducing the input resistance of the LMC at the extremities of the response range. However, my recent measurements of LMC membrane current-voltage curves suggest that the membrane resistance is largely unaffected by polarization over the response range, i.e. voltage-sensitive conductances have a negligible effect on the contrast-response curve. This holds both for dark-adapted cells (Laughlin 1974; Guy and Srinivasan

1988) and, more relevantly, for light-adapted cells. Measurements and analysis of the characteristic curve that relates receptor input amplitude to LMC response amplitude suggest that the non-linear properties of synaptic transmission are sufficient to account for both the shape and the slope of the contrast-response curve (Laughlin et al. 1987). The characteristic curve is sigmoidal and follows a simple model of synaptic transmission in which the release of transmitter is an exponential function of pre-synaptic potential (Falk and Fatt 1972). According to this model, the slope of the mid-region of the characteristic curve is solely determined by the sensitivity of the transmitter release mechanism. Our analysis suggests that the receptor-LMC synapses have a high gain, in the sense that little more than a 1.5 mV change in pre-synaptic potential produces an e-fold change in postsynaptic response. This compares with values of from 1.6 to 8.7 mV in other systems (Siegler 1985). Thus, our analysis of signal transmission suggests that a high transmitter release sensitivity has evolved to match contrast coding to input levels. Experiments are currently under way to see if this matching is entirely "hard-wired" or is susceptible to environmental factors, such as contrast levels encountered in early life (McCook and Laughlin 1986).

9 Factors Underlying the Design of Processing

Our work suggests that synaptic amplification reduces the limitations imposed by limited neuronal response range and intrinsic noise. Two procedures, built into the circuitry of the lamina, promote amplification and reduce the risk of saturation. Predictive coding removes an estimate of the background signal using antagonism that is carefully tailored to signal statistics. Matched amplification then operates on the remainder to ensure full use of the LMC response range.

   Why has this system evolved? It is true that amplification is usually beneficial when noise in higher-order circuits sets a minimum level of resolvable LMC response. However, measurements of noise in receptors and LMC suggest a more compelling reason (Laughlin et al. 1987). Amplification reduces the effects of the synaptic noise generated at the photoreceptor terminals. In other words, it is specifically directed towards reducing the limitations imposed by an identified neural constraint. Noise accompanies photoreceptor and LMC responses at all light levels. Receptor noise comes from two sources. Photon noise is generated by the random nature of photon absorptions. Transducer noise is generated during the chemical amplification process that converts a photon absorption into a change in receptor potential (Lillywhite and Laughlin 1979). The effects of receptor noise on visual acuity depends upon the relative amplitudes of noise and signal. We have compared the amplitude of the noise with the amplitude of an intensity signal (Fig. 5a), in this case the response to an intensity step of known contrast. The resulting signal-to-noise ratio (S-N-R) gives a direct measure of the effective amplitude of the noise. The receptor S-N-R rises with intensity (Fig. 5), as expected from the combined effects of photon and transducer noise, but at high intensities it reaches the stable asymptote that is set by the finite number of transduction units in a photoreceptor (Howard and Snyder 1983; Howard et al. 1987). The S-N-R for

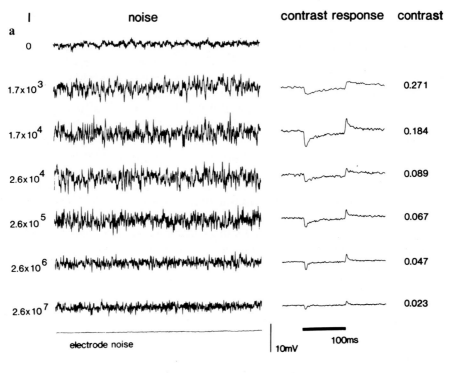

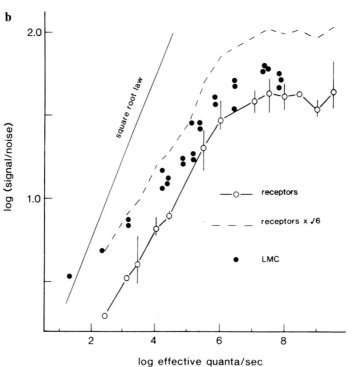

LMC's, measured by the same method (Fig. 5) shows a similar intensity dependence. At low intensities the S-N-R is $\sqrt{6}$ better than in the receptors, as expected from the convergence of the receptors R1-6 onto each LMC. However, at high intensities the improvement falls far short of the expected $\sqrt{6}$. Extra noise must have been introduced during the passage of signal from the receptor cell body to the LMC. This extra noise is apparent only at high intensities, when levels of photoreceptor noise are relatively low. A comparison between the kinetics of signal transfer from receptor to LMC and the noise power spectra suggests that much of this LMC intrinsic noise is generated at the photoreceptor-LMC synapses.

The observation that synaptic noise limits signal quality comes as no surprise. The formidable array of 200 synapses connecting each photoreceptor to each LMC is designed to reduce the levels of synaptic noise. A major component of synaptic noise will result from the random release and binding of transmitter molecules. Just as increasing the number of photons reduces the effect of photon noise, increasing the transmitter released per unit change in pre-synaptic voltage will reduce the effects of synaptic noise. A large area of synaptic contact is one way of improving the synaptic signal-to-noise ratio (Laughlin 1973; Shaw 1979). Decreasing the conductances of post-synaptic channels, the load resistance of the post-synaptic membrane or the transmitter content of vesicles so as to increase the numbers of molecules and vesicles involved will have the same type of effect (Laughlin et al. 1987). However, there is a second and more effective method; increasing the sensitivity of the transmitter release mechanism. We used the model of synaptic transmission that accounts for the characteristic curve to derive the signal-to-noise ratio for a chemical synapse (Laughlin et al.1987). The signal-to-noise ratio is directly proportional to the release sensitivity but only proportional to the square root of the number of synapses (i.e. transmitter dose or number of vesicles). Because the slope of the characteristic curve defines the gain of the synapse and is proportional to the transmitter release sensitivity, a high gain improves the signal-to-noise ratio for synaptic transmission. Photoreceptor contrast signals are small, so that the need for noise protection is pressing. Thus the effect of synaptic noise is reduced by using a coding procedure that maximizes synaptic amplification: hence the need for predictive coding and matched amplification.

Fig. 5a,b. Signal-to-noise ratios in receptors and LMC's at different mean intensities. *a* The method of measuring the signal-to-noise ratio, illustrated by recordings from an LMC. Noise is recorded during application of a background of intensity I, given in effective photons per receptor per second. The signal is the contrast response, defined as the average of 200 responses to a small step increment of known contrast, applied to the background. For each background the stimulus contrast is chosen to generate a contrast response of amplitude similar to the noise level. Because the cells are operating in their linear range, the S-N-R is defined as (contrast response/r.m.s. noise fluctuations) and the S-N-R is normalized with respect to an input signal of unit contrast by dividing by the contrast used. *b* The S-N-R's of photoreceptors and LMC's rise with background intensity. At high light levels the LMC S-N-R falls short of the values expected from the convergence of 6 photoreceptors (receptors x $\sqrt{6}$). This shortfalll indicates that synaptic transmission has added functionally significant amounts of intrinsic noise. (Laughlin et al. 1987)

10 Time Resolution and Information Capacity

A neuron's information capacity is determined by a number of factors. We have seen that noise and limited response range determine the amount of information that can be associated with a particular value of response level. A second factor is the number of different response levels that can be transmitted per second. This is analogous to the data transmission rate or baud for the link between a computer and a printer. For the LMC, each level of membrane potential conveys a certain amount of information, or number of bits. The maximum transmission rate is given by the number of bits per reading of membrane potential multiplied by the number of times that the membrane potential can change value in a second. As an aside, note that these two parameters are related. Large excursions in membrane potential will take longer to execute than small ones. Fortunately, a rigorous analysis is not required to understand the limitations imposed by LMC properties on information capacity. Synaptic transmission is much faster than phototransduction. The hyperpolarizing component of synaptic transmission has a time constant of 0.5 ms (Laughlin et al. 1987) and this compares with a time constant of at least 2 ms for phototransduction (Howard et al. 1987). Thus the receptors simply do not produce data at a high enough rate for synaptic transmission to be a limiting factor. The possibility remains that the passive propagation of signals along the LMC axon limits the transmission rate and synaptic transmission may boost high frequencies to compensate for this. This problem is being investigated (van Hateren and Laughlin in preparation), using measurements of LMC impedance and a cable model (van Hateren 1986) to estimate the effects of cable properties on signal transmission.

11 Comparisons with Other Systems

The analysis of the transfer of signals from photoreceptors to LMC's in the lamina suggests that this particular neural circuit is designed to maximize the amount of pictorial information transmitted to higher centres. The design involves amplification by a large parallel array of high gain synapses to minimize the effects of synaptic noise, and lateral and self-inhibition to remove redundant signal components. Performance is enhanced by adjusting antagonism to take account of the signal-to-noise ratio, and by matching amplification to the statistics of natural scenes. Where else might similar procedures operate?

The most obvious systems to examine are those with similar sensory functions. There are indications that certain parts of the vertebrate retina use similar coding principles, but the data are incomplete (rev. Attwel 1986). The reasons for this are largely technical, and remind one of the fundamental advantages of working on the fly lamina. The LMC's are accessible to stable high-quality recordings in an intact preparation. Moreover, the projection of receptors to LMC's is known. By comparison, bipolar cells are notoriously difficult to record from in intact, or eye-cup preparations. The most accessible are probably photon buckets, collecting rod signals over a large area (Ashmore and Falk 1980). Only cone bipolars with small receptive fields are comparable to LMC's because they code fine spatial detail in bright light. Data from cone bipolars operating at relatively low intensities suggest that the essential elements of design are present (rev. Laughlin 1981a). As constraints, receptors produce small contrast signals superimposed on a background level, synaptic noise is added, and bipolar cells have a limited dynamic range. As solutions, the receptors carry out a logarithmic transform, antagonism operates on bipolar cells to set their operating range to the mean light level, and there is a synaptic amplification of signal. However, parameters that are essential for a convincing analysis of design have not been measured. These include the precise receptor projection, the exact temporal and spatial profile of antagonism at different signal-to-noise ratios, and contrast-response curves. Nonetheless, predictive coding and matched amplification provide a good working hypothesis for understanding retinal coding (e.g. Attwell 1986), and they improve coding efficiency by reducing redundancy, as suggested for the vertebrate retina by Barlow (1961).

Because all neurons have a limited response range and generate noise, similar coding principles may operate in other sensory systems, such as ears, and in other neural circuits. The performance of a sensory system is limited by three factors, the physical properties of the sense organ, the sophistication of neural processing (i.e. the algorithms used), and the accuracy of neural processing. Improved accuracy allows behavioural performance to approach the limits imposed by the sense organ and the algorithms used by the brain. Thus an improvement in accuracy increases the return that an animal gets from its investment, and on this basis it is reasonable to suppose that there is strong selection pressure for computational accuracy in neural circuits.

An example of neural adaptation mechanisms operating to enhance the accuracy of coding at a higher level is the movement sensitive neuron of the fly lobula plate, H1. This unique cell (revs. Hausen 1984; Hausen and Egelhaaf this Vol.; Franceschini et al. this Vol.) integrates movement signals over a wide eye region and generates a spike train that depends upon the contrast, spatial pattern and angular velocity of the retinal image. Instantaneous spike frequencies seldom exceed 400/s. Very simple experiments reveal a pronounced adaptation in H1's response to sustained movement (Fig. 6). Adaptation improves the efficiency with which velocity is coded by H1. When the retinal image has been stationary for some time, H1 is extremely sensitive to low velocities but moderate velocities of 70°/s saturate its response. In the presence of a sustained movement, adaptation lowers H1's sensitivity so that the mean velocity now falls in the mid-region of the velocity-response curve, where the slope is higher. This patently enhances H1's sensitivity to changes in velocity about the mean level (Fig. 6). Adaptation in H1 was independently discovered by de Ruyter and colleagues (1986), who found that it produced an increase in temporal resolving power. These simple properties of H1 resemble those of photoreceptors (Fig. 6). Indeed, the same constraints could well operate on the two cell types. At moderate and high intensities, the photoreceptor response is generated by the summation of many elementary conductance events, each produced by a single photon hit. This means that the relationship between intensity and response must follow a hyperbolic function. This coding function produces the highest sensitivity to changes in intensity when the photoreceptor operates at half maximal response. Adaptation mechanisms hold the receptor

a

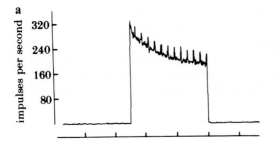

b

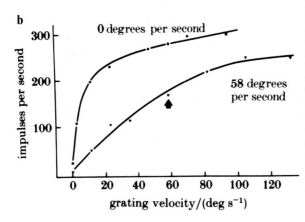

c

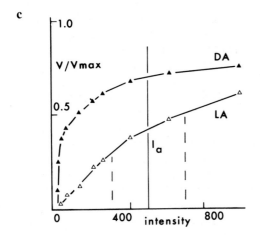

Fig. 6a-c. Adaptation increases the relative sensitivities of the fly lobula plate neuron, H1, and a fly photoreceptor in a similar manner. *a* The average response of H1 to movement of a grating for 5 s at 58°/s. The initial strong response to movement decays as movement continues. This adaptation is accompanied by an increase in sensitivity to brief (38 ms) increments in velocity, applied during the period of otherwise constant movement. The time markers show 2 s intervals. *b* Velocity-response curves for a neuron adapted to a stationary background, and adapted to continuous movement of 58°/s. Adaptation releases H1 from saturation and allows increments and decrements to be coded in a lower region of the velocity-response curve, where the slope is greater (*arrow* indicates background velocity of 58 degrees per second). *c* The intensity-response curves of a photoreceptor show remarkably similar behavior. Response amplitude (normalized to a fraction of the maximum, V_{max}) is plotted against intensity for the dark adapted state (DA) and for the light-adapted state (LA) with a constant background of intensity $I_a = 500$ arbitary intensity units. Light adaptation reduces the response to the adapting light but increases the sensitivity to increments and decrements by increasing the gradient of intensity-response curve in the vicinity of the adapting intensity, I_a

response close to this optimum level at high intensities (Laughlin 1981a). In H1 it is thought that the response is produced by the summation of many parallel synaptic inputs from elementary movement detectors. As in the photoreceptor, the H1 generator potential is probably constrained to follow a hyperbolic function. In this case the generator potential is maximally sensitive to changes in velocity at its half maximum amplitude. A second similarity between photoreceptor and H1 is that adaptation is a local phenomenon, restricted to the areas of strong activation (Hagins et al. 1962; Maddess and Laughlin 1985). Such local effects are advantageous because they preserve signals by preventing local sub-units from saturating. The net result of adaptation is that H1 codes a time average of the velocity with a low sensitivity, and codes changes in velocity about this mean with high sensitivity. In this sense it resembles photoreceptors coding the background intensity with low sensitivity and coding fluctuations about the background with a higher value (rev. Laughlin 1981a). Indeed, one can think of H1 coding velocity relative to the mean, or velocity contrast (Maddess and Laughlin 1985). In addition, local adaptation will tend to pick out boundaries where velocity changes, so aiding the detection of objects with different relative motions (Maddess and Laughlin 1985; Horridge 1986), much as a contrast gain control will pick out areas where texture changes (Osorio et al. 1988).

12 Coding Efficiency and the Evolution of Neural Circuits

This discussion of adaptation in H1 suggests that coding constraints play an important role in the evolution of neural circuits. At first sight it seems difficult to see how the complicated, yet orderly neural interactions required for neural processing have evolved by the selection of random aberrations. In the absence of a functioning network, a small change would probably be of no selective advantage. How then are whole new orders of complexity and levels of interaction added? One way of circumventing this problem is for an existing function to pre-adapt the network for its new role. Let me return to optics for an example of pre-adaptation.

How might the bee's sky compass have evolved? The compass uses receptors that are sensitive to polarized light. If there are no receptors with polarization sensitivity (PS) there is no selection pressure for the compass circuitry, but if there is no circuitry why would PS evolve? It turns out that microvilli have, by virtue of their circular cross section, an inherent sensitivity to polarized light (Moody and Pariss 1961; Rossel this Vol.). Furthermore, microvilli are much more efficient at absorbing light if their visual pigment molecules are aligned so that they are at right angles to the incident light beam. This molecular alignment produces a photoreceptor with greatly increased polarization sensitivity. Thus we can suggest a solution to the evolutionary problem. PS evolved to promote the efficiency of an established mechanism, photon absorption (Shaw 1969), and this PS formed the substrate for the evolution of the neural circuitry required for a sky compass.

Similarly, the efficiency with which a neural circuit executes an existing function may provide the selection pressure for changes that pre-adapt the system for a novel role. Take the evolution of a hypothetical movement sensitive cell similar to H1. One might suppose that a primitive neuron summed velocity inputs over a large retinal area to compute the mean slip speed. An adaptation mechanism then evolved to maximize the sensitivity of this cell over a wide range of input velocities. The selection pressure operating here is simply the enhancement of accuracy. However, the adaptation processes have altered the nature of the information that is coded. The seed of abstraction has been sown. Velocity contrast replaces velocity and, if one small component of this gain control involves local adaptation, selection pressure can operate to enhance higher order image characteristics such as boundaries. A functional argument for such a progression has been independently formulated by Horridge (1987). Similarly the analysis of coding in LMC's demonstrates how the biophysical constraints upon coding efficiency are minimized by adding new levels of neural interaction to the basic mechanism, the transfer of signal from receptor to interneuron via a single type of chemical synapse. To remove the mean signal, neural interactions have evolved that compute this mean over defined areas of space and time. This computation integrates inputs from different ommatidia and introduces interactions with different time constants. These new interactions now open up new functional possibilities, either by changing and enhancing these interactions to perform new functions or by combining outputs from this new cell type. As an example of the latter case, Hubel and Wiesel (1962) suggested that the patterns of lateral interaction seen in cortical simple cells could, in principle, be generated by adding the outputs of a linear array of geniculate cells.

In conclusion, our analysis of coding in the fly lamina suggests that two neural interactions, namely antagonism and amplification, enhance coding efficiency. This enhancement minimizes the effects of inviolable neural constraints, much as good optical design minimizes the effects of diffraction and photon noise. These interactions increase the amount of information that is generated by a sense organ by improving the accuracy with which nerve cells operate. Thus considerations of economy provide a ready explanation for the evolution of such interactions. However, in reducing the effects of such a basic constraint, a new level of processing can emerge which can then form the substrate for the evolution of the complex circuitry required for tasks such as pattern recognition.

References

Ashmore J, Falk G (1980) Responses of rod bipolar cells in the dark-adapted retina of the dogfish, *Scyliorhinus canicula*. J Physiol 300:115-150

Attwell D (1986) Ion channels and signal processing in the outer retina. Q J Exp Physiol 71:497-536

Autrum H (1958) Electrophysiological analysis of the visual systems in insects. Exp Cell Res Suppl 5:426-439

Autrum H (1984) Comparative physiology of invertebrates: hearing and vision. In: Dawson WW, Enoch JM (eds) Foundations of sensory science. Springer, Berlin Heidelberg New York, pp 1-23

Autrum H, Stöcker M (1952) Über optische Verschmelzungfrequenzen und stroboskopisches Sehen bei Insekten. Biol Zentralbl 71:129-152

Autrum H, Zettler F, Järvilehto M (1970) Post-synaptic potentials from a single monopolar neuron of the ganglion opticum I of the blowfly *Calliphora*. Z Vergl Physiol 70:414-424

Barlow HB (1961) The coding of sensory messages. In: Thorpe WH, Zangwill OL (eds) Current problems in animal behaviour. Univ Press, Cambridge, pp 331-360

Braitenberg V, Strausfeld NJ (1973) Principles of the mosaic organisation in the visual system's neuropil of *Musca domestica* L. In: Jung R (ed) Handbook of sensory physiology, vol VII/3A. Springer, Berlin Heidelberg New York, pp 631–659

Burkhardt D (1962) Spectral sensitivity and other response characteristics of single visual cells in the arthropod eye. Symp Soc Exp Biol 16:86–109

Cajal SR, Sanchez D (1916) Contribucion al conocimiento de los centros nerviosos de los insectos. Pt 1: Retina y centros opticos. Trab Lab Invest Biol Univ Madrid 13:1–164

Caveney S, McIntyre P (1981) Design of graded-index lenses in the superposition eyes of scarab beetles. Philos Trans R Soc London Ser B 294:589–632

Coombe PE (1986) The large monopolar cells L1 and L2 are responsible for ERG transients in *Drosophila*. J Comp Physiol A 159:655–665

de Ruyter van Steveninck RR, Zaagman WH, Mastebroek HAK (1986) Adaptation of transient responses of a movement-sensitive neuron in the visual system of the fly *Calliphora erythrocephala*. Biol Cybernet 53:451–463

Dubs A (1982) The spatial integration of signals in the retina and lamina of the fly compound eye under different conditions of luminance. J Comp Physiol A 146:321–334

Falk G, Fatt P (1972) Physical changes induced by light in the rod outer segment of vertebrates. In: Dartnall HJA (ed) Handbook of sensory physiology, vol VII/1. Springer, Berlin Heidelberg New York, pp 200–244

Gonzalez RC, Wintz P (1977) Digital Image Processing. Addison-Wesley, Reading, Mass

Guy RG, Srinivasan MV (1988) Integrative properties of second-order visual neurons: A study of large monopolar cells in the dronefly *Eristalis*. J Comp Physiol A 162:317–332

Hagins WA, Zonana HV, Adams RG (1962) Local membrane current in the outer segment of squid photoreceptors. Nature (London) 194:844–847

Hardie RC (1985) Functional organization of the fly retina. In: Ottoson D (ed) Progress in sensory physiology, vol. 5. Springer, Berlin Heidelberg New York, pp 1–79

Hartline HK (1969) Visual receptors and retinal interaction. Science 164:270–278

Hateren JH van (1986) An efficient algorithm for cable theory, applied to blowfly photoreceptor cells and LMCs. Biol Cybernet 54:301–311

Hausen K (1973) Die Brechungsindices im Kristallkegel der Mehlmotte *Ephestia kühniella*. J Comp Physiol 82:365–378

Hausen K (1984) The lobula-complex of the fly: structure, function and significance of visual behavior. In: Ali MA (ed) Photoreception and vision in invertebrates. Plenum, New York, pp 523–559

Heisenberg M (1971) Separation of receptor and lamina potentials in the electroretinogram of normal and mutant *Drosophila melanogaster*. J Exp Biol 55:85–100

Helmholtz H von (1924) Treatise on physiological optics, vol 2. The Sensations of vision. Optic Soc Am, pp 172–174 Transl from 3rd German edn, Southall JPC (ed)

Horridge GA (1986) A theory of insect vision: velocity parallax. Proc R Soc Lond Ser B 229:13–27

Horridge GA (1987) The evolution of visual processing and the construction of seeing systems. Proc R Soc London Ser B 230:279–292

Horridge GA, Giddings C, Stange G (1972) The superposition eye of skipper butterflies. Proc R Soc Lond Ser B 182:457–495

Howard J, Snyder AW (1983) Transduction as a limitation on compound eye function and design. Proc R Soc London Ser B 217:287–307

Howard J, Blakeslee B, Laughlin SB (1987) The intracellular pupil mechanism and the maintenance of photoreceptor signal to noise ratios in the blowfly *Lucilia cuprina*. Proc R Soc London Ser B 231:415–435

Hubel DH, Wiesel TN (1962) Receptive fields, binocular interaction and functional architecture in the cat's visual cortex. J Physiol 160:106–154

Järvilehto M, Zettler F (1971) Localized intracellular potentials from pre- and post-synaptic components in the external plexiform layer of an insect retina. Z Vergl Physiol 75:422–440

Kunze P (1979) Apposition and superposition eyes. In: Autrum H (ed) Handbook of sensory physiology, vol VII/6A. Springer, Berlin Heidelberg New York, pp 441–502

Land MF (1981) Optics and vision in invertebrates. In: Autrum H (ed) Handbook of sensory physiology, vol VII/6B. Springer, Berlin Heidelberg New York, pp 472–592

Land MF (1984) The resolving power of diurnal superposition eyes measured with an ophthalmoscope. J Comp Physiol A 154:515–533

Laughlin SB (1973) Neural integration in the first optic neuropile of dragonflies. I. Signal amplification in dark-adapted second order neurons. J Comp Physiol 84:335–355

Laughlin SB (1974) Resistance changes associated with the response of insect monopolar neurons. Z Naturforsch 29c:449–450

Laughlin SB (1981a) Neural principles in the peripheral visual systems of invertebrates. In: Autrum H (ed) Handbook of sensory physiology, vol VII/6B. Springer, Berlin Heidelberg New York, pp 135–280

Laughlin SB (1981b) A simple coding procedure enhances a neuron's information capacity. Z Naturforsch 36c:910–912

Laughlin SB (1982) Matching coding to scenes to enhance efficiency. In: Braddick OJ, Sleigh AC (eds) The physical and biological processing of images. Springer, Berlin Heidelberg New York, pp 42–52

Laughlin SB, Hardie RC (1978) Common strategies for light adaptation in the peripheral visual systems of fly and dragonfly. J Comp Physiol 128:319–340

Laughlin SB, Howard J, Blakeslee B (1987) Synaptic limitations to contrast coding in the retina of the blowfly *Calliphora*. Proc R Soc London Ser B 231:437–467

Lillywhite PG, Laughlin SB (1979) Transducer noise in a photoreceptor. Nature (London) 277:569–572

Maddess T, Laughlin SB (1985) Adaptation of the motion sensitive neuron H1 is generated locally and governed by contrast frequency. Proc R Soc London Ser B 225:251–275

Martin GR (1983) Schematic eye models in vertebrates. Prog Sens Physiol 4:43–81

McCook L, Laughlin SB (1986) Origin of the electroretinogram and determination of contrast coding in the compound eye in Diptera. Soc Neurosci Abstr 12:856

Meinertzhagen IA, Fröhlich A (1983) The regulation of synapse formation in the fly's visual system. Trends Neurosci 6:223–228

Moody MF, Pariss JR (1961) The discrimination of polarised light by *Octopus*: a behavioral and morphological study. Z Vergl Physiol 44:268–291

Osorio D, Snyder AW, Srinivasan MV (1988) Bi-partitioning and boundary detection in natural scenes. Spatial Vision (in press)

Pumfrey RJ (1961) Concerning vision. In: Ramsay JA, Wigglesworth VB (eds) The cell and the organism. Univ Press, Cambridge

Shannon CE, Weaver W (1949) The mathematical theory of communication. Univ Ill Press, Urbana

Shapley RJ, Enroth-Cugell C (1984) Visual adaptation and retinal gain controls. Prog Retinal Res 3:263–346

Shaw SR (1969) Sense cell structure and inter-species comparisons of polarised light absorption in arthropod compound eyes. Vision Res 9:1031–1040

Shaw SR (1979) Signal transmission by slow potentials in the arthropod peripheral visual system. In: Schmitt FO, Wordern FG (eds) The neurosciences: 4th study programme. MIT, Cambridge, Mass, pp 275–295

Shaw SR (1981) Anatomy and physiology of identified non-spiking cells in the photoreceptor-lamina complex of the compound eye of insects, especially Diptera. In: Roberts A, Bush BMH (eds) Neurons without impulses. Univ Press, Cambridge, pp 61–116

Shaw SR (1984) Early visual processing in insects. J Exp Biol 112:225–251

Siegler M (1985) Nonspiking interneurons and motor control in insects. Adv Insect Physiol 18:249–304

Snyder AW (1979) The physics of vision in compound eyes. In: Autrum H (ed) Handbook of sensory physiology vol VII/6A. Springer, Berlin Heidelberg New York, pp 225–314

Snyder AW, Laughlin SB, Stavenga DG (1977) Information capacity of eyes. Vision Res 17:1163–1175

Srinivasan MV, Bernard GD (1975) The effect of motion on visual acuity of the compound eye: a theoretical analysis. Vision Res 15:515–525

Srinivasan MV, Laughlin SB, Dubs A (1982) Predictive coding: a fresh view of inhibition in the retina. Proc R Soc London Ser B 216:427–459

Strausfeld NJ, Nässel DR (1981) Neuroarchitectures serving compound eyes of crustacea and insects. In: Autrum H (ed) Handbook of sensory physiology, vol VII/6B. Springer, Berlin Heidelberg New York, pp 1–32

Zettler F (1969) Die Abhängigkeit des Übertragungsverhaltens von Frequenz und Adaptationszustand, gemessen am einzelnen Lichtrezeptor von *Calliphora erythrocephala*. Z Vergl Physiol 64:432–449

Zettler F, Järvilehto M (1972) Lateral inhibition in an insect eye. Z Vergl Physiol 76:233–244

Chapter 12

Neurotransmitters in Compound Eyes

ROGER C. HARDIE, Cambridge, England

1 Introduction

Although chemical transmission had been suggested as early as 1848 by Du
Bois-Reymond, it was 1904 before Elliot made the specific suggestion that
adrenaline was a chemical mediator released by sympathetic nerve endings, and the
concept of a chemical synapse did not become firmly established until the 1930's.
It is thus no wonder that Exner did not consider the role of neurotransmitters in his
treatise on the compound eye, and indeed, except, perhaps, with reference to the
problem of the mechanisms of light- and dark-adaptation, neurotransmission was
not relevant to his study. In recent years, however, neuropharmacology has become
such an important part of any neurobiological study that it is relevant to include a
survey of neurotransmitters in the optic lobes in the present volume.

   Why should we be interested in neurotransmitters? There are at least four
major reasons: (a) first, simply for the intrinsic pharmacological interest, e.g. the
identity of transmitters, nature of the receptors, mechanisms of release and
inactivation, coupling of receptors to response (be it directly via channels or via
second messenger systems), and the question of co-localization of two or more
neurotransmitters in one neuron; (b) secondly, histochemical techniques — particularly those employing antibodies — apart from providing evidence for the
nature of the neurotransmitter in specific neurons, can also provide very useful
anatomical information in their own right: specific subsets of neurons are stained,
often identifying neuronal strata, tracts or even novel neurons which had not been
identified by other techniques (see e.g. the serotonergic cells described later). Such
stains have also often been used with success for following the development of
identified populations of neurons during embryonic and larval life (e.g. Nässel et
al. 1987; Budnik and White 1988); (c) thirdly, a knowledge of the neurotransmitters in identified pathways allows us to test theories of neural processing by
using selective antagonists to block specific pathways (e.g. Autrum and Hoffman
1957; Bülthoff and Bülthoff 1987); (d) finally, there are some practical implications: in vertebrate studies we usually think in terms of clinical spin-offs, whilst for
invertebrates in terms of pesticides.

   The first pharmacological study of the insect optic lobes appears to have been
an investigation of the effects of some neurotoxins on the electroretinogram (ERG)
(Autrum and Hoffmann 1957). Both nicotine and picrotoxin had the effect of
inhibiting the ERG transients in *Calliphora*, thus supporting the hypothesis that

Stavenga/Hardie (Eds.) Facets of Vision
© Springer-Verlag Berlin Heidelberg 1989

these fast components of the ERG were derived from postsynaptic elements (see Burkhardt this Vol.). Notwithstanding this early physiological approach, the development of the field owes largely to the implementation of various histo-chemical methods (e.g. Elofsson et al. 1966; Lee et al. 1973; revs. Klemm 1976; Nässel 1987). In recent years there has been a minor explosion of papers exploiting the availability of antibodies raised against either putative transmitters themselves, or enzymes involved in their metabolism (rev. Nässel 1987). In addition, more classical biochemical assays have been used to detect the presence and synthesis of putative transmitters (e.g. Maxwell et al. 1978; Evans 1978) and receptor-binding studies have identified potential postsynaptic sites (e.g. Schmidt-Nielsen et al. 1977; Scheidler et al. 1986). Evidence for transmitter identity has also been sought by autoradiographic uptake studies, however, with limited success in optic lobe tissue (Campos-Ortega 1974; Buchner and Rodrigues 1984). Useful information has also been obtained from *Drosophila* mutants with deficiencies in enzymes involved in transmitter metabolism (rev. Hall 1982). Only a very few studies have investigated the effects of exogenously applied agonists or antagonists on phys-iological responses (e.g. Zimmermann 1978; Bülthoff and Bülthoff 1987; Hardie 1987).

2 Biogenic Amines

The biogenic amines include the catecholamines (noradrenaline, dopamine and adrenaline), the phenolamines (octopamine and tyramine), the indolealkyl-amine (serotonin or 5-HT) and the imidazole-amine (histamine). The evidence for 5-HT and histamine will be reviewed first, separately, since the information is much clearer, whilst the other amines, for which our knowledge in the optic lobes is still very sketchy, will be considered together.

2.1 Serotonin (5-HT)

The description of optic lobe neurons showing immunoreactivity to 5-HT antisera is better than for any other putative transmitter antiserum. This is thanks largely to the availability of reliable antibodies to 5-HT (Steinbusch et al. 1978) and the fact that only a very limited number of 5-HT immunoreactive (5-HTi) neurons are found (e.g. 24 in the optic lobes of the fly) so that the anatomical details of individual cells can be clearly resolved. Nässel (1987) has written an excellent review of the subject, so only a few points will be highlighted here. For the historical record, the Falck-Hillarp method and the glyoxylic acid reaction (see below) have also been used to demonstrate 5-HT-containing neurons (rev. Klemm 1976); however, the immunocytochemical data are so far superior that the former results will not be further considered here.

In all insects yet studied (22 spp., Nässel 1987) 5-HTi neurons are found in all neuropile regions of the optic lobes. In general, 5-HTi neurons are characterized, firstly by their paucity and secondly by extensive arborizations, typically in several

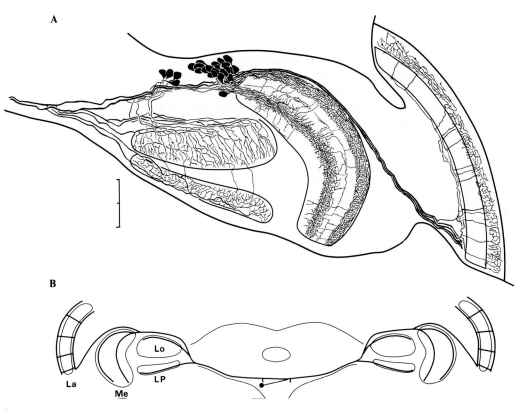

A

B

La   Me   Lo   LP

Fig. 1a,b. 5-HT immunoreactive neurons in optic lobes of the fleshfly, *Sarcophaga. a* Tracings in horizontal section. Two groups of cell bodies have processes invading the medulla and lobula respectively. The lamina is innervated only by the LBO-5-HT neurons, whose cell bodies lie in the caudal midbrain. Scale bar: 100 μm. *b* Schematic tracing of one of the two LBO-5-HT neurons, with arborizations in every neuropile region (lamina, medulla, lobula and lobula plate) of both optic lobes. (Nässel et al. 1987)

neuropiles. This is dramatically exemplified by the so-called LBO-5-HT neurons in fly optic lobes (Fig. 1). Each of two such neurons has its cell body in the protocerebrum, and projects bilaterally to every optic lobe neuropile (lobula, lobula plate, medulla and lamina) where its arborizations cover the entire retinotopic field (Nässel et al. 1987).[1]

This example demonstrates the anatomical impact of immunocytochemical techniques, since despite exhaustive anatomical studies in the fly, particularly of the lamina (rev. Strausfeld and Nässel 1981), this cell class remained unrecognized until the implementation of the 5-HT antibody.

The varicose terminals of the LBO-5-HT neurons reside in the most distal part of the lamina in a non-synaptic layer occupied only by receptor axons, monopolar

---

[1]The neuron is actually the same cell which was originally designated as tan3 and which was believed only to project to the medulla and lamina (Nässel et al. 1983).

cell bodies and glial cells. Ultrastructural studies show no indication of synaptic specializations, ns, although the varicosities contain large granular vesicles and small clear vesicles, both of which are immunoreactive to 5-HT (Nässel et al. 1985). In other insects, e.g. *Apis*, 5-HTi neurons in the lamina do in fact form synapses with, as yet, unidentified interneurons. Again, both types of vesicle are present (Nässel et al. 1985). The 5-HTi neurons in the bee lamina are restricted to the most proximal synaptic layer (epl-C), where they overlap extensively with GABAi neurons. Double stainings have demonstrated that the 5-HTi and GABAi neurons are in fact distinct (Fig. 7; Schäfer and Bicker 1986a).

The medulla typically shows several tangential 5-HTi layers ranging from one in *Apis* (Schürmann and Klemm 1984), three in the fly, six in the locust (Nässel and Klemm 1983) to eight layers in the dragonfly (Nässel 1987). Most of these are large neurons connecting to the brain, but amacrine neurons may also be present.

The lobula complex also shows extensive 5-HT immunoreactivity in all species studied but in a more diffuse pattern than the clearly stratified medulla pattern.

Very little is known about the function of serotonergic neurons in the optic lobes. No recordings have been reported from any of them, and there are very few reports of the effects of 5-HT on any aspect of visual function in the compound eye. Nevertheless, the striking morphology of 5-HTi neurons has inevitably led to speculation that they may in some way be involved in modulating overall sensitivity or activity related, perhaps, to arousal states or circadian activity (e.g. Nässel et al. 1985). In other invertebrates (e.g. the molluscs *Aplysia* and *Hermissenda*) evidence does in fact implicate 5-HT in a circadian rythm of visual sensitivity (Eskin and Maresh 1982; Crow and Bridge 1985). 5-HT has also been reported to modulate the sensitivity of *Limulus* photoreceptors (Barlow et al. 1977); however, efferent control of sensitivity in this species is now believed to be effected via octopamine (see Barlow et al. this Vol.).

2.2 Histamine (HA)

Histamine is one of the more rarely encountered neurotransmitter candidates. Only in so-called C2 cells in *Aplysia* is its role as a neurotransmitter clearly established (e.g. McCaman and Weinreich 1985), although it is strongly suspected of being a neurotransmitter or neuromodulator in vertebrate CNS (revs. Prell and Green 1986; Schwartz et al. 1986). It is therefore of considerable interest that Elias and Evans (1983) demonstrated the presence of large amounts of histamine in the retinae of *Locusta*, *Periplaneta* and *Manduca*. The amount of HA measured in the locust retina (736 pmol per eye) would actually be equivalent to a concentration of ca. 2 mM were it to be restricted to the photoreceptor cytoplasm. Significant quantities of HA were also found in the lamina and minor amounts in other regions of the optic lobes (Table 1). In the same study Elias and Evans (1983) also demonstrated the ability of retina and other optic lobe tissue both to synthesize histamine (from histidine) and to metabolize histamine (to N-acetyl histamine and imidazole-4-acetic acid). Maxwell et al. (1978) have also shown the ability of optic lobe tissue in *Manduca* to synthesize and store histamine.

Further attempts to characterize the distribution and pharmacology of HA in the locust via autoradiographic binding and uptake studies (Elias and Evans 1984; Elias et al. 1984) met with limited success. Although radioactive mepyramine (a ligand supposedly specific for the so-called H1 histamine receptor in vertebrates) was found to bind, for example, to lamina neuropile, the binding was not strictly specific to histamine receptors (possibly cross-reacting with octopamine sites). Uptake of radioactive HA was found only around the borders of the medulla and the lobula and not in any neuropile region.

The high concentration of histamine in the retina suggests a role as the neurotransmitter of the photoreceptors since, with the exception of sensory cells innervating corneal hairs in some species, these are the only neurons present in the retina. This suggestion receives strong support from a recent study in which the effects of a wide variety of transmitter candidates were tested on the intracellularly recorded responses of cells directly postsynaptic to the photoreceptors, the large monopolar cells (LMC's), in the fly lamina (Hardie 1987). After testing virtually all the "classical" neurotransmitter candidates (which were applied via ionophoresis from a multi-barrelled electrode glued onto the recording micropipette), only histamine was found to mimic the action of the natural transmitter. Like the response to light, the response to an ionophoretic pulse of histamine is a fast hyperpolarization associated with a conductance increase to chloride ions (Fig. 2). The LMC's are very sensitive to histamine, saturating in response to pulses of as little as 1nC (which corresponds to ca. 10^9 molecules). Similar results have now also be obtained from the second order L-neurons of the locust ocellus, suggesting that histamine may also be the neurotransmitter of ocellar photoreceptors (Simmons and Hardie 1988).

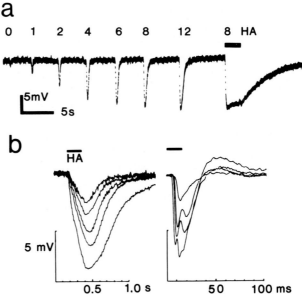

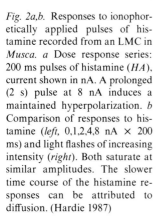

Fig. 2a,b. Responses to ionophoretically applied pulses of histamine recorded from an LMC in *Musca. a* Dose response series: 200 ms pulses of histamine (*HA*), current shown in nA. A prolonged (2 s) pulse at 8 nA induces a maintained hyperpolarization. *b* Comparison of responses to histamine (*left*, 0,1,2,4,8 nA × 200 ms) and light flashes of increasing intensity (*right*). Both saturate at similar amplitudes. The slower time course of the histamine responses can be attributed to diffusion. (Hardie 1987)

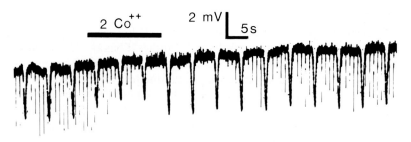

Fig. 3. Responses to light (rapid downward deflections) and ionophoretic histamine pulses (8 nA 0.5 s, delivered every 6 s) during a synaptic blockade induced by ionophoresis of cobalt ions (2 nA). Whereas the responses to light are abolished, the responses to histamine are only slightly affected. (Hardie 1987)

If histamine is the photoreceptor neurotransmitter, one would predict that the effects of histamine on the LMC's should be direct (rather than being mediated via other neurons). In order to test this, a general synaptic blockade was induced using cobalt ions (which block the presynaptic calcium channels). Under these conditions, although the responses to light were blocked, the responses to ionophoretic pulses of histamine were largely unchanged (Fig.3).

A number of drugs have been found which antagonize both the effects of exogenously applied histamine and also the responses to light (Hardie 1987, 1988a,b). Whilst these included known histamine antagonists, the most effective drugs (namely: hexamethonium, benzoquinonium and gallamine) were those which are usually considered as cholinergic agents. The histamine-sensitive receptors in the insect lamina may thus represent a novel class of receptor.

Most recently, specific immunolabelling of fly photoreceptors has been demonstrated using antibodies raised against histamine conjugates (Nässel et al. 1988). The axons of photoreceptors R1-6, and also one of the long visual fibres (R8) were stained along with the photoreceptors of the ocelli.

To summarize, there is evidence for the presence, synthesis and enzymatic inactivation of histamine in the insect retina and optic lobes, and immunocytochemical evidence indicates localization in the receptor terminals. Exogenously applied histamine mimics the action of the photoreceptor transmitter, and certain pharmacological agents antagonize both the response to light, and the response to ionophoretically applied histamine. There is thus a strong case for suggesting that histamine is the neurotransmitter of some insect photoreceptors; however, more evidence — particularly demonstration of release from the photoreceptor terminals — will be required to establish this firmly. Significant quantities of histamine are also present in the rest of the optic lobes, raising the possibility of further classes of histaminergic neurons (Elias and Evans 1983).

2.3 Catecholamines and Phenolamines

The distribution of catecholamines (CA) in the optic lobes has largely been inferred from histochemical techniques exploiting the fluorescence of aldehyde derivatives (Falck-Hillarp method) or the glyoxylic acid reaction product (Falck et al. 1962;

Bjorklund et al. 1972). In principle the identity of the original amine can be inferred from the fluorescence spectra. Currently, there is no reliable histochemical test for octopamine,[2] although radioactive binding studies have indicated a characteristic pattern of putative octopamine receptors in bee medulla and lobula (Scheidler et al. 1986). More recently, antibodies raised against dopamine and also enzymes involved in catecholamine synthesis have been used, although the specificity has not been clearly established. In addition, direct biochemical assays have been made of the concentrations of various amines (including octopamine) in different compartments of the optic lobes (Table 1). Some data are also available from *Drosophila* mutants deficient in catecholamine metabolism (rev. Hall 1982).

Typically, in many species, both the Falck-Hillarp method and the glyoxylic acid reaction reveal prominent tangential layers in the medulla, the dopamine fluorophore in particular being implicated. Less ordered marking is found in the lobula complex, and also the lamina of some species. Conspicuously, of all species

Table 1. Quantities of putative neurotransmitters in optic lobe tissue. All values expressed as pmol/structure

a) Whole optic lobes (one side minus retina)							
	OA	DA	NA	5-HT	HA	GABA	ACh
Schistocerca	8.5				23		
Periplaneta	3.6	1.2	0.2		62.5		
Apis	2.5	7.6	0.5	9.7			
Calliphora	1.6	0.9		0.8			
Manduca					25	12.7	850

b) Compartments	OA	DA	NA	5-HT	HA
(i) *Schistocerca*					
retina	0.33				736
lamina	0.34				19
medulla	6.1				2.3
lobula	2.0				0.9
(ii) *Periplaneta*					
retina	0.6				2525
lamina	0.35				
medulla	1.9				
lobula	0.75				
(iii) *Apis*					
retina	0.23	5.1	nd	3.0	
lamina	0.45	0.44	nd	1.4	
medulla	1.07	nd	nd	1.5	
lobula	0.53	0.31	0.25	2.6	

Data from: Evans (1978); Dymond and Evans (1979); Elias and Evans (1983) Mercer et al. (1983); Kingan and Hildebrand (1985); Nässel and Laxmyr (1983). nd — none detected.

---

[2] Actually the dye neutral red should stain all amine-containing cells, and thus cells which stain with neutral red, but do not fluoresce with the Falck-Hillarp or glyoxylic acid methods, may contain octopamine (e.g. Evans 1985).

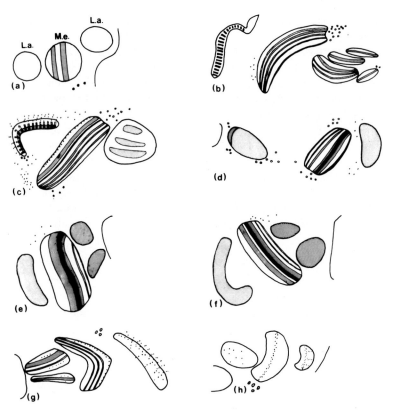

Fig. 4a-h. Schematic representation of the distribution of amine containing neuropile and cell bodies in insect optic lobes. *Stippled areas* presumed 5-HT fluorescence (yellow); *hatched and black areas* presumed catecholamine (DA or NA) fluorescence (green); *black areas* being particularly intense. *Open circles* 5-HT cell bodies; *filled circles* CA cell bodies. *a*) Silverfish (*Lepisma*); *b* Dragonfly nymph (*Aeshna*); *c* locust (*Schistocerca*); *d* cricket (*Acheta*); *e* caddisfly (*Limnephilidae*); *f* Noctuid moth (*Spodoptera*); *g* fly (*Calliphora*); *h* honeybee (*Apis*). (Klemm 1976)

investigated, only the bee showed no CA fluorescence in the optic lobes (Klemm 1976), although both 5-HT and octopamine are present (Mercer et al. 1983; Table 1). Figure 4 (from Klemm 1976) gives an idea of the variety reported. It has not yet been possible to identify any of the stained neurons, although presumably tangential and amacrine neurons are represented.

Unfortunately, recent results with newly developed antibodies have not clarified the issue. For example, a dopamine antibody gave a staining pattern that only partially overlapped with the Falck-Hillarp method in the locust optic lobes (Vieillemaringe et al. 1984), in particular failing to stain the lamina and medulla neuropile, which show the spectral characteristics of the dopamine fluorophore with the Falck-Hillarp method (Elofsson and Klemm 1972). An antibody raised against dopamine-β-hydroxylase (required for noradrenaline synthesis) stained a total of ca. 60 cell bodies in the optic lobes of *Calliphora*, none of which, however, fluoresce with the Falck-Hillarp method (Klemm et al. 1985). Note however, that in *Drosophila*, several hundred medullary cell bodies show CA fluorescence with the glyoxylic acid method (Budnik and White 1988).

The picture is thus rather unclear, at best one can say that there appear to be a limited number of catecholaminergic neurons in the optic lobes (maybe several hundred in the fly, for example) but none has been correlated with identified neurons, and in most cases there is also doubt as to the identity of the amine. Although octopamine is present in the optic lobes of all species investigated (see Table 1) there is virtually no indication of its detailed distribution.

There is also very little information as to the possible function of aminergic neurons in the optic lobes; however, some intriguing results have come from studies of *Drosophila* mutants with deficient catecholamine metabolism (revs. Hall 1982; Wright 1987). Two mutants in particular, *tan* and *ebony*, lack both "on" and "off" transients in the electroretinogram (ERG). Mosaic studies (Hotta and Benzer 1970) also showed that the transients were lacking when the mutation was restricted to the retina. Biochemically, the *tan* mutant is associated with low levels of dopamine and β-alanine (Konopka 1972; Wright 1987) due to a lack of β-alanyl dopamine hydroxylase activity, whilst the *ebony* mutant has raised levels of these two substances. Since the ERG transients are derived from cells postsynaptic to the photoreceptors (Autrum and Hoffmann 1957; Heisenberg 1971; Coombe 1986; Burkhardt this Vol.), these results have led to the suggestion that dopamine, for example, could be the photoreceptor neurotransmitter (Hall 1982). However, in the larger flies neither dopamine nor β-alanine, when ionophoretically applied, had any effect on the LMC's (Hardie 1987). Furthermore, neither the retina nor the lamina show any CA fluorescence in either *Calliphora* (Klemm 1976) or *Drosophila* (Budnik and White 1988). In view of the finding that histamine can mimic the photoreceptor neurotransmitter (Hardie 1987, see above) it would be of interest to see if these *Drosophila* mutants also have abnormal levels of histamine.

In crustaceans, the migration of the distal retinal pigment is controlled by antagonistic dark- and light-adapting hormones (rev. Rao 1985; see also Stavenga this Vol.). Recently evidence has been obtained implicating DA in the release of the dark-adapting, and NA in the release of the light-adapting hormone (Kulkarni and Fingerman 1986).

3 γ-Amino Butyric Acid (GABA)

GABA is widely believed to be the inhibitory neurotransmitter at the arthropod neuromuscular junction and a number of studies suggest that it has a similar role in the insect CNS (Pitman 1971; rev. Usherwood 1978).

Several studies have shown the presence (Kingan and Hildebrand 1985), synthesis (Maxwell et al. 1978), or uptake (Campos-Ortega 1974) of GABA by insect optic lobe tissue; however, the most detailed information available comes from a number of recent investigations using apparently reliable antibodies raised against GABA conjugates. Data are available from the flies *Musca* and *Calliphora* (Meyer et al. 1986; Datum et al. 1986), the bee, *Apis* (Schäfer and Bicker 1986b;

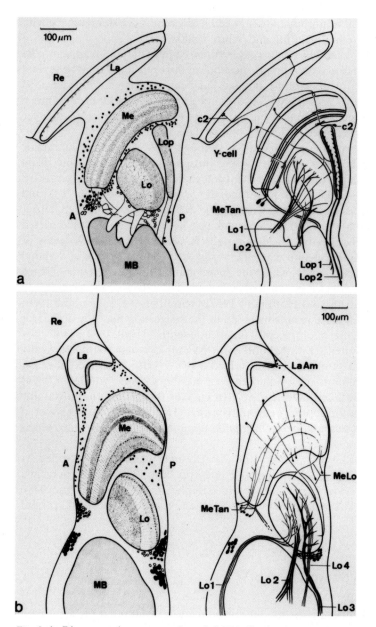

Fig. 5a,b. Diagrammatic representation of GABA-like immunoreactivity in the optic lobes of *a Calliphora* and *b Apis. Left.* cell bodies and immunoreactive layers; *right.* representative neurons. *a c2* columnar centrifugal cells connecting medulla and lamina; *Y-cell* columnar neurons connecting medulla with lobula plate and lobula; *MeTan* large tangential neurons running to mid-brain; *Lo1, Lo2* tracts containing thin columnar neurons and a smaller number of large lobula neurons; *Lop1* possibly corresponding to CH cells; *Lop2* contralaterally projecting neurons of lobula plate. *b* Amacrines (*LaAm*) of the lamina epl-C layer; *MeLo* columnar neurons connecting medulla and lobula; *MeTan* large tangential medullary neurons with possible connections to mid-brain; *Lo1* ca. six thin neurons with possible contralateral connections; *Lo2* thicker neurons of inner lobula layers connecting to the mid-brain; *Lo3* and *Lo4* outer lobula neurons connecting to the mid-brain. (Meyer et al. 1986)

Meyer et al. 1986) and the moth, *Manduca* (Homberg et al. 1987). Conveniently the results are sufficiently similar to be treated together (see also Fig. 5). Approximately 5% of the neurons in the optic lobes are GABA immunoreactive (GABAi), with ca. 9000 cell bodies counted in the bee, 10,000 in the fly and 18,000 in the moth (figures for one side only).

In each species the lamina contains only one or two classes of GABAi cells. In both moth and fly these are columnar centrifugal elements: the so-called C2 cell in fly (Fig. 6), and probably the t1 cell (after Strausfeld and Blest 1970) in the moth. In the bee the most proximal layer of the lamina (eplC) contains a set of tangential fibres originating from a small group of cell bodies in the outer and inner chiasm (Fig. 7). Datum et al. (1986) also report that one of the photoreceptors (the R7 cell, which projects to the medulla) reacts with anti-GABA and anti-GAD, although this was not confirmed in an independent study (Meyer et al. 1986).

In the medulla there is a more extensive staining pattern organized in tangential layers (9 in the bee, 7 in *Manduca* and 4 in the fly). These correspond to the arborizations of a combination of tangential, amacrine and columnar elements (including some connecting the medulla to the lobula and lobula plate in the fly). Note that if the cell body counts are reliable, then there is only scope for the involvement of maybe two strictly periodic columnar elements, one of which in the fly is the centrifugal C2 cell.

Extensive staining is also found in the lobula complex. Apart from a number of intrinsic elements (possibly amacrines) a limited number of giant tangential fibres are found in all three species. These have been tentatively identified as the CH cells and the V2 or V3 cells (after Hausen 1981) in the fly lobula plate (Meyer et al. 1986), and one of the fibres in the bee lobula is reminiscent of the HR (horizontal regressive) neuron (after DeVoe et al. 1982).

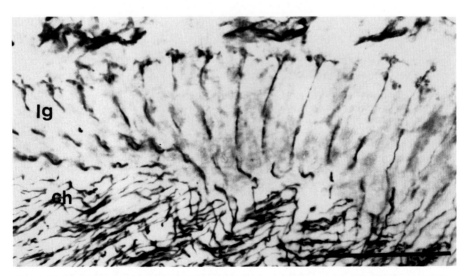

Fig. 6. Horizontal section through *Calliphora* lamina (*lg*), stained with an antibody against a GABA synthetic enzyme, GAD. C2 terminals arising from fibres in the chiasm (*ch*) are labelled in practically every cartridge (Datum et al. 1986)

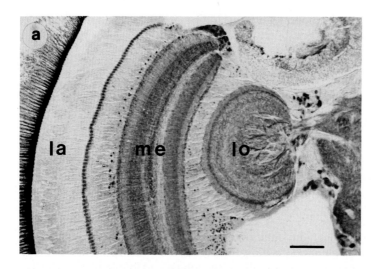

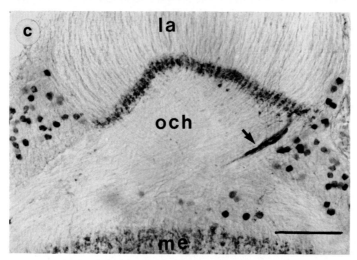

Fig. 7. a Frontal section through optic lobes of honeybee showing the stratification of GABA-like immunoreactivity in the lamina (*la*), medulla (*me*) and lobula (*lo*). Labelled cell bodies are seen in and around the chiasmata. *b* Comparable section showing 5-HTi-like immunoreactivity *c* Horizontal section through outer chiasm (*och*). Labelled GABAi fibres (*arrowhead*) approach the lamina. Their arborizations are confined to the most proximal stratum of the lamina (epl-C). Numerous cell bodies are also visible *d* Comparable section showing 5-HT immunoreactivity. Despite the similarity to the GABAi fibres, the 5HTi fibres are different neurons. Scale bars: (1,2) 100 μm; (3,4) 50 μm. (Schäfer and Bicker 1986b)

It is generally assumed that GABA's role in the optic lobes is inhibitory (i.e. causing hyperpolarizations via chloride or potassium conductance increases), but there is no hard evidence for this. In the lamina neuropile of the flies *Musca* and *Calliphora*, ionophoretic application of GABA actually leads to a depolarization of

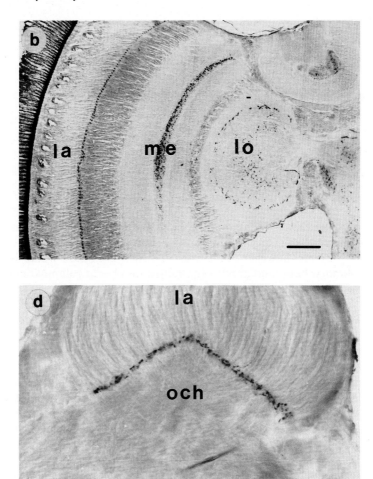

the LMC's (Fig. 8, Hardie 1987), a conclusion reached on indirect evidence by Zimmerman (1978). Since the response of the LMC's to light increments is actually a hyperpolarization, this can still be thought of as an inhibitory action, and indeed responses to light flashes are also reduced by the action of GABA. The LMC's response to GABA is associated with a conductance increase to ion(s) unknown (but almost certainly not chloride).

In a series of experiments, Bülthoff and co-workers have examined the effects of picrotoxin on motion-sensitive neurons and behavior in *Drosophila* and *Calliphora*. In general, directional selectivity of the H1 neuron and also the optomotor turning reaction is abolished and even transitorily reversed (Bülthoff and Schmid 1983; Bülthoff and Bülthoff 1987). Picrotoxin is known to block inhibitory chloride channels (especially, though not exclusively, those activated by GABA), and

10 GABA

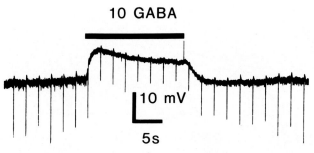

Fig. 8. Responses to ionophoretically applied GABA in an LMC recorded in *Musca*. Prolonged (16 s) ionophoresis of GABA (10 nA) causes a small depolarization and a reduction in both membrane noise and the response to light (rapid downward deflections). (Hardie 1987)

therefore the results support models of motion detection where directionality is conferred by a shunting inhibition (e.g. Barlow and Levick 1965; Torre and Poggio 1978). Note, however, that in the fly, such inhibition would be required on a columnar basis, and only a very limited number of columnar neurons (including the C2 cell), were found to react with the GABA antibody.

4 Acetylcholine (ACh)

As in the rest of the insect CNS (rev. Sattelle 1985), ACh is likely to be a major transmitter in the optic lobes. For example, in the optic lobes of *Manduca*, ca. 40x more ACh was synthesized and stored than any other transmitter candidate tested (Maxwell et al. 1978). The nicotinic ligand, α-bungarotoxin, binds strongly to all optic lobe tissue except the lamina in both *Drosophila* (Schmidt-Nielsen et al. 1977) and *Manduca* (Maxwell and Hildebrand 1981). Enzyme histochemistry indicates the presence of cholinesterase in all optic lobe tissue in *Drosophila* (Greenspan et al. 1980) and also crickets (Lee et al. 1973), though this is not necessarily specific for acetylcholinesterase.

Nevertheless, evidence for specific, identified cholinergic pathways in the optic lobes is very sparse. What we do know comes mainly from studies in *Drosophila*, either using immunocytochemistry for choline acetyl transferase (ChAT) antibodies (Salvaterra et al. 1983; Buchner et al. 1986; Gorczyca and Hall 1987) or mutants deficient in either synthetic (ChAT) or degradative (acetylcholinesterase = AChE) enzymes.

The ChAT antibody stains several layers in both the medulla and lobula of *Drosophila* together with a rather diffuse staining of the lobula plate. Stained neurons probably include both columnar and tangential or amacrine type elements. Staining in the lamina is less extensive. Buchner et al. (1986) describe only one structure, repeated in every cartridge, which was "highly reminiscent of C2" – a centrifugal element with cell body and further arborizations in the medulla[3].

---

[3]The homologous cell in the larger flies, *Calliphora* and *Musca* stains with antibodies to GABA (Meyer et al. 1986; see above). Whether this represents a species difference, a unique example of colocalization of ACh and GABA, or simply highlights the problems of immunocytochemical evidence remains to be seen.

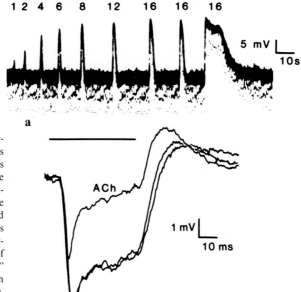

Fig. 9a,b. Responses to ionophoretically applied ACh in LMC's recorded in *Musca. a* 1 s pulses (current indicated in na) cause large depolarizations. A saturating dose (16 na) also reduces the response to light (downward deflections). *b* Response to a 45 ms light flash in another LMC. During continued ionophoresis of ACH (subsaturating) the "off" transient is enhanced and both transients become more phasic. (Hardie 1987)

Gorczyca and Hall (1987) report that lamina interneuron cell bodies were stained in addition to a centrifugal element and both they and Salvaterra et al. (1983) interpret the immunoreactive layers in the medulla as the terminals of the large monopolar cells, L1 and L2. Note, however, that the LMC's in the bee show glutamate-like immunoreactivity (Schäfer 1987; see Sect. 5).

Further evidence for a major class of cholinergic neurons in the lamina comes from studies on *Drosophila ChAT* and *AChE* mutants. Mosaics in which the mutation is restricted to the lamina have abnormal ERG's, in particular the "off" transient is reduced or absent (Greenspan et al. 1980; Greenspan 1980). A similar defect in the ERG can also be induced by feeding flies the AChE inhibitor neostigmine (Dudai 1980). *AChE* mosaics with mutant tissue in either the lamina or the lobula plate are also often optomotor blind (Greenspan et al. 1980).

Ionophoretic studies in the lamina of larger flies (*Musca*) show that ACh has a pronounced effect on the large monopolar cells, resulting in a conductance increase and depolarization of the cells − possibly consistent with a role in generating the depolarizing "off" transient (Hardie 1987). Unlike GABA (which also depolarizes the LMC's), ACh had the effect of increasing the size of the "off" transient and making both transients more phasic (Fig. 9), thus mimicking light adaptation (Laughlin and Hardie 1978). The mechanism for these actions is unknown. Note that the putative receptors in the lamina may be of a different class to ACh receptors in the rest of the optic lobes, since the lamina shows virtually no specific binding of the nicotinic ligand α-bungarotoxin (Schmidt-Nielsen et al. 1977).

Evidence for cholinergic function in the lamina also exists in *Manduca*: developmental manipulations leading to only vestigial lamina tissue (sectioning the stemmatal nerve in the pupal stage) result in a greatly reduced capacity of the optic lobes as a whole to synthesise and store ACh (Maxwell and Hildebrand 1981). These authors also interpret the binding pattern of radioactive α-bungarotoxin in the medulla as corresponding to the terminals of the laminar interneurons.

5 Glutamate

Although glutamate is the best-characterized neurotransmitter in arthropods, being generally accepted as the excitatory transmitter at the neuromuscular junction (rev. Usherwood 1978), very little is known about its function in the CNS. One reason for this has been the lack, until very recently, of antibodies against glutamate, another is that glutamate is a ubiquitous metabolite with multiple pathways for synthesis and metabolism (e.g. it is directly used for protein synthesis, and is the major precursor for GABA) so that, alone, the presence of glutamate or associated enzymes is poor evidence of neurotransmitter function.

Nevertheless, an attempt has recently been made in *Drosophila* to identify neural structures with possible glutamatergic function (Chase and Kankel 1987). A variety of techniques were used including radioactive glutamate binding, enzyme histochemistry for a range of key enzymes and a genetic analysis of one of these (glutamate oxaloacetic transaminase = GOT, a major glutamate synthetic enzyme) which was found to have a differential distribution within the CNS.

As far as the optic lobes are concerned, the most significant findings were: (a) GOT activity (determined via enzyme histochemistry) was more intense in the lamina than in virtually any other tissue; (b) a second enzyme (glutamate dehydrogenase) was also strongly active in both retina and lamina; (c) double mutants (*GOT-1/GOT-2*) lacking both forms of GOT had abnormal ERG's, in particular both "on" and "off" transients were reduced or absent. Although this suggests the possible involvement of glutamate in normal visual function, as the authors point out, the disruption of neural function may not necessarily be caused by the lack of glutamate per se, but, for example, by the concomitant reduction in the GABA pool.

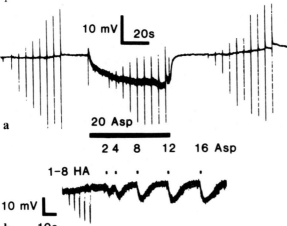

Fig. 10. a Prolonged iono- phoresis (20 na) of aspartate induces a noisy hyperpola- rization in an LMC (*Musca*). Responses to light flashes of increasing intensity are also monitored, showing a reduc- tion in amplitude during aspartate application. *b* Comparison of responses to histamine and L-aspartate applied from different barrels of the same ionophoretic pipette. Responses to aspar- tate are much slower and require greater doses. HA pulses 0.5 s, aspartate pulses 1 s long. (Hardie 1987)

Another line of evidence implicating glutamate as a lamina neurotransmitter comes from ionophoretic studies in the housefly lamina (Fig. 10; Hardie 1987). Application of either glutamate or aspartate hyperpolarized the LMC's. Compared to the responses to the putative photoreceptor transmitter, histamine, however, the responses were slower and smaller (for similar doses). Furthermore, the responses were typically associated with large noise levels and could be blocked by cobalt ions, implying that the effect is not on the LMC directly but via another neuron or neurons. Interestingly, hyperpolarizations were also observed when the recording electrode was in the extracellular space (e.c.s.) of the lamina cartridge. Such a hyperpolarization of the e.c.s. would be expected to result in neurotransmitter release from the photoreceptors, since they would then become effectively depolarized, thus possibly accounting for the responses observed in the LMC's.

Which neurons in the lamina might actually use glutamate as a neurotransmitter remains an open question. A speculative idea is that the responses just referred to might derive from synaptic receptors on the glial cells, in which case amacrine and/or so-called T1 cells would come into question as the putative glutamatergic neurons since these have been reported to form synapses with the glial cells (rev. Shaw 1984). Most recently, Schäfer (1987) has found that the large monopolar cells of the bee lamina, and also the second order L-neurons of the ocelli, react with antibodies raised against glutamate.

As a general comment it should be pointed out that, although optic lobe tissue has been screened with antibodies to the majority of the known "classical" neurotransmitter candidates, only a relatively small proportion of the cells have been accounted for. By far the largest number of cells were counted with GABA antibodies, but even these account for only ca. 5% of the cells in the optic lobes, whilst the various amines account for only a handful of neurons (maybe a few hundred not counting the photoreceptors). Unless a completely unexpected neurotransmitter plays a major role in the optic lobes, one can predict that a significant proportion of the neurons not yet accounted for will turn out to be either cholinergic or glutamatergic.

6 Miscellaneous

a) Taurine. Although detected in many insect tissues (e.g. Whitton et al. 1987) the status of taurine as a neurotransmitter is still dubious. Nevertheless, ionophoretically applied taurine has been shown to have inhibitory actions in insect CNS (e.g. Hue et al. 1979; Wafford and Sattelle 1986).

High levels of taurine have been reported in the retinae of both the locust (7μmol/gm wet weight: Whitton et al. 1987) and also the blowfly, where it is the most abundant free amino acid (29% of the total free amino-acid pool: Langer et al. 1976). Recently Schäfer et al. (1988) found that all photoreceptors in both the compound eyes and the ocelli of the bee are immunoreactive to an antibody raised against taurine, and suggested that taurine might be the photoreceptor neurotransmitter. However, my own experiments on the housefly failed to show any effect of taurine when ionophoresed onto the postsynaptic elements, the LMC's

(Hardie 1987). Taurine is actually very abundant in nervous tissue in general (e.g. Oja and Kontro 1983) and in many cases some function other than that of a neurotransmitter may be indicated.

b) Peptides. At present there are only a few reports of peptide immunoreactivity from insect optic lobes. White et al. (1986) describe FMRFamide-like immuno-reactive fibres in the medulla and lobula of *Drosophila*, though there were no cell bodies in the optic lobes. Schoofs et al. (1987) describe a group of small MSH-i (melanophore stimulating hormone-immunoreactive) neurons with fine axons entering the lamina in *Locusta*. Johansson and Nässel (unpubl. cited in Nässel 1987) found several gastrin/CCK-i cell bodies distal to the medulla in *Calliphora* with processes in the medulla, other immunoreactive neurons with cell bodies in the mid-brain invade the lobula. The lobula is also invaded by processes from proctolin-like immunoreactive cell bodies at the base of the lobula (Nässel and O'Shea 1987). In crickets, both somatostatin-i and gastrin/CCK-i neurons are found with fibres in the lamina and medulla. In both species the pattern of immunoreactive fibres was broadly similar to that of 5-HTi fibres, though the cell body populations were entirely different.

More substantial peptide immunoreactivity has been reported from crus-taceans. For example, in the lobster (*Panulirus*) all the retinula cells react with an antibody to enkephalin, whilst substance P immunoreactivity is found in tangential fibres in the lamina, and also the medulla interna and externa (Mancillas et al. 1981). The lamina neurons could be tentatively identified with the neurosecretory cells described by Hamori and Horridge (1966), whilst those in the medullae corresponded to Hanström's (1924) cell types 11 and 15 respectively. In the fiddler crab, *Uca*, enkephalin-like immunoreactivity was found in retinula cells and the lamina, substance P-like immunoreactivity in all optic lobes and all retinula cells and FMRFFamide-like immunoreactivity was seen particularly in the three optic chiasmata (Fingerman et al. 1985).

No information is available for the function of neuropeptides in insect optic lobes. A primary photoreceptor transmitter function for enkephalin (or a related peptide) was suggested by Mancillas et al. (1981) for the lobster, but it may be more realistic to consider the possibility of co-localization with a more classical neu-rotransmitter. Some of the putative peptide-containing neurons in crustaceans may also be involved in modulating the release of hormones involved in pigment migration. Indeed Quackenbush and Fingerman (1984) found that met-enke-phalin stimulated the release of pigment-concentrating hormones from isolated eyestalks in *Uca*.

7 Conclusions and Outlook

Clearly the study of neurotransmitters in the optic lobes is still in its infancy, and none of the candidates considered has satisfied all the pharmacological criteria for acceptance as a neurotransmitter. Particularly conspicuous is the absence of any evidence for transmitter release — certainly one of the most crucial criteria for neurotransmitter function. In the immediate future, probably the transmitter most likely to be successfully characterized is the photoreceptor neurotransmitter, for which histamine is currently the leading candidate.

Whilst there is direct biochemical evidence for the presence, and in many cases synthesis, of most of the putative optic lobe transmitter candidates, evidence for their distribution has come largely from immunocytochemistry, and here a *caveat* is in order: whilst the GABA and 5-HT antibodies are reasonably well characterised, to a greater or lesser extent, results from other antibodies must still be treated with caution until their specificity has been corroborated by independent means. Nevertheless, in the near future we can look forward to increasingly reliable antibodies and hence a more detailed and reliable description of the distribution of different classes of immunoreactive neurons. Because of the, already, excellent anatomical description, the ease of interpretation afforded by the insect optic lobes will make them a favourite tissue for further investigations.

Finally, there are conspicuously few studies of transmitter function and receptor pharmacology in the optic lobes. Hopefully, the slowly unravelling picture afforded by immunocytochemical techniques, together with the excellent physiological and anatomical framework provided by the optic lobes, will generate an impetus for more physiological and pharmacological studies.

References

Autrum H,Hoffmann E (1957) Die Wirkung von Pikrotoxin und Nikotin auf das Retinogramm von Insekten. Z Naturforsch 12b:752–757

Barlow HB, Levick WR (1965) The mechanism of directionally sensitive units in rabbit's retina. J Physiol 178:477–504

Barlow RB jr, Chamberlain SC, Kaplan E (1977) Efferent inputs and serotonin enhance the sensitivity of the *Limulus* lateral eye. Biol Bull 153:414 (Abstr)

Bjorklund A, Lindvall O, Svensson LA (1972) Mechanisms of fluorophore formation in the histochemical glyoxylic acid method for monoamines. Histochemie 32:113–131

Buchner E, Rodrigues V (1984) Autoradiographic localisation of [^{3}H]choline uptake in the brain of *Drosophila melanogaster*. Neurosci Lett 42:25–31

Buchner E, Buchner S, Crawford G, Mason WT, Salvaterra PM, Sattelle DB (1986) Choline acetyltransferase-like immunoreactivity in the brain of *Drosophila melanogaster*. Cell Tissue Res 246:57–62

Budnik V, White K (1988) Catecholamine-containing neurons in *Drosophila melanogaster:* distribution and development. J Comp Neurol 268:400–413

Bülthoff H, Bülthoff I (1987) Combining neuropharmacology and behavior to study motion detection in flies. Biol Cybernet 55:313–320

Bülthoff H, Schmid A (1983) Neuropharmakologische Untersuchungen bewegungsempfindlicher Interneurone in der Lobula-Platte der Fliege. Verh Dtsch Zool Ges 273

Campos-Ortega JA (1974) Autoradiographic localization of ^{3}H-γ-aminobutyric acid uptake in the lamina ganglionaris of *Musca* and *Drosophila*. Z Zellforsch 147:415–431

Chase BA, Kankel DR (1987) A genetic analysis of glutamatergic function in *Drosophila*. J Neurobiol 18:15–41

Coombe PE (1986) The large monopolar cells L1 and L2 are responsible for ERG transients in *Drosophila*. J Comp Physiol A 159:655–665

Crow T, Bridge MS (1985) Serotonin modulates photoresponses in *Hermissenda* type B photoreceptors. Neurosci Lett 60:83–88

Datum K-H, Weiler R, Zettler F (1986) Immunocytochemical demonstration of γ-amino butyric acid and glutamic acid decarboxylase in R7 photoreceptors and C2 centrifugal fibres in the blowfly visual system. J Comp Physiol A 159:241–249

DeVoe RD, Kaiser W, Ohm J, Stone LS (1982) Horizontal movement detectors of honeybees: directionally selective visual neurons in the lobula and brain. J Comp Physiol A 147:155–170

Du Bois-Reymond E (1848) Untersuchungen über thierische Electricität. Reimer, Berlin

Dudai Y (1980) Cholinergic receptors of *Drosophila*. In: Sattelle DB, Hall LM, Hildebrand JG (eds) Receptors for Neurotransmitters, hormones and pheromones in insects. Elsevier, New York North Holland, pp 93–110

Dymond GR, Evans PD (1979) Biogenic amines in the nervous system of the cockroach *Periplaneta americana*: association of octopamine with mushroom bodies and dorsal unpaired median (DUM) neurons. Insect Biochem 9:535–545

Elias MS, Evans PD (1983) Histamine in the insect nervous sytem: distribution, synthesis and metabolism. J Neurochem 41:562–568

Elias MS, Evans PD (1984) Autoradiographic localization of ^{3}H-histamine accumulation by the visual system of the locust. Cell Tissue Res 238:105–112

Elias MS, Lummis SCR, Evans PD (1984) [^{3}H] Mepyramine binding sites in the optic lobes of the locust: autoradiographic and pharmacological studies. Brain Res 294:359–362

Eliot TR (1904) On the action of adrenalin. J Physiol 31:20P

Elofsson R, Klemm N (1972) Monoamine-containing neurons in the optic ganglia of crustaceans and insects. Z Zellforsch 133:475–499

Elofsson R, Kauri T, Nielsen S-O, Strömberg J-O (1966) Localization of monoaminergic neurons in the central nervous system of *Astacus astacus* (Crustacea). Z Zellforsch 74:464–473

Eskin A, Maresh RD (1982) Serotonin or electrical nerve stimulation increases the photosensitivity of the *Aplysia* eye. Comp Biochem Physiol 73C:27–31

Evans PD (1978) Octopamine distribution in the insect nervous system. J Neurochem 30:1009–1013

Evans PD (1985) Octopamine. In: Kerkut GA, Gilbert LI (eds) Comprehensive insect physiology biochemistry and pharmacology vol 9. Pergamon, Oxford New York, pp 499–530

Falck B, Hillarp NA, Thieme G, Torp A (1962) Fluorescence of catecholamines and related compounds with formaldehyde. J Histochem Cytochem 10:348–354

Fingerman M, Hanumante MM, Kulkarni GK, Ikeda R, Vacca LL (1985) Localisation of substance P-like, leucine-enkephalin-like, methionine-enkephalin-like, and FMRF-amide-like immuno-reactivity in the eyestalk of the fiddler crab, *Uca pugilator*. Cell Tissue Res 241:473–477

Gorczyca MG, Hall JC (1987) Immunohistochemical localization of choline acetyltransferase during development and in Cha[ts] mutants of *Drosophila melanogaster*. J Neurosci 7:1361–1369

Greenspan RJ (1980) Mutations of choline acetyltransferase and associated neural defects in *Drosophila melanogaster*. J Comp Physiol A 137:83–92

Greenspan RJ, Finn JA, Hall JC (1980) Acetylcholinesterase mutants in *Drosophila* and their effects on the structure and function of the central nervous system. J Comp Neurol 189:741–774

Hall JC (1982) Genetics of the nervous system in *Drosophila*. Q Rev Biophys 15:223–479

Hamori J, Horridge GA (1966) The lobster optic lamina. II Types of synapses. J Cell Sci 1:257–270

Hanström B (1924) Untersuchungen über das Gehirn, insbesondere die Sehganglien der Crustaceen. Ark Zool 16:1–119

Hardie RC (1987) Is histamine a neurotransmitter in insect photoreceptors? J Comp Physiol A 161:201–213

Hardie RC (1988a) The use of local ionophoresis to identify neurotransmitter candidates in the housefly, *Musca domestica*. J Physiol 396:7P

Hardie RC (1988b) Effects of antagonists on putative histamine receptors in the first visual neuropile of the housefly (*Musca domestica*). J Exp Biol 138:221–241

Hausen K (1981) Monocular and binocular computation of motion in the lobula plate of the fly. Verh Dtsch Zool Ges 1981:49–70

Heisenberg M (1971) Separation of receptor and lamina potentials in the electroretinogram of normal and mutant *Drosophila*. J Exp Biol 55:85–100

Homberg U, Kingan TG, Hildebrand JG (1987) Immunocytochemistry of GABA in the brain and suboesophageal ganglion of *Manduca sexta*. Cell Tissue Res 248:1–24

Hotta Y, Benzer S (1970) Genetic dissection of the *Drosophila* nervous system by means of mosaics. Proc Natl Acad Sci USA 73:4154–4158

Hue B, Pelhate M, Chanelet J (1979) Pre- and postsynaptic effects of taurine and GABA in the cockroach central nervous system. J Can Sci Neurol 6:243–250

Kingan JG, Hildebrand JG (1985) GABA in the CNS of metamorphosing and mature *Manduca sexta*. Insect Biochem 15:667–675

Klemm N (1976) Histochemistry of putative transmitter substances in the insect brain. Prog Neurobiol 7:99–169

Klemm N, Nässel DR, Osborne NN (1985) Dopamine-β-hydroxylase like immunoreactive neurons in two insect species, *Calliphora erythrocephala* and *Periplaneta americana*. Histochemistry 83:159–164

Konopka RJ (1972) Abnormal concentrations of dopamine in a *Drosophila* mutant. Nature (London) 239:281–282

Kulkarni GD, Fingerman M (1986) Distal retinal pigment of the fiddler crab, *Uca pugilator*: evidence for stimulation of release of light-adapting and dark-adapting hormones by neurotransmitters. Comp Biochem Physiol 84C:2–9–224

Langer H, Lues I, Rivera ME (1976) Arginine phosphate in compound eyes. J Comp Physiol A 107:179–184

Laughlin SB, Hardie RC (1978) Common strategies for light adaptation in the peripheral visual systems of fly and dragonfly. J Comp Physiol A 128:319–340

Lee AN, Metcalf RL, Booth GM (1973) House cricket acetylcholine esterase: histochemical localization and in situ inhibition by O,O-dimethyl s-aryl phosphorothiates. Ann Entomol Soc Am 66:333–343

Mancillas JR, McGinty JF, Selverston A, Karten H, Bloom FE (1981) Immunocytochemical localization of enkephalin and substance P in retina and eyestalk neurons of lobster. Nature (London) 293:576–578

Maxwell GD, Hildebrand JG (1981) Anatomical and neurochemical consequences of deafferentiation in the development of the visual system of the moth *Manduca sexta*. J Comp Neurol 195:667–680

Maxwell GD, Tait JF, Hildebrand JG (1978) Regional synthesis of neurotransmitter candidates in the CNS of the moth *Manduca sexta*. Comp Biochem Physiol 61C 109–119

McCaman RE, Weinrich D (1985) Histaminergic synaptic transmission in the cerebral ganglion of *Aplysia*. J Neurophysiol 53:1016–1037

Mercer AR, Mobbs PG, Davenport AP, Evans PD (1983) Biogenic amines in the brain of the honeybee, *Apis mellifera*. Cell Tissue Res 234:655–677

Meyer EP, Matute C, Streit P, Nässel DR (1986) Insect optic lobe neurons identifiable with monoclonal antibodies to GABA. Histochemistry 84:207–216

Nässel DR (1987) Serotonin and serotonin-immunoreactive neurons in the nervous system of insects. Prog Neurobiol 30:1–85

Nässel DR, Klemm N (1983) Serotonin-like immunoreactivity in the optic lobes of three insect species. Cell Tissue Res 232:129–140

Nässel DR, Laxmyr L (1983) Quantitative determination of biogenic amines and DOPA in the CNS of adult and larval blowflies *Calliphora erythrocephala*. Comp Biochem Physiol 75C:259–265

Nässel DR, O'Shea M (1987) Proctolin-like immunoreactive neurons in the blowfly central nervous system. J Comp Neurol 265:437–454

Nässel DR, Hagberg M, Seyan HS (1983) A new, possibly serotonergic neuron in the lamina of the blowfly optic lobe: an immunocytochemical and Golgi-EM study. Brain Res 280:361–367

Nässel DR, Meyer EP, Klemm N (1985) Mapping and ultrastructure of serotonin-immunoreactive neurons in the optic lobes of three insect species. J Comp Neurol 232:190–204

Nässel DR, Ohlsson L, Sivasubramanian P (1987) Postembryonic differentiation of serotonin-immunoreactive neurons in fleshfly optic lobes developing in situ or cultured in vivo without eye discs. J Comp Neurol 255:327–340

Nässel DR, Holmqvist MH, Hardie RC, Hakånson R, Sundler F (1988) Histamine-like immunoreactivity in photoreceptors of the compound eyes and ocelli of flies. Cell Tissue Res 253:639–646

Oja SS, Kontro P (1983) Taurine. In: Lathja A (ed) Handbook of Neurochemistry, vol 3, pp 501–533

Pitman RM (1971) Transmitter substances in insects: a review. Comp Gen Pharmacol 2:347–371

Prell GD, Green JP (1986) Histamine as a neuroregulator. Annu Rev Neurosci 9:209–254

Quackenbush LS, Fingerman M (1984) Regulation of neurohormone release in the fiddler crab, *Uca pugilator*: effects of gamma-aminobutyric acid, octopamine met-enkephalin and beta-endorphin. Comp Biochem Physiol 79C:77–84

Rao KR (1985) Pigmentary effectors. In: Bliss DE (ed) The biology of Crustacea, vol 9. Academic Press, Orlando New York London, pp 395–462

Salvaterra PM, Crawford GD, Klotz JL, Ikeda K (1983) Production and use of monoclonal antibodies to biochemically defined insect neuronal antigens. In: Breer H, Miller TA (eds) Neurochemical techniques in insect research. Springer, Berlin Heidelberg New York, pp 223–242

Sattelle DB (1985) Acetylcholine receptors. In: Kerkut GA, Gilbert LI (eds) Comprehensive Insect Physiology Biochemistry and Pharmacology. Pergamon, Oxford New York, pp 395–434

Schäfer S (1987) Immunocytologische Untersuchungen am Bienengehirn. Phd Thesis Free Univ Berlin

Schäfer S, Bicker G (1986a) Common projection areas of 5-HT and GABA-like immunoreactive fibres in the visual system of the honeybee. Brain Res 380:368–370

Schäfer S, Bicker G (1986b) Distribution of GABA-like immunoreactivity in the brain of the honeybee. J Comp Neurol 246:287–300

Schäfer S, Bicker G, Ottersen OP, Storm-Mathiesen J (1988) Taurine-like immunoreactivity in the brain of the honeybee. J Comp Neurol 268:60–70

Scheidler A, Kaulen P, Bruning G, Erber J (1986) Autoradiographic localisation of octopamine and serotonin-binding sites in the brain of the honeybee. Verh Dtsch Zool Ges 79:293

Schmidt-Nielsen BK, Gepner JI, Teng NNH, Hall LM (1977) Characterisation of an α-bungarotoxin binding component from *Drosophila melanogaster*. J Neurochem 29:1013–1031

Schoofs L, Jegou S, Vaudry H, Verhaert P, De Loof A (1987) Localization of melanotropin-like peptides in the central nervous system of two insect species, the migratory locust, *Locusta migratoria*, and the fleshfly, *Sarcophaga bullata*. Cell Tissue Res 248:25–31

Schürmann FW, Klemm N (1984) Serotonin-immunoreactive neurons in the brain of the honey bee. J Comp Neurol 225:570–580

Schwartz J-C, Arrang J-M, Garbarg M, Korner M (1986) Properties and roles of the three subclasses of histamine receptors in brain. J Exp Biol 124:203–224

Shaw SR (1984) Early visual processing in insects. J Exp Biol 112:225–251

Simmons PJ, Hardie RC (1988) Evidence that histamine is a neurotransmitter of photoreceptors in the locust ocellus. J Exp Biol 138:205–219

Steinbusch HWM, Verhofstad AAJ, Joosten HWJ (1978) Localization of serotonin in the central nervous system by immunocytochemistry: description of a specific and sensitive technique and some applications. Neuroscience 3:811–819

Strausfeld NJ, Blest AD (1970) Golgi studies on insects. Pt 1. The optic lobes of lepidoptera. Philos Trans R Soc London Ser B 258:81–134

Strausfeld NJ, Nässel DR (1981) Neuroarchitecture of brain regions that subserve the compound eyes of Crustacea and Insects. In: Autrum H (ed) In: Handbook of sensory physiology, vol VII/6B. Springer, Berlin Heidelberg New York, pp 1–134

Torre V, Poggio T (1978) A synaptic mechanism possibly underlying directional selectivity to motion. Proc R Soc London Ser B 202:409–416

Usherwood PNR (1978) Amino acids as neurotransmitters. Adv Comp Physiol Biochem 7:227–309

Vieillemaringe J, Duris P, Geffard M, Moal Ml, Delaage M, Bensch C, Girardie J (1984) Immunohistochemical localisation of dopamine in the brain of the insect *Locusta migratoria migratorioides* in comparison with the catecholamine distribution determined by the histofluorescence technique. Cell Tissue Res 237:391–394

Wafford KA, Sattelle DB (1986) Effects of amino-acid neurotransmitter candidates on an identified insect motoneurone. Neurosci Lett 63:135–140

White K, Hurteau T, Punsal P (1986) Neuropeptide-FMRFamide-like immunoreactivity in *Drosophila*: Development and distribution. J Comp Neurol 247:430–438

Whitton PS, Strang RHC, Nicholson RA (1987) The distribution of taurine in the tissues of some species of insects. Insect Biochem 17:573–577

Wright TRF (1987) Genetics of Biogenic Amine Metabolism, sclerotization and melanization in *Drosophila melanogaster*. Adv Genet 25:00–00

Zimmerman RP (1978) Field potential analysis and the physiology of second-order neurons in the visual system of the fly. J Comp Physiol A 126:297–317

Chapter 13

Circadian Rhythms in the Invertebrate Retina

ROBERT B. BARLOW jr., STEVEN C. CHAMBERLAIN and HERMAN K. LEHMAN,
Syracuse, New York, USA

1 Introduction

Sunlight is 100 million times more intense than starlight. Survival often requires an ability to see specific features of the environment over this wide range of illumination. Animals achieve this extraordinary feat by adapting their visual sensitivity to the ambient level of illumination.

Visual adaptation begins at the retina. It involves numerous biochemical and cellular mechanisms, many of which are activated by changes in ambient illumination. They adapt the sensitivity of the visual system by responding to changes in light intensity. However, some mechanisms of adaptation are endogenous. They adapt visual sensitivity by anticipating changes in light intensity. These adaptation mechanisms are controlled by circadian oscillators which appear to be closely associated with most, if not all, visual systems (Aschoff 1981; Takahashi and Zatz 1982). They can be located in the retina or in the central nervous system.

Circadian oscillators modulate retinal function in a wide range of animals. In vertebrates, they change the structure and sensitivity of the fish retina (Levinson and Burnside 1981; Dearry and Barlow 1987), the activity of serotonin N-acetyltransferase in chicken and *Xenopus* retinas (Hamm and Menaker 1980; Besharse and Iuvone 1983), the sensitivity of the lizard and rabbit retinas (Brandenburg et al. 1983; Fowlkes et al. 1984), and the rod-cone dominance in the pigeon retina (Barattini et al. 1981). In the rat, circadian oscillators modulate the synthesis of dopamine in the retina (Wirz-Justice et al. 1984), the shedding of outer segments of rod photoreceptors (La Vail 1976), the transmission of information in the central visual systems (Hanada and Kawamura 1984), and sensitivity to light (Rosenwasser et al. 1979).

Circadian rhythms are particularly widespread among invertebrate visual systems (see Page 1981; Barlow 1983). They have been detected in the compound eyes of many arthropods and are commonly found in the visual systems of crustaceans, arachnids, and insects (Welsh 1938; Jahn and Wulff 1943; Aréchiga and Wiersma 1969; Page 1981). Kiesel (1894) first described the persistent daily movements of screening pigments in the eyes of arthropods. Demoll (1911) confirmed Kiesel's observation and suggested that the rhythmic movements may be controlled by the nervous system. Such circadian movements of screening pigments mirror the well known photomechanical movements of screening pigments in invertebrate eyes. Both have prominent roles in adapting visual sensitivity

Stavenga/Hardie (Eds.) Facets of Vision
© Springer-Verlag Berlin Heidelberg 1989

(Jahn and Crescitelli 1940; Autrum 1981). Another example of the joint control of visual sensitivity by a circadian oscillator and light is the hormonal system of the crustacean eyestalk (see Aréchiga et al. 1985).

Why does visual adaptation require both circadian and photon-triggered mechanisms in some animals? Both mechanisms often influence the same adaptation processes. What advantage does one have over the other? Why do the visual systems of some animals need to anticipate changes in light intensity rather than respond directly to them? Recent studies of the horseshoe crab, *Limulus polyphemus* suggest some answers.

The *Limulus* visual system provides a clear example of the circadian modulation of retinal function (Barlow et al. 1977a; Barlow 1983). At night, a clock in the brain transmits neural activity via efferent optic nerve fibers to all major photoreceptor organs of the animal (Eisele et al. 1982). The efferent impulses impinge directly on retinal neurons (Fahrenbach 1973; Barlow and Chamberlain 1980) and exert multiple circadian changes in their anatomical and physiological properties.

Table 1 lists the circadian changes detected thus far in the lateral compound eye. The first one, the efferent input, mediates the others. All except "photomechanical movements" and "membrane shedding" are endogenous; that is, they continue unabated when the animal remains in constant darkness. They all combine to increase visual sensitivity at night.

The large number of changes in Table 1 is remarkable. *Limulus* appears to have discovered "every trick in the book" to increase its retinal sensitivity at night.

Table 1. Circadian rhythms in the *Limulus* lateral eye

Retinal property	Day	Night	Reference
Efferent input	Absent	Present	Barlow et al. (1977a); Barlow (1983)
Gain	Low	High	Renninger et al. (1984); Barlow et al. (1987)
Noise	High	Low	Barlow et al. (1977a); Kaplan and Barlow (1980); Barlow et al. (1987)
Quantum bumps	Short	Long	Kaplan et al. (1986)
Frequency response	Fast	Slow	Batra (1983); Renninger (1983)
Dark adaptation	Fast	Slow	Kass (1985)
Cell position	Proximal	Distal	Chamberlain and Barlow (1977); Barlow et al. (1980)
Aperture	Constricted	Dilated	Chamberlain and Barlow (1977, 1987)
Acceptance angle	6°	13°	Barlow et al. (1980)
Photon catch	Low	High	Barlow et al. (1980)
Photomechanical movements	Trigger	Prime	Chamberlain and Barlow (1981, 1987)
Screening pigment	Clustered	Dispersed	Barlow and Chamberlain (1980)
Membrane shedding	Trigger	Prime	Chamberlain and Barlow (1979, 1984)
Lateral inhibition	Strong	Weak	Renninger and Barlow (1979); Batra and Barlow (1982)
Visual sensitivity	Low	High	Powers and Barlow (1985)

This may be easy to understand when we realize that the animal uses its eye to find mates both day and night (see below) and it does so with just a single type of photoreceptor, the retinular cell, whereas most vertebrate retinas, for example, possess two types of photoreceptors, rods and cones. The multiple changes in Table 1 may combine to extend the operating range of the retinular cells. Why then are they driven by an endogenous clock rather than by light? One possible answer suggested by our recent studies is that the retina must be adapted before exposure to high intensities to prevent light damage.

In this chapter we review briefly the circadian rhythms in the anatomy and physiology of the *Limulus* eye and the neurotransmitter mechanisms that mediate them. We also discuss their possible role in the animal's behavior. Where appropriate, we call attention to results from other animals, particularly invertebrates, that may shed light on the general role of circadian rhythms in the retina.

2 Circadian Rhythms in Retinal Anatomy

2.1 The *Limulus* Eye and its Efferent Innervation

The lateral eye of an adult *Limulus* contains about 800 ommatidia arranged in a domed oval yielding a hemispherical visual field. The cuticular cones (Fig. 1) protrude inward from the corneal surface and are immediately apposed internally to clusters of photoreceptors and pigment cells. The distal pigment cells form a cylindrical aperture around the ommatidial axis between the cuticular cones and photoreceptors and function with the proximal pigment cells to optically isolate each ommatidium. About a dozen photosensitive retinular cells are arranged around the axial dendrite of the eccentric cell like the segments of an orange (Fig. 2). The plasma membrane of the rhabdomeral segment of each photoreceptor is specialized into a microvillar rhabdomere, and abutting rhabdomeres of adjacent cells are fused to form a rhabdom that is shaped like an asterisk. At the center of the asterisk, the circular arrangement of the rhabdomeres forms an interdigitated microvillar array which couples the photoreceptors to the second-order cell that transmits optic nerve activity to the brain. The ommatidial array is underlain by a lateral plexus of branches of eccentric cell axons (Fahrenbach 1985) that mediate inhibitory interactions among ommatidia (Hartline 1972). Axons of both the photoreceptors and eccentric cells exit the lateral plexus and join together to form the optic nerve (Fahrenbach 1971) which curves anteriorly toward the midline and then plunges posteroventrally to enter the optic lobe of the brain (Patten 1912; Chamberlain and Barlow 1980).

Two efferent pathways innervate the lateral eye. One pathway leaves the brain via the lateral optic nerve, branches in the lateral plexus, and terminates on all three cell types of each ommatidium (Fahrenbach 1973, 1981; Barlow and Chamberlain 1980). These fibers are part of a general efferent system that transmits circadian activity to other photoreceptor organs (Fahrenbach 1975; Calman and Chamberlain 1982; Eisele et al. 1982; Evans et al. 1983). The lateral eye also receives a general epidermal innervation by efferent fibers with co-localized substance P-like

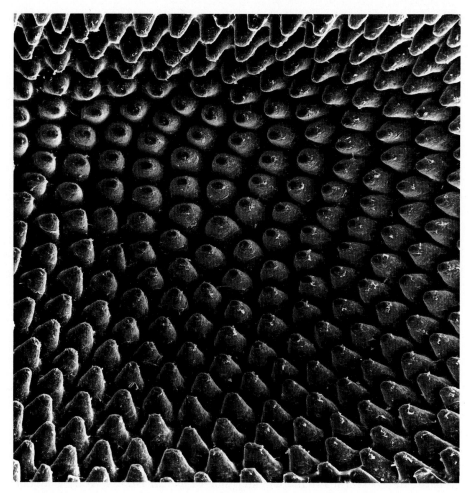

Fig. 1. Scanning electron micrograph of the interior surface of the cornea of the *Limulus* lateral eye. All retinal tissue was removed to reveal the proximal tips of the corneal lenses whose optical properties were first studied by Exner in 1891. The corneal lenses are 200 μm in diameter

Fig. 2. Circadian changes in the structure and field of view of a *Limulus* ommatidium. Longitudinal reconstructions are based on serial cross-sections of dark-adapted ommatidia. Schematic efferent terminals are included to indicate that each retinal cell receives neural input from a circadian clock located in the brain. Curves at the *top* give the relative sensitivity of the response of a single optic nerve fiber as a function of the angle of incidence of a light stimulus. Diameter of corneal lens is 200 μm. (Barlow et al. 1980)

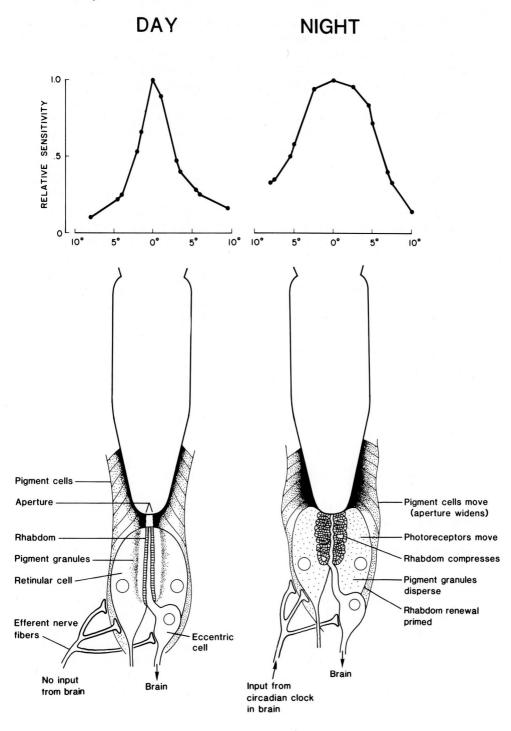

and FMRFamide-like immunoreactivity (Chamberlain and Engbretson 1982; Chamberlain and Lewandowski 1986). These fibers leave the brain via a dorsal segmental nerve trunk and join the lateral optic nerve through an anastomosis behind the eye. These fibers do not innervate the ommatidia, as reported by Mancillas and Selverston (1984), but pass distally to innervate the corneal epidermis.

The efferent fibers that transmit circadian activity are characterized ultra-structurally by large, paracrystalline cylindrical granules with a cylindrical indentation in one end and smaller lucent vesicles. The distinctive granules serve as natural markers permitting the identification of small processes in electron micrographs. The lateral optic nerve trunk contains as few as a dozen efferent fibers (Fahrenbach 1973; Evans et al. 1983) which ramify in the lateral plexus to produce as many as 100,000 distal processes. The distal processes synapse with collaterals of eccentric cell axons in the plexus (Fahrenbach 1985), enter ommatidia, and synapse repeatedly with every cell encountered, including the distal pigment cells that form the aperture (Fahrenbach 1981).

The central circuitry of the efferent system has proven difficult to unravel. A variety of physiological evidence, however, suggests that the somata of the efferent fibers which transmit circadian activity to photoreceptors reside among the ganglion cells of the medullar group and, furthermore, that the overall efferent control system may involve a hierarchy of interconnected circadian oscillators in both the brain and subesophageal ganglion. The central pathways of the circadian system receive inputs from the median, lateral and ventral eyes (Chamberlain and Barlow 1980; Barlow et al. 1981; Batra and Chamberlain 1985; Horne and Renninger 1988). Photoreceptors in the tail transmit information to the circadian system to synchronize the phase of the oscillator(s) with the solar cycle (Hanna et al. 1988).

2.2 Efferent Induced Circadian Changes in Retinal Anatomy

Efferent input exerts both real time and delayed changes in retinal anatomy. Real time changes begin with the evening onset of efferent activity and end with the morning offset. Delayed changes are primed by the nighttime efferent input but triggered by daylight.

The real time changes in anatomy in Table 1 are illustrated in Fig. 2. Circadian changes in cell position, aperture, and screening pigment combine to increase photon catch and acceptance angle of each ommatidium and thereby adapt the retina to function under dim levels of nocturnal illumination (Barlow et al. 1980; Chamberlain and Barlow 1987). This coordinated set of changes is initiated by the onset of efferent activity near dusk. The aperture formed by the distal pigment cells opens and shortens. The rhabdom in the retinular cells moves toward the tip of the cuticular cone and the screening pigment moves away from the rhabdom. The acceptance angle of an individual ommatidium increases from about 6° in the daytime dark adapted eye to about 13° at night (Barlow et al. 1980). The size of the acceptance angle is determined by the optical properties of the cuticular cone, the width of the aperture, and the distal cross section of the rhabdom. Quantitative modeling of the optical properties of these structures (Land 1979; Chamberlain and Fiacco 1985) confirms Exner's hypothesis (Exner 1891), that the cuticular cone functions as a lens cylinder rather than an ideal light collector, as suggested by Levi-Setti et al. (1975).

The real time changes in anatomy reverse in early morning hours at the offset of efferent activity. The aperture narrows and lengthens, the rhabdom elongates and moves away from the tip of the cuticular cone, and the retinular screening pigment aggregates in a cylindrical band beyond the radial array of rhabdomeres. These changes occur in complete darkness and adapt the retina for exposure to bright light. In natural lighting, photomechanical effects amplify the circadian changes in structure and further adapt the eye for daytime function. For example, the cessation of efferent activity near dawn allows the aperture to lengthen about 8 times. Exposure to natural lighting enhances the lengthening to 20 times. The same is true for the arrangement of the rhabdom which is equally long and wide at night. At the offset of efferent activity, the rhabdom lengthens and narrows ($l/w = 2.5$) in darkness, a process that is enhanced by exposure to daylight ($l/w = 4$).

Exner (1891) hypothesized these anatomical changes and their functional effects in *Limulus* nearly 100 years ago. In Chapter II of his treatise, he wrote:

"I describe below how the pigment in insects migrates under the influence of light, occupying a different position depending on whether the animal is in the light or dark. This may also be the case in *Limulus*; however, as I have made no relevant observations, I can only speculate. It is possible that, on illumination, the pigment sheath around the retinula changes its width, or varies its distance from the apical surface (proximal tip of cone) as a result of a reduction in the width and/or height of the unpigmented cone which adjoins the apical surface. A narrowing of the pigment sheath would correspond to the light-adapted state, a broadening would increase the brightness of the retinal image, but probably at the cost of its sharpness."

In sum, *Limulus* trades acuity for sensitivity at night. The circadian changes in anatomy increase photon catch as much as 100 times, increasing visual sensitivity by an equal amount (Barlow 1983; Chamberlain and Fiacco 1985). The corresponding increase in the field of view of each ommatidium reduces the acuity of the eye as anticipated by Exner (see above quote). These changes occur in an eye which is already designed for high sensitivity and low resolution as indicated by its very high value for the eye parameter p (Snyder 1977; see Land this Vol.).

Efferent input during the night primes the retina for delayed changes in anatomy triggered by the dawn's early light. The delayed changes in Table 1 are the shedding of microvillar membrane and photomechanical movement of cells (Chamberlain and Barlow 1979, 1984, 1987). These changes occur at dawn under natural environmental lighting, but they can be triggered by light later in the day under controlled laboratory conditions.

Membrane shedding is a vigorous transient breakdown of the rhodopsin-containing membrane of the retinular cells (Fig. 3). Whorls of the microvillar membrane pinch off from the rhabdom and move into the cytoplasm of the rhabdomeral (R) segment of the cell. Within minutes the rhabdom is rebuilt and

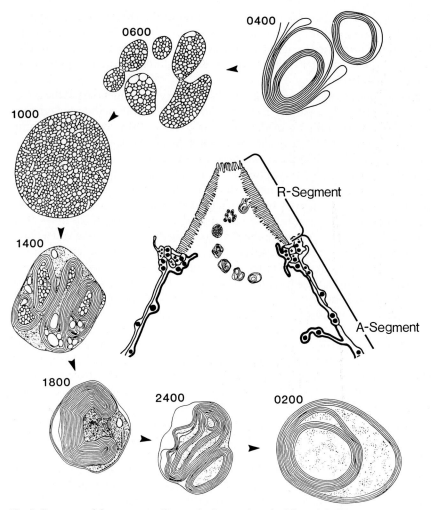

Fig. 3. Summary of the sequence of events in the transient shedding of rhabdomeral membrane in a *Limulus* retinular cell. The drawing in the center shows the approximate positions of the organelles during the day. Whorls of membrane form at about 0400 h, and over the next 22 h they convert to multivesicular bodies (1000 h) combination bodies (1400 h) and ultimately lamellar bodies (1800 to 0200 h) which move further into the A-segment and become lost in the general population of residual bodies. (Chamberlain and Barlow 1984)

the shed membrane migrates to the arhabdomeral (A) segment of the cell where it is degraded through a lysosomal pathway. This burst of shedding occurs only once each day and can be blocked by eliminating efferent input the preceding night.

Eliminating efferent input at night also blocks photomechanical movements the following day. To sum up, efferent input not only adapts the overall structure of the retina in anticipation of diurnal changes in ambient illumination, but it also primes the further adaptation of retinal structure triggered by daylight.

2.3 Comparison with Other Species

Although Exner did not describe circadian rhythms in compound eyes, his pioneering studies of pigment migration set the stage for current research in the field. Stavenga (1979) and Autrum (1981) have written excellent and comprehensive reviews of recent work. Three years after Exner's treatise, Kiesel (1894) observed the rhythmic movement of screening pigment in the eye of a noctuid moth, but more than 30 years elapsed before the full impact of his observation was appreciated (see Welsh 1938). It is now clear that rhythmic movements of screening pigment characterize many arthropod eyes; however, in most cases, the cellular mechanisms that control them are not well understood. Thus far, apart from *Limulus*, the most thoroughly studied invertebrate systems are those of the scorpion *Androctonus australis*, crayfish *Procambarus clarkii*, and the cockroach.

Circadian rhythms in retinal sensitivity and pigment migration in the median eye of the scorpion share characteristics with those of the *Limulus* eye (Fleissner and Fleissner 1978, 1985). Retinal sensitivity is increased at night by efferent optic nerve activity transmitted from a circadian clock located in the brain. The ultrastructural appearance of efferent terminals on retinal cells (Fleissner and Schliwa 1977) is remarkably similar to that in *Limulus*. The somata of these efferent fibers are arranged in two bilaterally symmetrical clusters in the supraesophageal ganglion near the circumesophageal connectives and separate from the central terminations of the visual afferents (Fleissner and Heinrichs 1982). Each cluster of cells sends an equal number of fibers to both median eyes. A circadian pacemaker may be located in the subesophageal ganglion of the scorpion, *Euscorpius flavicaudis*, but its relationships to the efferent optic nerve fibers is not known (Carricaburu and Muñoz-Cuevas 1986). Since scorpions and horseshoe crabs are closely related chelicerates, similarities in the nature of their retinal circadian rhythms would not be surprising. It is interesting that their retinal circadian rhythms also share common features with a more distant relative, the orb weaving spider (Yamashita and Tateda 1978).

The compound eye of the isopod crustacean, *Ligia exotica*, exhibits circadian rhythms in pigment position, rhabdom structure, and retinal sensitivity (Hariyama et al. 1986). The rhythms in pigment position and retinal sensitivity match the overall form of those rhythms in *Limulus*. The rhythm in rhabdom structure, large at night and small during the day, is different from that observed in *Limulus* but the conditions of the experiment are also different.

The circadian rhythm of migration of distal screening pigment in crayfish (Welsh 1941) generates a corresponding rhythm in retinal sensitivity (Aréchiga and Wiersma 1969; Aréchiga and Fuentes 1970). The control mechanisms for this retinal rhythm are complex. As discussed below, they involve multiple circadian oscillators and both neural and humoral components. Migration of screening pigment may also generate the circadian rhythm in the sensitivity of the retina of the cockroach (Wills et al. 1985). The underlying oscillator, which is located in the optic lobe near the lobula, may also control the rhythmic locomotor activity in this animal (Wills et al. 1985).

In the molluscs, *Aplysia* and *Bulla*, the organization of the circadian system is completely different. Circadian pacemaker cells are located in the retina (Jacklet and Rolerson 1982; Block and Wallace 1982). The pacemaker cells are weakly coupled to retinal photoreceptors for entrainment to the light cycle and may themselves be sensitive to light (Block and McMahon 1983), but have little or no direct effect on the function of the eye (Jacklet 1984). The axons of the pacemaker cells widely innervate the central nervous system (Olson and Jacklet 1985) and are undoubtedly involved in the circadian rhythm in locomotor activity. In *Aplysia*, efferent fibers innervate the retinal pacemakers from cells in the brain (Luborsky-Moore and Jacklet 1976; Olson and Jacklet 1985) and modulate their activity (Eskin 1971). In *Bulla*, the efferent fibers innervating the retinal pacemakers from the brain show FMRFamide-like immunoreactivity and FMRFamide suppresses pacemaker activity (Jacklet et al. 1987).

Thus it appears that invertebrates show a diversity in retina-circadian pacemaker arrangements. In some, such as *Limulus*, scorpion, *Ligia*, and crayfish, circadian activity is expressed in the structure and function of the retina itself. In others, such as *Aplysia* and *Bulla*, the retina serves as an input, but not a target, of the circadian pacemaker system.

3 Circadian Rhythms in Retinal Physiology

3.1 The *Limulus* Eye

The circadian rhythms in the anatomy of the *Limulus* retina produce corresponding rhythms in response. Figure 4 shows that the electroretinogram (ERG) of the lateral and median eyes exhibit synchronous circadian rhythms and that cutting the optic nerve abolishes them. In each case cutting the optic nerve at

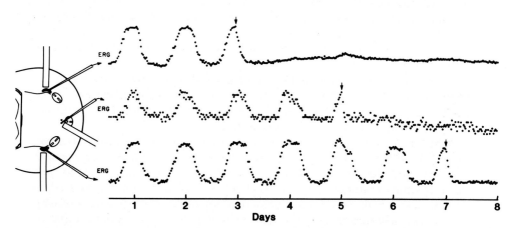

Fig. 4. Circadian rhythms in the ERG's of the lateral and median eyes of *Limulus*. The diagram on the *left* shows the placement of corneal electrodes and fiber optic light pipes as well as the access holes in the shell for sectioning the optic nerve trunks. On the *right* are plotted the peak-to-peak amplitudes of the ERG during the 8 days the animal remained in darkness. Day-night changes in the ERG were 40 to 200 μV for the lateral eyes and 35 to 120 μV for the median eyes. Sectioning the optic nerve at times indicated by *arrows* abolished the circadian rhythms. (Barlow 1983)

midnight reduced the amplitude to the daytime state and abolished further circadian changes in ERG amplitude. Synchronous bursts of efferent activity (Eisele et al. 1982; Barlow 1983) could be recorded form the proximal stumps of the three cut optic nerves, but only at night. The efferent activity exhibits a circadian rhythm in phase with the changes in ERG amplitude (Barlow et al. 1977a). Stimulating the distal stumps of the cut optic nerve *in situ* with pulses of current that mimic efferent activity increases the amplitude of the ERG to the nighttime state (Barlow 1983). It is clear that efferent optic nerve activity mediates the circadian rhythms in retinal response.

The circadian rhythms in retinal response in Fig. 4 reflect changes in retinal sensitivity. Figure 5 plots the ERG amplitudes on log-log coordinates. The "Day" and "Night" curves were fitted with the same function shifted 1.3 log units on the abscissa and 0.1 log units on the ordinate. The major portion of the increased sensitivity at night for brief flashes (lateral shift) can be attributed to an increase in photon catch resulting from changes in the structure of pigment and retinular cells shown in Fig. 2. Such changes can produce up to 100-fold increases in retinal sensitivity (Barlow et al. 1980).

Intracellular recordings show that the circadian clock in the *Limulus* brain modulates retinal sensitivity in part by acting directly on the most peripheral cell in the visual system, the photoreceptor (Kaplan and Barlow 1980). Although the clock innervates all types of retinal cells, the major physiological rhythms in Table 1 originate in the photoreceptor cells. Specifically, the circadian changes in gain, noise, bump shape, and frequency response occur in the photoreceptor cells and significantly modulate their response (Renninger 1983; Kaplan et al. 1986; Barlow et al. 1987). It is evident that the circadian clock in the brain can reach out to

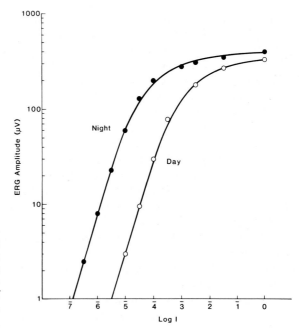

Fig. 5. Intensity-response functions of the lateral eye ERG at night and during the day. Each response was evoked by a 100-ms flash presented under dark-adapted conditions. Both sets of data were fitted by the same curve. (Barlow 1983)

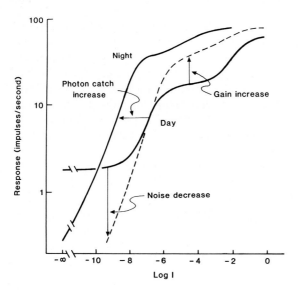

Fig. 6. Scheme of how circadian rhythms in noise, gain and photon catch change the intensity coding properties of the photoreceptor cell. The onset of efferent activity causes a rapid increase in gain and decrease in noise producing an intermediate function (*dashed line*). Continued efferent input slowly increases photon catch which moves the function to the left to produce the nighttime state. (Barlow et al. 1987)

influence very early events in the process of visual transduction. This may not be an isolated case. Circadian oscillators may also influence transduction processes in other photoreceptors such as mammalian rods (Brann and Cohen 1987).

Figure 6 presents a scheme for how some of the components in Table 1 modulate the response characteristics of a single photoreceptor cell (Barlow et al. 1987). Plotted are the intensity-response functions for the mean receptor potential recorded from a dark-adapted retinular cell. The "Day" function is graded over about 9 log units and has a characteristic shape with a plateau at intermediate light levels (Barlow and Kaplan 1977). The onset of efferent activity at night causes a rapid increase in gain (upward arrow) and a rapid decrease in noise (downward arrows) producing an intermediate function (dashed line). The decrease in noise in darkness (log $I = -\infty$) reflects a reduction in the rate of spontaneous quantum bumps. The cellular mechanism underlying this reduction in noise is not known, mainly because the source of the noise itself has not been identified. The increase in gain, prominent at intermediate intensities, represents an increase in membrane depolarization per absorbed photon. Voltage-clamp studies of ventral photoreceptors (Pepose and Lisman 1978) suggest that the efferent input may increase gain by reducing the efficacy of voltage-dependent mechanisms that repolarize the membrane potential during light exposure. In short, the efferent input appears to increase the amplitude of the receptor potential by influencing a membrane mechanism of light adaptation.

Prolonged efferent input then slowly shifts the intermediate photoreceptor function in Fig. 6 to the left by 1.2 log units just as it did in the ERG function in Fig. 5. Both lateral shifts reflect increases in photon catch resulting from the anatomical changes shown in Fig. 2.

Circadian rhythms in the response properties of single photoreceptors strongly influence the information transmitted to the brain. Specifically, the circadian changes in the intensity-response (I-R) function of the retinular cell in Fig. 6 are reproduced in the I-R functions of single optic nerve fibers (Barlow et al. 1977a). In both cases, response amplitudes increase and noise levels decrease at night, yielding about a 50-fold increase in the signal-to-noise characteristics of the dark-adapted retina (Barlow et al. 1987). This represents a lower limit of the overall circadian increase in retinal sensitivity. Circadian changes in photon catch, together with changes in the shapes of the I-R curves, can produce as much as a 100,000-fold increase in retinal sensitivity (Barlow et al. 1987).

3.2 Comparison with Other Species

Although no other animals have been analyzed in comparable detail at the cellular level, it is interesting that the invertebrates which have been studied in most detail, scorpion and orb-weaving spider, happen to be close relatives of *Limulus*, and their circadian rhythms appear to be closely related to those of *Limulus*. Both scorpion and orb-weaving spider generate circadian rhythms in the response of their median ocelli with efferent optic nerve activity transmitted at night from a circadian oscillator in the brain as in *Limulus* (Fleissner and Fleissner 1978, 1985; Yamashita and Tateda 1978, 1981). Retinal sensitivity, as measured by changes in ERG amplitude, is increased by about 4 log units for scorpion and 1 log unit for orb-weaving spider as compared to 2 log units for *Limulus*. Other studies show increases of 3 log units for beetles (Fleissner 1982), 4 log units for noctuid moths (Bennitt 1932) and up to 0.6 log units for the cockroach (Wills et al. 1985). The circadian rhythm in the cockroach retina also appears to be mediated by efferent input from an oscillator in the brain. Such circadian changes in retinal sensitivity are generally attributed to the migration of screening pigments or the movement of pigment cells, but additional mechanisms such as those acting on phototrans-duction in *Limulus* (Table 1) cannot be ruled out.

In an elegant study of *Bulla*, McMahon (1985) focused on the cells in the retina that generate circadian rhythms rather than the cells influenced by them. He found that changing the membrane potential of putative pacemaker cells by current injection shifted the phase of their endogenous rhythm. This is the first direct observation of a possible mechanism for how a light stimulus can shift phase of a cellular oscillator.

4 Neurotransmission of Circadian Rhythms in the Retina

What neurotransmitter mediates the circadian rhythms in the *Limulus* eye listed in Table 1? Is more than one involved? Several laboratories have investigated these questions and together their results strongly point to octopamine as a transmitter of the clock's action on the retina. Although other molecules may be involved as discussed below, only octopamine satisfies all of the criteria of a neurotransmitter: localization, synthesis, release, physiological mimicry, and pharmacological blockade.

4.1 Efferent Neurotransmitter(s) in the *Limulus* Eye

Octopamine was first detected in the *Limulus* visual system in 1980 by Battelle using radioenzymatic techniques. She and her colleagues then localized newly synthesized octopamine in the efferent processes of the retina and found it could be released by elevated levels of potassium ions (Battelle et al. 1982). Moreover, the release mechanism appears to be both calcium- and sodium-dependent (Battelle et al. 1982; Battelle and Evans 1986). Pharmacological studies by Kass and Barlow (1980, 1982, 1984) showed that topical application of octopamine and its agonists increased the amplitude of the ERG of the lateral eye and changed its structure toward the nighttime state. They also found that clozapine, an antagonist of octopamine, blocked the effects of both octopamine and the clock's input on ERG amplitude. Schneider et al. (1987) then showed that octopamine and its potent agonist NC-5 partially reproduced the clock's effects on the responses of single optic nerve fibers by increasing their gain and photon catch (see Fig. 5) but had no effect on their spontaneous activity. They also found that metaclopramide was a more potent antagonist than either phentolamine or yohimbine suggesting that the octopamine receptor was of the type II A class (see Evans 1980). Finally, Pelletier et al. (1984) found that octopamine and its agonists, when applied to isolated slices of retina, could modulate the physiological properties of gain and noise in single photoreceptor cells but not their anatomical properties (see below). To sum up, a wide range of biochemical, anatomical, and physiological evidence supports the role of octopamine in the neurotransmission of circadian rhythms to the *Limulus* eye.

Octopamine appears to modulate retinal properties via the second messenger, cAMP. Exposure of the retina to analogs of cAMP replicate in part the effects of the clock on the ERG, photoreceptor potential and optic nerve responses (Pelletier et al. 1984; Kass et al. 1988; Schneider et al. 1987). This is also true for forskolin, a potent activator of the enzyme, adenylate cyclase. Edwards and Battelle (1987) reported that octopamine, forskolin and cAMP stimulate the phosphorylation of a 122 kDa protein in the lateral eye. In an electrophysiological study, Kaplan et al. (1986) found that the clock's input modulated the anatomical and physiological properties of retinal cells without producing detectable changes in either their membrane resistance or potential. These biochemical and electrophysiological results suggest the following scheme: (1) efferent impulses depolarize synaptic terminals in the retina and cause the release of octopamine, (2) octopamine diffuses across the extracellular space and couples to specific receptors in the membranes of postsynaptic cells, (3) the receptors in turn activate the membrane bound enzyme, adenylate cyclase, which increases the intracellular levels of cAMP, and (4) the cAMP diffuses throughout the cell affecting a range of cellular organelles (membrane channels, cytoskeletal elements, etc.) via the activation of specific phosphokinases.

Octopamine may not act alone in the *Limulus* eye. Several lines of evidence point to multiple efferent neurotransmitters. First, terminals of the efferent fibers within the retina contain two types of organelles, dense granules and clear vesicles (Fahrenbach 1973). In general, dense granules are indicative of neurosecretory cells; however, it should be noted that those found in the *Limulus* retina (see above) are of an uncommon variety. Second, octopamine does not completely mimic the action of the clock's input regardless of whether it is assayed with ERG, optic nerve or photoreceptor responses (Kass and Barlow 1984; Schneider et al. 1987; Kass et al. 1988). Third, separation of retina and brain extracts by gel filtration, ion exchange and HPLC yields three components that increase ERG amplitude when injected in the lateral eye during the day (Lehman and Barlow 1987). Thus far, the identity of these components is not known, but the loss of activity with protease suggests that one or more are peptides. With regard to the identity of neuro-transmitters in the *Limulus* eye, Battelle et al. (1986) recently found that veratridine elicited a slow release of γ-glutamyltyramine and γ-glutamyloctopamine from the retina. What role these molecules have in the circadian rhythms of the eye is not known. Substance P has also been suggested as a possible transmitter (Mancillas and Selverston 1984), but the physiological results of this study have not been replicated in other laboratories (R. Barlow and L. Kass. unpublished results). High concentrations of serotonin can reproduce several actions of the clock (Barlow et al. 1977b) but neither can serotonin be found in the retina nor can the retina synthesize it (Chamberlain et al. 1986). To sum up, there is substantial evidence for octopamine as an efferent neurotransmitter and considerable evidence for addi-tional transmitter(s), but none as yet has been identified.

4.2 Push-Pull Mechanism

The structure of the *Limulus* retina appears to move between two active states, daytime and nighttime. Onset of efferent activity "pushes" the retina to the nighttime state and tonic activity maintains it in that state. Blocking the input at midnight causes an immediate movement back to the daytime state. Blocking the input for several days produces an extreme daytime state (Barlow and Cham-berlain 1980) as if the retina were under a constant "pull". This does not occur if the efferent input is chronically removed by excising the retina and maintaining it in organ culture. Instead, the retina drifts toward the nighttime state, but the drift can be stopped by adding an extract of *Limulus* blood to the culture medium. Moreover, maintaining the retina in a culture medium containing the extract preserves the retina in the daytime state (Chamberlain et al. 1987).

Retinal structure appears to be controlled by a push-pull mechanism (Fig. 7). Efferent neurotransmitter(s) "push" the retina to the nighttime state. Light and a circulating hormone "pull" it back to the daytime state. The actions appear coordinated and interdependent. The "pull" action of light requires the preceding "push" action of efferent input (Chamberlain and Barlow 1987).

The existence of two active structural states is also suggested by the action of microtubule and actin inhibitors (Barlow and Chamberlain 1980). Injecting such inhibitors beneath the cornea during the night, decreases retinal sensitivity and moves its structure to an intermediate state. The same intermediate state can be achieved by injecting the inhibitors during the day which increases sensitivity. We conclude that the two extreme states require the integrity of the cytoskeletal structures. A circulating hormone constantly pulls the cytoskeleton in one direction and the nighttime efferent input pushes it in the opposite direction.

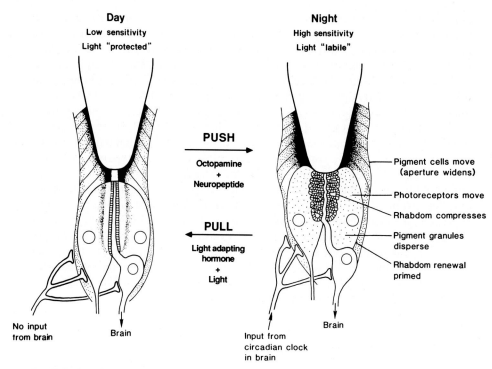

Fig. 7. Push-pull model of the circadian changes in the *Limulus* retina. Efferent neurotransmitters "push" the ommatidia to the high sensitivity nighttime state which is labile to the damaging effects of bright light. After the offset of efferent activity, light and a circulating hormone "pull" the retina back to the low sensitivity daytime state which is protected against the effects of bright lights

4.3 Comparison with Other Species

Efferent neurotransmission in the *Limulus* eye appears to be another example of an octopamine adenylate cyclase-linked second messenger system. They have been found in other invertebrates, especially insects, but not in the retina (Nathanson 1979; Evans 1980, 1981; Lingle et al. 1982), and none are associated with circadian oscillators. However, it is interesting that a similar dopamine adenylate cyclase-linked system is intimately related with circadian processes in at least two vertebrate retinas. Specifically, dopamine-sensitive adenylate cyclase has a role in the circadian regulation of retinomotor movements in both *Xenopus* (Pierce and Beshare 1985) and green sunfish (Dearry and Burnside 1986). The same second messenger system also mediates interplexiform-horizontal cell interactions in the carp retina, but thus far no circadian component has been detected in this control system (Mangel and Dowling 1987).

Neurohormones mediating retinal circadian rhythms have been studied extensively in crustaceans. The rhythmic migration of screening pigments in crustaceans shares several features with that of pigment cells in *Limulus*. The distal pigment of the crustacean retina migrates in a circadian fashion between the distal and proximal retina in response to hormones released from the sinus gland. The proximal pigment does not respond to hormones, but is thought to move in response to changes in photoreceptor potential (Aréchiga et al. 1985).

The hormone responsible for the movement of distal pigments to the daytime state is distal pigment light adapting hormone (DPLH). DPLH is an octadecapeptide (Fernlund 1976), originally described by its actions on the distal pigments. The concentration of DPLH in the sinus gland is low during the subjective day and high at night (Aréchiga and Mena 1975). The concentration in the sinus gland decreases during the day because the hormone is released into the circulation by an endogenous mechanism. Light can also trigger the release of DPLH (Aréchiga et al. 1985) giving the crayfish two mechanisms to ensure adequate levels of DPLH during the day. The light driven effect may also function to entrain the rhythm (Page and Larimer 1976).

Another sinus gland hormone, neurodepressing hormone (NDH), is released during the day from the crustacean sinus gland (Aréchiga and Huberman 1980). Neurodepressing hormone was first described by its role in circadian activity patterns and neural depressing activity, yet the molecule remains elusive and unsequenced. NDH activity released from cultured sinus glands peaks during the day and has a rhythm with period length of 22 h which can be phase-shifted (Aréchiga et al. 1985). When injected into an intact, dark-adapted, nighttime animal, extracts containing NDH decrease ERG amplitude and both the driven and spontaneous firing rates of the sustaining fibers (Aréchiga 1974). The first and second effects can be explained by well-known changes in the retinal screening pigments, but not the third. Changes in spontaneous activity require physiological changes in photoreceptors cells.

Red pigment concentration hormone (RPCH) was originally identified by its actions on the red integumentary chromatophores of the crustacean *Palaemon*, but it is now known to have effects on retinal pigments (Fernlund and Josefsson 1972; Josefsson 1973). PCH has the opposite actions of DPLH; the distal pigments of the retina move from the proximal to the distal regions of the retina in response to the hormone. PCH activity can be released in greater quantities at night by electrical stimulation of isolated sinus glands (Aréchiga et al. 1985).

Several common themes emerge from the crustacean and *Limulus* studies which may have general application: (1) multiple efferent actions may require multiple neurotransmitters, (2) generation of both daytime and nighttime states of the retina is an active process driven by complementary "push-pull" mechanisms, and (3) the full range of adaptation changes in the retina requires both endogenous (circadian) and exogenous (light) inputs.

5 Role of Circadian Rhythms in Behavior

How are the circadian rhythms in the *Limulus* eye related to the animal's behavior? Before answering this question it is first necessary to establish the role of vision in the animal's behavior. The *Limulus* visual system has been studied intensely since 1928 (Hartline 1972), but for many years no one understood what the animal uses

Fig. 8. Above A pair of horseshoe crabs nesting at night in Buzzards Bay near Woods Hole, Massachusetts. *Below* Male *Limulus* attempting to mate with a black cement casting of a female

its eyes for. In fact, Hartline often joked that he was "studying vision in a blind animal". Finally, in 1982 we discovered that *Limulus* uses its eyes to locate mates along the water's edge (Barlow et al. 1982, 1986a).

The animals migrate in the spring to protected beaches along the eastern coast of North America and build nests near the water's edge at high tide (Fig. 8). This behavior follows both the daytime and nighttime high tides, but most activity takes place at night (Barlow et al. 1986b). We uncovered a role for vision by observing the behavior of males in the vicinity of cement castings of female shells and other forms placed in the water. The males approached and attempted to mate with the castings (Fig. 8, bottom). Their degree of attraction depended on the form and contrast of the objects and the time of day.

How well can *Limulus* see during the day and at night? We explored this question by videotaping the animal's behavior in the vicinity of an underwater object and then measuring how close the animals must be to the object to see it. We found that their "sighting distance" was about the same both day (1.2 m) and night (1.0 m) (Barlow et al. 1986a). Apparently the nighttime increase in the sensitivity of their lateral eyes nearly compensates for the decrease in ambient light intensity. Seeing a female at a distance of 1.2 m involves only about four ommatidia (Barlow et al. 1986a). Detection by so few receptors suggests that the animal may be operating near the optical limit of its compound eyes. At night the animal sacrifices acuity to increase photon catch, but the loss does not appear to degrade its visual performance.

To sum up, *Limulus* has evolved elegant circadian mechanisms to adapt its visual sensitivity to the large day-night changes in environmental lighting. Although retinal circadian rhythms have been studied in a number of other animals as discussed in this chapter, to our knowledge none of the studies relate the retinal rhythms to well-defined behavior. We are further refining our field studies of *Limulus* behavior with the hope of relating specific aspects of its visual performance with the circadian rhythms in its retina.

Acknowledgments. Research supported by NIH Grants EY-00667, EY-03446, EY-05861 and NSF Grant BNS-8320315.

References

Aréchiga H (1974) Circadian rhythm of sensory input in the crayfish. In: Schmidt FO (ed) The neurosciences 3rd study program. FO Schmitt (ed) MIT, New York, pp 517–523

Aréchiga H, Fuentes B (1970) Correlative changes between retinal shielding pigments position and electroretinogram in crayfish. Physiologist 13:137

Aréchiga H, Huberman A (1980) Hormonal modulation of circadian rhythmicity in crustaceans. In: Valverde C, Aréchiga H (eds) Frontiers of hormone-research, vol 6. Comparative aspects of neuroendocrine control of behavior. Basel, pp 16–34

Aréchiga H, Mena F (1975) Circadian variations of hormonal content in the nervous system of the crayfish. Comp Biochem Physiol 52A:581–584

Aréchiga H, Wiersma CAG (1969) Circadian rhythm of responsiveness in crayfish visual units. J Neurobiol 1:71–85

Aréchiga H, Cortes JL, Garcia U, Rodriguez-Sosa L (1985) Neuroendocrine correlates of circadian rhythmicity in crustaceans. Am Zool 25:265–274

Aschoff J (1981) A survey on biological rhythms. In: Aschoff J (ed) Handbook of behavioral neurobiology, vol 4. Biological rhythms. Plenum, New York, pp 3–10

Autrum H (1981) Light and dark adaptation in invertebrates. In: Handbook of sensory physiology, vol VII/6C. Springer, Berlin Heidelberg New York, pp 1–91

Barattini S, Battisti B, Cervetto L, Marroni P (1981) Diurnal changes in the pigeon electroretinogram. Rev Can Biol 40:133–137

Barlow RB, jr. (1983) Circadian rhythms in the *Limulus* visual system. J Neurosci 3:856–870

Barlow RB, jr., Chamberlain SC (1980) Light and a circadian clock modulate structure and function in *Limulus* photoreceptors. In: Williams TP, Baker BN (eds) The effects of constant light on visual processes. Plenum, New York, pp 247–269

Barlow RB, jr., Kaplan E (1977) Properties of visual cells in the lateral eye of *Limulus* in situ: Intracellular recordings. J Gen Physiol 69:203–220

Barlow RB, jr., Bolanowski SJ, Brachman ML (1977a) Efferent optic nerve fibers mediate circadian rhythms in the *Limulus* eye. Science 197:86–89

Barlow RB, jr., Chamberlain SC, Kaplan E (1977b) Efferent inputs and serotonin enhance the sensitivity of the *Limulus* lateral eye. Biol Bull 153:414 (Abstr)

Barlow RB, jr., Chamberlain SC, Levinson JZ (1980) *Limulus* brain modulates the structure and function of the lateral eye. Science 210:1037–1039

Barlow RB, jr., Chamberlain SC, Bolanowski SJ, jr., Galway LA, jr., Joseph DP (1981) One eye modulates the sensitivity of another in *Limulus*. Invest Opthalmol Visual Sci (Suppl) 20:180

Barlow RB, jr., Ireland LC, Kass L (1982) Vision has a role in *Limulus* mating behavior. Nature (London) 296:65–66

Barlow RB, jr., Powers M-K, Kass L (1986a) Vision and mating behavior in *Limulus*. In: Popper AN, Fay R, Atema J, Travolga W (eds) Symp sensory biology of aquatic animals. Springer, Berlin Heidelberg New York Tokyo

Barlow RB, jr., Powers M-K, Howard H, Kass L (1986b) Migratory behavior of *Limulus* for mating: relation to lunar phase, tide height, and sunlight. Biol Bull 171:320–329

Barlow RB, jr., Kaplan E, Renninger GH, Saito T (1987) Circadian rhythms in *Limulus* photoreceptors. I. Intracellular studies. J Gen Physiol 89:353–378

Batra R (1983) Efferent control of visual processing in the lateral eye of the horseshoe crab. Ph D Diss, Inst Sens Res, Syracuse Univ

Batra R, Barlow RB, jr. (1982) Efferent control of pattern vision in *Limulus* lateral eye. Soc Neurosci Abstr 8:49

Batra R, Chamberlain SC (1985) Central connections of *Limulus* ventral photoreceptors revealed by intracellular staining. J Neurobiol 16:435–441

Battelle B-A (1980) Neurotransmitter candidates in the visual system of *Limulus polyphemus*: Synthesis and distribution of octopamine. Vision Res 20:911–922

Battelle B-A, Evans JA (1984) Octopamine release from centrifugal fibers of the *Limulus* peripheral visual system. J Neurochem 42:71–79

Battelle B-A, Evans JA (1986) Veratridine-stimulated release of amine conjugates from centrifugal fibers in the *Limulus* peripheral visual system. J Neurochem 46:1464–1472

Battelle B-A, Evans JA, Chamberlain SC (1982) Efferent fibers to *Limulus* eyes synthesize and release octopamine. Science 216:1250–1252

Battelle B-A, Edwards SC, Maresch HM, Pierce SK (1986) Synthesis of gamma-glutamyltyramine and gamma-glutamyloctopamine in the nervous system of *Limulus polyphemus*. Soc Neurosci Abstr 12:1292

Bennitt R (1932) Diurnal rhythm in the proximal pigment cells of the crayfish retina. Physiol Zool 5:65–69

Besharse J, Iuvone M (1983) Circadian clock in *Xenopus* eye controlling retinal serotonin N-acetyltransferase. Nature (London) 305:133–135

Block GD, McMahon D (1983) Localized illumination of the *Aplysia* and *Bulla* eye reveals relationships between retinal layers. Brain Res 265:134–137

Block GD, Wallace SF (1982) Localization of a circadian pacemaker in the eye of a mollusc, *Bulla*. Science 217:155–157

Brandenburg J, Bobbert AC, Eggelmeyer F (1983) Circadian changes in the response of the rabbit's retina to flashes. Behav Brain Res 7:113–123

Brann MR, Cohen LV (1987) Diurnal expression of transducin mRNA and translocation of transducin in rods of rat retina. Science 235:585–587

Calman BG, Chamberlain SC (1982) Distinct lobes of *Limulus* ventral photoreceptors. II. Structure and ultrastructure . J Gen Physiol 80:839–862

Carricaburu P, Muños-Cuevas A (1986) Spontaneous electrical activity of the suboesophageal ganglion and circadian rhythms in scorpions. Exp Biol 45:301–310

Chamberlain SC, Barlow RB, jr. (1977) Morphological correlates of efferent circadian activity and light adaptation in the *Limulus* lateral eye. Biol Bull 153:418–419 (Abstr)

Chamberlain SC, Barlow RB, jr. (1979) Light and efferent activity control rhabdom turnover in *Limulus* photoreceptors. Science 206:361–363

Chamberlain SC, Barlow RB, jr. (1980) Neuroanatomy of the visual efferents in the horseshoe crab (*Limulus polyphemus*). J Comp Neurol 192:387–400

Chamberlain SC, Barlow RB, jr. (1981) Modulation of retinal structure in *Limulus* lateral eye: Interactions of light and efferent inputs. Invest Ophthalmol Visual Sci (Suppl) 20:75

Chamberlain SC, Barlow RB, jr. (1984) Transient membrane shedding in *Limulus* photoreceptors: control mechanisms under natural lighting. J Neurosci 4:2792–2810

Chamberlain SC, Barlow RB, jr. (1987) Control of structural rhythms in the lateral eye of *Limulus*. Interactions of diurnal lighting and circadian efferent activity. J Neurosci 7:2135–2144

Chamberlain SC, Engbretson GA (1982) Neuropeptide immunoreactivity in *Limulus*. I. Substance P-like immunoreactivity in the lateral eye and protocerebrum. J Comp Neurol 208:304–315

Chamberlain SC, Fiacco PA (1985) Models of circadian changes in *Limulus* ommatidia: Calculations of changes in acceptance angle, quantum catch, and quantum gain. Invest Ophthalmol Visual Sci Suppl 26:340

Chamberlain SC, Lewandowski TJ (1986) Colocalization of FMRFamide and substance P-like immunoreactivities in neurons of the horseshoe crab brain and retina. Soc Neurosci Abstr 12:628

Chamberlain SC, Pepper J, Battelle B-A, Wyse GA, Lewandowski TJ (1986) Immunoreactivity in *Limulus*. II. Studies of serotonin-like immunoreactivity, endogenous serotonin, and serotonin synthesis in the brain and lateral eye. J Comp Neurol 251:363–375

Chamberlain SC, Lehman HK, Schuyler PR, Vadasz A, Calman BG, Barlow RB, jr. (1987) Efferent activity and circulating hormones: Dual roles in controlling the structure and photomechanical movements of the *Limulus* lateral eye. Invest Opthalmol Visual Sci Suppl 28:186

Dearry A, Barlow RB, jr. (1987) Circadian rhythms in the green sunfish retina. J Gen Physiol 89:745–770

Dearry A, Burnside B (1986) Dopaminergic regulation of cone retinomotor movement in isolated teleost retinas. I. Induction of cone contraction is mediated by D2 receptors. J Neurochem 46:1006–1021

Demoll R (1911) Über die Wanderung des Irispigments im Facettenauge. Zool Jahrb Physiol 30:159–180

Edwards SC, Battelle B-A (1987) Octopamine- and cyclic AMP-stimulated phosphorylation of a protein in *Limulus* ventral and lateral eyes. J Neurosci 7:2811–2820

Eisele LE, Kass L, Barlow RB, jr. (1982) Circadian clock generates optic nerve activity in the excised *Limulus* brain. Biol Bull 163:382 (Abstr)

Eskin A (1971) Properties of the *Aplysia* visual system: In vitro entrainment of the circadian rhythm and centrifugal regulation of the eye. Z Vergl Physiol 74:353–371

Evans JA, Battelle B-A, Chamberlain SC (1983) Audioradiographic localization of newly synthesized octopamine to retinal efferents and the *Limulus* visual system. J Comp Neurol 219:369–383

Evans PD (1980) Biogenic amines in the insect nervous system. Adv Insect Physiol 25:317–473

Evans PD (1981) Multiple receptor types for octopamine in the locust. J Physiol 318:99–122

Exner S (1891) Die Physiologie der facettierten Augen von Krebsen und Insecten. Deuticke, Leipzig

Fahrenbach WH (1971) The morphology of the *Limulus* visual system IV. The lateral optic nerve. Z Zellforsch 114:532–545

Fahrenbach WH (1973) The morphology of the *Limulus* visual system. V. Protocerebral neurosecretion and ocular innervation. Z Zellforsch 144:153–166

Fahrenbach WH (1975) The visual system of the horseshoe crab *Limulus polyphemus*. Int Rev Cyto 41:285–349

Fahrenbach WH (1981) The morphology of the *Limulus* visual system VII. Innervation of photo-receptor neurons by neurosecretory efferents. Cell Tissue Res 216:655–659

Fahrenbach WH (1985) Anatomical circuitry of lateral inhibition in the eye of the horseshoe crab, *Limulus polyphemus*. Proc R Soc London Ser B 225:219–249

Fernlund P (1976) Structure of a light-adapting hormone from the shrimp *Pandalus borealis*. Biochim Biophys Acta 439:17–35

Fernlund P, Josefsson L (1972) Crustacean color-change hormone: amino acid sequence and chemical synthesis. Science 177:173–175

Fleissner G (1982) Isolation of an insect circadian clock. J Comp Physiol A 149:311–316

Fleissner G, Fleissner G (1978) The optic nerve mediates the circadian pigment migration in the median eyes of the scorpion. Comp Biochem Physiol A 61:69–71

Fleissner G, Fleissner G (1985) Neurobiology of a circadian clock in the visual system of scorpions. In: Barth FG (ed) Neurobiology of arachnids. Springer, Berlin Heidelberg New York Tokyo, pp 351–375

Fleissner G, Heinrichs S (1982) Neurosecretory cells in the circadian clock system of the scorpion, *Androctonus australis*. Cell Tissue Res 224:233–238

Fleissner G, Schliwa M (1977) Neurosecretory fibres in the median eyes of the scorpion, *Androctonus australis* L. Cell Tissue Res 178:189–198

Fowlkes D, Karwoski C, Proenza L (1984) Endogenous circadian rhythm in electroretinogram of free-moving lizards. Invest Opthalmol Visual Sci 25:121–124

Hamm H, Menaker M (1980) Retinal rhythms in chicks: circadian variation in melatonin and serotonin N-acetyltransferase activity. Proc Natl Acad Sci USA 77:4998–5002

Hanada Y, Kawamura H (1984) Circadian rhythms in synaptic excitability of the dorsal lateral geniculate nucleus in the rat. Int J Neurosci 22:253–262

Hanna WJ, Horne JA, Renninger GH (1988) Circadian photoreceptor organs in *Limulus*. II. The talson. J Comp Physiol A 162:133–140

Hariyama T, Meyer-Rochow VB, Eguchi E (1986) Diurnal changes in structure and function of the compound eye of *Ligia exotica* (Crustacea, Isopoda). J Exp Biol 123:1–26

Hartline HK (1972) Visual receptors and retinal interaction. In: Nobel Lectures: Physiology or Medicine 1963–1970. Elsevier, New York, pp 269–288

Horne JA, Renninger GH (1988) Circadian photoreceptor organs in *Limulus*. I. Ventral, median, and lateral eyes. J Comp Physiol A 162:127–132

Jacklet JW (1984) Neural organization and cellular mechanisms of circadian pacemakers. Int Rev Cytol 89:251–294

Jacklet JW, Rolerson C (1982) Electrical activity and structure of retinal cells of the *Aplysia* eye. II. Photoreceptors. J Exp Biol 99:381–395

Jacklet JW, Klose M, Goldberg M (1987) FMRFamide-like immunoreactive efferent fibers and FMRF-amide suppression of pacemaker neurons in eyes of *Bulla*. J Neurobiol 18:433–449

Jahn TL, Crescitelli F (1940) Diurnal changes in the electrical response of the compound eye. Biol Bull 78:42–52

Jahn TL, Wulff VJ (1943) Electrical aspects of a diurnal rhythm in the eye of *Dytiscus fasciventris*. Physiol Zool 16:101–109

Josefsson L (1973) Invertebrate neuropeptide hormones. Int J Peptide Protein Res 21:459–470

Kaplan E, Barlow RB, jr. (1980) Circadian clock in *Limulus* brain increases response and decreases noise of retinal photoreceptors. Nature (London) 286:393–395

Kaplan E, Barlow RB, jr., Renninger GH (1986) The circadian clock in the *Limulus* brain modifies the electrical properties of the photoreceptor membrane. Biol Bull 171:495 (Abstr)

Kass L (1985) Circadian alteration of dark adaptation in *Limulus* lateral eye. Soc Neurosci Abstr 11:474

Kass L, Barlow RB, jr. (1980) Octopamine increases the ERG of the *Limulus* lateral eye. Biol Bull 159:487 (Abstr)

Kass L, Barlow RB, jr. (1982) Efferent neurotransmission of circadian rhythms in the *Limulus* lateral eye: specificity for octopamine. Invest Ophthalmol Visual Sci Suppl 22:179

Kass L, Barlow RB, jr. (1984) Efferent neurotransmission of circadian rhythms in *Limulus* lateral eye. I. Octopamine-induced increases in retinal sensitivity. J Neurosci 4:904–917

Kass L, Pelletier JL, Renninger GH, Barlow RB, jr. (1983) cAMP: A possible intracellular transmitter of circadian rhythms in *Limulus* photoreceptors. Biol Bull 165:540 (Abstr)

Kass L, Renninger G, Barlow RB, jr. (1987) In preparation

Kaupp UB, Malbon CC, Battelle B-A, and Brown JE (1982) Octopamine stimulated rise in cAMP in *Limulus* ventral photoreceptors. Vision Res 22:1503–1506

Kiesel A (1894) Untersuchungen zur Physiologie des facettierten Auges. Sitzungsber Kais Akad Wiss Wien Math Nat Kl 103:97–139

Land MF (1979) The optical mechanism of the eye of *Limulus*. Nature (London) 280:396–397

Larimer JL, Smith JTF (1980) Circadian rhythm of retinal sensitivity in crayfish: Modulation by the cerebral and optic ganglia. J Comp Physiol A 136:313–326

La Vail MM (1976) Rod outer segment disk shedding in rat retina: relationship to cyclic lighting. Science 194:1071–1074

Lehman HK, Barlow RB, jr. (1987) An efferent neuropeptide in the eye of *Limulus*. Soc Neurosci Abstr 13:237

Levi-Setti R, Park D, Winston R (1975) The corneal cones of *Limulus* as optimised light concentrators. Nature (London) 253:115–116

Levinson G, Burnside B (1981) Circadian rhythms in teleost retinomotor movements. Invest Ophthalmol Vis Sci 20:294–303

Lingle CE, Marder E, Nathanson JA (1982) The role of cyclic nucleotides in invertebrates. In: Kebabian JW, Nathanson JA (eds) Handbook of experimental pharmacology, vol 58II. Springer, Berlin Heidelberg New York, pp 787–845

Luborsky-Moore JL, Jacklet JW (1976) *Aplysia* eye: Modulation by efferent optic nerve activity. Brain Res 115:501–505

Mancillas JR, Brown MR (1984) Neuropeptide modulation of photosensitivity. I. Presence, distribution, and characterization of a Substance P-like peptide in the lateral eye of *Limulus*. J Neurosci 4:832–846

Mancillas JR, Selverston AI (1984) Neuropeptide modulation of photosensitivity. II. Physiological and anatomical effects of Substance P on the lateral eye of *Limulus*. J Neurosci 4:847–859

Mangel SC, Dowling JE (1987) The interplexiform-horizontal cell system of the fish retina: effects of dopamine, light stimulation and time in the dark. Proc R Soc London Ser B 231:91–121

McMahon DG (1985) Cellular mechanisms of circadian pacemaker entrainment in the mollusc *Bulla*. PhD Diss, Univ Virginia

Nathanson JA (1979) Octopamine receptors, adenosine 3', 5'-monophosphate and neural control of firefly flashing. Science 203:65–68

Olson LM, Jacklet JW (1985) The circadian pacemaker in the *Aplysia* eye sends axons throughout the central nervous system. J Neurosci 5:3214–3227

Page TL (1981) Neural and endocrine control of circadian rhythmicity in invertebrates. In: Aschoff J (ed) Handbook of behavioral neurobiology, vol 4. Biological rhythms. Plenum, New York, pp 145–172

Page TL, Larimer JL (1976) Extraretinal photoreception in entrainment of crustacean circadian rhythms. Photochem Photobiol 23:245–251

Patten W (1912) The evolution of the vertebrates and their kin. Blakiston, Philadelphia

Pelletier JL, Kass L, Renninger GH, Barlow RB, jr. (1984) cAMP and octopamine partially mimic a circadian clock's effect on *Limulus* photoreceptors. Invest Ophthalmol Visual Sci Suppl 25:288

Pepose JS, Lisman JE (1978) Voltage-sensitive potassium channels in *Limulus* ventral photoreceptors. J Gen Physiol 71:101–120

Pierce ME, Besharse JC (1985) Circadian regulation of retinomotor movements. I. Interaction of melatonin and dopamine in the control of cone length. J Gen Physiol 86:671–689

Powers MK, Barlow RB, jr. (1985) Behavioral correlates of circadian rhythms in the *Limulus* visual system. Biol Bull 169:578–591

Renninger GH (1983) Circadian changes in the frequency response of visual cells in the *Limulus* compound eye. Soc Neurosci Abstr 9:217

Renninger GH, Barlow RB, jr. (1979) Lateral inhibition, excitation, and the circadian rhythm of the *Limulus* compound eye. Soc Neurosci Abstr 5:804

Renninger GH, Kaplan E, Barlow RB, jr. (1984) A circadian clock increases the gain of photoreceptor cells of the *Limulus* lateral eye. Biol Bull 167:532 (Abstr)

Rosenwasser AM, Raibert M, Terman JS, Terman M (1979) Circadian rhythm of luminance detectability in the rat. Physiol Behav 23:17–21

Schneider M, Lehman HK, Barlow RB, jr. (1987) Efferent neurotransmitters mediate differential effects in the *Limulus* lateral eye. Invest Ophthalmol Visual Sci Suppl 28:186

Snyder AW (1977) Acuity of compound eyes: Physical limitations and design. J Comp Physiol A 116:161–182

Stavenga DG (1979) Pseudopupils of compound eyes. In: Autrum H (ed) Handbook of sensory physiology, vol VII/6A. Springer, Berlin Heidelberg New York, pp 357–439

Takahashi JS, Zatz M (1982) Regulation of circadian rhythmicity. Science 217:1104–1111

Welsh JH (1938) Diurnal rhythms. Q Rev Biol 13:123–139

Welsh JH (1941) The sinus gland and 24-hour cycles of retinal pigment migration in the crayfish. J Exp Zool 86:35–49

Wills SA, Page TL, Colwell CS (1985) Circadian rhythms in the electroretinogram of the cockroach. J Biol Rhythms 1:25–37

Wirz-Justice A, Prada M, Reme C (1984) Circadian rhythm in rat retinal dopamine. Neurosci Lett 45:21–25

Yamashita S, Tateda M (1978) Spectral sensitivities of the anterior median eyes of the orb web spiders, *Argiope bruennichii* and *A. amoena*. J Exp Biol 74:47–57

Yamashita S, Tateda H (1981) Efferent neural control in the eyes of orb weaving spiders. J Comp Physiol A 143:477–483

Chapter 14

Color Vision Honey Bees: Phenomena and Physiological Mechanisms

RANDOLF MENZEL and WERNER BACKHAUS, Berlin, FRG

1 Introduction

Colors are meaningful signals for honey bees. The information value originates from the phylogenetic experience of the species and the ontogenetic experience of each individual with significant objects in its world such as flowers and the immediate surroundings of the hive entrance. The historical controversy concerning color vision in bees arose between the young Karl von Frisch and the physiologist C. von Hess (von Frisch 1914; von Hess 1913) from a misunderstanding of the context specificity of color vision. Von Frisch demonstrated that bees see object colors as a visual quality different from grey shades in the behavioral context of feeding and homing, whilst von Hess showed that bees are color blind in their escape runs towards the light. Von Hess incorrectly generalized that bees are color blind in all behavioral contexts, and von Frisch did not realize that color vision in bees may be limited to certain behaviors.

At the feeding place, bees are exceptionally sensitive to the chromatic properties of targets that mark the nectar source (Kühn 1927; Daumer 1956; von Helversen 1972). They also show the characteristics of lower colorimetry, such as the mixing of two different spectral lights to form a new chromatic or achromatic signal (Daumer 1956). Higher colorimetric phenomena (Backhaus and Menzel 1987), color constancy (Neumeyer 1981; Werner et al. 1988) and the perceptual uniqueness of color signals (Menzel 1967, 1985; rev. Backhaus et al. 1987b) have also been demonstrated, and will be discussed in this chapter.

As in humans, monkeys, and many other animal species, color vision in honey bees is trichromatic. The physiological basis of trichromacy as developed by Young (1802), Maxwell (1860), and von Helmholtz (1896) for humans, was experimentally verified for humans, monkeys, and bees in the same year, namely 1964. MacNichol and his co-workers (Marks et al. 1964) measured the spectral absorbance of single primate cones, Brown and Wald (1964) that of the rods and cones in the human retina, and Autrum and von Zwehl (1964) performed the first spectral sensitivity measurements by electrophysiological recordings from single photoreceptors in the compound eye of the bee.

The Young-Helmholtz theory of trichromacy states that three independent spectral inputs are necessary to provide the brain with the information for the perceptual representations of hue, saturation and brightness. These three qualities determine, via introspection, our experience of color. We cannot test whether the honey bee (or any other animal) has such sensations, but we can infer the logical

Stavenga/Hardie (Eds.) Facets of Vision
© Springer-Verlag Berlin Heidelberg 1989

structure of the hypothetical representations of color by testing the similarity judgments of animals in behavioral discrimination tests. With these limitations in mind, we shall nevertheless argue that the honey bee's perception of color is only slightly different from that of man.

In human color vision, certain colors cancel each other when mixed and give rise to white light, whereas simultaneously or successively presented colors enhance each other if they belong to certain pairs of colors. The first kind of phenomenon was the major issue of color research in the last century and culminated in the formulation of the Graßmann color-mixing rules (Graßmann 1853). The latter phenomena were explained by Hering (1920) in his theory of mutually inhibitory colors (opponent colors), whereby three primary color pairs are coupled in a mutually inhibitory fashion, i.e., red-green, yellow-blue, and white-black. The neural basis for color mixing are the three photoreceptor types. Those for the color-opponent processes are spectrally opponent visual neurons, which have been worked out to some extent for primates, and have also been extended to include the combined spectral and spatial contrast phenomena (Hubel and Wiesel 1972; DeValois and DeValois 1975). Once again, color vision in honey bees is basically similar to that of primates, in that certain pairs of colors mix to give "bee-white" ("color cancellation"). Furthermore, simultaneous and successive color contrast and color opponent spatial contrast are also perceptual phenomena in bees. The neural correlates of the contrast phenomena are spectrally opponent visual interneurons that code spectral lights in opponent processes (Kien and Menzel 1977b; Hertel 1980). As in primates and humans, the major strategies of color coding in bees appear to be trichromacy at the retinal level, and spectral opponency at the level of visual neurons.

2 Trichromacy of the Bee's Visual System

2.1 Spectral Receptor Types

The accurate determination of a photoreceptor's spectral sensitivity by intracellular recordings is difficult since, even with a stable voltage baseline, a test flash of medium intensity can change the experimentally obtained spectral sensitivity (Menzel et al. 1986). However, new methods have helped to collect more reliable data. In the worker bee, it now appears that the three spectral receptors each contain a single rhodopsin photopigment and that there is little, if any, electrical coupling between neighboring photoreceptors. The sensitivity peaks of the spectral receptor types are at 335 nm for the "ultraviolet" (UV) receptor, 435 nm for the "blue" (B) receptor, and 540 nm for the "green" (G) receptor (Fig. 1).

Functional side band sensitivities originate only from the beta-absorption peak of the respective rhodopsin (see B and G in Fig. 1). It should be noted that Autrum and von Zwehl (1964) originally reported two kinds of B-receptors, one absorbing maximally at 420 nm and the other at 470 nm (Fig. 1). Although we have recorded from hundreds of receptors we have not yet found any indication of more than one population of B-receptors.

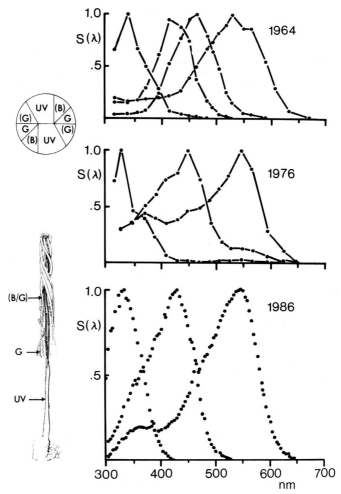

Fig. 1 Spectral sensitivity functions of the photoreceptors in the bee's compound eye as measured by intracellular recordings. *Upper graph* four selected recordings of four single cells from Autrum and von Zwehl (1964); *middle graph* average S(λ) functions as determined by the old method, where flashes of spectral light at different intensities were used to stimulate the eye. The spectral sensitivity is subsequently calculated from response/intensity functions for the different spectral lights (from Menzel and Blakers 1976). *Lower graph* average S(λ) functions measured by the new spectral scan method. The spectrum (300–700 nm) is scanned quickly (15 s) and a feedback system is used to continuously adjust the intensity in such a way that the light response of the intracellularly recorded cell is clamped to a preselected value (Menzel et al. 1986). The correspondence between the nine receptor cells in each ommatidium and the spectral type was determined by intracellular marking after recording the spectral responses. The upper left figure gives a schematic cross-section of an ommatidium just below the lens, where eight of the nine receptors are found. Each median frontal ommatidium contains two UV (plus one proximal UV cell), two blue (B) and four green (G) cells. The axonal projections of these receptors are sketched in the *lower left drawing*. The UV receptors terminate in the second visual neuropile, two green cells with fat axons terminate in the proximal layer of the medulla, and the thin green and the blue cells terminate in the median lamina layer. The spectral peaks of the UV, B and G cells are 335, 435 and 540 nm respectively

Each median frontal ommatidium contains the three receptor types in a characteristic arrangement (Fig. 1). Thus, each point of view is simultaneously analyzed in the three spectral bands "ultraviolet", "blue", and "green". Since the three pigments are combined in a fused rhabdom they mutually act as spectral filters and counterbalance self-screening in the long and highly absorbing rhabdom (Snyder et al. 1973). It is at present unknown whether ommatidia in other eye regions are organized in a similar or different manner. Preliminary evidence indicates a deviation in the extreme ventral (with more B receptors) and posterior (only G receptors) eye region (Milde 1978).

2.2 Lower Colorimetry and a Receptor-Based Chromaticity Diagram

The same color sensation can be produced by different mixtures of chromatic lights. This visual phenomenon is the basic paradigm of the Young-Helmholtz trichromacy theory. This theory states that there are only three independent spectral receptors; and so any two spectral mixtures which stimulate each receptor type equally will be indistinguishable. Furthermore, Graßmann's (1853) linear mixture rules provide a quantitative description even for those conditions in which the mixtures do not match for brightness differences. The results of the mixture calculations can be graphically represented in a chromaticity diagram, where the three basic lights (called primaries) define the edges of a symmetric triangle (Fig. 2). The locus of a mixture is constructed by dividing the straight line between the loci of the two lights used in the mixture into the same but inverse proportion of the respective light intensities. Using such a "center-of-gravity" construction one can represent and predict graphically the results of mixture experiments. Daumer (1956) found that such a procedure can successfully be applied in quantitatively describing the results of matching experiments in bees, where the primaries are UV (360 nm), blue (440 nm) and green (588 nm) lights.

Three phenomena characterize the mixture triangle:
1. "Bee-white" results from a mixture of 15% 360 nm + 85% 490 nm or from a mixture of 15% 360 nm + 30% 440 nm + 55% 590 nm.
2. The resulting mixture of two spectral lights between 350 nm and 440 nm or 440 nm and 580 nm appears identical to other spectral lights positioned between the two lights (see table in Fig. 2).
3. The mixture of the two ends of the visible spectrum appears different from any spectral light and from each of the components of the two mixed lights ("bee-purple").

The primaries can be changed to other lights. The loci differ when the lights are represented in another color space, but the Graßmannian mixture rules still hold (Krantz 1975). This allows the construction of a physiological color stimulus space (tristimulus space) (Le Grand 1948, 1971; Cornsweet 1970; Rushton 1972; Rodieck 1973), where the primaries correspond to the photon fluxes absorbed by each of the three receptor types (tristimulus values). Since the spectral sensitivities of the receptor types of the bee are so well known, one can construct a tristimulus space

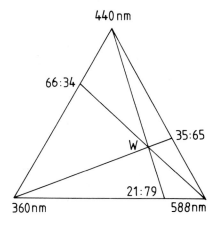

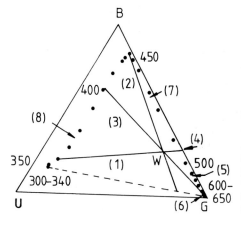

´Bee white´= 15% 360 + 30% 440 + 55% 590

(1) 360 ⟶ 490 (= 65% 590 + 35% 440)
(2) 440 ⟶ 79% 590 + 21% 360
(3) 590 ⟶ 66% 440 + 34% 360

(4) 490 = 65% 590 + 35% 440
(5) 530 = 60% 616 + 40% 490
(6) 588 = (20–80)% 616 + (80–20)% 530
(7) 474 = (40–60)% 440 + (60–40)% 490
(8) 375 = 20% 440 + 80% 360

Fig. 2. Comparison of Daumer's chromaticity diagram (*upper graph*) with the physiological (receptor) chromaticity diagram (*lower graph*). The table gives the average proportions of spectral lights as calculated by Daumer (1956) for complementary color stimuli (*1–3*) from the measured colorimetric equation for "bee-white", and the independently measured intensity mixing proportions of noncomplementary spectral lights (*4–8*). The numbers denote wavelength in nm. The loci of the color stimuli in both graphs are determined by the "center-of-gravity" construction from Daumer's equations. The *edges of the upper diagram* represent three basic spectral lights as primaries. The *edges of the lower diagram* represent the photon fluxes absorbed by the three receptor types. The comparison shows that the mixture rules hold in the physiological diagram as well, since the absorbed quantum fluxes are linear functionals of the light intensities. (Krantz 1975)

and examine whether it predicts the experimental results obtained by Daumer (1956). Since "bee-luminance" is assumed to be proportional to the sum of the tristimulus values, the "bee-brightness" of any spectral mixture is only dependent on the spectral sensitivities of the receptors. Therefore, the color stimulus space corresponds to the color luminance space. The corresponding physiological chromaticity diagram represents a plane of constant "bee-brightness" (constant totally absorbed photon flux) (see Backhaus and Menzel 1987, for details). Figure 2 (lower graph) demonstrates how well Daumer's color-mixing experiments are predicted by the receptor based chromaticity diagram. However, it should be pointed out that Daumer's matching experiments did not perfectly match for equal brightness, as the diagram in Fig. 2 assumes. Daumer was concerned about this problem, but nevertheless realized that brightness differences did not affect, or only slightly affected, the matching between color signals. We shall come back to this point later.

3 Chromatic and Achromatic Vision in Honey Bees

Primates lose color vision under scotopic light conditions. This is due to two factors: (1) At low light levels the cones in the fovea pool their excitation in such a way that the differential processing of spectral input is lost, and (2) the rods in the periphery are more sensitive than the cones (Bouman 1969; Walraven 1973). Bees do not appear to possess specialized low light receptors. Although two of the four green receptors have a slightly larger rhabdomere volume and project with thick axons into the proximal layer of the lamina, they are not more sensitive to low light than the other receptor types (Menzel and Blakers 1976). How then is scotopic vision organized in bees? Are they able to see colors down to the threshold of absolute light sensitivity, or is there a neuronal mechanism which gains sensitivity by summing the excitation from all receptors of one ommatidium? Behavioral experiments have addressed this question by determining the thresholds for color vision and for absolute light sensitivity (Menzel 1981).

Bees were trained in a T-maze to turn left on viewing spectral light (e.g., 533 nm) or right when viewing white light, both lights being shone at the choice point of the T-maze. The bees learned this double task within a few hours, and responded to the two light signals with more than 80% correct choices. Since the bees were trained individually to run straight from the dark hive through a dark tube into the dark experimental chamber, they were all well adapted to darkness. The critical experiment consisted of a series of tests where the quantum flux of the spectral light was changed over a wide range (5–6 log steps) in a random sequence. Figure 3 shows that there are two thresholds. The higher threshold is for the response to the side, which was trained to the spectral light, and would seem to indicate that, below this threshold, the bees lose the ability to identify a spectral light as a chromatic stimulus. The other, lower threshold is for the response to the side trained to the white light, and is a particularly important finding since it shows that bees interpret a certain range of quantum flux of a spectral light as "bee-white" (see further Menzel 1981).

This "achromatic" intensity range is equally large (ca. 1.5 log units wide) for each of the three spectral lights (413, 440, 533 nm) so far tested. Furthermore, the two thresholds for 413 nm are higher than those of 440 nm, and these in turn are higher than those of 533 nm (Fig. 3). It is likely, therefore, that "achromatic" dim light vision in bees is the result of an addition of the three spectral inputs. Visual interneurons with broad-band spectral properties, and receiving input from all three receptor types, have been recorded at all levels of visual integration in the bee brain (Kien and Menzel 1977a).

Exner, to whom this Volume is dedicated, was of the opinion that "the compound eye functions in a similar fashion to the periphery of the human retina" (Exner 1891, p. 185). His reasons for saying so were that (p. 183) "the compound eye is more perfectly suited to detect the (temporal) changes of figures", as the periphery of the vertebrate eye does. As far as color vision is concerned, it is indeed possible that the temporal succession of chromatic stimuli is more important for insect vision than for primate vision (see below). However, as far as the receptor and neuronal mechanisms underlying fine grain color vision are concerned, there is a

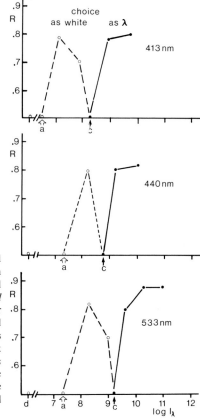

Fig. 3. Thresholds for color vision (*dark arrow c*) and achromatic vision (*light arrow a*) are separated by an achromatic interval. Dark adapted bees were trained and tested in a T-maze as described in the text. The *full line* and *dark symbols* indicate the choice of the bee for the spectral light seen as color, whilst the *dotted line* and *open symbols* indicate the choice of the spectral light as an "achromatic white" light. If the bees run in a dark T-maze, they choose the arms randomly (*left points* marked with *d*). The abscissa gives the quantal flux of the spectral test light in log steps, and the ordinate gives the response value of the dual forced choice. (After Menzel 1981)

clear correspondence between the fovea of the vertebrate eye and the compound eye of the honey bee. The equivalent of ommatidia in the vertebrate fovea are groups of cones which analyze more or less the same point in space with respect to its chromatic properties. This group of cones is functionally connected under low light conditions to improve sensitivity, but such a connection is at the expense of color vision. The structural ommatidia in the bee eye operate in the same way.

4 Higher Colorimetry and Perceptual Distances in the Color Space

The resolution of the color vision system in bees has been measured in training experiments dealing with wavelength discrimination (see Fig. 4) (von Helversen 1972), and with first color saturation discrimination threshold for equally bright spectral lights that surrounds the white point (Lieke 1984). Both of these discrimination tasks are wavelength-dependent. Since the concept of the chromaticity diagram and color space is based on color matching judgments

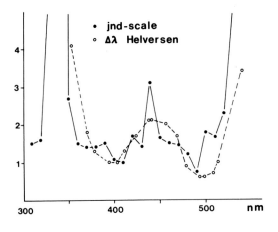

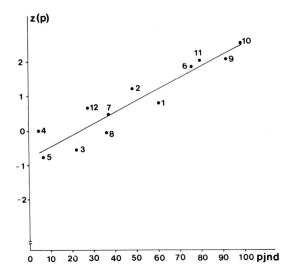

Fig. 4. Upper graph A comparison between the $\Delta\lambda/\lambda$-function as determined by von Helversen (1972) in behavioral experiments (o---o) and the predictions of the model calculation as described in the text. A unit at the ordinate corresponds to 8 nm for the von Helversen data, and to 38 pjnd for the model calculation. *Lower graph* Correlation between discrimination values determined by dual forced choice tests (ordinate, the percent of correct choices is expressed in z-values, $z = 0$ corresponds to 75% correct choices) and the jnd-values as calculated for the perceptual distances (pjnd) in the receptor plane. The pjnd values are calculated, as outlined in the text, by ignoring any perceptual distances resulting from brightness differences. The pjnd-values for the plane correlate better with the behavioral discrimination values than the spatial pjnd-values, which include perceptual distances resulting from brightness differences. The *numbers at the points* indicate colors which differed in their spectral components, proportion of white light and intensity (see Backhaus and Menzel 1987)

(identity judgments), the results of color difference judgments in discrimination experiments cannot be predicted in a simple way from the differences of tristimulus values or chromaticity coordinates.

Von Helmholtz (1896) overcame this problem by measuring the dissimilarity of colors in steps which refer to a subjective, perceptual measure, namely the smallest just noticeable distance (jnd-step). In the chromaticity diagram, the lines with the smallest number of steps and the lines of most similar colors are curved and not equidistantly spaced. The shortest lines of most similar colors can be derived mathematically as geodesics (line elements) in a non-Euclidean (Riemannian) space. Schrödinger (1920) worked out the concept of line elements and showed that the color discrimination and dissimilarity judgments of "higher colorimetry" have no simple relation to the "lower colorimetry" of color mixture. In accordance with von Helmholtz, he said: "Two colors are assigned to be more similar if the shortest distance in the Riemannian space involves fewer jnd-steps". We follow Schrödinger's proposal and assume that in the case of "small field color vision" (one ommatidium, three receptor types — three ultraviolet, two blue, and four green receptors), the resolution of the color vision system is ultimately determined by the photoreceptors, and that the resolution of the photoreceptors is limited by random fluctuations in the receptor potential. The graded receptor potential is the total number of units of transmembrane potential related to the number of absorbed photons. The photon absorption process and the transduction process is thus responsible for fluctuations in the receptor potential (shot noise and trans-ducer noise) (Laughlin 1981). To a first approximation these fluctuations were found to be independent of the magnitude of the potential. This was also the case in our intracellular recordings (Backhaus and Menzel 1987). The mean fluctuation amplitude is converted into a perceptual jnd (pjnd) by assuming that one pjnd is reached if the variation in intensity, wavelength and white-light proportion, with respect to a reference light, causes a significant change in at least one of the three receptor types (rjnd). In other words, if two targets with the same brightness are viewed one after another by an ommatidium, the subjective color difference is determined by the total number of rjnd steps in the three receptors.

When the pjnd-scale is compared with the dissimilarity judgment of the bee, a very good agreement is found. Two examples are given in Fig. 4, which shows the comparison of the pjnd scale for spectral lights in 10 nm steps with the $\Delta\lambda/\lambda$-function of von Helversen (1972). Figure 4 (lower) gives the correlation between the pjnd-values of non-spectral color plates in daylight and the probability transformed choice proportions z(p) of dual choice training experiments. The correlation is optimal if any differences in intensity are ignored and the pjnds are calculated for the chromaticity plane of equal brightness. This observation also applies to colored objects which differ markedly in their brightness components (see Backhaus et al. 1987a). Therefore, one can conclude that bees ignore differences in brightness when trained to colors. This is in good agreement with the observations of Daumer (1956), Menzel (1967), and von Helversen (1972), who were concerned about the possible influence of brightness differences on the discriminability of hue and saturation (chromaticity). This result is important, in that it suggests that the color vision system of bees may be two-dimensional, as opposed to the three perceptual dimensions of man's color vision.

Backhaus et al. (1987a) used multidimensional scaling procedures to analyze similarity judgments of bees in a test situation with 12 simultaneous alternative color plates. Although the light differed in spectral composition, intensity, and the proportion of white light, only two scales were found to be necessary in order to reconstruct the experimental data (Fig. 5). In an attempt to interpret these scales on a physiological basis, we analyzed various correlations between these two scales and the parameters of the color vision system. Color difference is calculated as the sum of the absolute differences of the two perceptual coordinates (city-block metric). Neither of the two scales correlates with one of the receptor inputs from the 12 color signals and must, therefore, correlate with a combination involving all three spectral inputs. As can be seen in Fig. 5, the arrangement of the color loci in

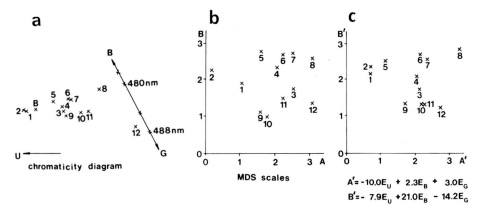

Fig. 5a-c. Results of the multidimensional scaling procedure (MDS) applied to twelve matrices of 12 × 12 discrimination values. The twelve color stimuli (Nos. *1* to *12*) are given with their respective color loci in the chromaticity diagram (*a*). MDS procedures provide two scales *A, B* (*b*) which are related by a city-block metric. A comparison between the relative position of the respective color loci indicate a very good reconstruction on the basis of the two scales *A* and *B*. The result of an interpretation of the scales *A* and *B* as two spectrally opponent axes is shown in *c*. *E* denotes the excitation of the respective receptors (see text, and Backhaus and Menzel 1987)

a chromaticity diagram (Fig. 5a) and the color loci drawn in the two orthogonally arranged perceptual coordinates (Fig. 5b) are very similar (see Backhaus et al. 1987a).

This would suggest that the two scales correlate with either the von Helmholtz coordinates of hue and saturation or the color-opponent axes blue/purpleness and UV/blue-greenness. The latter possibility was further examined by assuming that the two scales are determined by a linear combination of the receptor potentials (color opponent model). The gain coefficients are calculated for the best fit of the excitations in the color opponency model to the corresponding scale values derived by multidimensional scaling analysis. The result is given as: scale $A = -10.0 E_U + 2.3 E_B + 3.0 E_G$, and scale $B = -7.9 E_U + 21.0 E_B - 14.2 E_G$ (where E denotes the excitation of the three different receptor types UV: E_U, B: E_B, and G: E_G. Thus it is possible that scale A corresponds to a UV/blue-green opponency, and that scale B corresponds to a blue/purple opponency.

5 Color Opponent Processes

In humans, most phenomena of color vision are best explained by a combination of trichromacy at the retinal level and color opponency at the level of visual neurons. This also holds for color vision in honey bees. Hue "cancellation" within the confines of an object supports Hering's (1920) theory of color opponency. The physiological substrates for color opponency, as well as simultaneous and successive color contrast phenomena, are spectrally opponent visual interneurons with overlapping receptive fields of antagonizing receptor inputs. In the honey bee such neurons have been described for two sets of spectral opponencies: type I: UV[+] B[-] G[-] (or UV[-] B[+] G[+]) and type II: UV[+] B[-] G[+] (or UV[-] B[+] G[-]) (Kien and Menzel 1977b, Hertel 1980), (see Fig. 6). These neurons offer a straight-forward qualitative explanation of the results of Daumer's "cancellation" experiments: a mixture of 360 and 490 nm in proper proportion would cancel a response in type I, and similarly a mixture of 360 + 440 + 588 nm would cancel a response in type II, while a combination of type I and type II would explain the cancellation of 400 + 588 nm. A quantitative description cannot be derived from the recording experiments, since the excitatory and inhibitory portions of the spectral response functions are very different for different neurons. As already pointed out above, the interpretation of the two perceptual dimensions derived from a multidimensional scaling procedure suggests two color opponent systems with a specific set of coefficients. It should be noted that the results of recording experiments agree with the predictions of these two color opponent systems.

Simultaneous and successive color contrast, as well as color constancy, are well established perceptual phenomena in the bee. A blue target on a yellow back-

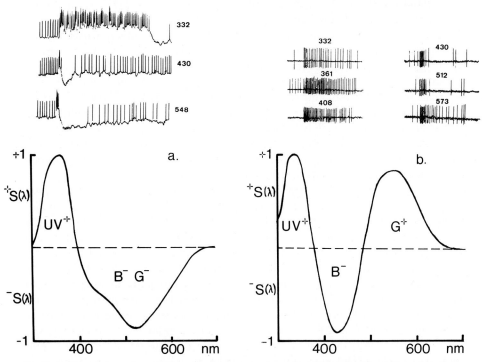

Fig. 6a,b. Two classes of spectrally opponent neurons were found that developed their opponency relatively slowly and which then expressed this opponency in their tonic response to prolonged flashes of spectral light. Type I neurons (*a*) respond antagonistically to UV and long wavelength light (either UV[+] B[-] G[-] or UV[-] B[+] G[+]) whilst type II (*b*) are excited by UV and G and inhibited by B or vice versa (Kien and Menzel 1977b; Hertel 1980). The spectral functions are generalized from several recordings of each type, both of which differ quite considerably with respect to their relative sensitivities at various wavelengths. The excitatory and inhibitory parts are normalized independently to their respective optima

ground looks more blue to the bee than a blue target on a grey background, and a yellow target on a blue background more yellow than if it were on a grey background (Neumeyer 1980). A change in the overall chromatic illumination has little or no effect on the choice behavior towards reflecting color marks (Neumeyer 1981). Bees were also tested in a multicolored display similar to that of the "Mondrian" arrangement used by Land (1977). The display is illuminated by three spectral lights of differing wavelength that correspond to the respective spectral receptor types (Werner et al. 1988). After the bee has been trained to one of the differently colored areas, the illumination is changed in such a way that the light flux emitted formerly by the trained area is now emitted by another area. The bees still choose the trained area, thus indicating that, like humans, their color vision is independent of the spectral content of the illumination.

Simultaneous color contrast and color constancy in vertebrates are thought to be under the control of double opponent neurons in the cortex that have a combined spectral and spatial opponency in their concentric receptive fields (Gouras 1974; Zeki 1984). Although concentric double opponent neurons have not been found in the bee's visual system, a wide variety of neurons that show a combination of spectral and spatial contrast have been observed (Kien and Menzel 1977b; Hertel 1980). These neurons generally have separated receptive fields for different spectral compositions, e.g. a UV sensitive field looking upwards, a green sensitive field looking forwards and downwards, or a UV^+ B^- G^- field in one eye and a UV^- B^+ G^+ in the other eye (Hertel and Maronde 1987). The receptive subfields are small or large, partially overlapping or completely separated, and are characterized by sometimes having very complicated structures of their spectral-spatial sensitivities for the different temporal components of their responses to light flashes (see complex neurons in Kien and Menzel 1977b). Perhaps such neurons are specialists for particular color patterns, rather than those multipurpose color coding neurons that subserve properties such as color contrast and color constancy independent of the actual arrangement of colored patterns.

Since the type I and II spectrally opponent neurons, i.e., those with overlapping receptive fields of relatively large size (40°-60° visual angle) and no spatial antagonism, seem to be the most uniform and frequently found color coding cells in the visual system of the bee, it is worthwhile asking whether such neurons are sufficient to explain the observed perceptual phenomena. So far all behavioral experiments testing color contrast and constancy were carried out with free flying bees that scan over the colored targets and transpose the spatial arrangement into a temporal sequence of stimulation. Since spectrally opponent neurons have no spatial antagonism, their response is unambiguous for color and location or color and brightness. This is also the case for the single opponent neurons (e.g., G^+ centre, R^- surround) that are so common in the primate retina and the lateral geniculate nucleus (DeValois and DeValois 1975).

However, in the bee, the fact that the receptive fields of the spectrally opponent mechanisms completely overlap excludes the possibility of simultaneous contrast enhancement and suggests a temporal mechanism in establishing color constancy. For the general processes of local contrast enhancement and wide field color constancy, the bee may not need double opponent neurons with concentric receptive fields if appropriate adaptation mechanisms exist peripheral to or in these neurons. In that case the temporal sequence of stimulation would adjust each spectrally opponent neuron to a dynamic working point against which the positive and negative going fluctuations of excitation resulting from the sequential stimulation are expressed. The advantage of such a system is that spatial and temporal information is not confounded in a single, spectrally coding neuron. Most interestingly, such neurons have complex temporal response properties to spectral light flashes. Many neurons develop spectral opponency over considerable length of time (100 ms or more), others respond antagonistically in their fast ON and OFF responses (Kien and Menzel 1977b; Hertel 1980; Riehle 1981). Further experiments are necessary in order to examine the question of whether these properties are related to the temporal integration of successive stimulation, as suggested above.

6 Wavelength Selective Behavior, Color Categories and the Uniqueness of Colors

In bees, color vision in its true sense is well documented only for orientation at the immediate surround of the food source and the entrance of the colony housing. Other behaviors and visual orientation tasks are also guided by the spectral components of the light stimuli, but color vision is not involved. This was recently confirmed by Menzel and Greggers (1985), who found that the strength of the phototactic response to a mixture of spectral lights follows a strict additive rule of effective quanta without any indication of color effects. We already pointed out in the introduction that bees are color blind in their phototactic runs. This is also true for the optomotor response (Kaiser and Liske 1974) and for their visual scanning behavior in front of vertical bars (Lehrer et al. 1985). The spectral sensitivity functions for these two behaviors differ considerably. The optomotor and scanning response is dominated by the green receptors (Kaiser and Liske 1974), while the phototactic response is controlled by all photoreceptors, although under certain conditions the strongest contribution is from the UV-receptors (Kaiser et al. 1977; Menzel and Greggers 1985). Orientation to polarized light is a particularly interesting example of wavelength selective behavior. Extended light sources emitting short-wavelength light with or without polarization are interpreted as part of the sky, whereas point light sources emitting unpolarized long-wavelength light are interpreted as the sun (von Helversen and Edrich 1974; Edrich et al. 1979; Brines and Gould 1979; Rossel and Wehner 1984). Behavioral test situations can be arranged in such a way that the emission spectrum of the dorsally arranged light source is the only parameter guiding the bee during its dance behavior on a horizontal comb. In such a situation, bees categorize a light stimulus as either sun or sky, depending on the spectrum of the light source. Wavelengths around the transition point between the two categories (around 400 nm) confuse the bee. There is, however, no indication yet that bees use color vision in their celestial orientation apart from this categorial separation along the wavelength scale.

The special characteristics of certain colors become evident when bees are trained to colored targets at the feeding place. For example, the learning rate of spectral colors is fastest for violet and slowest for bluish-green, with the other wavelengths lying in between (Menzel 1967). Any perceptual parameter such as subjective brightness, saturation, or the wavelength dependence of spectral discrimination, cannot be the reason for these differences. It is more likely that certain evolutionary predispositions enable the animal to categorize colors as potential food sources. On the basis of these and other observations, we can conclude that the memory for colors induces distortions in the perceptual representations of remembered colors, which in turn lead to biased generalizations. Such color biases may be interpreted as being an indication of the existence of unique colors and color categories.

7 Conclusion: An Integrative View of Color Vision in Honey Bees

The analogy between color vision in primates and honey bees has been stressed with good reason (e.g., Autrum 1963; Daumer 1956; Menzel 1979) and the functional similarities are indeed striking. For example, trichromacy at the input stage, applicability of the Graßmannian mixture rules, different thresholds for achromatic and chromatic vision (particularly when compared with foveal vision in humans), color contrast effects, color constancy, and the possibility of the experience of unique colors. However, there are also important differences. In humans the luminance function $[V(\lambda)]$ is highly tuned, the $S(\lambda)$ functions of the three receptor types are unevenly spaced throughout the visual range, and the receptors contribute unequally to color vision (in particular the blue cone has lowest weight). In bees, however, the $V(\lambda)$ function is broad and the $S(\lambda)$ functions are evenly spaced and contribute about equally. Furthermore, humans have three perceptual color dimensions (von Helmholtz's coordinates hue, saturation, brightness, or the opponent axes blue/greenness, green/redness, black/whiteness depending on the experiment) whereas the bee has two perceptual dimensions (UV/blue-greenness, blue/purpleness) — at least if the animals are not specifically trained to brightness differences.

The neuronal mechanisms of color coding in bees and primates are basically similar but differ in important details. The two-zone theory put forward by von Kries (1905) for man is equally applicable to the bee. The Young-Helmholtz first stage, with three spectral classes of receptors as inputs, explains the color mixing and hue cancellation phenomena (lower colorimetry). With extensions for dissimilarity judgments, the structure of color distance is also explained on this level (higher colorimetry). The second Hering stage is based on spectrally opponent neurons, and accounts for the phenomena of successive and simultaneous color induction. In both humans and bees, color distance can be explained as a function of the absolute differences in the excitations of two color opponent cell types.

Simultaneous color contrast and color constancy in primates — the two additional and most important phenomena of color vision — can be explained in the second stage of Hering's theory with additional assumptions about the lateral interactions of color opponent cells. Center-surround double opponent neurons at the level of the visual cortex are well suited to serving simultaneous color contrast by short distance local computations, and color constancy by long distance interactions (Daw 1984; Zeki 1984; Land 1986). Such double opponent neurons have not been found in the bee brain or any other insect brain, and may, therefore, not exist. Combined spatial and spectral opponency would appear to be a characteristic of visual neurons with specially structured, and sometimes highly complex, receptive fields (Kien and Menzel 1977b; Hertel 1980). These neurons are probably involved in a kind of feature-detecting system for particular color patterns. Therefore, simultaneous color contrast may exist only for particular patterns of color arrangements, whilst other color contrast phenomena may be based solely on successive contrast effects. Even the phenomenon of color constancy may be based upon temporary sampling of contrast values alone. To determine the color of an object independently of the chromatic illumination, the nervous system must compare the wavelength composition of light reflected by an object with respect to the wavelength composition of the light reflected from other areas in the surrounding. If the bee collects the necessary information by scanning over the objects, as opposed to instantaneous comparison of the neural representations of the visual objects, then a sensory memory in the range of seconds is needed to adjust for any changes in illumination. At the present time we lack experimental evidence that would allow us to speculate about the neural basis of the constancy phenomenon.

The combined efforts of behavioral, neurophysiological, and modeling analysis have elucidated many details of the honey bee's color vision. The next stage is the rationalization of the processes that are involved and which control color contrast and color constancy phenomena — two subjects that are behaviorally documented but which nevertheless are little understood.

References

Autrum H (1963) Wie nimmt das Auge Farben wahr? Umschau 63:332–336

Autrum H, Zwehl V von (1964) Spektrale Empfindlichkeit einzelner Sehzellen des Bienenauges. Z Vergl Physiol 48:357–384

Backhaus W, Menzel R (1987) Color distance derived from a receptor model of color vision in the honeybee. Biol Cybernet 55:321–331

Backhaus W, Menzel R, Kreißl S (1987a) Multidimensional scaling of color similarity in bees. Biol Cybernet 55:331–333

Backhaus W, Werner A, Menzel R (1987b) Color vision in honeybees: metric, dimensions, constancy, and ecological aspects. In: Menzel R, Mercer A (eds) Neurobiology and behavior of the honeybee. Springer, Berlin Heidelberg New York Tokyo

Bouman MA (1969) My image of the retina. Q Rev Biophys 2:25–64

Brines ML, Gould JL (1979) Bees have rules. Science 206:571–573

Brown PK, Wald G (1964) Visual pigments in single rods and cones of the human retina. Science 144:45–52

Cornsweet TN (1970) Visual perception Academic Press, New York London

Daumer K (1956) Reizmetrische Untersuchungen des Farbensehens der Bienen. Z Vergl Physiol 38:413–478

Daw NW (1984) The psychology and physiology of colour vision. Trends Neurosci 6:330–335

DeValois RL, DeValois KK (1975) Neural coding of color. In: Caterette EC, Friedman MP (eds) Handbook of Perception, vol 5. Academic Press, New York London, pp 117–166

Edrich W, Neumeyer C, Helversen O von (1979) "Anti-sun orientation" of bees with regard to a field of ultraviolet light. J Comp Physiol A 134:151–157

Exner S (1891) Die Physiologie der facettierten Augen von Krebsen und Insecten. Deuticke, Leipzig

Frisch K von (1914) Der Farbensinn und Formensinn der Biene. Zool J Physiol 37:1–238

Gouras P (1974) Opponent-colour cells in different layers of foveal striate cortex. J Physiol 238:583–602

Graßmann H (1853) Zur Theorie der Farbenmischung. Ann Phys Chem 89:69–84

Gribakin FG (1972) The distribution of the long wave photoreceptors in the compound eye of the honey bee as revealed by selective osmic staining. Vision Res 12:1225–1230

Helmholtz H von (1896) Handbuch der physiologischen Optik, 2 edn. Voß, Hamburg

Helversen O von (1972) Zur spektralen Unterschiedsempfindlichkeit der Honigbiene. J Comp Physiol A 80:439–472

Helversen O von, Edrich W (1974) Der Polarisationsempfänger im Bienenauge: Ein Ultraviolett-rezeptor. J Comp Physiol A 94:33–47

Hering E (1920) Grundzüge der Lehre vom Lichtsinn. Springer, Berlin

Hertel H (1980) Chromatic properties of identified interneurons in the optic lobes of the bee. J Comp Physiol A 137:215–231

Hertel H, Maronde U (1987) Processing of visual information in the honeybee brain. In: Menzel R, Mercer A (eds) Neurobiology and behavior of the honeybee. Springer, Berlin Heidelberg New York Tokyo

Hess C von (1913) Experimentelle Untersuchungen über den angeblichen Farbensinn der Bienen. Zool Jahrb Abt Physiol 34:81–106

Hubel DH, Wiesel TN (1972) Laminar and columnar distribution of the geniculocortical fibres in the macaque monkey. J Comp Neurol 146:421–450

Kaiser W, Liske E (1974) Die optomotorischen Reaktionen von fixiert fliegenden Bienen bei Reizung mit Spektrallichtern. J Comp Physiol A 89:391–408

Kaiser W, Seidl R, Vollmar J (1977) The participation of all three colour receptors in the phototactic behavior of fixed walking honeybees. J Comp Physiol A 122:27–44

Kien J, Menzel R (1977a) Chromatic properties of interneurons in the optic lobes of the bee. I. Broad band neurons. J Comp Physiol A 113:17–34

Kien J, Menzel R (1977b) Chromatic properties of interneurons in the optic lobes of the bee. II. Narrow band and colour opponent neurons. J Comp Physiol A 113:35–53

Krantz DH (1975) Color measurement and color theory: I. Representation theorem for Graßmann structures. J Math Psychol 12:283–303

Kries J von (1905) Die Gesichtsempfindungen. In: Nagell W (ed) Handbuch der Physiologie des Menschen, vol 3. Vieweg, Braunschweig, p 269

Kühn A (1927) Über den Farbensinn der Bienen. Z Vergl Physiol 5:762–800

Land EH (1977) The retinex theory of color vision. Sci Am 108–128

Land EH (1986) Recent advances in the retinex theory. Vision Res. 26:7–21

Laughlin S (1981) Neural principles in the peripheral visual systems of invertebrates. In: Autrum H (ed) Handbook of sensory physiology, vol VII/6B. Springer, Berlin Heidelberg New York, pp 133–280

Le Grand Y (1948) Optique physiologique, vol 2. Luminière et couleurs. Rev Optique (Paris)

Le Grand Y (1971) Light, colour and vision. Chapman, London

Lehrer M, Wehner R, Srinivasan M (1985) Visual scanning behavior in honeybees. J Comp Physiol A 157:405–415

Lieke E Farbensehen bei Bienen: Wahrnehmung der Farbsättigung. Diss. Freie Universität Berlin.

Marks WB, Dobelle WM, MacNichol EF (1986) Visual pigments of single primate cones. Science 143:45–52

Maxwell JC (1860) On the theory of compound colours of the spectrum. Philos Trans R Soc London 150:57–84; Sci Pap 1890, vol 1. Univ Press, Cambridge, pp 410–444

Menzel R (1967) Untersuchungen zum Erlernen von Spektralfarben durch die Honigbiene, *Apis mellifica*. Z Vergl Physiol 56:22–62

Menzel R (1979) Spectral sensitivity and color vision in invertebrates. In: Autrum H (ed) Handbook of sensory physiology. vol VII/6A Berlin, Heidelberg, New York, pp 504–580

Menzel R (1981) Achromatic vision in the honeybee at low light intensities. J Comp Physiol A 141:389–393

Menzel R (1985) Learning in honeybees in an ecological and behavioral context. In: Hölldobler B, Lindauer M (eds) Experimental behavioral ecology. Fischer, Stuttgart, pp 55–74

Menzel R, Blakers M (1976) Colour receptors in the bee eye — morphology and spectral sensitivity. J Comp Physiol A 108:11–33

Menzel R, Greggers U (1985) Natural phototaxis and its relationship to colour vision in honeybees. J Comp Physiol A 157:311–321

Menzel R, Ventura DF, Hertel H, de Souza J, Greggers U (1986) Spectral sensitivity of photoreceptors in insect compound eyes: comparison of species and methods. J Comp Physiol A 158:165–177

Milde J (1978) Bestimmung der spektralen Empfindlichkeit in verschiedenen Augenregionen der Biene. Dipl Arbeit, Freie Univ Berlin

Neumeyer C (1980) Simultaneous color contrast in the honeybee. J Comp Physiol A 139:165–176

Neumeyer C (1981) Chromatic adaptation in the honeybee: successive color contrast and color constancy. J Comp Physiol A 144:543–553

Riehle A (1981) Color opponent neurons of the honey bee in a hetero-chromatic flicker test. J Comp Physiol A 142:81–88

Rodieck RW (1973) The vertebrate retina. Freeman, San Francisco

Rossel S, Wehner R (1984) Celestial orientation in bees: the use of spectral cues. J Comp Physiol A 155:605–613

Rushton WAH (1972) Pigments and signals in colour vision. J Physiol 220:1–31

Schrödinger E (1920) Grundlinien einer Theorie der Farbenmetrik im Tagessehen: Der Farbenmetrik II. Teil: Höhere Farbenmetrik (eigentliche Metrik der Farbe). Ann Phys 63:481–520

Snyder AW, Menzel R, Laughlin SB (1973) Structure and function of the fused rhabdom. J Comp Physiol A 87:99–135

Walraven PL (1973) Theoretical models of the colour vision network. Colour 73. Hilger, London

Werner A, Menzel R, Wehrhahn C (1988) Color constancy in the honeybee. J Neurosci 8:156–159

Young T (1802) On the theory of light and colours. Philos Trans R Soc London Ser B 92:12–48

Zeki S (1984) Colour pathways and hierarchies in the cerebral cortex. In: Ottoson D, Zeki S (eds) Central and peripheral mechanisms of colour vision. MacMillan, London, pp 19–44

Chapter 15

Polarization Sensitivity in Compound Eyes

SAMUEL ROSSEL, Freiburg, FRG

Introduction

The compound eyes of insects and many other arthropods can detect a basic property of light which normally remains invisible to us: the plane of polarization. Arising by scattering in air and water, as well as by reflection from surfaces (Fig. 1), linearly polarized light (characterized by the E-vector direction) is widespread in nature and provides the sensitive eye with a great deal of extra optical information. How polarized light is analyzed and used by arthropods is the theme of this chapter.

The study of polarization sensitivity in compound eyes began 40 years ago when von Frisch (1949) discovered that bees can use the polarization patterns in the sky as a compass. At that time von Frisch investigated how bees orient their waggle dances on horizontal combs, i.e., when they could not use gravity as an orientational cue. Previous experiments had already shown that under these circumstances the bees could orient by means of the sun, but now it turned out that the bees danced correctly towards the foraging site, even when they could see only a patch of blue sky instead of the sun itself. In an elegant series of experiments, von Frisch proved that this compass orientation depended on the celestial polarization patterns. Briefly, he placed a polarizing filter above the hive and as he rotated the filter, the bees changed their dance direction by a predictable amount (rev. von Frisch 1965).

This demonstration of polarization orientation came as a surprise and stimulated a great deal of interest in polarization sensitivity. As a result, evidence rapidly accumulated showing that not only bees, but a wide variety of invertebrates can perceive polarized light (von Frisch 1965). In most of them this sensory capacity seemed to be involved in celestial orientation, but recently an alternative use has been demonstrated by Schwind (1983a, 1984). He found that backswimmers, while flying from one pond to another, detect their water habitat by virtue of the horizontally polarized light which is reflected from the water surface. Other uses of polarization sensitivity have also been proposed. For example, dragonflies may use water surface polarization as some kind of artificial horizon, an implied ability which would help these insects to stabilize flight (Laughlin 1976). Water striders, on the other hand, may use polarization sensitivity to reduce glare, thus improving their ability to detect prey objects on the water surface (Schneider and Langer 1969; Bohn and Täuber 1971). Finally, the ability to detect polarized light may be useful in underwater vision of crustaceans and other invertebrates by increasing the

Stavenga/Hardie (Eds.) Facets of Vision
© Springer-Verlag Berlin Heidelberg 1989

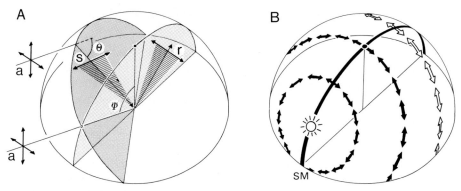

Fig. 1A,B. Light polarization by scattering and reflection. *A* General principles. Unpolarized sunlight (*a*) streaming into the atmosphere is scattered by air molecules and reflected from surfaces (scattered and reflected light beams are denoted by *s* and *r*, respectively). Scattered light in the forward direction (scattering angle $\theta = 0°$) is unpolarized, off this axis it is partially polarized, reaching a maximum degree of polarization when $\theta = 90°$. The direction of polarization (E-vector direction, *double arrow*) is normal to the plane defined by a and s. The degree of polarization of reflected light depends on the angle of incidence ψ. If light strikes a water surface at an angle of 53°, maximal polarization occurs with the E-vector direction normal to the plane of incidence and therefore parallel to the water surface. *B* The celestial polarization pattern. According to the rules outlined in *A*, E-vector directions are concentrically arranged around the sun (or its opposite, the anti-sun). Direct sunlight is not polarized. Maximal polarization occurs along the great circle 90° from the sun (marked by white E-vectors). Note that for a terrestrial observer the solar meridian (*SM*) and the antisolar meridian form a plane of symmetry, i.e., the E-vector pattern to the left and to the right of these meridians are mirror images to each other. For further details of the principles of light polarization in nature see Waterman (1981) and Brines and Gould (1982)

visibility of objects (Lythgoe and Hemmings 1967) or by providing positional information under conditions when the habitat is otherwise devoid of appropriate orientational cues (Waterman 1981).

Apart from its biological significance, polarization sensitivity raises a number of questions in relation to the underlying mechanisms. For example, what are the analyzing structures in the compound eyes? How is polarized light detected by the visual system? What features of the polarized world can a particular animal discriminate and use appropriately in behavior? These and related questions represent a fascinating challenge to sensory neurobiology and ecology, and a brief review of our present understanding is given in this chapter. We look first at how rhabdomeric photoreceptors of arthropods act as basic polarization analyzers, and then at their distribution within compound eyes. This is followed by a case study, demonstrating how bees use their photoreceptors to analyze the polarization patterns in the sky for navigational means. Finally, the results from the bee experiments are compared with neurobiological data obtained from other insects. The picture emerging from all these studies is that the operational principles underlying polarization vision are not as complex as it originally seemed.

The field of polarization vision is rich and diverse and for further documentation of the topic the reader is referred to the reviews by von Frisch (1965), Laughlin et al. (1975), Waterman (1981, 1984), Wehner and Rossel (1985) and Rossel and Wehner (1987).

2 Photoreceptors: the Basic Analyzers for Polarized Light

Shortly after von Frisch's discovery of polarization sensitivity in bees, Autrum and Stumpf (1950) suggested that the analyzer for polarized light is situated in the photoreceptors themselves rather than in the overlying dioptric system. Impressed by the striking radial arrangement of photoreceptors in ommatidial cross-sections, they proposed that each photoreceptor is sensitive to a specific plane of polarization owing to some structural property of its photoreceptive organelle. This hypothesis was a subject of much debate until Stockhammer (1956, 1959) provided compelling optical evidence, demonstrating that neither the cornea lenses nor the crystalline cones of compound eyes could act as analyzers. This finding was later confirmed by physiological measurements reported by Shaw (1967). He succeeded in recording simultaneously from pairs of locust retinula cells located in the same ommatidium. Such pairs of receptors could differ in the plane of polarization to which they were optimally sensitive. Clearly this result could not be explained by some analyzing function of the dioptric system, for this should be equally effective in all photoreceptors of a single ommatidium. Apart from this, Shaw's experiments directly proved the original proposal of Autrum and Stumpf, namely that each ommatidium contains photoreceptors which are sensitive to different E-vector directions.

Exactly how can a photoreceptor analyze polarized light? The answer to this question lies in the dichroic absorption properties of both the visual pigments and the membrane structures in which the pigment molecules are embedded. This idea was proposed as early as 1953 by de Vries and colleagues. Nevertheless, all efforts to prove the hypothesis met with failure until the mid 1960's when Langer (1965) first demonstrated dichroic absorption in the photoreceptors of flies.

2.1 Dichroism and Polarization Sensitivity

The visual pigment chromophore is a dipole and as such exhibits maximal absorption when the E-vector of incident light is parallel to the long axis of the molecule. Dichroic chromophores, however, are not themselves sufficient to explain the receptor's polarization sensitivity. Rather the photoreceptive organelle as a whole must be dichroic. In the rhabdomeric photoreceptors of compound eyes (Fig. 2A and B), this is indeed the case. As first reported by Fernández-Morán (1956), Miller (1957), Goldsmith and Philpott (1957) and Wolken et al. (1957), the rhabdomeres are composed of parallel arrays of microvilli, oriented perpendicular to the optical axis of the receptor and hence to the direction of incident light. The visual pigment molecules are embedded within the membranes of the microvilli. Optical examination of the disc membranes of vertebrate photoreceptors suggests that the visual pigment molecules maintain a fixed angle relative to the membrane surface, but are free to rotate within the tangent plane of the membrane (rev. Laughlin et al. 1975). The resulting random orientation in this plane explains why vertebrate photoreceptors exhibit no dichroic absorption under normal conditions of illumination. However, with the tubular structure of the microvilli of invertebrate photoreceptors, dichroic absorption will automatically result, even when the

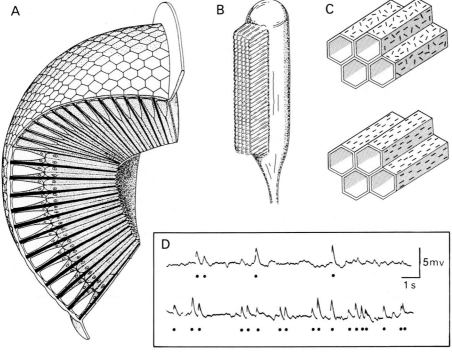

Fig. 2A-D. Basic anatomy of compound eyes and rhabdomeric photoreceptors. *A* Compound eyes consist of a large number of units, called ommatidia. Each ommatidium has its own dioptric apparatus and retinula. The latter consists of only a few photoreceptors, radially arranged relative to the ommatidial axis. (After von Frisch 1960). *B* Single photoreceptor with parallel array of microvilli, collectively called rhabdomere. *C* Schematic representation of microvilli with dipole visual pigments within the membranes (the *black lines* indicate the axis of preferential absorption). *Top* the dipole molecules are randomly oriented within the plane of the membrane; *bottom* they are aligned with the long axis of the microvilli. *D* Quantum bumps recorded from a retinula cell of a crab (bumps are marked with *dots*). In the upper and lower record the E-vector of incident light was perpendicular and parallel to the microvilli, respectively. (After Doujak 1984)

visual pigments rotate freely within the membrane plane. This is because the pigments localized in the side parts of the microvilli can be activated only when plane polarized light is parallel to the microvillar axis (Fig. 2C, top). Moody and Parriss (1961) were the first to realize this, predicting a dichroic ratio of 2 for a random array of dipoles in the tangent plane of the membrane. Any higher dichroic ratios require that the chromophores are preferentially aligned relative to the long axis of the microvilli (Fig. 2C, bottom), with a theoretical maximum value of 20, if perfect alignment is assumed (Laughlin et al. 1975).

How do optical and physiological data fit in with these theoretical estimates? Direct microspectrophotometric measurements on isolated crustacean rhabdoms first seemed to conform to the Moody-Parriss model (Waterman et al. 1969). However later measurements showed that the dichroic ratios can be significantly higher than 2 (e.g., 2–3.5 in crayfish photoreceptors: Goldsmith 1975), indicating that some dipole alignment with the microvillar axis does indeed occur. The same conclusion can be derived from experiments reported by Goldsmith and Wehner (1977). They were able to alter the dichroic ratios of crayfish photoreceptors by bleaching the visual pigments selectively with polarized light of different E-vector directions. The theoretical analysis of their results led to the prediction that crustacean photoreceptors should have a dichroic ratio of 5–7 with the visual pigments making an angle no greater than 50° relative to the long axis of the microvilli.

The case of alignment is supported by physiological measurements. Many dark-adapted invertebrate photoreceptors produce discrete membrane potentials (bumps) in response to dim illumination, with each bump representing the response to a single photon capture of a visual pigment molecule (Lillywhite 1977). When polarization sensitivity (PS) is measured directly from bump frequencies obtained when the E-vector is parallel and then perpendicular to the microvilli, it gives a direct measure of the receptor's dichroic ratio. Appropriate measurements in locusts (Lillywhite 1978) and more recently in crabs (Doujak 1984), yielded dichroic ratios of around 5 (Fig. 2D). This value is in good agreement with the predictions derived from the photo-induced dichroism in the Goldsmith-Wehner experiments, but it is significantly higher than the values measured directly by microspectrophotometry. The reason for this discrepancy is still unknown. It has been suggested, however, that some degradation of pigment orientation may occur during in vitro microspectrophotometry due to the lability of the microvillar cytoskeleton which may be critical for the alignment of visual pigment molecules (Blest et al. 1982; Stowe 1983).

More commonly, PS is determined by measuring the two intensities necessary to produce graded potentials of equal amplitude when plane polarized light is parallel and perpendicular to the microvilli. Using this criterion reliable PS data have been obtained from a wide variety of insect photoreceptors, starting with the first single cell recordings (Kuwabara and Naka 1959; Burkhardt and Wendler 1960). More recent examples include dragonflies (Laughlin 1976), bees (Labhart 1980), butterflies (Horridge et al. 1983), flies (Hardie 1984), crickets (Labhart et al. 1984) and desert ants (Labhart 1986). In all these insects cells are found with PS values as high as 6–19, and in the butterfly as high as 50. This latter value in particular is hard to reconcile with current concepts of dichroism in rhabdomeres which predict polarization sensitivities of 20 at best (Laughlin et al. 1975). In this context it should be noted that theoretical possibilities exist for achieving polarization sensitivities beyond the capacity of dichroism. These include, for example, optical screening by an overlying photoreceptor (Snyder 1973), or inhibitory electric field interactions between receptors with different microvillar directions (Mueller 1973; Shaw 1975; Horridge et al. 1983; Hardie 1984). Of the two, electrical interactions are the most plausible explanation for the large PS values of butterflies and probably other insects as well, but their actual role in amplifying polarization sensitivities of receptors still awaits thorough experimental investigation.

Several mechanisms also exist which may degrade or even eliminate PS. One is twisting the microvillar directions along the visual axis of the receptor, as is the case, for example, in bees (Wehner et al. 1975) and flies (Smola and Tscharntke 1979). Other mechanisms include self-screening in long rhabdomeres (Snyder 1973; Vogt 1987) or electrical coupling between cells with different microvillar directions (Lillywhite 1978; van Hateren 1986). Finally, PS of the whole visual system may be reduced by synaptic summation of the output signals of receptors with phase-shifted PS functions. All these possibilities stress the need for a careful identification and localization of the effective polarization analyzers within compound eyes. Thereby the spectral sensitivities of the analyzers as well as their retinal distribution are of the utmost importance.

2.2 Spectral Sensitivity and Retinal Distribution

It appears reasonable to assume that polarization detection should rely upon one spectral type of photoreceptor. Otherwise visual processing would be complicated by the need to separate color and polarization information. Physiological measurements from various insects support this hypothesis. All were shown to possess different color receptors, but in each case only one of them (the UV receptor in bees (Menzel and Snyder 1974, Labhart 1980), ants (Labhart 1986) and flies (Hardie 1984); the blue receptor in crickets (Labhart et al. 1984) was highly sensitive to polarized light. Behavioral observations in backswimmers (Schwind 1983a, 1984), bees (von Frisch 1965; von Helversen and Edrich 1974), and ants (Duelli and Wehner 1973) yielded similar conclusions. Particularly instructive in this respect are the experiments of von Helversen and Edrich. They were able to reconstruct the bee's spectral sensitivity function of E-vector orientation, by correlating the bee's waggle dance performance with the intensity and the wavelength of the polarized light source. The result conformed exactly to the wavelength sensitivity profile of the UV photoreceptor inferred from intracellular recordings (Fig. 3A). Clearly, this convincingly proves that only one spectral type of receptor is involved in the detection of polarized light.[1]

Although polarization-sensitive photoreceptors have been found to be distributed over most of the compound eye, the picture now emerging is that receptors responsible for the detection of a particular aspect of environmental polarization are grouped within specialized eye regions. In the backswimmer, such a region exists in the ventral parts of the eyes and serves to detect polarized light from the water surface (Schwind 1983a, b, 1984). In the bee and the desert ant, on the other hand, selective blinding experiments have identified the most dorsal rim of the eyes, containing only a few percent of the total ommatidia, as being responsible for the detection of polarized skylight (Wehner 1982; Wehner and Strasser 1985). In appreciating their role in polarization vision, such distinct groups of specialized

---

[1]It is not yet decided why most insects use UV receptors for polarization detection. One may expect the detecting system to be optimally sensitive to the predominant wavelength of the polarization source or to the wavelength where maximal polarization occurs. However, with regard to both celestial and water surface polarization, such a match would best be achieved if the analyzers were sensitive in the blue, rather than in the ultraviolet range of the spectrum (Waterman 1981, Brines and Gould 1982). This apparent paradox can perhaps be resolved by assuming that the use of UV improves the insect's ability to discriminate polarization from other optical cues (Brines and Gould 1982; Schwind 1983a). This idea is based on the fact that the contrast between a patch of sky or water surface and its surround is greatest in the UV. In fact, it has already been shown that bees use this effect to distinguish between sun and sky (Edrich et al. 1979; Brines and Gould 1979; Rossel and Wehner 1984b).

A
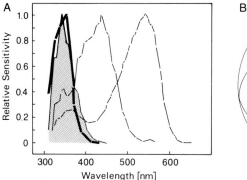

B
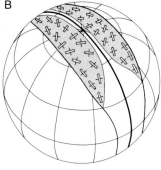

Fig. 3A. Spectral sensitivity of E-vector orientation of bees (*heavy black line*), as determined from behavioral experiments. In addition, intracellular measurements of the spectral sensitivities of the three· classes of receptors found within the eyes of bees are plotted (gray shading: the UV photoreceptor). (Combined from von Helversen and Edrich 1974, and Menzel et al. 1986). *B* The spatial layout of polarization-sensitive photoreceptors within the POL areas of the eyes of bees. The sphere shows the visual field of the left and the right eye (separated by *heavy black line*). The POL areas are marked by *gray shading*. Anatomical and physiological studies have identified two types of analyzers with orthogonal microvillar directions (*white* and *thin black bars*). Notice that the POL areas have contralateral fields of view. (After Sommer 1979)

ommatidia have been called "POL areas" of compound eyes (Wehner and Strasser 1985).

The POL areas of insects have a number of anatomical, optical, and physiological peculiarities in common (bees: Schinz 1975; Sommer 1979; Labhart 1980; Meyer and Labhart 1981; ants: Herrling 1976; Wehner 1982; Labhart 1986; flies: Wada 1974; Wunderer and Smola 1982; Hardie 1984; moths: Hämmerle and Kolb 1986; butterflies: Kolb 1986; crickets: Burghause 1979; Labhart et al. 1984; backswimmers: Schwind 1983b). Specifically, their ommatidia contain sets of receptors with high polarization sensitivity and orthogonally arranged microvillar directions. The receptors are enlarged in diameter and do not twist (in bees and flies such a twist does occur in the remainder of the eye with a resultant reduction in polarization sensitivity). In bees and crickets (but not in flies and desert ants) the receptors have extremely wide visual fields, owing to some structural specializations of the corneal lenses or to the lack of screening pigments in the ommatidia. This allows the receptors to sample polarization information far beyond the small area around the sky's zenith to which their optical axes are directed. In the ventral POL regions of the backswimmer, one of the two crossed analyzers is arranged horizontally throughout the visual field, whenever the insect assumes its normal flying posture. This contrasts to insects with POL regions at the dorsal margin of the eyes. Here the polarization-sensitive photoreceptors have contralateral fields of view, with their microvillar directions rotating from the back to the front, so that a fan-like pattern of crossed retinal analyzers results (Fig. 3B). This peculiar pattern provides an important clue to an understanding of the mechanisms involved in E-vector detection and orientation (for anatomical reconstructions of interneurons in the polarization-sensitive pathways of flies see Strausfeld this Vol.).

3 Polarized Light Detection and Orientation

Photoreceptors of compound eyes are sensitive to polarized light, but, individually, they are not discriminators of this parameter. As quantum detectors, photo-receptors are unable to distinguish polarization from wavelength and intensity, unless they are allowed to rotate about their visual axis (Kirschfeld 1972). If this cannot be achieved, the visual system must compare the responses of several photoreceptors, each being maximally sensitive to a different E-vector direction, in order to detect polarized light. In the backswimmer, the output signals of single pairs of analyzers may suffice to accomplish this task, because this insect must detect only one distinct plane of polarization (Schwind 1984). However, a more complicated situation may be expected in insects which use the polarization patterns in the sky as a compass, for these patterns are composed of many different E-vector directions. To account for this, it is tempting to speculate that each individual ommatidium is able to detect the E-vector direction of a patch of light and that an array of similar ommatidia perceives a polarization pattern very much as it does a spatial intensity or chromatic pattern. Although propagated by many workers in the field, this hypothesis must be rejected because its basic requirements are not met by the structural organization of the specialized dorsal margin of insect eyes. Their ommatidia contain only two analyzer directions, whereas three would be needed if each ommatidium were able to assess and report independently on the E-vector direction in a patch of light falling within its receptive field of view (Kirschfeld 1972). Furthermore, the analyzer directions rotate from the back to the front of the eyes, so that no reference system exists for a local analysis of E-vector directions. Exactly how then can an insect detect the polarization patterns in the sky and how can it use these patterns to steer compass courses? To date this question has only been studied extensively in bees, but a number of supportive data are also available from other insects.

3.1 The Bee's E-Vector Compass

Following von Frisch's demonstration that bees could navigate by means of a patch of polarized light as small as 10°, it has been widely believed that bees have a precise knowledge of the E-vector distributions in the sky. In fact, it was this premise that led to the sort of detection model mentioned above, but work over the past decade has shown that it is incorrect. This became evident from experiments designed to test whether bees were able to cope with the rather complex changes of the polarization patterns as they occur when the sun moves across the sky. It turned out that bees make consistent orientation errors depending on both the elevation of the sun and the E-vector direction in the small patch of skylight presented to them. Thus the position assumed by bees for a particular E-vector direction does not necessarily correspond to its actual position in the sky, but deviates from the latter by an angle which equals the orientation error. On this assumption, the bee's version of the polarization pattern has been reconstructed. The result shows that bees ignore the sun's elevation altogether and instead rely on an invariant celestial map in which each E-vector direction has a fixed bearing relative to the sun (Rossel et al. 1978; Rossel and Wehner 1982). Later, this finding led to a new and remarkably simple model of E-vector detection and orientation (Rossel and Wehner 1984a). The essence of this model is that the invariant E-vector map deduced from the orientation errors is embodied in the array of polarization analyzers, and that the bee uses this array to scan and match the polarization patterns in the sky (for review see Wehner and Rossel 1985).

The bee's invariant E-vector map happens to be a close approximation to the sky's polarization pattern when the sun is on the horizon. Thus it can be assumed that this particular pattern is mimicked by the distribution of polarization-sensitive photoreceptors within the eyes, such that the microvillar directions rotate from the back to the front in the same fashion as do the E-vectors in the sky (Fig. 4A). To use this array in celestial orientation, the bee must simply turn until the analyzers match the E-vectors visible in the sky. When this is achieved, the receptor array produces a maximal signal that tells the bee she is aligned with the pattern's principal plane of symmetry, the solar, and the antisolar meridian. Once aligned, the bee discriminates between these two meridians and uses either of them as a reference to select the definitive compass course.[2] Because the analyzer array is invariant and corresponds only to the E-vector pattern at dawn and dusk, mistakes are bound to occur when the bee navigates underneath a daytime polarization pattern (Fig. 4B). It is from these mistakes that the scanning model and the spatial layout of retinal analyzers have been deduced (Rossel and Wehner 1984a). For the moment we can ignore the fact that the analyzer array inferred from behavior differs significantly from the actual receptor mosaic in the bee's POL areas (Fig. 3B), but we shall come back to this point later.

Bees still make mistakes when presented with more than one patch of skylight; however, these mistakes are exactly as large as predicted from the scanning model. The prediction is that each E-vector visible in the sky induces an error angle and that the mean of the error angles induced by all E-vectors determines the bee's deviation from the correct compass course. In this respect one type of experiment is very instructive: substantial errors occur, when a patch of a daytime E-vector pattern is presented to the left and then to the right of the pattern's symmetry plane, but none are observed when bees are allowed to view both patches simultaneously (Fig. 4C). Obviously, this is because the E-vectors in each celestial half are mirror images to each other, so that the error angle induced by each E-vector in the left and the right half of the sky are opposite in sign and thus cancel each other out. This observation demonstrates that the scanning mechanism normally provides quite

---

[2]To distinguish between the solar and the antisolar meridian the bee uses both polarization and wavelength (Rossel and Wehner 1984b). When the sun is above the horizon, the antisolar half of the sky is more strongly polarized than its opposite (Fig. 1B). In addition, there is a distinct spectral gradient in the sky, with the relative UV content increasing from the solar towards the antisolar meridian. The bee fails to distinguish between the two meridians when only a small patch of sky is available and when the E-vector in this patch occurs at two different positions in the sky (one lying in the solar half, another in the antisolar half of the sky). In this case, of the two possibilities, the bee interprets the patch as being the further from the solar meridian (Brines and Gould 1979, Rossel and Wehner 1982). This results in either of the two meridians being mistakenly identified half of the time. It still remains to be answered how the bee, once she has determined the reference meridian, selects the proper compass course (but see Rossel and Wehner 1987).

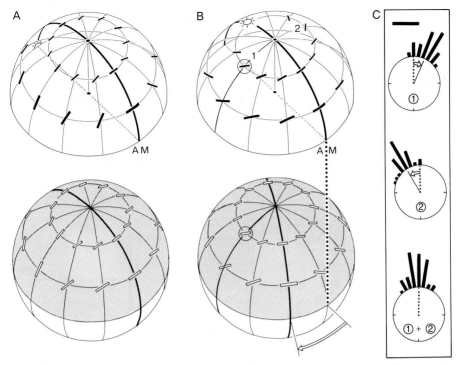

Fig. 4A-C. Scanning model of E-vector orientation in bees. *A* The *upper diagram* illustrates the E-vector pattern in the sky when the sun is on the horizon (*black bars* E-vector directions; *A M* antisolar meridian). The *lower diagram* shows model eyes with array of polarization sensitive photoreceptors (*white bars* microvillar directions). The analyzer array matches the E-vector pattern when the bee's longitudinal plane (marked by *heavy black line*) is aligned with the solar and the antisolar meridian. *B* A different polarization pattern is realized when the sun is high in the sky (50° in the figure). The bee has turned so as to achieve a match between an appropriately oriented analyzer and the E-vector encircled in the right celestial half. The resulting deviation (*white arrow*) of the bee from the pattern's symmetry plane is manifested by orientation errors. *C* In the experiments shown from top to bottom, dancing bees were first presented with a patch of skylight in the right celestial half, then with another patch in the left celestial half (the patches are labeled by *1* and *2* in B). Finally, both patches were presented simultaneously. The *dotted* and *solid* lines in the diagrams show the correct compass course (direction of foraging site) and the mean dance direction, respectively. The bee's mistakes are indicated by *open arrows*. Scale: 20 waggle runs

accurate compass bearings. It is only under experimental conditions when patches of skylight are presented in one celestial half that substantial orientation errors may occur.

Besides predicting the general compass strategy, the scanning model specificly formulates how polarized light is detected by the bee. When the bee rotates about its vertical body axis, a patch of light with a given E-vector successively stimulates a ring of orthogonal photoreceptors (Fig. 5A). As the angle between the microvilli of the receptors and the E-vector varies gradually during the turn, the differential output of the array of crossed analyzers is a modulating signal passing through a peak as the microvilli of one type of analyzer (the X-type in Fig. 5A) become parallel to the E-vector in the patch of skylight. A similar modulation is detected

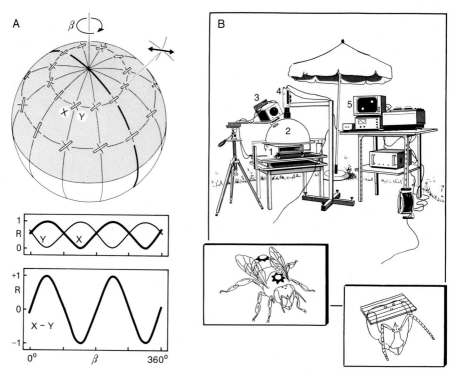

Fig. 5. A The detection of polarized skylight. In addition to the receptors shown in Fig. 4 (here called X-type receptors) there is an additional array of Y-type receptors with orthogonally arranged microvillar directions (evidence for such crossed analyzers comes from anatomy and physiology). The bee views a small patch of skylight with a single E-vector direction (*double arrow*). She is first aligned with the E-vector ($\beta = 0°$) and then performs a full turn about the vertical axis ($\beta = 0°-360°$). The output of the X- and Y-type receptors (R; X,Y) are assumed to interact so as to produce a differential output (R; X − Y). The latter is a modulating signal reaching a peak as the microvilli of the X-type receptors become parallel to the E-vector in the patch of light. *B* Experimental set-up used to test the scanning model. Bees are trained to a feeding station some hundred meters from the hive. Upon their return they perform directional dances on a horizontal comb (*1*) underneath a Plexiglas hemisphere (*2*). A patch of unpolarized UV light from a Xenon arc (*3*) is presented as an orientational cue. The intensity of the light source modulates in intensity according to the bee's horizontal orientation. This is achieved as follows: The dancing bee, which carries two circular reflectors on its body (see inset to the left), is filmed from above by means of a video camera (*4*). Only the reflectors are visible on the TV monitor (*5*), where they indicate the bee's longitudinal axis. By using the appropriate electronic equipment, the orientation of this axis is continuously registered and the resulting signal is used to control the intensity of the Xenon arc. Note that the bees are equipped with overhead filters parallel to the transverse axis of the eyes (see *inset to the right*), such that only the X-type receptors are stimulated

by the bee when the E-vector direction is changed, except that the modulation now rises to a peak in a different position on the eyes. Thus the model predicts that the perceived direction of a particular E-vector and the azimuthal angle associated with that E-vector is determined by what part of the eyes produces the largest output.

On these assumptions, the scanning model has recently been tested (Rossel and Wehner 1986, for review see Rossel 1987, Rossel and Wehner 1987). In the experiments the bee's dance orientation with respect to a particular E-vector direction was mimicked by a beam of unpolarized ultraviolet light which was modulated in intensity according to the bee's horizontal orientation (Fig. 5B). For example, the model predicts that a vertical E-vector is represented in the lateral region of the eyes (compare Fig. 5A). Accordingly, the modulation of the light source was arranged so that the beam was of peak intensity when it stimulated the lateral parts of the eyes. When this was achieved the bees were expected to dance in the same direction as when presented with a patch of vertically polarized light.

The experiment was complicated by the existence of crossed analyzers, which were assumed to interact so as to produce a differential output. When presented with polarized light, such opponent interactions provide the bee with a high-contrast E-vector signal. However, a modulated beam of unpolarized light stimulates pairs of crossed analyzers equally and in phase, so that the differential output will be zero throughout the modulation circle. This was in fact shown in the experiments, i.e., the bees danced in the same direction irrespective of whether the beam of unpolarized light was modulated or not.

The solution to this problem was to neutralize the opponency mechanism by stimulating selectively one of the two crossed analyzer arrays on its own. This was achieved by covering the eyes with a piece of a polarizing filter oriented parallel to the transverse axis of the eyes (Fig. 5B). Geometrical considerations show that under these conditions incoming light is polarized by the filter so as to stimulate only one type of analyzer throughout the eyes (the X-type analyzer in Fig. 5A). In consequence one analyzer array will dominate as the bee rotates, and the differential output of the crossed analyzer array will modulate in response to the stimulus light. The model then predicts that this signal will be interpreted by the bee as polarized light. The experimental data completely confirm this prediction (Rossel and Wehner 1986, 1987). One example is documented in Fig. 6A.

One might argue that E-vector orientation in the experiments was not affected by the intensity modulations of the light source, but was caused by the E-vectors projected onto the eyes by the overhead filter. This possibility, however, can be excluded on the basis of experiments in which the transmission axis of the filter is 45° from the long axis of the head. In this case no E-vector orientation can be observed (Fig. 6B). This result can hardly be explained by the argument mentioned above, but it fully conforms to the scanning model: With the transmission axis of the filter oriented obliquely, the crossed analyzers are stimulated equally and in phase, in which case no modulating signal will be detected by the rotating bee (Rossel and Wehner 1986).

In summary, behavioral experiments suggest that bees use a scanning strategy in order to read compass information from the polarization patterns in the sky. While rotating about their vertical body axis, the bees transform the polarization patterns into a modulating signal, which rises to a peak when the body is aligned with the solar and the antisolar meridian. The bees then discriminate these two meridians and use either of them as a reference to select their proper compass course.

3.2 Evidence from Other Insects

It seems likely that the principles uncovered in bees are generally applicable to insects that detect and use polarized skylight for orientation. As mentioned earlier, the mosaic of polarization-sensitive photoreceptors at the dorsal margin of

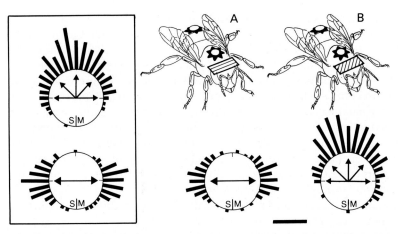

Fig. 6. Simulation of E-vector orientation by means of modulated beam of unpolarized light — results.' Dance directions are calculated to indicate where the navigating bee expects the light stimulus to occur in the sky (*SM* solar meridian). In the control experiments (see *inset from top to bottom*) the bees are presented with a patch of unpolarized nonmodulated UV light and then with a patch of polarized light with a vertical E-vector. The former is interpreted by the bee to lie anywhere within the antisolar half of the sky, the latter to lie 90° to the left and to the right of the solar meridian (for details see Rossel and Wehner 1982, 1984b). *A* Bees are presented with an unpolarized beam of light which modulates in intensity so as to simulate a vertical E-vector. The overhead filter stimulates selectively the X-type receptors (see Fig. 5). The results show that the bees treat the modulated beam of light as a patch of polarized light with a vertical E-vector (compare with *lower diagram in the inset*) *B* The same stimuli conditions as in *A*, except that the transmission axis of the overhead filter is rotated by 45°. In this case the X- and Y-type receptors are stimulated equally and the bees behave as when presented with an unpolarized nonmodulated beam of light (compare with *upper diagram in the inset*). Scale: 50 waggle runs

compound eyes is similar in all insects studied so far, and — at least in the desert ant *Cataglyphis* (Fent 1986) — this similarity is directly manifested by orientation errors analogous to those observed in bees. A further piece of evidence is the recent discovery of polarization-opponent interneurons in the visual system of crickets (Labhart 1987, 1988). These interneurons are spontaneously active and this activity only changes in the presence of polarized light. Depending on the E-vector direction the neurons can vary their firing rates about the level of spontaneous activity, with alternating 90°-phases of excitation and inhibition when the E-vector is rotated about the neuron's optical axis (Fig. 7). This antagonistic mode of response can readily be explained if one assumes that the neurons receive excitatory and inhibitory inputs from two types of photoreceptors with orthogonally arranged microvillar directions (Labhart 1987, 1988). It seems likely, then, that this sort of neuron represents the physiological basis of the antagonism demonstrated in bees by means of behavioral experiments.

The physiological measurements in crickets yielded another interesting result: namely that there are three different classes of POL neurons, each specified by a particular E-vector direction at which maximal spike discharge occurs. To account for this, Labhart (1987, 1988) assumes that the receptor outputs from three separate parts of the POL regions converge onto single opponent interneurons. In fact, it is this sort of convergence that could solve one major remaining problem in con-

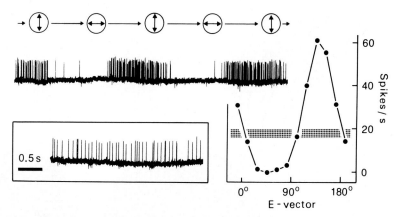

Fig. 7. Polarization-opponent interneurons in the eyes of crickets. The record shows the response of a neuron when the E-vector direction (*double arrow*) in a patch of light is continuously rotated through 360° (the spontaneous activity of the neuron is shown in the inset). (After Labhart 1988). In the graph the spike frequency of another interneuron is plotted as a function of the E-vector direction (the level of spontaneous activity is marked by *dotted band*). (After Labhart 1987)

nection with the E-vector compass of bees. The problem is how the array of polarization analyzers deduced from the bee's orientation errors relates to the actual receptor mosaic within the POL areas of the eyes. In comparing Figs. 3B and 5A, we readily notice that major discrepancies exist in the frontal and caudal eye regions, where the microvillar directions are oriented obliquely or even vertically rather than horizontally as predicted from behavior. To resolve this discrepancy, take the photoreceptors in the middle range of the POL areas (Fig. 8A). Their microvillar directions are oriented more or less parallel to the transverse plane of the eyes. Together with the fact that individual photoreceptors have large visual fields (Labhart 1980), neural convergence then results in a single large-field detector which is also oriented parallel to the tranverse plane (Fig. 8B). Interestingly enough, such a large-field detector has exactly the same analyzing properties as the whole array of small-field detectors assumed in the model eyes of Figs. 4 and 5. For example, the detector is maximally active with a vertical E-vector in the lateral field of view, a horizontal E-vector in the anterior and posterior field of view, and so on. To use this large field detector in orientation, the bee again rotates about its vertical axis and assumes, whenever there is a peak in the output of the detector, that the body is aligned with the solar and the antisolar meridian. Because bees are able to discriminate these two meridians solely on the basis of E-vector information (Rossel unpublished data), an extra set of detectors would be needed to accomplish this task. Obviously, such detectors could result from the remaining photoreceptors located in the anterior and posterior range of the POL regions (cf. Fig. 8A). Altogether, the bee would then possess three different types of large-field detectors, as shown by electrophysiology in the eyes of crickets.

Finally, note that a set of three global detectors with overlapping receptive fields could provide an insect with an alternative and rather direct method of determining compass bearings (Fig. 8C). According to this method, the insect would measure the ratio of the output signals of the detector triplet and would use this

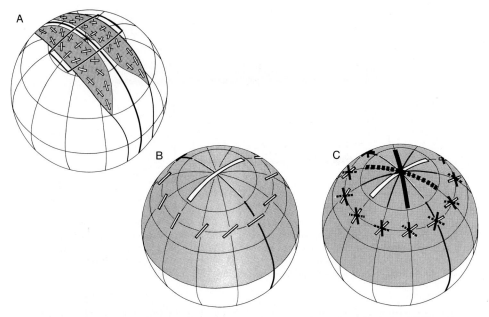

Fig. 8A-C. Neural convergence within the POL areas of compound eyes (hypothetical). *A* POL areas of the bee as in Fig. 3B. The crossed photoreceptors *framed by the black line* are assumed to converge onto a single polarization opponent interneuron. *B* Resulting large-field detector. The *small bars* mark E-vector directions in different positions of the receptive field of the detector, oriented to stimulate the detector maximally. *C* The remaining photoreceptors in the anterior and posterior eye could also converge onto single opponent interneurons. The effective analyzer direction of these detectors would depend not only on the orientation of the photoreceptors in each eye but also on the sort of binocular interactions. In the figure the binocular interactions are assumed to be diametrical. The result is a triplet of large-field detectors which intersect in the zenith of the visual field (*large open, solid* and *dotted bars*; the *small bars* indicate the corresponding E-vector directions which stimulate the detectors maximally). In principal the bee could use such a detector arrangement to determine the compass course directly

ratio as a direct measure of the body angle relative to the sun's bearing. In other words, a particular ratio of detector outputs would determine the insect's compass course, such that no prior alignment of the body with the pattern's symmetry plane would be needed. If the receptive fields of the detectors are sufficiently large, this method would work even when the insect can view only a small patch of polarized skylight. Furthermore, it would result in the same orientation errors as described above, provided that the three detectors intersect at the most dorsal part of the visual field. At present, however, there is no evidence for this possibility. In the experiments reported above, a polarizing filter was fixed on the bee's head, so that the ratio of the detector outputs would always have been the same, irrespective of the bee's orientation relative to the sun. Furthermore, one would have expected the dance behavior to be the same whether the transmission axis of the filter was parallel to the transverse axis of the eyes or oblique to it. The results show that this was not the case (Fig. 6).

In conclusion, we may revise the scanning model described at the outset by assuming that the principal design features of the polarization patterns in the sky are mimicked by the on- and off-axis analyzing properties of single large-field detectors (Fig. 8), rather than by an appropriate geometrical pattern of individual pairs of crossed photoreceptors (Fig. 5). In fact it is not important how the mosaic of crossed photoreceptors is oriented in space as long as neuronal convergence results in large-field detectors appropriately oriented within the bee's field of view. This concept may simplify the issue in terms of evolution. It is well known that the genetically controlled formation of the compound eye retina results in a highly regular pattern of microvillar directions as the one shown in Fig. 8A. Evolution could have modified this pattern so as to achieve the best possible match with the E-vector pattern in the sky. However, widening the visual fields of the photoreceptors in combination with neural convergence may have been a much simpler strategy in order to provide the insect with a workable E-vector compass.

4 Concluding Remarks

The work described in this brief review sheds some light onto the once magical phenomenon of polarization sensitivity in compound eyes. Anatomical, optical, and physiological studies from various invertebrates identify the rhabdomeric photoreceptors as basic polarization analyzers, and behavioral observations in bees provide instructive insights how E-vector information extracted from these receptors is utilized in celestial orientation. Some progress has also been made in unraveling the neuronal processing of polarized light. In fact, this analysis has a realistic chance of success, because the number of receptor inputs to the detecting system is relatively small and because the general principle of operation of this system is already evident from behavior. A more difficult undertaking is to obtain information on how polarized light detection is integrated into the overall visual processing and orientation behavior of a particular animal of interest; yet it seems obvious that only with such knowledge, will the study of polarization sensitivity in compound eyes be capable of complete interpretation.

References

Autrum H, Stumpf H (1950) Das Bienenauge als Analysator für polarisiertes Licht. Z Naturforsch 5b:116–122

Blest AD, Stowe S, Eddey W (1982) A labile, Ca^{2+}-dependent cytoskeleton in rhabdomeral microvilli of blowflies. Cell Tissue Res 223:553–573

Bohn H. Täuber U (1971) Beziehungen zwischen der Wirkung polarisierten Lichtes auf das Elektroretinogramm und der Ultrastruktur des Auges von Gerris lacustris. Z Vergl Physiol 72:32–53

Brines LB, Gould JL (1979) Bees have rules. Science 206:571–573

Brines LB, Gould JL (1982) Skylight polarization patterns and animal orientation. J Exp Biol 96:69–91

Burghause FMHR (1979) Die strukturelle Spezialisierung des dorsalen Augenteils der Grillen (Orthoptera, Grylloidea). Zool Jahrb Physiol 83:502–525

Burkhardt D, Wendler L (1960) Ein direkter Beweis für die Fähigkeit einzelner Sehzellen des Insektenauges, die Schwingungsrichtung polarisierten Lichtes zu analysieren. Z Vergl Physiol 43:687–692

de Vries H, Spoor A, Jielof R (1953) Properties of the eye with respect to polarized light. Physica 19:419–432

Doujak FE (1984) Electrophysiological measurement of photoreceptor membrane dichroism and polarization sensitivity in a Grapsid crab. J Comp Physiol A 154:597–605

Duelli P, Wehner R (1973) The spectral sensitivity of polarized light orientation in *Cataglyphis bicolor* (Formicidae, Hymenoptera). J Comp Physiol A 86:37–53

Edrich W, Neumeyer C, Helversen O von (1979) Anti-sun orientation of bees with regard to a field of ultraviolet light. J Comp Physiol 134:151–157

Fent K (1986) Polarized skylight orientation in the desert ant Cataglyphis. J Comp Physiol A 158:145–150

Fernández-Morán H (1956) Fine structure of the insect retinula as revealed by electron microscopy. Nature (London) 177:742–743

Frisch K von (1949) Die Polarisation des Himmelslichtes als orientierender Faktor bei den Tänzen der Bienen. Experientia 5:142–148

Frisch K von (1960) "Sprache" und Orientierung der Bienen. Dr. Albert Wander-Gedenkvorlesung 3. Huber, Bern Stuttgart, pp 1–43

Frisch K von (1965) Tanzsprache und Orientierung der Bienen. Springer, Berlin Heidelberg New York

Goldsmith TH (1975) The polarization sensitivity-dichroic absorption paradox in arthropod photoreceptors. In: Snyder AW, Menzel R (eds) Photoreceptor optics. Springer, Berlin Heidelberg New York, pp 392–409

Goldsmith TH, Philpott DE (1957) The microstructure of the compound eyes of insects. J Biophys Biochem Cytol 3:429–440

Goldsmith TH, Wehner R (1977) Restrictions on rotational and translational diffusion of pigment in the membranes of a rhabdomeric photoreceptor. J Gen Physiol 70:453–490

Hämmerle B, Kolb G (1986) Rhabdomstruktur im dorsalen Augenbereich des Apfelwicklers *Adoxophyes reticulana* (Tortricidae, Lepidoptera). Verh Dtsch Zool Ges 79:364

Hardie RC (1984) Properties of photoreceptors R7 and R8 in dorsal marginal ommatidia in the compound eyes of *Musca* and *Calliphora*. J Comp Physiol A 154:157–167

Hateren JH van (1986) Electrical coupling of neuroommatidial photoreceptor cells in the blowfly. J Comp Physiol A 158:795–811

Helversen O von, Edrich W (1974) Der Polarisationsempfänger im Bienenauge: ein Ultraviolettrezeptor. J Comp Physiol A 94: 33–47

Herrling PL (1976) Regional distribution of three ultrastructural retinula types in the retina of *Cataglyphis bicolor* Fabr. (Formicidae, Hymenoptera). Cell Tissue Res 169:247–266

Horridge GA, Marčelja L, Jahnke R, Matič T (1983) Single electrode studies on the retina of the butterfly *Papilio*. J Comp Physiol A 150:271–294

Kirschfeld K (1972) Die notwendige Anzahl von Rezeptoren zur Bestimmung der Richtung des elektrischen Vektors linear polarisierten Lichtes. Z Naturforsch 30c:88–90

Kolb G (1986) Rhabdome der dorsalen Augenrandzone und anderer Augenbereiche des Tagfalters *Aglais urticae* L. (Nymphalidae). Verh Dtsch Zool Ges 79:368

Kuwabara M, Naka K (1959) Response of a single retinula cell to polarized light. Nature (London) 184:455–456

Labhart T (1980) Specialized photoreceptors at the dorsal rim of the honey bee's compound eye: polarization and angular sensitivity. J Comp Physiol A 141:19–30

Labhart T (1986) The electrophysiology of photoreceptors in different eye regions of the desert ant, *Cataglyphis bicolor*. J Comp Physiol A 158:1–7

Labhart T (1987) The physiology of polarization-opponent interneurons in the visual system of crickets. In:Elsner NC, Creutzfeld O (eds) Neue Wege in der Hirnforschung. Beiträge zur 15. Göttinger Neurobiologentagung. Thieme, Stuttgart New York, p 141

Labhart T (1988) Polarization-opponent interneurons in the insect visual system. Nature (London) 331:435–437

Labhart T, Hodel B, Valenzuela I (1984) The physiology of the cricket's compound eye with particular reference to the anatomically specialized dorsal rim area. J Comp Physiol A 155:289–296

Langer H (1965) Nachweis dichroitischer Absorption des Sehfarbstoffes in den Rhabdomeren des Insektenauges. Z Vergl Physiol 51:258–263

Laughlin SB (1976) The sensitivity of dragonfly photoreceptors and the voltage gain of transduction. J Comp Physiol 111:221–247

Laughlin SB, Menzel R, Snyder AW (1975) Membranes, dichroism and receptor sensitivity. In: Snyder AW, Menzel R (eds) Photoreceptor optics. Springer, Berlin Heidelberg New York, pp 237–259

Lillywhite PG (1977) Single photon signals and transduction in an insect eye. J Comp Physiol A 122:189–200

Lillywhite PG (1978) Coupling between locust photoreceptors revealed by a study of quantum bumps. J Comp Physiol A 125:13–27

Lythgoe JN, Hemmings CC (1967) Polarized light and underwater vision. Nature (London) 213:893–894

Menzel R, Snyder AW (1974) Polarized light detection in the bee, *Apis mellifera*. J Comp Physiol 88:247–270

Menzel R, Ventura DF, Hertel H, Souza JM de, Greggers U (1986) Spectral sensitivity of photoreceptors in insect compound eyes: Comparison of species and methods. J Comp Physiol A 158:165–177

Meyer EP, Labhart T (1981) Pore canals in the cornea of a functionally specialized area of the honey bee's compound eye.Cell Tissue Res 216:491–501

Miller WH (1957) Morphology of the ommatidia of the compound eye of *Limulus*. J Biophys Biochem Cytol 3:421–428

Moody MF, Parriss JR (1961) The discrimination of polarized light by *Octopus:* a behavioral and morphological study. Z Vergl Physiol 44:268–291

Mueller KJ (1973) Photoreceptors in the crayfish compound eye: electrical interactions between cells as related to polarized light sensitivity. J Physiol 232:573–595

Rossel S (1987) Das Polarisationssehen der Biene. Naturwissenschaften 74:53–62

Rossel S, Wehner R (1982) The bee's map of the E-vector pattern in the sky. Proc Natl Acad Sci USA 79:4451–4455

Rossel S, Wehner R (1984a) How bees analyze the polarization patterns in the sky. Experiments and model. J Comp Physiol A 154:607–615

Rossel S, Wehner R (1984b) Celestial orientation in bees: The use of spectral cues. J Comp Physiol A 155:605–613

Rossel S, Wehner R (1986) Polarization vision in bees. Nature (London) 323:128–131

Rossel S, Wehner R (1987) The bee's E-vector compass. In:Menzel R, Mercer A (eds) Neurobiology and behavior of the honey bee. Springer, Berlin Heidelberg New York Tokyo, pp 76–93

Rossel S, Wehner R, Lindauer M (1978) E-vector orientation in bees. J Comp Physiol A 125:1–12

Schinz RH (1975) Structural specialization in the dorsal retina of the bee, *Apis mellifera*. Cell Tissue Res 162:23–34

Schneider L, Langer H (1969) Die Struktur des Rhabdoms im "Doppelauge" des Wasserläufers *Gerris lacustris*. Z Zellforsch 99:538–559

Schwind R (1983a) A polarization-sensitive response of the flying water bug *Notonecta glauca* to UV light. J Comp Physiol A 150:87–91

Schwind R (1983b) Zonation of the optical environment and zonation in rhabdom structure within the eye of the backswimmer *Notonecta glauca*. Cell Tissue Res 232:53–63

Schwind R (1984) Evidence for true polarization vision based on a two-channel analyser system in the eye of the water bug, *Notonecta glauca*. J Comp Physiol A 154:53–57

Shaw SR (1967) Simultaneous recordings from two cells in the locust retina.Z Vergl Physiol 55:183–194

Shaw SR (1975) Retinal resistance barriers and electrical lateral inhibition. Nature (London) 255:480–483

Smola U, Tscharntke H (1979) Twisted rhabdomeres in the dipteran eye.J Comp Physiol A 133:291–297

Snyder AW (1973) Polarization sensitivity of individual retinula cells. J Comp Physiol 83:331–360

Sommer E (1979) Untersuchungen zur topographischen Anatomie der Retina und zur Sehfeldtopologie im Auge der Honigbiene *Apis mellifera* (Hymenoptera). Diss, Univ Zürich

Stockhammer K (1956) Zur Wahrnehmung der Schwingungsrichtung linear polarisierten Lichtes bei Insekten. Z Vergl Physiol 38:30–83

Stockhammer K (1959) Die Orientierung nach der Schwingungsrichtung linear polarisierten Lichtes und ihre sinnesphysiologischen Grundlagen. Erg Biol 21:23–56

Stowe S (1983) A theoretical explanation of intensity-independent variation of polarization sensitivity in Crustacean retinula cells. J Comp Physiol A 153:435–441

Vogt K (1987) Chromophores of insect visual pigments. Photobiochem Photobiophysics Suppl 273–296

Wada S (1974) Spezielle randzonale Ommatidien der Fliegen (Diptera:Brachycera): Architektur und Verteilung in den Komplexaugen. Z Morphol Tiere 77:87–125

Waterman TH (1981) Polarization sensitivity. In: Autrum H (ed) Handbook of sensory physiology, vol VII/6C., Springer, Berlin Heidelberg New York, pp 281–469

Waterman TH (1984) Natural polarized light and vision. In: Ali MA (ed) Photoreception and Vision in Invertebrates. Plenum, New York, pp 63–114

Waterman TH, Fernández HR, Goldsmith TH (1969) Dichroism of photosensitive pigment in rhabdoms of crayfish *Orconectes*. J Gen Physiol 54:415–432

Wehner R (1982) Himmelsnavigation bei Insekten. Neurophysiologie und Verhalten. Neujahrsbl Naturforsch Ges Zürich 184:1–132

Wehner R, Rossel S (1985) The bee's celestial compass — A case study in behavioral neurobiology. In: Hölldobler B, Lindauer M (eds) Experimental behavioral ecology and sociobiology. Fischer, Stuttgart New York, pp 11–53

Wehner R, Strasser S (1985) The POL area of the honey bee's eye: behavioral evidence. Physiol Entomol 10:337–349

Wehner R, Bernard GD, Geiger E (1975) Twisted and non-twisted rhabdoms and their significance for polarization detection in the bee. J comp Physiol A 104:225–245

Wolken JJ, Capenos J, Turano A (1957) Photoreceptor structures. J Biophys Biochem Cytol 3:441–448

Wunderer H, Smola U (1982) Fine structure of ommatidia at the dorsal eye margin of *Calliphora erythrocephala* Meigen (Diptera: Calliphoridae): an eye region specialized for the detection of polarized light. Int J Morphol Embryol 11:25–38

Chapter 16

Beneath the Compound Eye: Neuroanatomical Analysis and Physiological Correlates in the Study of Insect Vision

NICHOLAS J. STRAUSFELD, Tucson, Arizona, USA

1 Introduction

Students of the compound eye deserve to be pleased with the scope and depth of their knowledge about the diversity of physiological optics and receptor structure. The same may be said for modern research on receptor transduction, the structure and kinetics of visual pigments, and the molecular biology of the photoreceptor membrane. Now, 100 years after Exner published his views on the compound eye, our understanding of the molecular and biophysical properties of the receptor layer is certainly comparable to knowledge of vertebrate photoreceptors.

Beneath the compound eye, it is altogether another story; for here the intricate organization of many thousands of neurons has relinquished comparatively few of its secrets. Despite this, the insect's visual system has definite technical advantages over its vertebrate counterpart, the foremost being that it is highly accessible for combining intracellular physiological studies with sophisticated neuroanatomy. However, for insect vision research to resonate amongst those working on vertebrates, it is necessary to identify in insects so-called "model" systems that can be used for studying general phenomena that, in vertebrates, are less accessible. Likely systems can be identified, such as neuroanatomical substrates for parallel processing of visual information and for integrating vision with other sensory modalities. In this chapter I shall review the research trends in neuroanatomy, reflecting on the impact of past and present strategies that relate the organization of the compound eye to behavior.

2 Research Strategies

Hitherto, most structural studies on the insect visual nervous system have followed one of two lines: (1) intracellular recording and dye filling of uniquely identifiable single neurons whose synaptic relationships are obscure and (2) detailed neuroanatomical studies on the relationships of neurons within restricted areas of the central nervous system where functional interpretations rely on reasoned conjecture rather than on direct experimentation. The only exception to this dichotomy has been research on the peripheral lamina. There, complex synaptic relationships between photoreceptors, interneurons, and specialized glial cells are known in such detail as to allow debate about the significance of synaptic organization with respect to a wealth of physiological data. But even in the lamina

Stavenga/Hardie (Eds.) Facets of Vision
© Springer-Verlag Berlin Heidelberg 1989

certain elements have been incompletely analyzed for reasons that are symptomatic of insect vision research: the extraordinary technical difficulty of recording from retinotopic neurons, many of which (in Diptera, Odonata and Orthoptera) have axon diameters of less than 1 μm.

The lamina's inaccessibility was once reason for speculating about the function of identified neurons solely on the basis of behavioral experiments, such as those derived from studies of the optomotor reaction (see also Sect. 2.1). For example, it was thought that the lamina might contain neural circuits for movement detection: certain geometrical arrangements amongst its L4 neurons (Sect. 2.3.1) matched theoretical circuits for optomotor responses (Braitenberg 1972). However, no intracellular recordings of lamina neurons have demonstrated motion sensitivity at that level and, despite the lamina's complex synaptology, the suggestion that the lamina contains elemental motion detectors (EMD's) is today virtually discounted. Yet, although behavioral evidence demonstrates the existence of small-field "elemental motion detectors," their identity and location as real neuronal circuits are still unknown. A reasonable supposition is that because so many computational tasks are based upon motion sensitivity, circuits for motion detection may be legion.

Until 1975 neuroanatomy of the optic lobes provided much information about the shapes of single nerve cells, but did little to explain their relevance in visual processing. Research was based on the precept that unravelling optic lobe anatomy from the periphery inward would reveal discrete neuronal arrangements, each representing a basic computational circuit underlying a behavior. The flaw in this notion is obvious: at each synapse there occurs a divergence to two or more interneurons. Already at the level of the lamina each photoreceptor terminal diverges to six relay neurons, which themselves diverge in the medulla to more than 40 parallel channels. At this level, riches for the light microscopist become a quagmire for the physiologist, and electron microscopy seems self-defeating.

Here we shall compare the results of this linear "receptor-downward" approach with three other strategies that have led to fundamental insights into the functional organization of the insect visual system. These are: systems analysis (biocybernetics), which relies almost exclusively on quantitative aspects of visual behavior to infer computational pathways or "neural networks"; predictive neuroanatomy, derived from resolving a novel subunit of optic lobe organization, the neuronal assembly; and the "muscle-upwards" approach, which relates neurons at the level of the behavioral output to identified interneurons of the visual system (Fig. 1, Table 1).

2.1 Systems Theory: Its Influence on Neuroanatomical Strategies

By the early 1950's the study of animal orientation had become a field of research in its own right, pursued most actively in post-war Germany under the intellectual leadership of Erich von Holst. Together with research on receptor physiology and insect behavior, which had been reestablished by Autrum at the University of Munich, quantitative ethology dominated research on the arthropod nervous system, leading to attempts to describe the relationships between sensory stimulation and behavior in terms of control systems. A major concern was not only to portray input-output relationships between the stimulus and stimulus-induced motor reactions, but also to address the question of how the nervous system discriminates between active external stimulation and apparent stimulation of the receptor system due to the animal's own movement.

In their famous "reafference principle", von Holst and Mittelstaedt (1950) suggested that a copy of motor instructions for an animal's movement was used to neutralize motor commands arising from sensory processing that resulted as a consequence of that movement. This model had found support from an experimental test using the highly regular receptor array of the hoverfly's compound eye whose geometry could be reversed, without rearrangement of the effector organs, by twisting the head around the neck and gluing it upside down to the body (Mittelstaedt 1949). In the control and test experiments, a syrphid (*Eristalis tenax*) was suspended in the axis of a vertically striped drum. When the drum was rotated around a normal fly, the fly turned in the direction of the moving stripes, attempting to stabilize the retinal image and maintain a constant orientation to the world around it. Since a fly does not exhibit "optomotor" reactions when it moves spontaneously in a stationary drum, von Holst suggested that either the optomotor circuit is switched off, or there is a feedback loop that cancels out perceived displacements of the retinal image induced by self-motion. In the fly with an upside-down head, the linear arrays of ommatidia are experimentally reversed through 180°. During spontaneous movement, stripes moving from right to left are perceived moving from left to right and motor commands are given to follow stripes moving from left to right. Under these conditions the more *Eristalis* attempts to compensate for retinal displacement the greater the apparent movement of gratings across the retina, and the more *Eristalis* tries to correct this.

The advantage offered by the compound eye for testing the validity of hypothetical pathways was underlined by Hassenstein and Reichardt (1956),who explained normal optomotor responses by the beetle *Chlorophanus* in terms of hypothetical cross-connections, linear filters, and multiplication elements between adjacent ommatidial channels. Their experiments led to the now famous "minimal model" for movement detection, published by Hassenstein in 1958, in which neighboring channels interact continuously to analyze the visual panorama (Reichardt 1969). The strength of the output from this "cross-correlation" circuit determines the appropriate strength and sign of the motor reaction.

Hassenstein and Reichardt's investigations initiated an entirely new approach to the study of the visual system, originating what was to become a major research program at the Max Planck Institute for Biological Cybernetics in Tübingen. Later, a second group of researchers following similar lines at the California Institute of Technology were to greatly influence the direction of insect vision research by electrophysiologically probing neurons involved in optomotor pathways. We shall return to this later.

From the outset, theoretical constructs that conceptualized neural processing emphasized the need for research on cellular arrangements beneath the compound eye. There is a revealing legend to the figure in Hassenstein and Reichardt's 1956 account: a micrograph showing photoreceptor projections out of the compound eye and, beneath it, a figure of retinotopic medulla neurons derived from Cajal and Sánchez's (1915) classic description of optic lobe anatomy. Freely translated, the legend reads: "The figures suggest the nature of the substrate in which the

Table 1. Neuroanatomical correlates to physiology (refer to Fig. 1)

Level 1. Morphology of receptors and projections to lamina/medulla. Basic anatomy of R1–R6, R7/8 and their projections: Trujillo-Cenoz (1965), Braitenberg (1967), Strausfeld (1971b). Electron microscopy of receptors: Shaw and Stowe (1982); types of receptors and distribution in retina: Wunderer and Smola (1982), Hardie et al. (1981), Hardie (1983), Strausfeld and Wunderer (1985)

Level 2. Organization of the lamina, including lamina amacrines. Light microscopy: Cajal and Sanchez (1915), Strausfeld (1970, 1971a), Strausfeld and Braitenberg (1970). Electron microscopy: Trujillo-Cenoz (1965), Boschek (1971), Strausfeld and Campos-Ortega (1977) [reviewed 1981 by Strausfeld and Nässel], Burkhardt and Braitenberg (1976), Hardie (1983), Fröhlich and Meinertzhagen (1982), Shaw (1984).

Level 3. Organization of the medulla. Light microscopy: Cajal and Sanchez (1915), Campos-Ortega and Strausfeld (1972), Strausfeld (1970, 1976a,b, 1984). Minimal electron microscopy: Trujillo-Cenoz (1969), Campos-Ortega and Strausfeld (1972a,b).

Level 4. Organization of the lobula plate. Light microscopy: Cajal and Sanchez (1915), Braitenberg (1970, 1972), Strausfeld (1970, 1976a,b, 1984), Strausfeld and Obermayer (1976), Hausen (1976, 1982a,b, 1984), Hengstenberg (1977, 1982), Egelhaaf (1985b), Strausfeld and Bassemir (1985a,b). Minimal electron microscopy: Pierantoni (1974, 1976), Hausen et al. (1980), Bishop and Bishop (1981).

Level 5. Structure of the lobula. Cajal and Sanchez (1915), Braitenberg (1970), Strausfeld (1970, 1976a,b, 1980), Strausfeld and Hausen (1977), Hausen and Strausfeld (1980), Strausfeld and Nässel (1981), Nässel and Strausfeld (1982). Locust: O'Shea and Williams (1974), Rind (1987).

Level 6. Organization of optic lobe outputs onto descending neurons in the lateral deutocerebrum. Light microscopy: Strausfeld (1976a,b), Strausfeld and Obermayer (1976), Strausfeld and Seyan (1985, 1987), Strausfeld and Bacon (1982), Bacon and Strausfeld (1986), Light and electron microscopy: Pierantoni (1976), Strausfeld and Bassemir (1983, 1985a,b). Locust: O'Shea and Williams (1974), Killmann and Schürmann (1985).

Level 7. Organization of visual descending neurons. Light microscopy: Strausfeld and Bacon (1982), Strausfeld and Singh (1980), Strausfeld et al. (1984), Bacon and Strausfeld (1986). Electron microscopy: King and Wyman (1980), King and Valentino (1983). Locust: O'Shea et al. (1974), Bacon and Tyrer (1978), Griss and Rowell (1986).

Level 8. Descending neuron output to neck motor. Strausfeld and Seyan (1985), Milde and Strausfeld (1986), Milde et al. (1987), Strausfeld et al. (1987). Moth: Rind (1983). Locust: Kien (1980).

Level 9. Descending neuron output to leg and flight motor. Bacon and Strausfeld (1986), King and Wyman (1980). Locust: O'Shea et al. (1974), Pearson (1983).

(computational) functions described in this paper are performed. Until now we know of no specialized role for any of the structures illustrated. But because they are so clear cut, by comparing these structures with the model circuit shown (in their Fig. 8) we hope to identify the physiological significance of anatomical structures in the insect visual system."

The photomicrograph of receptor axons figured in their paper came from a comparative histological study by the Tübingen histologist Günther Meyer (1951) on the effectiveness of the Gros-Schultze and Bodian reduced-silver stains on the brains of different phyla. This paper, which portrays the beautiful orderliness of the fly's optic lobe, must have greatly influenced later investment in neuroanatomy. For surely, embodied on a single slide and laid out in perfect order, were the biological networks that mapped visual space through the nervous system to behavior.

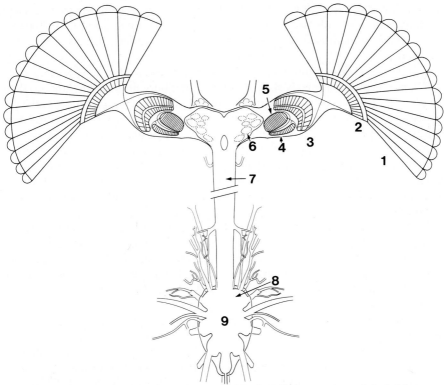

Fig. 1. Plan of the visual and central nervous system of the fly *Calliphora erythrocephala. 1–9* designate levels at which neuroanatomical studies have provided relevant structural correlates to physiology (see Table 1). *1* retina; *2* lamina; *3* medulla; *4* lobula plate; *5* lobula; *6* "optic foci" of lateral deutocerebrum; *7* cervical connective; *8* neck motor; *9* fused thoracic-abdominal ganglia

These early reduced-silver preparations provided the original data set for Braitenberg's 1966 and 1967 studies on the retina-to-lamina projections of receptor axons, which demonstrated that each unit of neuropil in the lamina (the "optic cartridge") obtained its six receptor axons from six different ommatidia. This corroborated Kirschfeld's 1967 studies on the physiological optics of the fly eye, which showed that each of a set of optically aligned receptors was in a different ommatidium. To obtain coherent information about a single point in space their axons should project to an exclusive target neuron.

For a short period the mainstay of neuroanatomical research was provided exclusively by Bodian reduced-silver preparations. But because the method selectively stained coagulated cytoskeletal proteins, these preparations provided a parsimonious representation of neurons whose regular arrangements and cross-connections were seductively similar to hypothetical wiring diagrams, derived from behavioral experiments. The two abstractions, neuroanatomical and behavioral, coexisted unperturbed until single nerve cells were revealed selectively.

In 1966 Collett and Blest used extracellular recordings from sphingid brains to show that single units in the brain and ventral cord respond selectively to "optomotor" stimuli. This research prompted a survey of the moth optic lobes, accomplished by reintroducing the Golgi method to insect neuroanatomy. Except for a little known study by Pflugfelder in 1937, Strausfeld and Blest's 1970 paper was the first comparative study on insect optic lobes since Cajal and Sánchez's 1915 publication.

Subsequent Golgi studies on *Calliphora* and then on *Musca* (Strausfeld 1970, 1971a,b) revealed complexly shaped neurons that bore little resemblance to their Bodian-stained counterparts. It also became obvious, that because many cell types were not stained by Bodian methods, confident identification of single neurons was only feasible using the Golgi technique. An optic cartridge, once thought to contain triplets of relay neurons postsynaptic to photoreceptor terminals, was shown to contain six interneurons, parts of two species of amacrine cells, and a variety of centrifugal fibers (Strausfeld 1971a; Strausfeld and Nässel 1981). Electron microscopy of Golgi-impregnated neurons eventually provided synaptic "wiring diagrams" that were incompatible with those derived from systems analysis of behavior. Instead, they offered plausible explanations for the phenomena of neural adaptation pools, receptor summation at monopolar neurons, enhancement of signal-to-noise ratio, and lateral inhibition (Strausfeld and Campos-Ortega 1977) – all at that time being demonstrated from intracellular recordings of dipterous and odonate lamina (Arnett 1972; Zettler and Järvilehto 1972; Laughlin 1975; see also Sect. 2.3.1).

The daunting complexity of the lamina, revealed by Golgi electron microscopy, threw into doubt whether neurons could ever be matched to computational elements envisaged by quantitative behavioral studies. Despite this, three major neuroanatomical achievements were inspired by systems analysis of compound-eye-mediated responses. These were: (1) descriptions of the geometrical layout of the optic lobes; (2) the experimental identification of neuropils involved in motion detection; and (3) iontophoretic dye-identification of tangential neurons that sum information about directional movement. The first of these achievements was a decisive paper by Braitenberg (1970) summarizing the main features of optic lobe architecture as portrayed in Bodian preparations. This paper served as a paradigm for subsequent work that attempted to relate single nerve cells and their dendritic arbors to the organization of the compound eye and its receptive fields. We shall return later to the significance of this study.

Although intracellular physiology and real neural circuits in the lamina could not be reconciled to models of motion detection circuitry, the abundance and crystalline regularity of cross-connections visible at any depth of the optic lobes suggested that there was no shortage of other possible correlates. Connections between retinotopic columns were generally oriented along two orthogonal (x,y) axes of neuropil, corresponding to the two oblique axes of the retinal lattice, described earlier by Braitenberg (1967). If "elemental motion detectors" were based on interconnections between adjacent retinotopic pathways (Reichardt 1961), then they should be optimally stimulated by gratings that moved in the appropriate direction across the ommatidial mosaic. Experiments using sequential stimulation of adjacent receptors, narrowed down all possible configurations of paired channels to adjacent retinotopic pathways aligned along either the x- and y-axes or subadjacent channels aligned along the horizontal (z-) axis of the compound eye (Buchner 1976; Pick and Buchner 1979).

This finding initiated a heroic effort to identify directly the cellular basis of elemental motion detection using radioactive deoxyglucose accumulation in visually activated neurons,which take up the glucose analogue at an enhanced rate. After initial glycolysis and phosphorylation, unmetabolized products are held captive in the cell and can be later revealed by autoradiography. Deoxyglucose "activity staining" was used by the Buchners (1983, 1984) to reveal specific optic lobe strata that had shown sustained activity during compound eye stimulation. The pattern of activity could be matched to specific depths in the lobula plate corresponding to the orientation selectivity of direction-sensitive neurons already known from intracellular recording and dye filling.

An important result of the Buchners' studies was that deoxyglucose accumulations designated specific strata of the medulla and lobula plate that were known to be connected by certain Golgi-impregnated small-field retinotopic neurons (Strausfeld 1970, 1976b, 1984). These are the bushy T-cells of which there are two types, T4 and T5 (see Fig. 2). Both are highly unusual in that they occur as identical pairs in each retinotopic column. The T4 pair projects to the lobula plate from the medulla, where its dendrites coincide with the ending of one of two highly specialized columnar interneurons (labeled S-ub; Fig. 2). The dendrites of S-ub enwrap the terminals of monopolar cells relaying data from a set of identically aligned photoreceptor axons. The second of the two columnar neurons (Tm1: Strausfeld 1970, 1976b) projects to an outer layer of the lobula where it terminates exclusively amongst the dendrites of the paired T5 cells (Fig. 2). These arrangements are repeated in every retinotopic column. The four terminals belonging to the T4 and T5 pairs in each column segregate out to the four discrete lobula plate strata, which show up as individual radioactive-labeled layers, depending on the direction and orientation of the gratings used to stimulate the eye during deoxyglucose uptake (Buchner and Buchner 1983). Because a preferred direction and orientation reveals one stratum in preference to the other three, the depths of T4 and T5 endings, corresponding to one or another of these activity layers, suggest that one population is associated with horizontal progressive motion, another with regressive motion, a third with upward, and a fourth with downward motion.

Electron microscopic studies on T4 neurons (selectively stained with silver chromate after filling motion-sensitive wide-field HS and VS neurons with cobalt-silver) demonstrate that one layer of T4 terminals is presynaptic to a similarly shaped profile from the neighboring and subneighboring retinotopic column (Strausfeld in preparation). In turn, these T4 terminals are presynaptic to the dendrites of the wide-field motion-sensitive elements (Strausfeld 1984). Hassenstein and Reichardt's 1956 model predicted that elemental motion detectors that compute the direction and orientation of visual flow fields involve cross-connection between neighboring retinotopic channels. It now seems likely that this model may include synapses amongst identifiable small-field retinotopic T4 or T5 neurons terminating on large-field "summatory" elements. The only difficulty is providing electrophysiological confirmation: if computation of motion is performed between neighboring T4 or T5 endings, this may not be accessible to intracellular recording if, as suggested by Torre and Poggio (1978), interactions are local subthreshold events. Motion-sensitive responses have, however, been

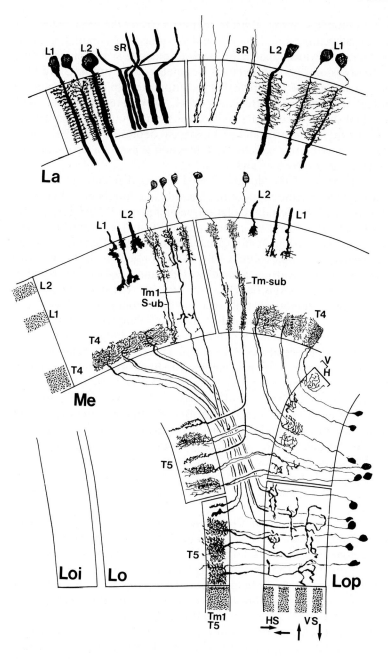

recorded in abundance from wide-field neurons postsynaptic to the T4 and T5 cells, and we shall now consider this major output from the optic lobes.

Braitenberg's 1970 account "Ordnung und Orientierung. . . ." (with figure captions in English!) identified two important features that characterize deeper levels of the system. First, the periodic arrangement of the lamina is represented by a similar mosaic of medulla columns. The linear arrangement of cartridges along the horizontal axis of the eye is inverted in the medulla by the first optic chiasma (Strausfeld 1970, 1971b) and re-inverted in the lobula and lobula plate by the second chiasma. Each medulla column, and its projection into the deeper neuropils of the optic lobes, represents a set of eight identically aligned rhabdomeres — the short visual fibers R1–R6 and the pair of long visual fibers R7 and R8.

Second, the retinotopic mosaic is mapped onto radically different cell arrangements in the lobula and lobula plate. The lobula consists of regularly spaced pyramidal cell-like dendritic trees that coarsen the original retina-lamina mosaic. Individual lobula neurons occur one to every n retinotopic columns ($n > 3$). In contrast, the lobula plate is characterized by tangential fibers, arranged orthogonally at two levels, one at the plate's outer surface and one deep in its neuropil (Braitenberg 1972). The outer group is arranged dorsoventrally, consisting of the horizontally oriented dendritic trees that cover the neuropil from top to bottom. The deeper group comprises a system of vertical fibers arranged from the front to the back of the neuropil. Serial section reconstructions of these fibers led Pierantoni (1974, 1976) to speculate that these neurons were cardinal to motion perception.

This prediction was shortly afterwards confirmed by McCann and his associates at the California Institute of Technology. Unlike their colleagues in Tübingen, the Pasadena group had been concerned with ascribing quantifiable behavior to histologically located motion-selective units that responded to optomotor stimuli (Bishop and Keehn 1967; Bishop et al. 1968), a strategy inspired

----

Fig. 2. Composite diagram showing "Common denominator neurons" of Diptera and Lepidoptera associated with the R1–R6 input. The left lamina (*La*) and left medulla (*Me*) and the lower part of the lobula and lobula plate (*Lo, Lop*) show neurons from *Calliphora erythrocephala.* The right lamina, medulla (and upper lobula complex) are from *Manduca sexta,* much reduced in size to conform with their dipteran counterparts. In both orders, the short photoreceptor terminals R1–R6 (indicated as *sR*) terminate onto spiny monopolar cells. In adult *Manduca* these have wide fields extending through several groups of sR's having optical alignments typical of superposition retinae. In Diptera, the *L1* and *L2* monopolars extend through six sR's (R1–R6) sharing the same optical alignment. L2 (and outer swellings of L1) in Me end on the small field relay neurons *Tm1* and *S-ub* which terminate, respectively, at the level of *T5* and *T4* neurons. The endings of T4, T5 define four direction and orientation levels in the Lop, as revealed by 2-deoxyglucose activity (layers *stippled,* directions indicated by *arrows*). In the medulla, activity levels (*stippled*) are correlated with flicker and motion but not direction or orientation (Buchner and Buchner 1984). In *Manduca,* L1, L2 neurons terminate at the small field relay neuron *Tm-sub,* which gives rise to axon collaterals and terminals, respectively, in the inner medulla and outer lobula coinciding with the dendritic trees of T4 and T5 cells. Tm-sub in moths appears to represent the dual Tm1 and S-ub channel of Diptera. In Lepidoptera, vertically and horizontally oriented wide-field neurons are situated at two superficial levels in the lobula plate (*V, H*) supplied by three levels of T5 terminals. Lobula plate neuropils are characteristic only of Diptera, Lepidoptera and Coleoptera. In other orders, the "undivided" lobula is characterized by a deep stratum (*Loi*) which receives terminals of T5 neurons that end at various depths, and gives rise to wide field neurons possibly involved in relaying information about wide-field motion. (Strausfeld 1976b)

by the work of Wiersma (also at Cal Tech) who had demonstrated that context-specific visual responses by interneurons of the crab *Podophthalmus* occupied characteristic positions in the optic nerve (Wiersma and Mill 1965; Wiersma 1966). Working on the green bottle fly, *Calliphora phaenicia,* Bishop et al. (1968) used extracellular electrodes and histochemistry to characterize a variety of motion-sensitive, direction-selective units, and to determine the approximate locations of the recording tip. They showed that such units occurred predominantly in the back part of the midbrain, in the lobula plate, in the inner medulla, and in an axon bundle of the frontal protocerebrum. Following Stretton and Kravitz's (1968) discovery of the fluorescent dye, procion yellow, and its employment in insect visual physiology (Autrum et al. 1970), the California group was amongst the first to use dye-filled glass capillary microelectrodes to intracellularly record and fill lamina monopolar cells (Arnett 1972) and lobula plate wide-field motion-sensitive neurons (Dvorak et al. 1975a,b). The latter corresponded to the horizontal and vertical elements reconstructed by Pierantoni (1974, 1976).

The accessibility of the lobula plate for intracellular studies has since been most fully exploited by the Tübingen group, the combined efforts of Hengstenberg (1977, 1982), Hausen (1976, 1982a,b, 1984), and Egelhaaf (1985a,b) providing an unsurpassed richness of data on the shapes and responses of its neurons. And although this topic is treated elsewhere in this book (Hausen and Egelhaaf this Vol.), a brief resume here should bring the salient points of this research into the perspective of functional neuroanatomy as a whole.

The lobula plate is dominated by "giant" direction-selective motion-sensitive neurons: three horizontal (HS) and eleven vertical (VS) cells, whose dendrites invade two discrete levels of the neuropil — HS at an outer layer and VS innermost. Both groups of cells terminate posteriorly and ipsilaterally in the lateral deutocerebrum with the axon prolongations of two HS cells extending into the subesophageal ganglion (Strausfeld and Bassemir 1985b).

Intracellular studies of lobula plate neurons have identified at least 20 morphologically and physiologically distinct classes of wide-field motion-sensitive cells (Hausen and Egelhaaf this Vol.). Those within the outer layers of the neuropil are successively activated by front-to-back, and back-to-front, movement of vertical gratings around the compound eye. Tangential neurons situated deep in the lobula plate respond to upward or downward movement of horizontal gratings. Certain tangentials have dendrites that invade inner and outer layers, and it would be expected that these respond to oblique orientations (Eckert and Bishop 1978; Hengstenberg 1982).

A characteristic feature of lobula plate organization is that none of its layers is uniform from front to back. Even amongst horizontal cells, which interact with the entire retinotopic mosaic, dendritic densities are much greater in the front part of the mosaic than elsewhere. Certain motion-sensitive neurons have dendritic fields restricted to the posterior, others to the anterior part of the retinotopic mosaic; others have unique dendritic fields that cover an area of the plate that is peculiar for that cell type (Hengstenberg 1977). This situtation is also characteristic of lobula neuropil, where such localized arrangements involve assemblies of small-field neurons. Their significance in behavior is discussed in the next section.

Two of the three HS cells, subtending the upper two-thirds of the retina, respond to bilateral stimulation, as do VS neurons subtending the lateral monocular field of the eye. Since HS and VS neurons terminate on the same side of the brain on which they originate, such heterolateral responses must be mediated by heterolateral connections. Hausen (1984) has demonstrated that several lobula plate neurons belong to this category, some directly linking the two lobula plates, others projecting to the dendrites of centrifugal neurons in the contralateral deutocerebrum, which then project distally into the appropriate lobula plate strata. Other heterolateral neurons project from the lobula plate contralaterally to terminate at the axon collaterals of HS and VS neurons.

One prominent heterolateral connection directly linking the two plates is provided by a unique pair of horizontal-motion-sensitive neurons (H1) whose axons are the exclusive occupants of a superficial and highly accessible anterior tract. This cell type has assumed importance as an experimental tool because it can be easily monitored during a variety of experimental situations. The most sophisticated of these is successive stimulation of optically isolated retinotopic channels from which can be derived the geometrical and biophysical properties of minimal retinotopic connections required for eliciting a "motion response" within a "motion-sensitive" neuron (Riehle and Franceschini 1984; Franceschini et al. this Vol.).

A variety of other wide-field lobula plate tangentials, obviously unrelated to HS or VS cells, arborize in a number of lateral and midbrain neuropils. However, their associations with other neurons are still obscure, partly because we know too little about the composition of these areas. Despite this, the observation that such cells distribute to a variety of centers suggests that the general function of the lobula plate is to collate data about wide-field panoramic motion, relaying this information to diverse targets in the brain. Only one target has been studied in detail: this is the class of nerve cells known as descending neurons, which supply motor neuropils in the thoracic and abdominal ganglia. These are described in the last section of this chapter.

2.2 "Receptor-Downward" Analysis: Classical Neuroanatomy

Working inward from the receptor level has been the classical approach to analyzing the central nervous system and was practiced on insects in the last century by Kenyon, and at the beginning of this century by Vigier. Although best known for his Golgi studies on the bee brain, Kenyon (1897) succeeded these by a short paper describing Golgi-impregnated photoreceptor endings and monopolar cells extending from the lamina to the medulla, forming a chiasma between these two regions. A few years later, the French physiologist Vigier (1907a,b, 1908) used Golgi impregnations to stain the pathways taken by receptor axons from fly ommatidia to lamina monopolar cells. He demonstrated that axons from an ommatidium did not project directly to the monopolar cell beneath them, as Kenyon described from the bee, but diverged to several other monopolar cells displaced laterally from the base of their ommatidium of origin.

As described by Dietrich in 1909, the seven observable rhabdomeres of a dipteran are distinct from each other, instead of being fused into a single coaxial unit as in the bee. Vigier (1907a) suggested that in flies the divergence of receptor axons from the base of an ommatidium reflected different viewing directions amongst the photoreceptors. He proposed that axons converging to a single monopolar cell arose from photoreceptors sharing the same optical axis but distributed in different ommatidia.

Sixty years later, these observations were elaborated by Kirschfeld (who subsequently rediscovered Vigier's original papers!). When the eye is illuminated antidromically and viewed through a stopped-down objective, rhabdomeres sharing the same optical axis can be observed as a rhomboidal configuration of seven receptors arranged in seven ommatidia. Trujillo-Cenóz (1965) had already demonstrated that discrete groups of receptor axons terminating in the lamina synapse onto at least two monopolar cells in a single cartridge. Extrapolating from this, Kirschfeld (1967) predicted that if the postsynaptic neurons were to receive information from a single point source, the axons from a set of identically aligned receptors must project laterally under the retina to a common target in the neuropil. Because there must be as many sets of identically aligned receptors as there are ommatidia, this projection pattern must be repeated across the eye with no variation except for local distortions due to the eye's curvature. The validity of Kirschfeld's model was demonstrated by Braitenberg's (1967) histological studies of receptor projections. Together, these two papers laid the foundation for all subsequent discussions (and much crucial research) about the configuration of parallel channels required for computing elemental motion detection.

The basic tenet of the fly's "neural superposition" retina is no different from that applying to apposition eyes (as in bees) in which receptors sampling the same point in space have fused rhabdomeres under a common lens: axons of identically aligned receptors terminate at a single site in the lamina containing a standard set of postsynaptic neurons enclosed by glial processes. In the fly, axons from an optic cartridge project together to a column in the medulla accompanied by two "long visual fibers". These arise from the optically corresponding seventh and eighth receptors, which contribute two tandemly arranged short rhabdomeres (one ultraviolet [termed R7], the other [termed R8] blue- or green-sensitive [Hardie 1985]) to the central element of a visual sampling unit (Melamed and Trujillo-Cenóz 1968; Campos-Ortega and Strausfeld 1972a).

Much of our knowledge about the variety of nerve cell shapes in insect optic lobes is derived from two seminal studies published in 1913 and 1915: one by the Russian anatomist Zawarzin, the other by the Spanish anatomists Cajal and Sánchez. Embedded in these beautifully illustrated papers are three sweeping generalizations, the first of which is only now finding resonance amongst vertebrate physiologists. This was the recognition that the basic layout of the insect visual system bears direct comparison to those of cephalopods and vertebrates. Zawarzin (1913, 1925), and Cajal and Sánchez (1915), proposed that analogies should be made between the insect lamina, medulla, and lobula and the mammalian external and internal plexiform layer, geniculate, and cortex. The second observation — which today may seem commonplace, but has strongly influenced our expectations

—was that neurons in the optic lobes are arranged in an exceedingly orderly fashion reflecting the hexagonal lattice of ommatidia. Indeed the precision of the neuronal arrangement so enchanted Cajal that he afterwards derided the vertebrate nervous system, calling it "coarse, rude, and deplorably elementary" (Cajal 1937). The third generalization was that neurons could be recognized time and again on the basis of their shape and their location. Basically, Cajal, Sánchez, and Zawarzin confirmed one of the dominant structural properties of arthropod and annelid nervous systems: namely, Retzius's (1891) discovery that they comprised "uniquely identifiable" neurons whose shapes and locations are typical of a species and are invariant. This feature would be emphasized some 30 years later by electrophysiologists who realized that response-specific units in cockroach and crayfish had characteristic locations in certain fiber bundles and tracts (Roeder 1948; Wiersma 1958). Later, yet another of Cajal and Sánchez's observations was confirmed: the similarities between neuronal shapes and dispositions amongst different insect species, indicating the existence of a common blueprint of central nervous system organization.

The previous section has already mentioned the negative impact made by Golgi studies and electron microscopy on the search for neural correlates to theoretical networks. However, even though Golgi studies provided a dismaying variety of neurons, they were useful in demonstrating the repertoire of cell shapes and the stratified organization of the optic lobes. It became clear that the medulla is composed of a staggering variety of small-field relay neurons — transmedullary neurons, T-cells, and Y-cells — characterized by complexly shaped dendrites that occupy discrete layers of neuropil containing stratified arrangements of amacrines (Strausfeld 1970, 1976a). An important finding was that whereas amacrines comprise a major class of neurons in the medulla, they are scarcely represented in the lobula and lobula plate. This suggests that the functional organization of these two inner neuropils reflects a system of summatory neurons that integrate elemental information about the visual world relayed to them from the medulla. More complex qualitative information is acquired at two levels: by the convergence of parallel relays from the medulla to lobula and lobula plate neurons and by the convergence of various combinations of lobula-complex neurons onto descending neurons. We shall return to this important aspect of optic lobe organization later, with reference to the structural basis for parallel processing by the visual system.

A conspicuous failure of the Golgi method was its inability to reveal lateral relationships between neurons. Because nerve cells are stochastically impregnated, their occurrence in the histological preparation could not reflect their relationships. Furthermore, it was virtually impossible to demonstrate if any cell type was representative of the whole eye or was limited to a specific zone serving a restricted area of the retina. To know if cells are distributed like this was of great importance because behavioral studies had demonstrated that certain zones of the compound eye are used for specific tasks, such as distance estimation and binocular vision (Maldonado and Barrós-Pita 1970; Collett and Land 1975). Only after cobalt staining was belatedly used for optic lobe anatomy could such questions be resolved.

2.3 Predictive Neuroanatomy: Central Representation of the Compound Eye

By 1975 cobalt "backfilling" into descending neurons had become an established technique for screening the shapes of dendrites in cricket, cockroach, and locust ganglia. The first cobalt backfills into the dipteran nerve cord revealed the entire populations of three species of optic lobe neuron: the HS and VS cells, and a palisade of small-field lobula neurons (Col A cells) spaced one to every three retinotopic columns (Strausfeld 1976a,b; Strausfeld and Obermayer 1976). Implantation of minute crystals into the neuropil was subsequently used to achieve stainings into axon bundles within the brain, revealing their dendrites and terminals. A refinement of the procedure employed blunt electrodes filled with cobalt solutions, later intensifying the precipitated cobalt profiles with silver (Strausfeld and Hausen 1977). These methods opened up an entirely new aspect of optic lobe organization.

Prior to the use of cobalt, evidence that a certain species of neuron occurred in each retinotopic column could be obtained only from many hundreds of Golgi impregnations. In contrast, direct application of cobalt into the neuropil directly reveals cell distributions. How this occurs is not fully understood, but the results are spectacular, demonstrating a new subunit of optic lobe organization: the collective arrangement of all of one particular species of nerve cell. When neurons are represented across the whole of the optic lobes, their summed organization is referred to as an isomorphic cell assembly. This is the arrangement seen most often. However, certain areas of the retina are represented by special groups of neurons, termed local nerve cell assemblies.

Six areas of retina have now been identified that are represented in the lobula. These are: (1) the whole of the monocular visual field; (2) the area of binocular overlap — a region determined by optical measurements (Beersma et al. 1977); (3) frontal-dorsal areas of high visual acuity that are found only in male flies (Land and Eckert 1985); (4) a dorsal marginal zone containing polarized-light-sensitive photoreceptors (Wunderer and Smola 1982; Hardie 1984); (5) a ventral area represented in both sexes; and (6) a posterior region represented in the lobula by a pair of neurons having asymmetrical dendritic trees on each side of the brain (Nässel and Strausfeld 1982). Each cell assembly sends its axons to a characteristic neuropil in the lateral deutocerebrum. As we shall discuss below, it is these areas from which originate the dendrites of descending neurons leading to thoracic motor centers.

2.3.1 Representation of the Whole Eye

Isomorphic and Heteromorphic Assemblies. Although the whole retinotopic mosaic is represented in the lobula plate by the projections of certain T- and Y-cells (the latter also sending a branch to the lobula), very few species of lobula plate output cells subtend the whole of the retina. Even in what appears from mass fillings to be a regular array of similarly shaped neurons, these may vary significantly with respect to their dendritic field shapes or their central projections. For example, the three HS neurons together subtend the entire retina; however, the lower of the three projects centrally to a location not shared by the other two (Strausfeld and Bassemir 1985b). The VS assembly is even more heteromorphic. Not only do its elements project to different targets in the brain (Strausfeld and Bassemir 1985a), but each species of VS neuron is obviously related to fields of view of discrete areas of the retina and the function of neck muscles (Strausfeld et al. 1987; Milde et al. 1987).

Whole-eye representation by isomorphic assemblies is, however, manifested by medulla columnar neurons and a variety of lobula neurons, some species of which also send dendritic branches into the lobula plate. Typically, each element of a lobula assembly has a relatively simple uni- or bistratified dendritic arbor, extending through a characteristic retinotopic domain. The following criteria distinguish each species of isomorphic columnar neuron: (1) dendritic tree morphology; (2) periodicity with respect to groups of retinotopic inputs from the medulla; (3) its relationship in depth with terminals from the medulla (Strausfeld and Hausen 1977); and (4) its target in the midbrain (Strausfeld and Bacon 1982).

The Organization of Parallel Processors into the Lobula. Anatomical evidence suggests that all the dendrites of an isomorphic assembly receive an identical combination of terminals from transmedullary cells, each type of medulla neuron penetrating to a characteristic depth in the lobula neuropil (Strausfeld 1976b). This organization suggests that specific combinations of medulla terminals are presynaptic onto characteristic assemblies.

If this is correct, then the lobula may hold the key to understanding the functional significance of the medulla's complexity. What amazed Cajal was the elaboration of cell shapes in this region. Indeed, this is what sets the medulla apart from other neuropils, and it is worth digressing briefly to review the neuron morphologies there. The medulla is composed of columns, each containing as many as 43 relay neurons (Campos-Ortega and Strausfeld 1972a,b). These project to the lobula, the lobula plate, or both. Each species of relay cell has a characteristic shape, many of which are very elaborate indeed. Whether singly or multistratified, dendrites extend through a characteristic domain of other columns at each level in the neuropil. These domains are often asymmetrical; some extend dorsoventrally, others extend along the horizontal axis of the retinotopic mosaic, others have more complex fields. Certain neurons have at least one set of dendrites that reside at the level of specific lamina-retina inputs. But this does not imply that all are postsynaptic to long visual fibers or lamina monopolar cells. Many probably receive indirect inputs from the lamina, routed through amacrines and other relay neurons, as suggested by layers of axon collateral reminiscent of L4 monopolar cell connections in the lamina (Strausfeld and Campos-Ortega 1973a; Strausfeld 1976b).

The second distinguishing feature of the medulla is its large variety of amacrine cells, possibly comprising as many as 35 distinct morphological types. These comprise three general classes of anaxonal neuron: those that extend in a diffuse and stratified manner through the neuropil, either within a single retinotopic column or visiting several columns and overlapping each other; and those that have overlapping flattened uni- or multistratified fields, designating prominent strata across the neuropil. The overall organization, portraying immense complexity, suggests a series of intricate but precisely engineered microcircuits.

Synaptic Organization in Retinotopic Neuropil. Some idea, both about the detail and the task of understanding such circuits, may be gained from the complex synaptic matrix of the lamina (summarized by Strausfeld and Campos-Ortega 1977; Strausfeld and Nässel 1981; and critically reviewed by Shaw 1984, this Vol.). Each set of R1-R6 photoreceptor terminals is associated with six relay neurons, the monopolar cells, L1-L5, and the T-cell, T1 (unrelated in function to the T4 and T5 neurons mentioned earlier). Of these, L3 receives a simple combination of inputs, from the R1-R6 receptors and from two types of centrifugal neurons (C2 and Tan 1) arising from the medulla (Campos-Ortega and Strausfeld 1973; Strausfeld and Campos-Ortega 1977). In contrast, the two largest relay cells, the monopolars L1 and L2, are involved in complex synaptic interactions, both within the retinotopic columns and amongst columns laterally, with the participation of the L4 and L5 monopolar cells (both third-order neurons that receive no direct inputs from R1-R6), the T1 neuron, and amacrines. The circuitry is complicated and certain crucial aspects of it are either disputed or are, as yet, unresolved. A network of axon collaterals of the L4 monopolar cells (upper diagram, Fig. 3) mediates lateral interactions between retinotopic outputs, carried by the L1, L2 pair from each cartridge (Strausfeld and Braitenberg 1970; Strausfeld and Campos-Ortega 1973a; Braitenberg and Debbage 1974). Collaterals of three L4 monopolars (that of the parent L1, L2 cartridge, and of two flanking cartridges, one aligned along the x- and one along the y-axis of the retinotopic mosaic) are presynaptic to the axon origin of L1, L2. The L4 neurons derive their input indirectly from many photoreceptors via the wide-field amacrines whose tuberous specializations (alpha processes) flank and are postsynaptic to the receptor terminals (Strausfeld and Campos-Ortega 1973a; Campos-Ortega and Strausfeld 1973).

A second lateral network (lower diagram, Fig. 3) involves amacrines, L4 neurons, and the T1 afferents whose dendrites (beta processes) climb up the outside of each cartridge where they are postsynaptic to alpha processes – the varicose specializations of amacrines. Golgi-EM studies showed the beta processes to be also apposed to receptor terminals (Campos-Ortega and Strausfeld 1973) to which they were interpreted as being postsynaptic. However, recently Shaw (1984) disputes this on the grounds that he could not observe R1-R6 presynaptic to T1. Rather, Shaw suggests that T1 is postsynaptic to L2 in a special layer at the base of the lamina where L2 is known to have presynaptic sites onto L1 and onto R1-R6. Interestingly, the ending of T1 in the medulla insinuates itself between that of the L2 monopolar and the follower transmedullary cell giving T1 the role of a local interneuron irrespective of its second- or third-order status (Strausfeld and Campos-Ortega unpublished).

The connections among receptors, amacrines, and T1 might provide a substrate for lateral inhibition onto T1, which, on the basis of one fill, has been reported to have an acceptance angle less than that of the receptors of its parent optic cartridge (Järvilehto and Zettler 1973). The network provided by connections amongst R1-R6 receptor terminals, amacrines, L4 neurons, and their collaterals to

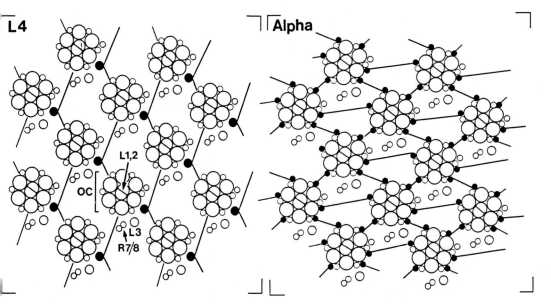

Fig. 3. Interconnections between retinotopic channels in the lamina. Optic cartridges are represented by two central elements (the *L1, L2* monopolars), surrounded by six receptor terminals and six smaller beta fibers belonging to T1 (represented here by six *small open profiles*). *L3, R7* and *R8* represent a separate output pathway from the retina and lamina. *Left* L4 output monopolars are represented by *filled profiles* with axon collaterals linking cartridges vertically. *Right* L4's receive inputs from amacrine cells whose alpha processes (*filled profiles*) surround the cartridges and are connected both obliquely and horizontally, providing a substrate that could account for lateral inhibition of L4 along the horizontal axis of the eye. Again, L3, R7, R8 are outside this pattern of connections

L1, L2 monopolar cells, suggests a second substrate for lateral interactions focused on the L1, L2 afferents (Strausfeld and Campos-Ortega 1977). As reviewed by Laughlin (1981) there is compelling physiological evidence to suppose that inter- and intracartridge inhibitory circuits tune monopolar cells to respond to contrast modulation within a narrow range of the ambient light intensity.

Shaw (1984) has forcefully argued against the role of either the amacrines or L4 being involved in lateral inhibition of L1, L2. His reasons are: (1) Amacrines may not behave as a wide-field unit. This objection relies on the assumption that an unidentified spiking neuron (Arnett 1972) having a narrow sustained "on" center field flanked by antagonistic (z-axis) "off" surrounds corresponds to the L4 monopolar. The orientation of its inhibitory fields does not correspond to a wide-field surround of amacrines. (2) L4 neurons are part of a low gain system, whereby their meager presynaptic sites onto L1, L2 are insufficient to override the many thousands of inputs derived from R1-R6.

Although entirely plausible, there is no evidence that Arnett's sustained unit was the L4 cell, beyond one procion fill of a spiking unit by Hardie (1978, cited by Laughlin 1981). In addition, EM studies (Strausfeld unpublished) show amacrine processes to be pre- and postsynaptic to each other within a discrete and superficial stratum of the lamina. Amacrine-amacrine fibers at this level are oriented along the horizontal (z-axis) and there is no compelling evidence to suppose that Arnett's spiking unit does not receive amacrine inputs from neighboring z-axis cartridges. Furthermore, L4 collaterals synapse onto L1 and L2 axons where the efficacy of the L4 synapse is entirely unknown.

Despite anatomical evidence for neuronally mediated lateral interactions, Shaw makes the point that synaptology cannot be simply reconciled with the response profiles of L1 and L2: transient hyperpolarizing to on-axis illumination, slow depolarizing responses to off-axis stimuli that resemble the positive response of local field potentials recorded in the lamina (Dubs 1982). It is these field potentials that Shaw (1975, 1984) suggests provide presynaptic electrical inhibition of flanking off-axis R1-R6 terminals. This idea, developed from studies of current flow in the locust and fly lamina and retina (Shaw 1975; Dubs 1982) is persuasively supported by a critical evaluation of the electrical and morphological compartmentalization of epithelial glial cells in Diptera, triplets of which provide an ionic compartmentalization of each optic cartridge (Shaw 1984). Intracellular receptor-induced currents are thought to travel between cartridges via epithelial cells (that are postsynaptic to alpha and beta processes) along the direction of least resistance (the z-axis).

There is tenuous evidence from a single fill into an L4 profile (Hardie 1978, cited by Laughlin 1984) that this is indeed a "sustained" spiking neuron. Considering that L4s provide a second regular network of collaterals, amongst L1, L2 terminals in the medulla, this cell type may play a cardinal role in elementary motion detection (EMD) circuits. L5, whose terminal enwraps those of the L1, L2 endings, is intimately associated with the L1, L2 pathway. In contrast, the parallel L3 channel is not involved in lateral connections and represents a separate channel into the medulla from the retina. The significance of this is discussed later.

2.3.2 The Binocularity Zone of the Lobula

Flies of both sexes exhibit abrupt body saccades towards small contrasty objects anywhere in the visual field. However, object fixation does not necessarily involve both eyes since it can be forced monocularly in tethered flies and seems, in any event, to be practiced monocularly in nature by cruising hoverflies (Collett and Land 1975). Nevertheless, other tasks common to both sexes, such as the ability of insects to land on small prominent objects like a flower, may involve true binocular vision employing depth perception by neurons responding to binocular disparity (see Schwind this Vol.).

In flies, both eyes are represented in the same neuropil by small-field retinotopic elements (Strausfeld 1979). These comprise heterolateral local neuron assemblies, representing the sex-unspecific frontal region of binocular overlap (Beersma et al. 1977), that exists in both males and females (Fig. 4). Obviously, such arrangements of neurons would be ideal for depth perception by binocular disparity judgements. The assembly is composed of columnar neurons arranged one to every six retinotopic columns, each having a relatively large vertically

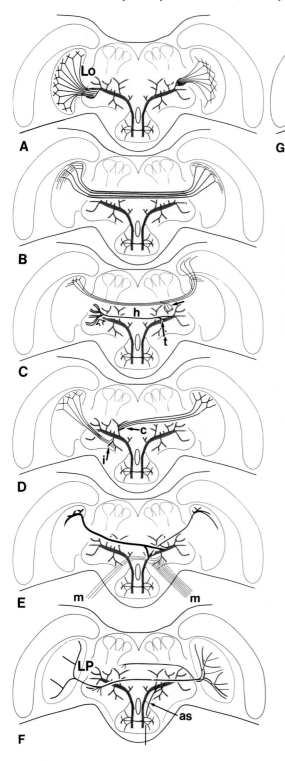

Fig. 4A-G. Organization of optic lobe outputs from the lobula (*A-E*) and the lobula plate (*F,G*). *A* Two isomorphic assemblies originating at two depths in the lobula (*Lo*) and terminating at two different dendritic arbors of a common descending neuron (shown as a *pair of stippled profiles at the center of each figure*). *B* Heterolateral connections between left and right lobulae by small-field retinotopic neurons representing the frontal region of binocular overlap (no output onto descending neurons). *C* Local assembly, representing the polarized-light-sensitive marginal zone of the retina, originates in left lobula and terminates in right lobula and medulla with axon collaterals onto a dendritic subfield of the descending neuron (*arrow*). Also shown: heterolateral neuron (*h*) sharing dendritic subfield with the DN (*open arrow*) and terminating contralaterally on the DN (*arrow t*). *D* Two local assemblies terminating on characteristic subfields of the DN. One, representing the male-specific eye zone terminates contralaterally (*c*), the other, representing the sex-unspecific binocularity zone, terminates ipsilaterally (*i*). *E* Wide-field male-specific neurons terminate on the main trunks of DN's, as do sex-unspecific wide-field tangentials of the lobula plate *(G)*. *E* also schematizes projections of mechanosensory afferents (*m*) onto special branches of the DN. *F* The majority of lobula plate (*LP*) tangential neurons project heterolaterally to a variety of targets in the brain and contralateral optic lobes. *F* also shows the typical site of termination of an ascending neuron (*as*) onto the trunk of the DN. *G* Although certain smaller wide-field tangentials terminate on DN's, the most pronounced relationship between the lobula and DN's is via the horizontal- and vertical-motion-sensitive neurons whose axon collaterals contact DN's at their main trunks

extended dendritic domain, which overlaps those of its neighbors. Each neuron of the assembly projects to the equivalent area in the contralateral lobula where its terminal arborizes amongst the dendritic tree of its contralateral counterpart. As a consequence, a frontal strip of the retina of each eye is represented by a mosaic of neurons in both lobulae. A second local assembly projects from this area into the lateral deutocerebrum of the same side where its axons terminate on descending neuron dendrites.

2.3.3 Sex-Specific Neuropil

In nature, a variety of male Diptera pursue and attempt to intercept other flies as a prelude to copulation. This behavior by the male fly is mediated visually (Land and Collett 1974; Zeil 1983a,b).

Most obviously in hoverflies, and to a lesser degree in *Calliphora* and *Musca*, the male compound eyes are distinguished by a frontal zone of large lenses, surmounting photoreceptors whose divergence angles are about half as wide as those elsewhere, imparting high acuity to this area (Collett and Land 1975; Land and Eckert 1985). This zone is represented in the lobula by prominent sex-specific neurons (Fig. 4). First described in *Calliphora* and *Musca* (Hausen and Strausfeld 1980; Strausfeld 1980) as comprising three unique tangential neurons and a local assembly of male columnar cells, a large variety of sex-specific neurons have since been resolved in male *Calliphora*, certain of which are "giant" versions of unique elements seen in females (Strausfeld 1980). So far 14 uniquely identifiable male-specific assemblies have been related to the high-acuity zone of *Calliphora* (Strausfeld 1987 and in preparation).

Each of the dendritic domains of male-specific assemblies circumscribes a characteristic zone of the retinotopic mosaic. These domains can be superimposed onto precise maps representing the penetration of each medulla column into the lobula. The maps can then be transformed to reveal the topographic relationships of sex-specific neurons with the retinal mosaic and hence the mosaic of visual sampling points. Conversion of these into putative receptive fields is done using Land and Eckert's (1985) optical projection maps, which relate the direction of view of the photoreceptors to coordinates of visual space. Such maps demonstrate that the male-specific assemblies are arranged in a roughly concentric fashion, subtending overlapping areas of the upper frontal visual field.

We can reasonably speculate that, together, the assemblies contribute to sex-specific behavior. This involves the detection of a small object somewhere in the visual field, turning towards it, tracking it, chasing it, and attempting to intercept it. The first two components, object detection followed by a body saccade, are also performed by female flies and are presumably a function of the whole eye. However, as demonstrated by Collett and Land (1975) and by Wehrhahn (1979), the tracking, pursuit, interception, and final approach are performed only by male flies and seem to involve the upper frontal retina. Crucial stimuli are: (1) the angle that the quarry subtends on the retina; (2) the eccentricity of this retinal image; and (3) its direction and velocity. Possibly the arrangement of sex-specific neurons, superimposed on the isomorphic assemblies of the lobula, represents a highly specialized visual processor in its own right. However, apart from an intriguing male-specific gradient amongst L3 monopolars (Braitenberg 1972) and specialized photoreceptor terminals underlying the high-acuity area of male house flies (Hardie 1983), there is no evidence so far that local assemblies of peripheral neurons represent specialized sex-specific pathways into the lobula. At present we must assume that, in cyclorraphan Diptera such as *Musca* and *Calliphora*, elemental computations for pursuit and interception, regarded as specifically male, are processed across the whole medulla. That only the upper frontal part of the retina is critical for pursuit and interception may be because in the lobula only this region is subtended by the appropriate postsynaptic elements. This suggestion may be put to the test by recording from descending neurons common to both sexes, which in males receive terminals of sex-specific neurons (Fig. 4).

In certain nematoceran Diptera, however, a male-specific retina is very obviously associated with a male-specific lamina, medulla and lobula complex. In male bibionids, each eye is completely divided into two separate components. One consists of large-apertured, upward looking ommatidia and has high spatial resolution. The other is a ventro-lateral retina, identical to the undivided eye of a female, which consists of small-apertured ommatidia having wide acceptance angles (Zeil 1979, 1983a). The upper male-specific eyes mediate detection and chasing of small targets (females) flying above the male. The small lateral eyes are used for optomotor flight control (Zeil 1983b). Using cobalt-silver staining into receptor axons and clusters of central neurons, Zeil (1983c) demonstrated that each of the optic lobe neuropils of the male *Bibio marci* is divided into two distinct components, one serving the dorsal eye, the other serving the ventro-lateral eye. Cobalt fills demonstrated that several of the neuronal assemblies revealed in the male bibionid have a morphological counterpart in *Musca* and *Calliphora*.

In bibionids and cyclorraphans, the lobula plate contains giant horizontal (HS) and vertical neurons (VS) which, in *Musca* and *Calliphora*, are presumed to mediate visually stabilized flight in response to movement of the whole visual panorama (see Hausen and Egelhaaf this Vol.). In these two species, HS and VS neurons are sexually isomorphic and are distributed across the entire lobula plate, thus serving the whole eye. But in the male bibionid *Dilophus febrilis*, HS and VS cells were observed only in the lobula plate serving the male-specific dorsal retina. A further interesting comparison between male bibionid and muscid flies concerns male-specific Col D neurons. In *Musca* Col D dendrites comprise a local assembly, the axons of which project to descending neurons (Hausen and Strausfeld 1980). In male bibionids, Col D neurons project to the contralateral lobula much in the same way as do binocularity neurons of male and female *Calliphora*, discussed in Section 2.3.2.

As yet, there are no data about the neuroanatomical organization of the smaller ventral eye of males or the undivided ventral eye of female bibionids. However, from the existing data it would seem that similarly shaped (and probably homologous) neurons, such as VS and HS cells, may have evolved to serve different functions in different genera: sex-specific behavior in bibionids and nonsex-specific visually stabilized flight in cyclorraphans.

2.3.4 Representation of the Marginal Eye Zone

One of the most intriguing orientation behaviors is the ability of bees and ants to return to their hive or nest after foraging over long distances, using polarization vectors as their navigational cues. However, despite extensive behavioral studies of this phenomenon (Rossel and Wehner 1987; Rossel this Vol.), little is known about the underlying neuronal organization. In Hymenoptera, behavioral studies have accredited the ability to orient to polarized light to a specialized dorsal strip of retina, and although there is only one study showing that Diptera exploit polarized light for course control (Wolf et al. 1980), there is direct evidence for a specialized system of optic lobe neurons that subtend a specialized dorsal "marginal zone" of the retina (Strausfeld and Wunderer 1985).

This zone is characterized by pairs of unusually large-diameter rhabdomeres provided by the outer segments of long visual fibers (Wada 1974). As elsewhere in the retina, each pair of these central rhabdomeres is arranged in tandem, but there the similarity stops. In the marginal zone of the retina, the microvilli of each of the central rhabdomeres are perfectly aligned and do not twist as in the rest of the eye (Wunderer and Smola 1982). In addition, the microvilli of the lower of the two receptors are arranged at right angles to those above it. Parallel microvilli alignment within a rhabdomere suggests a high degree of dichroism. This property, and sensitivity to ultraviolet light, have been convincingly demonstrated by intracellular recordings (Hardie 1984). Hardie (1984) and Melamed and Trujillo-Cenóz (1968) suggest that the orthogonal alignment of the microvillar axes between each pair of central rhabdomeres could maximize E-vector detection if the signals from the two receptors were subtracted from each other. A further refinement of the system is that the orientation of microvillar alignments change progressively from the front to the back of the marginal zone, going through a rotational shift of 180°. This fan-shaped arrangement seems ideally designed for detecting E-vector orientations, any stave in the fan being maximally excited by one orientation of linearly polarized light (see Rossel this Vol.).

The dorsal eye specialization is common to both sexes. In *Calliphora*, it is manifested in the optic lobes by three levels of structural specialization (Strausfeld and Wunderer 1985). In the retina, R1–R6 receptors of the marginal zone have much reduced rhabdomeres, or none at all. The lamina terminals are thin, and the complement of monopolar cells appears reduced. In contrast, the long visual fibers from the marginal zone penetrate deeply into the medulla where they are associated with a local assembly of small-field medulla Y-neurons that interact with a group of six or so R7/R8 receptor terminals. The Y-cells terminate amongst a unique assembly of marginal tangential cells in the lobula plate and a local assembly of closely spaced columnar neurons in the lobula. It is the latter that command special interest. Marginal columnar cells from one lobula project to the marginal neuropil of the contralateral lobula, some extending further into the marginal zone of the contralateral medulla (Fig. 4). Typically, these heterolateral neurons give rise to axon collaterals that arborize at descending neuron dendrites in the contralateral deutocerebrum.

2.3.5 Other Unique Neurons in the Lobula

So far, four general classes of optic lobe neurons that converge at the level of the descending neuron (Fig. 4) have been identified in flies: lobula isomorphic assemblies; sex-specific neurons; marginal zone neurons representing the polarized-light analyzer; and wide-field panoramic motion-sensitive neurons. In both sexes, a small tangential cell subserves the lower edge of the retina, implicating yet another subsystem whose role is entirely open to speculation.

Typically, such unique neurons occur in an identical area in the left and right lobulae, contributing to the beautiful symmetry of the insect brain. And even though certain tangential cells, such as the HS neurons, may have somewhat different patterns of bifurcations in the two lobes, such variations are not reflected in differences of dendritic domain: these are highly consistent.

There is, however, one pair of bizarre neurons that are typically asymmetric in each individual. The neurons have bilaterally symmetrical dendrite branches within the lateral midbrain which give rise to processes that asymmetrically invade a posterior strip of the lobula, and patches of neuropil in the lobula plate. The dimensions of their lobula complex domains always differ between right and left lobes, and between individuals. Interestingly, this neuron is the only optic lobe output to project directly to the thoracic ganglia (pro-and mesothorax) from where it can be filled retrogradely (Nässel and Strausfeld 1982). These asymmetric neurons subserve the rear of the retina, and it is amusing to speculate that such neurons may provide an unpredictable asymmetry to the flight motor during escape responses of a fly approached from behind.

2.4 From the Muscle Upwards: The Relationship Between Behavioral Pathways and Optic Lobe Organization

2.4.1 Splitting the Retinal Output

Neuroanatomical studies on the fly distinguish the structural differences between the three classes of photoreceptors: blue-green sensitive R1-R6 elements terminating in the lamina, and the UV-(R7) and blue or green-sensitive (R8) long visual fibers terminating at two levels in the medulla (Campos-Ortega and Strausfeld 1972a). The R1-R6 channel is split into two lamina output systems: (1) the simple L3 monopolar cell which has no lateral interactions in the lamina (Strausfeld and Campos-Ortega 1973b) and (2) the twin L1, L2 monopolar channel synaptically associated with amacrines, L4 collaterals, T1 cells, and a variety of centrifugal cells (Strausfeld and Campos-Ortega 1977; Strausfeld and Nässel 1981).

The divergence of R1-R6 outputs to two distinct channels, one complex the other simple, raises the intriguing possibility that this early segregation is related to the specialized function of the lobula plate: the L1, L2 channel, subject to complex lateral inhibition and neural adaptation pools within and between projection columns, providing duplicate monochromatic pathways to small-field elemental motion detectors (EMDs) supplying wide-field color blind motion-sensitive neurons. The simplicity of the L3 channel suggests that with R7 and R8 it contributes to a color coded trichromatic retinotopic pathway supplying the majority of medullary interneurons destined for the lobula (Fig. 5).

Wide-field motion-sensitive "collector" neurons are required for visually stabilized flight (Hausen and Egelhaaf this Vol.) responding to contrast modulation over a wide range of transient changes in ambient luminance, such as are encountered during fast locomotion within a textured visual environment. Collector neurons in the lobula plate that are activated by front-to-back (progressive)

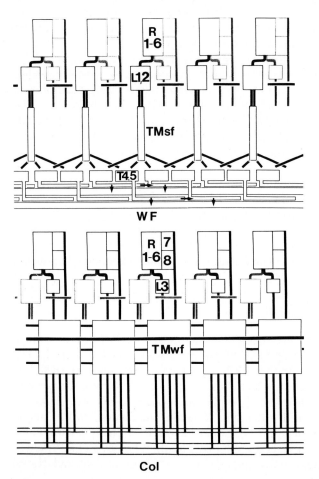

Fig. 5. Diagram showing the diverging outputs from each ommatidium. *Upper* R1–R6 via *L1, L2* diverge to small-field medullary relays (*TMsf upper diagram*), which provide inputs to neighboring bushy T-cells (*T4,5*) that synapse amongst themselves just distal to their synapses (*arrows*) onto wide-field directionally selective, motion-sensitive , color-insensitive neurons in the lobula plate (*WF*). *Lower* R1–R6 also synapse onto *L3*, which with *R7, R8* provides input to a large variety of wide-field transmedullary neurons (*TMwf*) that provide different levels of inputs to retinotopic columnar neurons in the lobula (*Col*). It is proposed that lateral interactions for the R1–6/L1, L2 channel occur mainly in the lamina, whereas those for the R7, 8/L3 channel occur in the medulla

or back-to-front (regressive) movement of the visual panorama are presumably supplied by retinotopic arrays of EMD's, which operate in an identical fashion but differ in their directional selectivity (Franceschini et al. this Vol.). Several of the above requirements are already fulfilled at the level of the lamina by networks associated with the L1, L2 monopolar cells: (1) pairs of first-order interneurons (L1, L2) duplicate the R1-R6 output in each retinotopic channel and (2) each pair is involved in a complex synaptic network interconnecting neighboring columns (Strausfeld and Campos-Ortega 1977; Shaw 1984). Interactions between neighboring optic cartridges serve to sharpen spatial acuity of L1, L2 by lateral inhibition. Lateral interactions over wider ranges, via specialized glia, or amacrines, T1, and L5 neurons (the terminals of which enwrap those of the L1 and L2 monopolar cells in the medulla; Strausfeld and Campos-Ortega unpublished), may provide local and wide-field neural adaptation channels liberating the responses of L1, L2 from local changes in ambient intensities (Laughlin 1975). This is a precondition for computations between neighboring channels that interact to determine a sequential change of intensity between one visual sampling point and the next.

Neuroanatomical observations of the relationships between the L1, L2 channel and medulla neurons provide further evidence that already at the level of the lamina there exists a profound segregation of retinotopic pathways to the lobula plate and the lobula (Strausfeld 1976b). Two types of small-field neurons (Tm1 and Sub1; Strausfeld 1976a), that have dendrites at the same level as the L2 terminal, originate from each column (Figs. 2, 5). The dendrites, axon collaterals and terminals of both of these second order interneurons are restricted to their parent retinotopic column (Strausfeld 1984) and their axons appear to be displaced from the other 40 or so columnar neurons accompanying them. The Sub1 relay terminates in the deepest stratum of the medulla at the level of the dendrites of bushy T-cells (T4; Strausfeld 1970, 1976a). Tm1 terminates in a superficial level of the lobula that contains, exclusively, T5 cells whose dendrites appear identical to those of T4 (Fig. 2). A pair of T4s and a pair of T5s originate from each column: their dendrites invade neighboring columns along the x- or y-axis (Strausfeld 1984). The four T-cell axons, together representing a single medullary column, terminate at four different levels in the lobula plate corresponding to the four motion orientation- and directionality-layers resolved by 2-deoxyglucose uptake (Buchner and Buchner 1984; see Sect. 2.1, Fig. 2).

Golgi studies have demonstrated that these cell types (L1, L2, Tm1, Sub1, T4 and T5) are "common denominators" of the optic lobes of several insect orders: Diptera and Lepidoptera (Fig. 2); Coleoptera and Hymenoptera (Strausfeld 1976a). All possess wide-field tangentials within the lobula plate or within a specialized layer of the undivided lobula, as in bees and ants. Further evidence that the T4 and T5 cells are specifically associated with HS- and VS-type neurons is provided by Golgi studies on lobula plate-less (*lop*) *Drosophila* in which HS neurons erroneously invade a deep layer of the medulla and send branches superficially into the lobula (Fischbach 1983). In the *lop* mutant, T4 and T5 neurons terminate at these two ectopic locations.

The unique functional role of the L3 monopolar cell finds support from studies on the layer relationships between lamina inputs and small field medulla interneurons. Golgi studies of the lepidopteran *Pieris brassicae,* and the Diptera *Calliphora, Eristalis,* and *Musca* (Strausfeld and Blest 1970; Strausfeld 1970, 1976a) demonstrated the precision in which each type of monopolar cell coincided with a specific type of retinotopic relay neuron. The accuracy of layering is reiterated between individuals and is independent of variations in the depth of the medulla (Strausfeld 1970, Campos-Ortega and Strausfeld 1972b). As originally demonstrated by Cajal and Sánchez (1915) this matching between specific neurons is highly suggestive of functional relationships between them.

Golgi studies determined three main projection systems between the medulla and the lobula complex: to the lobula via transmedullary cells; to the lobula plate via small T-cells arising in deep medulla strata or superficially in the lobula; and to both the lobula and lobula plate via transmedullary Y-cells (Strausfeld 1976b). Recently, Fischbach (1983) has provided convincing quantitative evidence from Golgi preparations of *Drosophila* that L3 terminals coincide in depth with transmedullary neurons destined exclusively for the lobula. In contrast, L2 terminals in *Drosophila* coincide with small-field retinotopic neurons closely associated with a deeper layer (stratum 10) containing T4 cells destined for the lobula plate, an organization that is identical to that of *Calliphora* (Strausfeld 1984).

In conclusion, Golgi studies (and mass filling with cobalt; Strausfeld, unpublished) demonstrate that in *Calliphora,* complex multistratified transmedullary neurons, whose dendrites coincide with the terminal levels of L3, R7 and R8, typically penetrate deep into the lobula (Strausfeld 1976a). This feature has been substantiated from quantitative studies of *Drosophila* (Fischbach 1983), and is also typical of two other investigated insect orders: Lepidoptera and Hymenoptera (Strausfeld, unpublished). Making allowances for topographical differences in insects having undivided lobulae, these cellular arrangements suggest that the great majority of medullary neurons may be part of a trichromatic visual system involving the R7, R8 and L3 retina-lamina outputs. If this interpretation of the neuroanatomical organization is correct, then medulla neurons contributing to elemental motion detector circuits should be color insensitive, like the optomotor reaction itself (Kaiser 1979), and should have very small receptive fields, a feature that has been demonstrated to be a property of the H1 lobula plate neuron (Franceschini et al. this Vol.). In comparison, lobula columnar cells should encode information about color, form, and position as a consequence of interactions amongst color- and contrast-sensitive transmedullary neurons that impinge upon them. The following section considers what is known about the functional properties of anatomically identified retinotopic neurons in the medulla, with special reference to the bee, locust, and fly.

2.4.2 Parallel Processing in the Medulla

Behavioral observations imply that the imago's visual world is rich in imagery: in certain species, such as hoverflies, form and pattern discrimination may be well developed (Collett and Land 1975). This ability is accomplished by the optic lobes, which concomitant with the postembryonic development of the compound eye (Meinertzhagen 1973) intervene between the new photoreceptor array and descending neurons that have undergone dendritic elaboration during pupal development.

Two lines of physiological evidence, in which recorded neurons have been structurally identified, suggest that there are two major functional classes of small field relay neurons in the medulla. In bees, intracellular recordings from the medulla and from that part of the undivided lobula complex that is equivalent to the dipteran lobula have identified a variety of multistratified retinotopic elements with dendritic fields spreading through only a few columns. In bees broad-band color-sensitive neurons, equivalent to the Y-cells of flies and projecting to the undivided lobula (Kien and Menzel 1977), have simply organized sharply bordered receptive fields dominated by UV, green or blue sensitivity. Hertel (1980) identified a number of similar neurons (albeit incompletely dye-filled) having stratified dendritic trees spreading through groups of retinotopic columns and having small receptive fields that exhibited center-surround color opponency and, occasionally, nondirectional motion sensitivity.

Only a few studies describe response characteristics of medulla neurons in flies, suggesting the existence of a variety of small field nondirectional or directional motion sensitive units (Mimura 1970, 1975; DeVoe and Ockleford 1976; DeVoe 1980) and only two incompletely dye-filled neurons have been tentatively matched to an electrophysiological response; a directionally sensitive unit with some structural features of a T2 neuron and a Y-cell triggered by light on.

From intracellular studies of the locust, Osorio (1986, 1987a,b) has persuasively argued that the peripheral position of the medulla makes it unlikely that the responses of any one of its neurons can be attributed to a specific behavioral subroutine, except possibly for those neurons that lead directly into the brain, bypassing deeper neuropils.

Assuming that certain principles of organization are common across orders, Osorio's (1986, 1987a,b) studies may, for the present, be taken as representing the types of encoding typical of that level even though only a small sample of medulla neurons have been anatomically and physiologically identified. Briefly summarized, small-field neurons, with receptive fields less than 20° in diameter, fell into two major functional classes. The first class, which may be called "sustained units," has approximately linear properties and respond to dimmings and brightenings with opposite polarities. These neurons received UV and green inputs and showed a variety of spatial, temporal and spectral interactions.

The second class, which may be called "transient units", corresponds approximately to non-directionally selective motion sensitive neurons recorded by other authors (Rowell 1971); for example, cells with larger receptive fields showing local adaptation. The neurons are highly non-linear, typically responding with a single precisely timed spike to both dimming and brightening above a sharply defined threshold. They are thus able to signal the timing and position of a suprathreshold stimulus, but not its polarity or amplitude. In contrast to the "sustained units", these non-linear cells receive purely green receptor input.

The receptive fields of both classes of neurons varied from about 2°, representing a single retinotopic column, up to 20°. However, apart from a correlation of dendritic field size with receptive field size, the small number of Lucifer yellow fills did not provide any basis for an anatomical classification.

Osorio's (1986) studies on the locust have best documented a dye-filled orientation-selective motion-sensitive element in the medulla (to upward movement of gratings, and viewing obliquely downward). One intriguing aspect of this neuron is, however, the relatively distal position at which inhibitory interaction between flicker-sensitive channels has already occurred. In common with several species of tangential neuron, this small-field retinotopic element projects centrally, bypassing the lobula.

Studies of identified small-field lobula output neurons are also sparse and describe their activity in relatively simplistic terms, noting approximate receptive fields, spectral sensitivities, "on-off" responses, and their directional or nondirectional motion sensitivity (Umeda and Tateda 1985). In part, the stimulus parameters used reflect our ignorance about which stimuli may be most effective for eliciting responses. In comparison, recent studies of descending neurons (DN's) that are postsynaptic to lobula outputs demonstrate that these convey data about specific events that are immediately relevant to behavior (Reichert and Rowell 1986; Milde 1987). As was suggested in the previous section, anatomy suggests that the lobula complex is the interface between parallel processors derived distally from the medulla and context-specific descending neurons centrally. This being so, the obvious strategy for exploring the functional properties of medullary interneurons (many too small for intracellular recording) would be to analyze the activity of neurons known to be postsynaptic to them. However, although this will be possible in the case of lobula plate tangentials (and sex-specific lobula tangentials) which have large axons, many of the columnar output neurons from the lobula have relatively small axon diameters. Suggestions about the role of medulla neurons into the lobula may have to be obtained by inference: (1) determining synaptic relationships between medulla neurons onto lobula output cells; (2) mapping lobula outputs onto uniquely identified DN's; (3) interpreting the functional significance of medulla-lobula-DN connections in the light of DN activation by specific visual stimuli.

2.4.3 The Uniquely Identifiable Neuron

Present strategies for functional neuroanatomy were anticipated and greatly influenced by the research of Roeder and Wiersma: descriptions of cockroach giant fiber projections, responses, and functional connections (Roeder 1948); and attempts to relate the location of specific fibers in crustacean optic nerves with electrophysiological responses to specific sensory stimuli (Wiersma 1958, 1966). These seminal accounts demonstrated that context-specific units have invariant locations in the cord or ganglion.

Visually induced responses from orthopteran ventral nerve cord demonstrated a variety of units activated by panoramic or small-field motion (Burtt and Catton 1954, 1959; Palka 1965, 1969). By the early 1970s extracellular electrodes could reliably pick out specific units including multimodal neurons responding to visual and mechanosensory/acoustic cues (Rowell 1971) and cells that discriminated between small-field motion and displacement of the visual world caused by the compound eye's own movement (Palka 1969).

As significant as the invention of the Golgi method was the introduction of two histological techniques designed specifically for use in conjunction with electrophysiology: iontophoresis of fluorescent dye (Stretton and Kravitz 1968) and cobalt iontophoresis (Pitman et al. 1972). The second method was used by O'Shea and Rowell (1975) to demonstrate for the first time the shape and connection of a lobula visual interneuron (LGMD) that terminated on a uniquely identifiable descending neuron (DCMD) contributing to an identified motor circuit in the metathoracic ganglion (O'Shea and Williams 1974; O'Shea and Rowell 1976).

The lobula of the locust *Schistocerca* contains a variety of wide-field motion-sensitive neurons. One of these, the LGMD (lateral giant motion detector) responds to novel abrupt movements of small contrasting objects anywhere in the visual surround. The LGMD does not, however, respond to wide-field movement. The fan-shaped dendritic tree, covering the whole retinotopic lobula, receives excitatory small-field motion-sensitive retinotopic inputs, which habituate rapidly. It is protected from wide-field stimulation by at least one inhibitory nonretinotopic input, activated by the onset of large-field motion (and possibly a second, which signals cessation of motion), which projects to a small dendritic tuft of the LGMD, occupying a lobe of neuropil adjoining the lobula. Originating in the medulla, three parallel processors (for which anatomical correlates exist) converge at specific target areas of the LGMD to control its activity.

A similar circuit has also been proposed for certain lobula plate neurons in flies. These are the large-field figure-detector (FD) neurons (Egelhaaf 1985a,b, Egelhaaf et al. 1988) which selectively discriminate between panoramic motion and regressive or progressive motion of an object subtending only a few degrees of arc on the retina. Assuming that such FD neurons receive the same kinds of elemental motion sensitive inputs as panoramic motion sensitive HS neurons, a feed-forward shunting inhibition by a system of large-field retinotopic motion detectors originating peripherally is required to protect the cell from activation by wide-field motion (Egelhaaf 1985c). Unlike the LGMD, which rapidly habituates and functions as a "novelty" detector, the FD neurons serve to monitor relative motion between foreground and background, thus contributing to visually stabilized flight (Hausen and Egelhaaf this Vol.).

Rowell and his colleagues (O'Shea and Williams 1974; O'Shea and Rowell 1975, 1976; O'Shea et al. 1974) demonstrated that the LGMD is presynaptic to at least two uniquely identifiable descending neurons, the DCMD and DIMD (descending contra- and ipsilateral motion detectors), the latter receiving LGMD and other types of visual inputs via chemical synapses. Electron microscopy (Killmann and Schürmann 1985) supports electrophysiological studies showing the LGMD electrically coupled to the DCMD (O'Shea and Rowell 1975) and also supports Rind's 1984 demonstration that the DCMD is driven by a chemical synapse from the LGMD. The DCMD is additionally activated by other sensory modalities, which are chemically presynaptic and which may gate LGMD-DCMD transmission. As we shall presently see, an almost identical organization exists for the dipteran giant descending neuron (GDN) which, like the DCMD, is involved in activation of an escape jump response.

From studies on the DCMD in particular (O'Shea et al. 1974) and from the segmental structure of descending neuron (DN) terminals in general, the evidence is that each species of DN targets onto a characteristic configuration of motor pools (Burrows and Rowell 1973; Pearson and Goodman 1979) each comprising the core elements of "local interneurons" and motor neurons. In the first and best-researched example (Robertson and Pearson 1985), the DCMD responds to mechanosensory cues as well as small-field "novelty" movements in the visual environment relayed to it by the wide-field optic lobe neuron, LGMD (O'Shea et al. 1974). The DCMD provides the visual component of several descending neuron channels (carrying tactile and auditory information) that converge onto two interneurons in the metathoracic ganglion – the "C" and "M" neurons. These are cardinal elements of a local multisensory circuit that, respectively, prime ("cock") and release hindleg extension. The DCMD and the accompanying DNs excite the "C-interneuron" which in turn co-excites the antagonistic extensor and flexor motor neurons (Steeves and Pearson 1982; Pearson 1983). The DCMD also converges with other multisensory DNs onto the M interneuron, which in conjunction with other interneurons (Gynther and Pearson 1987) can terminate the co-contraction of both extensor and flexor by inhibition of the flexor motor neuron. The extensor muscles suddenly shorten, resulting in the explosive extension of both hindlegs (Heitler and Burrows 1977). The analyses of this circuit did much to focus research on processing at the motor output (e.g., leg posture: Siegler 1984; Burrows and Siegler 1985).

A hiatus of several years separates research on the LGMD-DCMD-tibiae extensor system from studies on other multimodal visual descending neurons. However, recently another important class of these neurons has been discovered in the locust whose behavioral relevance is obviously associated with directional control of the flight motor.

The first of such cells was the "tritocerebral commissure giant" (TCG), which is wind-sensitive and sensitive to regressive wide-field motion during flight (Bacon and Tyrer 1978). Its activity suggests that inputs from the optic lobes, carrying information about visual flow fields, gate the response of the TCG to tactile stimulation. Recent studies (Bacon and Mohl 1983) suggest that the pair of TCG's respond to self-generated wind turbulence during flight. Asymmetry between the activity of the TCG pair implies a change in flight direction.

Recordings from locusts presented with conditions that simulate flying (an artificial horizon, visual texture, air current over the head) have demonstrated several other multimodal visual interneurons, which are thought to contribute to flight course correction by modulating the wing motor. Asymmetry in wind direction or tilting the horizon and visual surround signal a change that would occur if the animal deviated from a straight course. Certain descending neurons are vigorously activated by the combination of such stimuli, responding only feebly to single unimodal cues (Reichert and Rowell 1986).

The context in which these "course-deviation" neurons best respond suggests that they receive a variety of inputs: from the ocelli, compound eyes, and head hairs (Rowell 1988; Griss and Rowell 1986; Reichert and Rowell 1986). The synaptic relationships between afferents and DNs are largely obscure, however, because structural studies on locust descending neurons have largely taken second place to physiology. Exceptions are: early studies by Williams (1975), Goodman (1976), Goodman and Williams (1976) and, most recently, Rind's (1987) elegant reconstructions of locust motion-sensitive lobula output neurons, the terminals of which are comparable to those of wide-field lobula neurons in Diptera (Umeda and Tateda 1985). Both studies suggest a discretely partitioned organization of the lateral deutocerebrum.

Course deviation neurons respond to global changes associated with flying. In dragonflies, small-field "target detecting" descending neurons, which also modulate the wing motor, are activated by very different cues. Their receptive fields are usually directed forwards, and the neurons respond to movement of small objects contrasting against a uniform or textured background (Olberg 1986). Activity is orientation and direction selective, showing inhibition in the null direction. Patterned whole-field movement in the null direction of the target also activates these neurons, whereas whole-field pattern moving in the active direction of the target causes inhibition. Again, reconstructions of these Lucifer-yellow-filled neurons demonstrate that their main dendritic branches extend to several discrete areas of the lateral midbrain suggesting that they receive a variety of inputs from the lobula.

2.4.4 The Organization of Sensory Inputs onto Descending Neurons

Relatively few studies have attended to the integrative mechanisms amongst presynaptic elements onto complex dendritic trees of interganglionic interneurons. The notable exception is a history of research on ascending interneurons arising from the terminal ganglia of crickets (Murphey 1983; Walthall and Murphey 1986). Neuroanatomy, electrophysiology, development, and manipulation of identified dendritic trees (Jacobs et al. 1986) have combined to demonstrate the relevance of modality-specific and direction-sensitive receptor maps onto ascending neuron dendrites in eliciting behaviorally relevant activity.

Diptera is the only group in which structural relationships between descending neurons and their sensory inputs have been analyzed in any detail (Strausfeld and Bacon 1982; Strausfeld and Bassemir 1983, 1985a,b). Optic lobe outputs have been related to the convergence of other sensory systems onto common postsynaptic targets — the descending neurons (Fig. 4) or motor neurons that supply muscles involved in a specific behavioral routine (Strausfeld and Seyan 1985). Two systems will be summarized: the organization of the giant descending neuron supplying tergotrochanteral muscle and initiating midleg extension (Bacon and Strausfeld 1986) and the lobula plate connections to the head motor that mediate head saccades and compensatory head movements (Strausfeld and Bassemir 1985a,b; Milde and Strausfeld 1986; Strausfeld et al. 1987; Milde et al. 1987).

The organization of the lateral deutocerebrum. Existing neuroanatomical data provide a detailed background for studying interactions between assemblies of optic lobe outputs. To many of these can be ascribed general functional roles: polarized-light-sensitive pathways; orientation and direction selective, motion-sensitive wide-field neurons; high-acuity male-specific neurons; retinotopic small-field motion-sensitive neurons; intensity-sensitive and color-opponent elements; and binocular small-field elements. Each species of optic lobe output neuron projects to a characteristic glomerulus-like structure in the lateral deutocerebrum. These areas are termed the "optic foci": discrete neuropils invaded by descending neuron dendrites and having characteristic and invariant topographical relationships to each other in the same species. Optic foci are not, however, exclusively visual centers since the majority also receive primary mechanosensory afferents as well as chemosensory interneurons originating from protocerebral neuropils (Strausfeld et al. 1984; Strausfeld and Bacon 1982). Optic foci are also invaded by the terminals of ascending neurons originating in thoracic ganglia (Strausfeld et al. 1984). Together, this variety of pathways suggests that the lateral deutocerebrum is the destination of optic lobe outputs and the site of multimodal sensory convergence onto descending neurons.

Descending neurons (DN's) are uniquely identifiable nerve cells linking the brain with thoracic and abdominal neuropils. They represent the final link in complex cerebral pathways — a feature reflected by their ornately partitioned dendritic trees, which interact with many thousands of presynaptic terminals. DN's occur in discrete clusters shared by local interneurons and heterolateral neurons connecting the two halves of the brain. Each cluster of neurons shares some or all of the same sensory input. Although the number of clusters has not yet been determined, over 60 descending neurons originate from each side of the fly's brain. As a rule, each neuron consists of a stout initial segment, supplied by a thin neurite arising from a cell body. The initial segment gives rise to the descending axon proximally, and distally to several primary branches subdividing into secondaries. Each secondary branch repeatedly bifurcates forming a characteristic arbor of dendritic processes. These invade one or more foci into which project the terminals of one or more lobula assemblies (Fig. 4). Typically, large-field large-axon-diameter neurons of the lobula and lobula plate terminate on the main branches (primaries and secondaries) of the DN, on its initial segment, or on its axon. Terminals from small-field assemblies branch amongst the fine dendritic arbors (Strausfeld and Seyan 1987). A variety of other mechanosensory afferents terminate amongst the endings of visual interneurons in an optic focus or at specialized modality-specific branches of the DN itself (Strausfeld and Bassemir 1983; Bacon and Strausfeld 1986).

The segregation of wide-field and small-field visual interneurons onto distinct regions of a descending neuron may be of great functional importance with respect to neurons mediating "optomotor" and other visually guided actions. Typically, horizontal cell (HS) axon collaterals are presynaptic to a great variety of DN's, any single DN establishing relatively few synapses with the HS cells (Strausfeld and Bassemir 1985b). The current view is that HS and VS axons from the lobula plate diverge to many premotor channels, there being no single channel devoted to the flight motor. Terminations of male-specific tangentials on the main segments of sex-unspecific DN's suggest that the function of a DN common to both sexes may be modified in the male by events subtending the male-specific part of its compound eyes.

Descending neurons thus occupy a cardinal position in the chain between photoreceptor transduction in the compound eye and behavior. I shall now briefly review two descending systems in *Calliphora*, both of which illustrate the value of such elements in elucidating the functional organization of the optic lobes.

Jumping Flies and Uniquely Identifiable Circuits. Many Diptera have powerful dorso-ventral tergotrochanteral muscles (TTM's) broadly anchored to the tergum and tapering as a tendonlike attachment to the trochanter of the mesothoracic leg. TTM contraction results in the leg's sudden extension, reminiscent of the locust's hindleg extension. This action is controlled in part by a uniquely identifiable neuron analogous to the DCMD of locusts.

In flies, the existence of this neuron has been known since Power's classical description in 1948 of the thoracic abdominal nervous system of *Drosophila*. He described a pair of giant fibers (GF's), extending from the lateral deutocerebrum down the cord to the mesothoracic ganglion, from where Power suggested they innervated the TTM "jump" muscles. The hypothesis that the GF's were motor neurons received support from Levine and Tracy's (1973) physiological and anatomical studies: current injection into the GF axon elicited a muscle spike with no appreciable delay. Injection of cobalt in the GF appeared to prove its brain-to-muscle projection. However, combining electrophysiological recordings from the neck connective and behavioral studies using leg ablation, Mulloney (1969) proposed that the GF was an interneuron that played an important role in midleg extension (TTM muscle contraction) and the start of flight (activation of dorso-longitudinal muscles: DLM's). The plausibility of the GF being an interneuron was further strengthened by Hardie's (1977) intracellular recordings from the GF of *Lucilia cuprina* in conjunction with recordings from the DLM's. Spiking in the GF's was not invariably followed by DLM activity. When it was, a single spike in the DLM was preceded by one or more spikes in the GF, with a latency of 2–3ms suggesting the presence of an intervening motor neuron between the GF and DLM's.

Golgi studies on *Drosophila* demonstrated that the GF (now termed GDN: giant descending neuron) indeed terminated in the mesothoracic ganglion (Strausfeld and Singh 1980). Electron microscopy of the GDN (King and Wyman 1980) showed it presynaptic to a major branch of the tergotrochanteral muscle motor neuron (TTMMN) to which it is coupled by an electrical synapse (Tanouye and Wyman 1980). Doubtlessly, this explains why cobalt ions pass readily between the GDN and the TTMMN, leading Levine and Tracy to interpret the cobalt-filled structure as a single element rather than two contiguous ones. King and Wyman (1980) and Tanouye and Wyman (1980) demonstrated that the GDN is electrically presynaptic to a pair of interneurons, whose terminals are chemically presynaptic onto the axons of DLM motor neurons: the first power stroke of the wings follows midleg extension by one synaptic delay. Detailed anatomical studies of *Calliphora* and *Musca* (Strausfeld and Bassemir 1983; Strausfeld et al. 1984; Bacon and Strausfeld 1986) demonstrated that the GDN organization is essentially identical in these species. Recordings from the GDN and the TTMMN revealed negligible delay between them, providing additional evidence for electrical contiguity.

Anatomical studies have demonstrated that the GDN has a highly partitioned dendritic tree having characteristic synaptic relationships with primary afferents and interneurons from a variety of sensory systems. These include several lobula assemblies (one of which shares gap junctions with the GDN) and, in males, sex-specific lobula tangentials (Strausfeld and Bassemir 1983; Strausfeld and Bacon 1982; Nässel 1983). Clearly, the neuroanatomical organization onto its complexly shaped tree suggests this neuron is subject to sophisticated multisensory control.

Intracellular recordings from wild-type GDN's in *Drosophila* demonstrated its reluctance to spike to any unimodal stimulus (Tanouye and Wyman 1980). The most it would do was to spike occasionally in response to light flashes (but only in white-eyed mutants), or to pulses of current imposed across the head (Tanouye and Wyman 1980; Koto et al. 1981). In white-eyed *Lucilia*, GDN activity is elicted by flashes of monochromatic light within a spectral range conforming to that of the R1–R6 receptors (Hardie 1977). However, as in *Drosophila*, the GDN of *Calliphora* rarely spikes to a unimodal stimulus.

Bacon and Strausfeld (1986) demonstrated that a variety of visual and mechanosensory stimuli elicit subthreshold epsp's and ipsp's. These reflect dendritic activity and corroborate the general types of morphologically identified inputs: lateral displacement of the antenna registered by afferents from Johnston's organ; excitation in response to small-field movement frontally or in response to movement over the whole compound eye; and ipsp's in response to wind blown onto the thorax (Bacon and Strausfeld 1986). A single spike is occasionally provoked by sudden motion in the visual field or by displacement of the antennae. However, when these events occur simultaneously, antennal movement facilitates a tonic spiking response to visual motion (Milde 1987).

These studies demonstrate two important points. First, resolving DN activity and the neuroanatomical context in which it operates should identify appropriate stimulus parameters for investigating the functional nature of its presynaptic inputs. Second, as can be inferred from Milde's (1987) results, it is crucial to devise appropriate visual stimuli that are relevant to other modalities. So far, descending neurons appear to be activated by multimodal context specific events, presumably as a consequence of pre- and postsynaptic interactions between the great variety of sensory elements impinging on the descending neuron's dendritic tree.

Ungluing the Head: Saccades, Neck Muscles and the Lobula Plate. Quantitative studies of visually induced behavior required that responses to visual stimuli by flying female flies were derived exclusively from the flight motor (Reichardt and Poggio 1976). Suspended from a torque meter, the thorax of the fly was glued to the head preventing any movement between them. One consequence of this arrangement was that most discussions about giant motion-sensitive neurons related them solely to flight (Hausen 1984): a reasonable view considering that the most effective releasing stimuli for the optomotor responses were identical to those that activated HS and VS neurons.

In 1974 Land demonstrated that when the fly's head can move freely, an optomotor stimulus elicits a head saccade, which precedes a compensatory movement by the body. Head-body interaction was also demonstrated by Liske (1977), who showed that mechanical displacement of the head leading to a compensatory movement by the body is registered by a patch of mechanoreceptors on the ventral prothoracic segment, just beneath the neck, which detects movement of the prosternum. In addition, since walking flies also respond appropriately to optomotor stimuli, if lobula plate giant neurons mediate optomotor reactions, they should supply at least three channels: to the flight, to the neck, and to the leg motor.

Neuroanatomical studies have demonstrated that 21 pairs of muscles accomplish head movement: translation, rotation, inclination-supination, and extension-retraction (Strausfeld et al. 1987). Each muscle is innervated by an identifiable motor neuron connected in a specific manner to lobula plate neurons. The dendritic fields of the two most anterior VS neurons are mapped into the retinotopic mosaic representing the frontal binocular strip of retina. Their axons are connected to two pairs of motor neurons arising in the deutocerebrum, the terminals of which invest two pairs of muscles that pull the head downwards (Milde and Strausfeld 1986). As would be predicted by this arrangement, downward movement of horizontal gratings presented to the region of binocular overlap activate these motor neurons (Milde et al. 1987) and the VS cells supplying them (Hengstenberg et al. 1982; Hengstenberg 1982). In comparison, muscles that rotate the head around its long axis are supplied by a unique descending neuron whose Y-shaped dendrites in the brain are postsynaptic to VS cells subtending the lateral part of the compound eye (Strausfeld and Bassemir 1985a; Milde and Strausfeld 1986). Rotatory movement of horizontal gratings around the head activates these VS neurons and the motor neurons investing head rotation muscles (Milde et al. 1987).

Lateral movement of the head, typical of saccades, is mediated by two pairs of horizontal muscles (Strausfeld et al. 1987; Milde et al. 1987). Their motor neurons receive axon collaterals from complexly shaped descending neurons that also supply flight motor neuropils. Significantly, these DN's share gap junctions with HS neurons responding to translatory panoramic motion (Strausfeld and Bassemir 1985b). The same DN's also receive abundant inputs from small-field lobula assemblies, presumably activated by small-field events in the visual panorama. Although we do not yet know how small-field assemblies and HS neurons interact at these DN's, Olberg's (1986) studies suggest that such neurons in dragonflies carry figure-ground information ("target detection").

Like the flight motor, the neck motor involves other sensory modalities in addition to vision that play a role in compensatory head movements. Amongst these are mechanoreceptors: campaniform receptors of the organs of balance (halteres); the head-neck position receptors (prosternal organs); neck stretch receptors; and wind receptors on the shoulders of the prothoracic segment, to name but a few (Strausfeld and Seyan 1985; Liske 1977; Milde et al. 1987). The role of the ocelli as horizon detectors in flies is compatible with their convergence onto a common DN that is postsynaptic to VS neurons subtending the lateral part of the retina (Strausfeld and Bassemir 1985a). In an elegant series of experiments, Hengstenberg et al. (1986) demonstrated some of these neuroanatomical interactions behaviorally. When the body of the flying fly is made to oscillate, the head is maintained level if there is a visible horizon. Conversely, if the body is maintained horizontal during flight but the artificial horizon is made to oscillate around the long axis, the head will move accordingly to stabilize the retinal image. In addition, displacement of the halteres, simulating an imposed lateral movement of the body, results in a compensatory movement of the head laterally, correcting for anticipated retinal drift (Sandeman and Markl 1980; Hengstenberg 1984).

3 Conclusion

Neuroanatomical data complements modern behavioral and electrophysiological studies on insect vision, and suggests strategies for elucidating the function of the medulla, the most complex of neuropils. Clearly of great interest and value, is a comparative approach — studying the optic lobes and descending neurons supplying analogous motor circuits in a variety of insects: moths, locusts, dragonflies, and flies. However, only for flies do we have the wealth of neuroanatomical data that allows us to relate the incredible precision of the receptor mosaic to equally precise arrangements of interneurons. Similar studies are much needed on locusts, for the Orthoptera have dominated studies of sensorimotor integration and are likely to continue to do so. But despite these omissions, and despite many lacunae in our basic set of data, we can be optimistic that certain visual behaviors mentioned here will eventually be understood in terms of synaptic interactions amongst identified neurons. Then, possibly, principles of functional organization will emerge that can be compared to visual processing in other phyla, leading eventually to general models of neuronal interactions underlying visual perception.

Acknowledgments. I thank Drs. Camilla Strausfeld, Ed Arbas, and Brian Waldrop for offering criticism and advice on the manuscript. I am grateful to Ms Jennifer Lawrence for word processing. Supported by National Institutes of Health National Eye Institute Grant No. ROI EY 07151-01

References

Arnett DW (1972) Spatial and temporal integration properties of units in the first optic ganglion of dipterans. J Neurophysiol 35:429–444
Autrum H, Zettler F, Järvilehto M (1970) Postsynaptic potentials from a single monopolar neuron of the ganglion opticum I of the blowfly *Calliphora.* Z Vergl Physiol 70:414–424
Bacon JP, Mohl B (1983) The tritocerebral commissure giant (TCG) wind-sensitive interneuron in the locust. 1. Its activity in straight flight. J Comp Physiol A 150:439–452
Bacon JP, Strausfeld NJ (1986) The dipteran 'Giant fibre' pathway: neurons and signals. J Comp Physiol A 158:529–548
Bacon JP, Tyrer M (1978) The tritocerebral commissure giant (TCG): a bimodal interneurone in the locust, *Schistocerca gregaria.* J Comp Physiol A 126:317–325
Beersma DGM, Stavenga DG, Kuiper JW (1977) Retinal lattice, visual field and binocularities in flies. J Comp Physiol 119:207–220
Bishop CA, Bishop LG (1981) Vertical motion detectors and their synaptic relations in the third optic lobe of the fly. J Neurobiol 12:281–296
Bishop LG, Keehn DG (1967) Neural correlates of the optomotor response in the fly. Kybernetik 3:288–295

Bishop LG, Keehn DG, McCann GD (1968) Studies of motion detection by interneurons of the optic lobes and brain of the flies, *Calliphora phaenicia* and *Musca domestica*. J Neurophysiol 31:509–525

Boschek CB (1971) On the fine structure of the peripheral retina and lamina ganglionaris of the fly, *Musca domestica*. Z Zellforsch Mikrosk Anat 118:369–409

Braitenberg V (1966) Unsymmetrische Projektion der Retinulazellen auf die Lamina ganglionaris bei der Fliege *Musca domestica*. Z Vergl Physiol 50:212–214

Braitenberg V (1967) Patterns of projection in the visual system of the fly. I. Retina-Lamina projections. Exp Brain Res 3:271–298

Braitenberg V (1970) Ordnung und Orientierung der Elemente im Sehsystem der Fliege. Kybernetik 7:235–242

Braitenberg V (1972) Periodic structures and structural gradients in the visual ganglia of the fly. In: Wehner R (ed) Information processing in the visual system of arthropods. Springer, Berlin Heidelberg New York, pp 3–15

Braitenberg V, Debbage P (1974) A regular net of reciprocal synapses in the visual system of the fly *Musca domestica*. J Comp Physiol 90:25–31

Buchner E (1976) Elementary movement detectors in an insect visual system. Biol Cybernet 24:85–101

Buchner E, Buchner S (1983) Anatomical localization of functional activity in flies using 3H-2-deoxy-D-glucose. In: Strausfeld NJ (ed) Functional neuroanatomy. Springer, Berlin Heidelberg New York, pp 225–238

Buchner E, Buchner S (1984) Neuroanatomical mapping of visually induced neuron activity in insects by 3H-deoxyglucose. In: Ali MA (ed) Photoreception and vision in invertebrates. Plenum, New York, pp 623–634

Burkhardt W, Braitenberg V (1976) Some peculiar synaptic complexes in the first visual ganglion of the fly, *Musca domestica*. Cell Tissue Res 173:287–308

Burrows M, Rowell CHF (1973) Connections between descending visual interneurons and metathoracic motoneurons in the locust. J Comp Physiol 85:221–234

Burrows M, Siegler MVS (1985) The organization of receptive fields of spiking local interneurons in the locust with inputs from hair afferents. J Neurophysiol 53:1147–1157

Burtt ET, Catton WT (1954) Visual perception of movement in the locust. J Physiol 125:566–580

Burtt ET, Catton WT (1959) Transmission of visual responses in the nervous system of the locust. J Physiol 146:492–514

Cajal SR (1937) Recollections of my life (Recuerdos De Mi Vida). (Transl Horne EC, Cano J). MIT, Cambridge

Cajal SR, Sánchez y Sánchez D (1915) Contribucion al conocimiento de los centros nerviosos de los insectos. Parte I. Retina y centros opticos. Trab Lab Invest Biol Univ Madrid 13:1–168

Campos-Ortega JA, Strausfeld NJ (1972a) The columnar organization of the second synaptic region of the visual system of Musca domestica L. I. Receptor terminals in the medulla. Z Zellforsch 124:561–582

Campos-Ortega JA, Strausfeld NJ (1972b) Columns and layers in the second synaptic region of the fly's visual system: The case for two superimposed neuronal architectures. In:Wehner R (ed) Information processing in the visual systems of arthropods. Springer, Berlin Heidelberg New York, pp 31–36

Campos-Ortega JA, Strausfeld NJ (1973) Synaptic connections of intrinsic cells and basket arborisations in the external plexiform layer of the fly's eye. Brain Res 59:119–136

Collett TS, Blest AD (1966) Binocular, directionally selective neurones, possibly involved in the optomotor response of insects. Nature (London) 212:1330–1333

Collett TS, Land MF (1975) Visual control of flight behaviour in the hoverfly, *Syritta pipiens* L. J Comp Physiol A 99:1–66

DeVoe RD (1980) Movement sensitivities of cells in the fly's medulla. J Comp Physiol A 138:93–119

DeVoe RD, Ockleford EM (1976) Intracellular responses from cells of the medulla of the fly, *Calliphora erythrocephala*. Biol Cybernet 23:13–24

Dietrich W (1909) Die Facettenaugen der Dipteren. Z Wiss Zool 92:465–539

Dubs A (1982) The spatial integration of signals in the retina and lamina of the fly compound eye under different conditions of luminance. J Comp Physiol A 146:321–343

Dvorak DR, Bishop LG, Eckert HE (1975a) On the identification of movement detectors in the fly optic lobe. J Comp Physiol A 100:5–23

Dvorak DR, Bishop LG, Eckert HE (1975b) Intracellular recording and staining of directionally selective motion detecting neurons in fly optic lobe. Vision Res 15:451–453

Eckert HE, Bishop LG (1978) Anatomical and physiological properties of the vertical cells in the third optic ganglion of *Phaenicia serricata* (Diptera, Calliphoridae). J Comp Physiol A 126:368–419

Egelhaaf M (1985a) On the neuronal basis of figure-ground discrimination by relative motion in the fly. I. Behavioural constraints imposed on the neuronal network and the role of the optomotor system. Biol Cybernet 52:123–140

Egelhaaf M (1985b) On the neuronal basis of figure-ground discrimination by relative motion in the fly. II. Figure detection cells, a new class of visual interneurons. Biol Cybernet 52:195–209

Egelhaaf M (1985c) On the neuronal basis of figure-ground discrimination by relative motion in the visual system of the fly. III. Possible input circuitries and behavioural significance of the ED-cells. Biol Cybernet 52:267–280

Egelhaaf M, Reichardt W (1987) Dynamic response properties of movement detectors: theoretical analysis and electrophysiological investigation in the visual system of the fly. Biol Cybernet 55:1–19

Egelhaaf M, Hausen K, Reichardt W, Wehrhahn C (1988) Visual course control in flies relies on neuronal computation of object and background motion. Trends neurosci 11:351–358

Fischbach K-F (1983) Neurogenetik am Beispiel des visuellen Systems von *Drosophila melanogaster.* Habilitationsschr, Bay Julius-Maximilians-Univ, Würzburg

Fröhlich A, Meinertzhagen IA (1982) Synaptogenesis in the first optic neuropile of the fly's visual system. J Neurocytol 11:159–180

Goodman CS (1976) Anatomy of the ocellar interneurons of acridid grasshoppers. I. The large interneurons. Cell Tissue Res 175:183–202

Goodman CS, Williams JLD (1976) Anatomy of the ocellar interneurons of acridid grasshoppers. II. The small interneurons. Cell Tissue Res 175:203–225

Griss C, Rowell CHF (1986) Three descending interneurons reporting deviation from course in the locust. I. Anatomy. J Comp Physiol A 158:765–774

Gynther IC, Pearson KG (1987) Evaluation of a hypothesis for the neural control of jumping in the locust. Neurosci Abstr 13:1059

Hardie RC (1977) Flight initiation in the fly *Lucilia.* Proc Austral Physiol Pharmacol Soc 8:95P

Hardie RC (1983) Projection and connectivity of sex-specific photoreceptors in the compound eye of the male housefly (*Musca domestica*). Cell Tissue Res 233:1–21

Hardie RC (1984) Properties of photoreceptors R7 and R8 in dorsal marginal ommatidia in the compound eyes of *Musca* and *Calliphora.* J Comp Physiol A 154:157–165

Hardie RC (1985) Functional organisation of the fly retina. In: Ottoson D (ed) Progress in sensory physiology. Springer, Berlin Heidelberg New York Tokyo, pp 1–79

Hardie RC, Franceschini N, Ribi W, Kirschfeld K (1981) Distribution and properties of sex-specific photoreceptors in the fly *Musca domestica.* J Comp Physiol A 145:139–152

Hassenstein B (1958) Über die Wahrnehmung der Bewegung von Figuren und unregelmässigen Helligkeitsmustern. Z Vergl Physiol 40:556–592

Hassenstein B, Reichardt W (1956) Systemtheoretische Analyse der Zeit-, Reihenfolgen- und Vorzeichenauswertung bei der Bewegungsperzeption des Rüsselkäfers *Chlorophanus.* Z Naturforsch 11b:513–524

Hausen K (1976) Functional characterization and anatomical identification of motion sensitive neurons in the lobula plate of the blowfly *Calliphora erythrocephala.* Z Naturforsch 31c:629–633

Hausen K (1982a) Motion sensitive interneurons in the optomotor system of the fly. I. The horizontal cells: structure and signals. Biol Cybernet 45:143–156

Hausen K (1982b) Motion sensitive interneurons in the optomotor system of the fly. II. The horizontal cells: receptive field organization and response characteristics. Biol Cybernet 46:76–79

Hausen K (1984) The lobula complex of the fly: Structure, function, and significance in visual behavior. In: Ali MA (ed) Photoreception and vision in invertebrates. Plenum, New York, pp 523–599

Hausen K, Strausfeld NJ (1980) Sexually dimorphic interneuron arrangements in the fly visual system. Proc R Soc London Ser B 208:57–71

Hausen K, Wolburg-Buchholz K, Ribi WA (1980) The synaptic organization of visual interneurons in the lobula complex of flies. A light and electron microscopical study using silver-intensified cobalt impregnations. Cell Tissue Res 208:371–387

Heitler WJ, Burrows M (1977) The locust jump. II. Neural circuits of the motor programme. J Exp Biol 66:221–241

Hengstenberg R (1977) Spike response of a non-spiking visual interneurone. Nature (London) 170:338–340

Hengstenberg R (1982) Common visual response properties of giant vertical cells in the lobula plate of the blowfly *Calliphora erythrocephala*. J Comp Physiol A 149:179–193

Hengstenberg R (1984) Roll-stabilization during flight of the blowfly's head and body by mechanical and visual cues. In: Varju D, Schnitzler V (eds) Localization and orientation in biology and engineering. Springer, Berlin Heidelberg New York, pp 121–134

Hengstenberg R, Hausen K, Hengstenberg B (1982) The number and structure of giant vertical cells (VS) in the lobula plate of the blowfly *Calliphora erythrocephala*. J Comp Physiol A 149:163–177

Hengstenberg R, Sandeman DC, Hengstenberg B (1986) Compensatory head roll in the blowfly *Calliphora* during flight. Proc R Soc London Ser B 227:455–487

Hertel H (1980) Chromatic properties of identified interneurons in the optic lobes of the bee. J Comp Physiol A 137:215–232

Holst E von, Mittelstaedt H (1950) Das Reafferenzprinzip (Wechselwirkungen zwischen ZNS und Peripherie). Naturwissenschaften 37:464–476

Jacobs GA, Miller JP, Murphey RK (1986) Integrative mechanisms controlling directional sensitivity of an identified sensory interneuron. J Neurosci 6:2298–2311

Järvilehto M, Zettler F (1973) Electrophysiological-histological studies on some functional properties of visual cells and second order neurons of an insect retina. Z Zellforsch 136:291–306

Kaiser W (1979) The relationship between visual movement detection and colour vision in insects. In: Horridge GA (ed) The compound eye and vision of insects. Clarendon, Oxford, pp 358–377

Kenyon FC (1897) The optic lobes of the bee's brain in the light of recent neurological methods. Am Nat 31:369–376

Kien J (1980) Morphology of locust neck muscle motorneurons and some of their inputs. J Comp Physiol A 140:321–336

Kien J, Menzel R (1977) Chromatic properties of interneurons in the optic lobes of the bee. I. Broad band neurons. J Comp Physiol A 113:17–34

Killmann F, Schürmann FW (1985) Both electrical and chemical transmission between the "lobula giant movement detector" and the "descending contralateral movement detector" neurons of the locust are supported by electron microscopy. J Neurocytol 14:637–652

King DG, Valentino KL (1983) On neuronal homology: a comparison of similar axons in *Musca, Sarcophaga* and *Drosophila* (Diptera: Schizophora). J Comp Neurol 219:1–9

King DG, Wyman RJ (1980) Anatomy of the giant fibre pathway in *Drosophila*.I. Three thoracic components of the pathway. J Neurocytol 9:753–770

Kirschfeld K (1967) Die Projektion der optischen Umwelt auf das Raster der Rhabdomere im Komplexauge von *Musca*. Exp Brain Res 3:248–270

Koto ML, Tanouye MA, Ferris A, Thomas JB, Wyman RJ (1981) The morphology of the cervical giant neuron of *Drosophila*. Brain Res 221:213–217

Land MF (1974) Head movements and fly vision. In: Horridge GA (ed) The compound eye and vision of insects. Clarendon, Oxford, pp 469–489

Land MF, Collett TS (1974) Chasing behaviour of houseflies (*Fannia canicularis*). A description and analysis. J Comp Physiol 89:331–357

Land MF, Eckert HE (1985) Maps of the acute zones of fly eyes. J Comp Physiol A 156:525–538

Laughlin SB (1975) The function of the lamina ganglionaris. In: Horridge GA (ed) The compound eye and vision of insects. Clarendon, Oxford, pp 341–358

Laughlin SB (1981) Neural principles in the peripheral visual systems of invertebrates. In: Autrum H (ed) Handbook of sensory physiology, vol VII/6B. Springer, Heidelberg New York Berlin, pp 133–280

Laughlin SB (1984) The roles of parallel channels in early visual processing by the arthropod compound eye. In: Ali MA (ed) Photoreception and vision in invertebrates. Plenum, New York, pp 457–481

Levine JD, Tracy D (1973) Structure and function of the giant motorneiron of *Drosophila melanogaster*. J Morphol 140:153–158

Liske E (1977) The influence of head position on the flight behaviour of the fly *Calliphora erythrocephala*. J Insect Physiol 23:375–379

Maldonado H, Barrós-Pita JC (1970) A fovea in the praying mantis eye. I. Estimation of the catching distance. Z Vergl Physiol 67:58–78

Meinertzhagen IA (1973) Development of the compound eye and optic lobe of insects. In: Young D (ed) Developmental neurobiology of arthropods. Univ Press, Cambridge, pp 51–104

Melamed J, Trujillo-Cenóz O (1968) The fine structure of the central cells in the ommatidia of Dipterans. J Ultrastruct Res 21:313–334

Meyer GF (1951) Versuch einer Darstellung von Neurofibrillen im zentralen Nervensystem verschiedener Insekten. Zool Jahrb 71:413–426

Milde JJ (1987) GDNs and associated descending neurons in *Calliphora* have similar response characteristics and anatomical arrangement. In: Elsner N, Creutzfeld O (eds) New frontiers in brain research. Thieme, Stuttgart, p 70

Milde JJ, Strausfeld NJ (1986) Visuo-motor pathways in arthropods. Giant motion-sensitive neurons connect compound eyes directly to neck muscles in blowflies (*Calliphora erythrocephala*). Naturwissenschaften 73:151–153

Milde JJ, Seyan HS, Strausfeld NJ (1987) The neck motor system of the fly *Calliphora erythrocephala*. II. Sensory organization. J Comp Physiol A 160:225–238

Mimura K (1970) Integration and analysis of movement information by the visual system of flies. Nature (London) 226:964–966

Mimura K (1975) Units of the optic lobe, especially movement. In: Horridge GA (ed) The Compound Eye and Vision of Insects. Clarendon, Oxford, pp 421–436

Mittelstaedt H (1949) Telotaxis und Optomotorik von *Eristalis* bei Augeninversion. Naturwissenschaften 36:90

Mulloney B (1969) Interneurons in the central nervous system of flies and the start of flight. Z Vergl Physiol 64:243–253

Murphey RK (1983) Maps in the insect nervous system: their implications for synaptic connectivity and target location in the real world. In: Huber F, Markl H (eds) Neuroethology and behavioral Physiology, Springer, Heidelberg New York, pp 176–188

Nässel DR (1983) Horseradish peroxidase and other heme proteins as neuronal markers. In: Strausfeld NJ (ed) Functional Neuroanatomy. Springer, Heidelberg Berlin New York, pp 45–91

Nässel DR, Strausfeld NJ (1982) A pair of descending neurons with dendrites in the optic lobes projecting directly to thoracic ganglia of dipterous insects. Cell Tissue Res 226:355–362

Olberg RM (1986) Identified target-selective visual interneurons descending from the dragonfly brain. J Comp Physiol A 159:827–840

O'Shea M, Rowell CHF (1975) A spike-transmitting electrical synapse between visual interneurons in the locust movement detector system. J Comp Physiol A 97:143–158

O'Shea M, Rowell CHF (1976) Neuronal basis of a sensory analyzer, the acridid movement detector system. II. Response decrement, convergence, and the nature of the excitatory afferents to the LGMD. J Exp Biol 65:189–308

O'Shea M, Williams JLD (1974) The anatomy and output connection of a locust visual interneuron; the lobula giant movement detector (LGMD) neuron. J Comp Physiol 91:257–266

O'Shea M, Rowell CHF, Williams JLD (1974) The anatomy of a locust visual interneuron; the descending contralateral movement detector. J Exp Biol 60:1–12

Osorio D (1986) Directionally selective cells in the locust medulla. J Comp Physiol A 159:841–847

Osorio D (1987a) The temporal properties of non-linear, transient cells in the locust medulla. J Comp Physiol A 161:431–440

Osorio D (1987b) Temporal and spectral properties of sustaining cells in the medulla of the locust. J Comp Physiol A 161:441–448

Palka J (1965) Diffraction and visual acuity in insects. Science 149:551–553

Palka J (1969) Discrimination between movements of eye and object by visual interneurons of crickets. J Exp Biol 50:723–732

Pearson KG (1983) Neural circuits for jumping in the locust. J Physiol 78:765–771

Pearson KG, Goodman CS (1979) Correlation of variability in structure with variability in synaptic connections of an identified interneuron in locusts. J Comp Neurol 184:141–166

Pflugfelder O (1937) Vergleichende anatomische, experimentelle und embryologische Untersuchungen über das Nervensystem und die Sinnesorgane der Rhynchoten. Zoologica 34:1–102

Pick B, Buchner E (1979) Visual movement detection under light- and dark-adaptation in the fly, *Musca domestica*. J Comp Physiol A 134:45–54

Pierantoni R (1974) An observation on the giant fiber posterior optic tract in the fly. Biokybernetik 5:157–163

Pierantoni R (1976) A look into the cockpit of the fly: The architecture of the lobula plate. Cell Tissue Res 171:101–122

Pitman RM, Tweedle CD, Cohen JH (1972) Branching of central neurons: Intracellular cobalt injection for light and electron microscopy. Science 176:412–414

Power ME (1948) The thoraco-abdominal nervous system of an adult insect, *Drosophila melanogaster.* J Comp Neurol 88:347–409

Reichardt W (1961) Nervous integration in the facet eye. Biophys J 2:121–163

Reichardt W (1969) Movement perception in insects. In: Reichardt W (ed) Processing of optical data by organisms and by machines. Proc Int School Phys "Enrico Fermi." Academic Press, New York London, pp 465–493

Reichardt W, Poggio T (1976) Visual control of orientation behaviour in the fly. I. A quantitative analysis of neural interactions. Q Rev Biophys 9:311–375

Reichert H, Rowell CHF (1986) Neuronal circuits controlling flight in the locust: how sensory information is processed for motor control. Trends Neurosci 9:281–283

Retzius G (1891) Zur Kenntnis des centralen Nervensystems der Würmer, Das Nervensystem der Annulaten. Biol Untersuch NF 2:1–28

Ribi, WA (1978) Gap junctions coupling photoreceptors in the first optic ganglion of the fly. Cell Tissue Res 195:299–308

Riehle A, Franceschini N (1984) Motion detection in flies: A parametric control over on-off pathways. Exp Brain Res 54:390–394

Rind FC (1983) A directionally sensitive motion detecting neuron in the brain of a moth. J Exp Biol 102:253–271

Rind FC (1984) A chemical synapse between two motion detecting neurons in the locust brain. J Exp Biol 110:143–167

Rind FC (1987) Non-directional, movement sensitive neurons of the locust optic lobe. J Comp Physiol A 161:477–494

Robertson RM, Pearson KG (1985) Neural networks controlling locomotion in locusts. In: Selverston AI (ed) Model Neural Networks and Behavior. Plenum, New York, pp 21–35

Roeder KD (1948) Organization of the ascending giant fiber system in the cockroach (*Periplaneta americana*). J Exp Zool 108:243–262

Rossel S, Wehner R (1987) The bee's E-vector compass. In: Menzel R (ed) Honeybee Neurobiology. Springer, Berlin Heidelberg New York Tokyo, pp 76–93

Rowell CHF (1971) The orthopteran descending movement detector (DMD) neurons: A characterisation and review. Z Vergl Physiol 73:167–194

Rowell CHF (1988) Mechanisms of steering in flight by locusts. In: Camhi J (ed) Neuroethology: a multiauthored review. Experientia (in press)

Rowell CHF, O'Shea M (1976a) The neuronal basis of a sensory analyzer, the acridid movement detector system. I Effects of simple incremental and decremental stimuli in light and dark adapted animals. J Exp Biol 65:173–188

Rowell CHF, O'Shea M (1976b) The neuronal basis of a sensory analyzer, the acridid movement detector system. III. Control of response amplitude by tonic lateral inhibition. J Exp Biol 65:617–625

Rowell CHF, Reichert H (1986) Three descending interneurons reporting deviation from course in the locust. II. Physiology. J Comp Physiol A 158:775–794

Sandeman DC, Markl H (1980) Head movement in flies (*Calliphora*) produced by deflection of the halteres. J Exp Biol 85:43–60

Shaw SR (1975) Retinal resistance barriers and electrical lateral inhibition. Nature (London) 255:480–483

Shaw SR (1982) Synaptic gain control in insect photoreceptors. Soc Neurosci Abstr 8:44

Shaw SR (1984) Early visual processing in insects. J Exp Biol 112:225–251

Shaw SR, Stowe S (1982) Freeze-fracture evidence for gap junctions connecting the axon terminals of dipteran photoreceptors. J Cell Sci 53:115–141

Siegler MVS (1984) Interneurons and local interactions in arthropods. J Exp Biol 112:253–282

Steeves JD, Pearson KG (1982) Proprioceptive gating of inhibitory pathways to hind leg flexor motoneurons in the locust. J Comp Physiol A 146:507–515

Strausfeld NJ (1970) Golgi studies on insects. II. The optic lobes of Diptera. Philos Trans R Soc London Ser B 258:135–223

Strausfeld NJ (1971a) The organisation of the insect visual system (light microscopy). I. Projections and arrangements of neurons in the lamina ganglionaris of Diptera. Z Zellforsch 121:377–441

Strausfeld NJ (1971b) The organisation of the insect visual system (light microscopy). II. The projection of fibres across the first optic chiasma. Z Zellforsch 121:442–454

Strausfeld NJ (1976a) Atlas of an insect brain. Springer, Berlin Heidelberg New York

Strausfeld NJ (1976b) Mosaic organizations, layers and visual pathways in the insect brain. In: Zettler F, Weiler R (eds) Neural principles in vision. Springer, Berlin Heidelberg New York, pp 245–279

Strausfeld NJ (1979) The representation of a receptor map within retinotopic neuropil of the fly. Verh Dtsch Zool Ges 1979:167–179

Strausfeld NJ (1980) Male and female visual neurons in dipterous insects. Nature (London) 283:381–383

Strausfeld NJ (1984) Functional neuroanatomy of the blowfly's visual system. In: Ali MA (ed) Photoreception and vision in invertebrates. Plenum, New York, pp 483–522

Strausfeld NJ (1987) Sex-specific neurons in the visual system of blowflies (*Calliphora erythrocephala*) represent concentrically organized retinotopic domains and are segregated out to end at specific regions of premotor descending neurons. Soc Neurosci Abstr 13:137

Strausfeld NJ, Bacon JP (1982) Multimodal convergence in the central nervous system of insects. In: Horn E (ed) Multimodal convergence in sensory systems. Fischer, Stuttgart, pp 47–76

Strausfeld NJ, Bassemir UK (1983) Cobalt-coupled neurons of a giant fibre system in Diptera. J Neurocytol 12:971–991

Strausfeld NJ, Bassemir UK (1985a) Lobula plate and ocellar interneurons converge onto a cluster of descending neurons leading to neck and leg neuropil in *Calliphora erythrocephala*. Cell Tissue Res 240:617–640

Strausfeld NJ, Bassemir UK (1985b) The organisation of giant horizontal-motion-sensitive neurons and their synaptic relationships in the lateral deutocerebrum of *Calliphora erythrocephala* and *Musca domestica*. Cell Tissue Res 242:531–550

Strausfeld NJ, Blest AD (1970) Golgi studies on insects. I. The optic lobes of Lepidoptera. Phil Trans R Soc London Ser B 258:81–134

Strausfeld NJ, Braitenberg V (1970) The compound eye of the fly (*Musca domestica*): Connections between the cartridges of the lamina ganglionaris. Z Vergl Physiol 70:95–104

Strausfeld NJ, Campos-Ortega JA (1973a) The L4 monopolar neuron: a substrate for lateral interaction in the visual system of the fly *Musca domestica*. Brain Res 59:97–117

Strausfeld NJ, Campos-Ortega JA (1973b) L3, the 3rd 2nd order neuron of the 1st visual ganglion in the "neural superposition" eye of *Musca domestica*. Z Zellforsch 139:397–403

Strausfeld NJ, Campos-Ortega JA (1977) Vision in insects: pathways possibly underlying neural adaptation and lateral inhibition. Science 195:894–897

Strausfeld NJ, Hausen K (1977) The resolution of neuronal assemblies after cobalt-injection into neuropil. Proc R Soc London Ser B 199:463–476

Strausfeld NJ, Nässel DR (1981) Neuroarchitecture of brain regions that subserve the compound eyes of Crustacea and insects. In: Autrum H (ed) Handbook of sensory physiology, vol VII/6B. Springer, Berlin Heidelberg New York, pp 1–132

Strausfeld NJ, Obermayer M (1976) Resolution of intraneuronal and transsynaptic migration of cobalt in the insect visual and nervous systems. J Comp Physiol A 110:1–12

Strausfeld NJ, Seyan HS (1985) Convergence of visual, haltere·and prosternal inputs at neck motor neurons of *Calliphora erythrocephala*. Cell Tissue Res 240:601–615

Strausfeld NJ, Seyan HS (1987) Identification of complex neuronal arrangements in the visual system of *Calliphora erythrocephala* using triple fluorescence staining. Cell Tissue Res 247:5–10

Strausfeld NJ, Singh RN (1980) Peripheral and central nervous system projections in normal and mutant (*bithorax*) *Drosophila melanogaster*. In: Siddiqi O, Babu P, Hall LM, Hall JC (eds) Development and neurobiology of *Drosophila*. Plenum, New York, pp 267–291

Strausfeld NJ, Wunderer H (1985) Optic lobe projections of marginal ommatidia in *Calliphora erythrocephala* specialized for detecting polarized light. Cell Tissue Res 242:163–178

Strausfeld NJ, Bassemir U, Singh RN, Bacon JP (1984) Organizational principles of outputs from dipteran brains. J Insect Physiol 30:73–93

Strausfeld NJ, Seyan HS, Milde JJ (1987) The neck motor system of the fly *Calliphora erythrocephala*. I. Muscles and motor neurons. J Comp Physiol A 160:205–224

Stretton AO, Kravitz EA (1968) Neuronal geometry: determination with a technique of intracellular dye injection. Science 162:132-134

Tanouye MA, Wyman RJ (1980) Motor outputs of the giant nerve fibre in *Drosophila*. J Neurophysiol 44:405-421

Torre V, Poggio T (1978) A synaptic mechanism possibly underlying directional selectivity to motion. Proc R Soc London Ser B 202:409-416

Trujillo-Cenóz O (1965) Some aspects of the structural organization of the arthropod eye. Cold Spring Harbor Symp Quant Biol 30:371-382

Trujillo-Cenóz O (1969) Some aspects of the organization of the medulla of muscoid flies. J Ultrastruct Res 27:533-553

Umeda K, Tateda H (1985) Visual interneurons in the lobula complex of the fleshfly, *Boettcherisca peregrina*. J Comp Physiol A 157:831-836

Vigier P (1907a) Mécanisme de la synthèse des impressions lumineuses recueillies par les yeux composés des Diptères. C R Acad Sci 6:122-124

Vigier P (1907b) Sur les terminations photoréceptrices dans les yeux composés des Muscides. C R Acad Sci 145:532-536

Vigier P (1908) Sur l'existence réelle et le rôle des neurones. La neurone perioptique des diptères. CR Soc Biol 64:959-961

Wada S (1974) Spezielle randzonale Ommatidien von *Calliphora erythrocephala* Meig (Diptera: Calliphoridae): Architektur der zentralen Rhabdomeren-Kolumne und Topographie im Komplexauge. Int J Insect Morphol Embryol 3:397-424

Walthall WW, Murphey RK (1986) Positional information, compartments and the cercal system of crickets. Dev Biol 113:182-200

Wehrhahn C (1979) Sex-specific differences in the chasing behaviour of free flying houseflies (*Musca*). Biol Cybernet 21:213-220

Wiersma CAG (1958) On the functional connections of single units in the central nervous system of the crayfish *Procambarus clarkii* Girard. J Comp Neurol 110:421-472

Wiersma CAG (1966) Integration in the visual pathway of Crustacea. Symp Soc Exp Biol 20:151-177

Wiersma CAG, Mill PJ (1965) "Descending" neuronal units in the commissure of the crayfish central nervous system and their integration of visual, tactile, and proprioceptive stimuli. J Comp Neurol 125:67-94

Williams JLD (1975) Anatomical studies of the insect central nervous system: A ground-plan for the mid-brain and an introduction to the central complex in the locust, *Schistocerca gregaria* (Orthoptera). J Zool 176:67-86

Wolf R, Gebhardt B, Gademann R, Heisenberg M (1980) Polarization sensitivity of course control in *Drosophila melanogaster*. J Comp Physiol A 139:177-191

Wunderer H, Smola U (1982) Fine structure of ommatidia at the dorsal eye margin of *Calliphora erythrocephala* Meigen (Diptera: Calliphoridae): An eye region specialised for the detection of polarized light. Int J Insect Morphol Embryol 11:25-38

Zawarzin A (1913) Histologische Studien über Insekten. IV. Die optischen Ganglien der *Aeschna*-Larven. Z Wiss Zool 108:175-257

Zawarzin A (1925) Der Parallelismus der Strukturen als ein Grundprinzip der Morphologie. Z Wiss Zool 124:118-212

Zeil J (1979) A new kind of neural superposition eye: the compound eye of male Bibionidae. Nature (London) 278:249-250

Zeil J (1983a) Sexual dimorphism in the visual system of flies: the compound eyes and neural superposition in Bibionidae (Diptera). J Comp Physiol A 150:379-393

Zeil J (1983b) Sexual dimorphism in the visual system of flies: the free flight behaviour of male Bibionidae (Diptera). J Comp Physiol A 150:395-412

Zeil J (1983c) Sexual dimorphism in the visual system of flies: the divided brain of male Bibionidae (Diptera). Cell Tissue Res 229:591-610

Zettler F, Järvilehto M (1972) Lateral inhibition in an insect eye. Z Vergl Physiol 76:233-244

Chapter 17

Directionally Selective Motion Detection by Insect Neurons

NICOLAS FRANCESCHINI, ALEXA RIEHLE and AGNES LE NESTOUR,
Marseille, France

1 Introduction

Animals have several good reasons for detecting motion with their eyes. First, the motion of other animals – potential preys, mates, intruders or predators – provides essential information on which to base vital moves such as escape or chase. Secondly, information about self-motion is crucial, especially in the context of navigation, course stabilization, obstacle avoidance, and collision-free goal reaching. In fact, the wealth of information provided by passive, non-contact self-motion evaluation in visual systems has been likened to a kind of "visual kinaesthesis" (Gibson 1958). Even the 3D structure of the environment can be picked up by a moving observer (revs. Collett and Harkness 1982; Buchner 1984; Nakayama 1985; Hildreth and Koch 1987). Von Helmholtz (1867) was the first to clearly state the importance of this "motion parallax" in locomotion, and Exner (1891) proposed that arthropods make use of motion parallax as well as stereopsis to estimate distances (see also Horridge 1986).

When something moves in the surroundings of a motionless animal or when the animal itself moves through a stable environment, each contrast border of the optical image projected onto its retina inevitably produces an orderly sequence of excitations in the array of photoreceptor cells. This holds for the compound eyes of insects and crustaceans, just as it does for the camera eyes of vertebrates, because the imaging process of the eye, whatever its optical interface, results in an orderly mapping of the visual environment onto the receptor mosaic, as illustrated in Fig. 1. The sequence of electrical signals that is produced in the receptor array is processed by specialized neural microcircuits that inform the animal about its movements relative to the environment. In this context, the highly developed visual systems of vertebrates and arthropods are equipped with directionally selective (DS) motion-sensitive neurons. These "smart sensors" respond best to the movement of an object in a particular (preferred) direction. The opposite (nonpreferred or null-) direction elicits little or no response.

Although insects also possess nondirectional motion sensitive neurons (the best known example of which is perhaps the *lobula giant movement detector* which is to be found in various species: rev. O'Shea and Rowell 1977; see also Rind 1987), the scope of the present report is limited to the mechanisms underlying the responses of DS motion sensitive neurons. Special attention will be paid to the fly visual system,which has been the subject of the most extensive investigations. For a more general treatment, four excellent recent reviews cover all the aspects of

Stavenga/Hardie (Eds.) Facets of Vision
© Springer-Verlag Berlin Heidelberg 1989

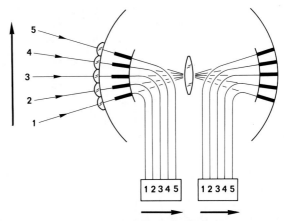

Fig. 1. Montage showing superimposed compound eye (*left*) and camera eye (*right*) having the same field of view. The scheme illustrates how different eye designs fundamentally serve common purposes in data acquisition: (i) *orderly* projection of the optical environment by a single lens (camera eye, *right*) or by a multilens array (compound eye, *left*); (ii) spatial sampling by a mosaic of discrete photo-transducing elements; (iii) *orderly* projection of the receptor electrical signals onto the first processing layer. The orderly projections that occur in both eye designs are such that motion of a contrast border (*vertical arrow*) in the optical environment gives rise to an orderly sequence (*1,2,3,4,5*) of electrical signals in the underlying neural layers. Directionally selective (DS) motion sensitive neurons are "smart sensors" that react to such sequences of input signals. They are present in both types of eyes and challenging logic and neuronal wiring problems are still being raised as to how they operate. (After Franceschini 1985)

motion vision in vertebrates, arthropods and machines (Ullman 1981; Buchner 1984; Nakayama 1985; Hildreth and Koch 1987).

   Although none of these recent reviews quote Sigmund Exner, the study of motion perception owes much to the genius of this Viennese physiologist. Thirty seven years before Wertheimer (1912) and the "Phi" phenomenon, Exner showed that sequential flashing of two neighboring, stationary light spots (still a commonly used stimulus today) creates a vivid impression of "apparent motion" in a human observer, providing the spacing and timing of the two flashes are chosen appropriately (Exner 1875a). From experiments of this kind, he concluded that motion was a primary sensory dimension. Thus in the early days Exner made it clear that DS motion detection was a spatio-temporal process. In his attempts to provide physiological explanations for psychological phenomena, he postulated that the sensation of directional motion might rely on specific "ganglion cells" located in some "motion perception centre" (Exner 1894). He even proposed a tentative wiring diagram involving "directionally selective motion sensitive cells", thought to drive corresponding eye muscles in the same direction as the perceived motion (Fig. 2). He recognized that directionally selective motion sensitive ganglion cells might solve their problem by using genuine delay lines that could result in signals from successively stimulated retinal elements coinciding in time. He suggested that the delays themselves might rely on the finite conduction velocity of excitation along nerve fibers (a point which had been demonstrated shortly before by von Helmholtz).

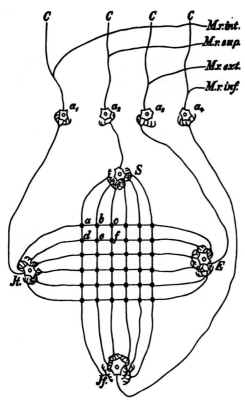

Fig. 2. The first neural scheme of a "centre for motion perception", as proposed by Exner about a century ago. Exner described his scheme as follows: "*a-f* and similar points are those locations into which the fibers from the retina come. *S,E,Jt,Jf* are centres into which excitations from each of these points arrive and where they can be summed. The time required for an excitation at one of these points to reach a centre is assumed to be approximately proportional to its distance as shown in the scheme. *a₁-a₄* centres that are closely related to the nuclei of the extraocular muscles (only four of them are shown here: *M.r.int., sup., ext., inf.*) or are themselves these nuclei. *c.* fibers to the cortex as the organ of consciousness". (Exner 1894)

Exner's studies on motion perception in man owe much to his reflections on the compound eye of arthropods (Exner 1875b, 1891), which was for him what it still is for us today: a convenient and particularly instructive model for analyzing fundamental problems in animal vision and the subtle ways in which nature has solved them. It is a humbling fact that more than one century of empirical vision research since Exner has not sufficed to fully elucidate the mechanisms of human motion detection. Today any animal model that can teach us something about this visual process is welcome. As O'Shea and Rowell (1977) put it, the likelihood of discovering general neural solutions to complex problems is high because complex integrative problems are liable to have only few neural solutions. Indeed, a recent lesson from studies in artificial intelligence is that even in the absence of neural constraints it is often exceedingly difficult to find one workable solution to a complex visual task (Ullman 1986).

2 The Correlation Model of Motion Perception

Developments in understanding motion detection have been slow, and it took 65 years to confirm Exner's idea about the existence of directionally selective motion sensitive ganglion cells in a visual system (vertebrates: Hubel and Wiesel 1959; Lettvin et al. 1959; arthropods: Waterman et al. 1964; Horridge et al. 1965). Meanwhile, a breakthrough had resulted from behavioral analyses of motion perception in the beetle *Chlorophanus*. Hassenstein (1951) analyzed the optomotor response of an insect whose visual feedback had been deliberately suppressed. This condition, ingeniously achieved by tethering the animal and recording its "open loop" walk on a "Y-maze globe", made sure that the moving stimulus presented to the eyes would not be affected by the motor reaction of the animal itself. By presenting the animal with sequences of light and dark steps that simulated motion in a given direction (cf. Exner's "apparent motion"), Hassenstein succeeded in eliciting a measurable optomotor response. From the responses to sequences presented with various angular separations between the two stimuli, he inferred that motion detection by the nervous system requires an interaction of signals from two directly neighboring or next neighboring ommatidia. From refined input-output analyses of this kind (Hassenstein 1958), it was inferred that the basic computation carried out in the underlying neural machine consisted of three steps (Hassenstein and Reichardt 1956, cf. Fig. 11):

— asymmetric linear filtering of the signals from the two sampling stations, one filter being of the low-pass type,
— multiplication of the resulting signals,
— time-averaging of the product obtained.

As the low-pass filter activated by the first stimulus acts as a delay, the overall process is a kind of "delayed comparison" between the two input signals, and the quantitative formulation leads to the classical concept of "correlation" (Reichardt 1957). It has been shown meanwhile that this correlation model of motion detection, originally deduced from analyses of *Chlorophanus* behavior, can satisfactorily account for the intensively studied, time-averaged optomotor response of walking or flying flies to continuously moving, periodic or nonperiodic stimulus patterns (*Musca:* Fermi and Reichardt 1963; McCann and MacGinitie 1965; *Drosophila:* Götz 1964, 1968; Buchner 1976). This topic has already been reviewed extensively (Reichardt 1961, 1969, 1987; Götz 1965, 1972; Poggio and Reichardt 1976; Reichardt and Poggio 1976; Buchner 1984).

Reichardt and Hassenstein succeeded in distilling the results of many behavioral experiments into a coherent and relatively simple functional diagram which was capable of predicting the reaction of the animal to other experimental paradigms. Moreover, their quantitative model showed that part of an animal's visual behavior might be related to a non-trivial mathematical computation that had to be carried out — in some way or another — by the underlying neuronal hardware. These authors identified a level of explanation for neural processes that transcends the neuronal hardware and concentrates on the underlying functional principles. In this sense the pioneering investigations by Hassenstein and Reichardt (1956) laid the foundations of modern research on computational vision and computational neuroscience (Marr and Poggio 1977; Marr 1982; Ullman 1986).

Interestingly, the correlation model of motion detection was born at a time when the mathematical concepts of auto- and cross-correlation were gaining great importance in signal processing (in particular for recovering faint signals buried in noise) and engineers were actively developing analogue methods of computing this function, such as the "optoelectric correlator" (e.g., Kharybin 1955, quoted by Solodovnikov 1965).

More recently, another low-level, intensity-based scheme for locally computing motion (the "gradient" scheme) was proposed in the field of engineering, namely for estimating the speed of moving objects from a television signal (Limb and Murphy 1975; Fennema and Thompson 1979). This scheme was suggested as a model for movement detection by vertebrate cortical cells (Marr and Ullman 1981). By confronting experimental data with predictions obtained using both schemes ("gradient" and "correlation"), Buchner (1984) provided evidence that DS motion detection in flies is based on correlation-like interactions rather than on a gradient-computing mechanism. Furthermore, the fundamental properties of the correlation model could be retraced at the level of the widefield, directionally selective H1 neuron in the fly visual system (Zaagman et al. 1978; Mastebroek et al. 1980) that will be the subject of the following section. Several variants of the correlation model have been proposed over the years (e.g., Thorson 1966; Kirschfeld 1972; van Santen and Sperling 1985), and its basic functional structure has even been used with great success for modelling DS motion detection in the visual system of vertebrates (Schouten 1967; Foster 1971; van Doorn and Koenderink 1976; Bülthoff and Götz 1979; van Santen and Sperling 1984; Adelson and Bergen 1985; Wilson 1985; van de Grind et al. 1986; Emerson et al. 1987).

It should be mentioned, however, that a correlation scheme is not the nec plus ultra for measuring angular velocities because its average output in response to a continuously moving, periodic grating varies not only with the speed but also with other stimulus dimensions, such as the contrast and spatial wavelength. Moreover, the response amplitude is not even unambiguously related to the speed or to the contrast frequency (product of angular velocity and spatial frequency) of the grating. These apparent "flaws" in a motion detector based on correlation (Hassenstein and Reichardt 1956) raise the question as to whether insects need these motion sensors for measuring speed or simply for indicating that motion has occurred in a given direction and within a given speed range (Buchner 1984): this information might be sufficient for some kinds of locomotor control such as corrective steering (Hassenstein and Reichardt 1956; Collett and King 1975).

3 The Directionally Selective H1-Neuron of the Fly as a Model Interneuron

Ever since directionally selective (DS) motion-sensitive neurons were discovered in the arthropod visual system (Waterman et al. 1964; Horridge et al. 1965; Collett and Blest 1966; Bishop and Keehn 1966), electrophysiological recordings have been adding considerably to our knowledge of motion vision. Within the last 20 years, directionally selective cells have been found to exist in the optic lobes of many insects (see Table 3 in the comprehensive review by Wehner 1981). Since the latter review was published, more DS cells have been discovered and stained in the optic lobes and brain of diverse insects such as the bee (DeVoe et al. 1982; Hertel et al. 1987; Hertel and Maronde 1987), the fly (Hausen 1981; Eckert 1982; Hengstenberg et al. 1982; Umeda and Tateda 1985; Egelhaaf 1985), the locust (Osorio 1986) and the moth (Rind 1983). The most thoroughly studied case is that of higher Diptera (flies). In these animals, the lobula complex (= third optic ganglion) is divided into two parts. The posterior one, called the lobula plate, is now known to house about 50 individually identifiable neurons, all of which are directionally selective and participate in some kind of optomotor control related to head or body movements (rev. Hausen 1981, 1984; Hausen and Egelhaaf this Vol.; Strausfeld this Vol.). Although most of these magnificent, fan-shaped neurons (Pierantoni 1976) are widefield DS cells, others with smaller excitatory fields, called "figure detection cells", have been discovered more recently (Egelhaaf 1985) and it seems that the guidance of a fly through its often cluttered visual environment essentially results from the co-operative interplay of these two types of DS neurons (rev. Hausen 1984; Wehrhahn 1984; Hausen and Egelhaaf this Vol.).

One of the widefield DS neurons of the fly lobula plate, called H1, has made a particularly significant contribution to our present understanding of directionally selective neurons, and all the results we will discuss here were obtained in experiments that took advantage of this neuron, which is unique in many respects. Only one specimen of H1 exists on each side of the midline. By means of intracellular staining, the shape of the H1 neuron has now been correlated with its response properties in three species of flies (Hausen 1976; Eckert 1980; Umeda and Tateda 1985) and the following statements can now be made on the basis of these and other studies:

1. H1 is a giant heterolateral neuron with an extensive dendritic arbor in the lobula plate (Fig. 4b). It is separated from the photoreceptor cells by three to four synapses at most. H1 sends its axon to the contralateral lobula plate, in which it terminates in a widely spreading arborization (Fig. 4b).
2. In keeping with the extensive dendritic arbor, the receptive field of H1 is broad and covers practically the whole panoramic field of view of the ipsilateral eye.
3. H1 is a spiking neuron with a low resting discharge (0–20 Hz). It responds with a conspicuous and tonic increase in firing rate to horizontal forward (inward, regressive) movements presented to the ipsilateral eye, and it is inhibited by motion presented in the backward (outward, progressive) direction (Fig. 3). The two bilaterally symmetrical H1 cells appear to be involved in the control of optomotor torque responses (Bishop et al. 1967; Eckert 1980; Hausen 1981). The two H1 neurons exert mutual inhibition, a feature that makes each particularly sensitive to either clockwise or anticlockwise rotatory (yaw) motion of the visual environment, as experienced by the fly when it performs a turn (McCann and Foster 1971; Hausen 1984).
4. H1-sensitivity to directional motion arises from the pooling of signals from smallfield units called "Elementary Movement Detectors" (EMD's, see Buchner 1976), which project perpendicularly to the dendritic fan in a retinotopic manner. This orderly mapping is such that the frontal EMD's project to the most distal part of the dendritic arbor, whereas the posterior EMD's project

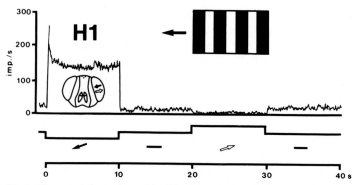

Fig. 3. Characteristic response of the H1-neuron to horizontal motion of a striped pattern (angular size: 100° width × 60° height; spatial wave-length: 33°; angular speed: 47°/s; contrast: 0.85) presented to the ipsilateral eye in the preferred (←) and nonpreferred (→) directions. (*Musca domestica*) PST histogram (bin width = 10 ms). Note the high and steady firing rate of the neuron in the preferred (back to front) direction and its rapid decay to a low resting level on cessation of motion (— = periods during which the illuminated pattern remains stationary). Note also the inhibition of the resting discharge during motion in the nonpreferred direction. (After Riehle and Franceschini 1984)

onto the proximal part of the dendrites (rev. Strausfeld 1976; Strausfeld and Hausen 1977; Strausfeld and Nässel 1980).
5. The array of EMD's feeding H1 must be wired down with fine grain because H1 readily responds to local motion between two points of the environment that are as little as one interommatidial angle apart (McCann and Arnett 1972; McCann 1973; Zaagman et al. 1977; Mastebroek et al. 1980).

Among the 50 DS motion-sensitive neurons of the lobula plate, H1 displays unique response properties, and it is relatively easy to locate, even with an extracellular microelectrode, in the head of the intact, unanesthetized animal. Since the class II-1 unit, discovered earlier in the fly visual system (Bishop et al. 1967, 1968), has been definitely identified as H1 by means of intracellular staining (Hausen 1976), this neuron is the only individually identifiable DS motion-sensitive cell known in the animal world that can pride itself on having been consistently investigated for 20 years, thereby giving rise to an impressive body of data (see also: Dvorak et al. 1980; Lillywhite and Dvorak 1981; Eriksson 1982; Srinivasan and Dvorak 1980; Srinivasan 1983; Bülthoff and Schmid 1983; Maddess 1986).

As in vertebrate retina and visual cortex, the detailed neuronal wiring of EMD's is still elusive. In those vertebrate retinae in which DS motion-sensitive ganglion cells have been mostly studied (fish, frog, turtle, pigeon and several rodents), there is evidence that a DS ganglion cell pools contributions from small-field sequence discriminating subunits (Barlow and Levick 1965; Levick et al. 1969), which may each result from a sophisticated synaptic microcircuit on the dendrites of the ganglion cell itself (Torre and Poggio 1978; Marchiafava 1979; Miller 1979; see also more recent contributions by Ariel and Daw 1982; Jensen and DeVoe 1983; Koch et al. 1983; Watanabe and Murakami 1984; Amthor et al. 1984; Masland and Tauchi 1986; Grzywacz and Koch 1987). However, neither the detailed neuronal implementation of these subunits (EMD's) nor their exact inputs in terms of individual cones or rods are known so far. In the fly it is still a major open question whether EMD's consist of single columnar neurons located in the more distal, second optic ganglion (the medulla), or whether they are built up on the H1-dendrites themselves (rev. Hausen 1981; Hausen and Egelhaaf this Vol.; Strausfeld 1984, this Vol.). In fact, small-field DS cells are to be found distally to H1 (in the medulla: McCann and Dill 1969; Mimura 1971, 1972; Collett and King 1975; DeVoe and Ockleford 1976; DeVoe 1980) but they do not necessarily drive the DS lobula plate neurons as there exist other, direct output pathways from the medulla to the brain (rev. Strausfeld 1976).

In the fly, comparisons between the spectral sensitivity of H1 and that of photoreceptor cells of the R1-6 type (i.e., the six peripheral receptors of an ommatidium, see inset of Fig. 4a) have suggested that these receptors are inputs to the motion-detecting circuitry feeding H1 (McCann and Arnett 1972). This conclusion tallies with the results of various optomotor experiments. First, the spectral sensitivity of the optomotor response (turning reaction around the yaw axis) reveals the existence of a dominant input from this receptor type (Eckert 1971). Secondly, a measurable, time-averaged optomotor response was induced in the walking fly by stimulating two photoreceptor cells of this type within a single ommatidium (Kirschfeld 1972; Kirschfeld and Lutz 1974). Thirdly, studies on the *Drosophila* mutants *ora* and *sevenless*, which lack receptors R1-6 and R7, respectively, have shown that the optomotor response is mediated predominantly by receptors of the R1-6 type (Heisenberg and Buchner 1977). The involvement of cells of the R1-6 type in DS motion detection is evidence that the neural columns (called "cartridges") of the underlying ganglion (the lamina), onto which these cells are known to project, are themselves involved in the motion detection circuitry driving H1. It should therefore be possible to learn more about the intrinsic properties of a single EMD by recording the H1 response when a pair of neighboring cartridges is activated.

4 Intrinsic Properties of an Elementary Movement Detector (EMD)

4.1 Optical Microstimulation of Single Photoreceptor Cells

Closer insight into the properties of an EMD has recently resulted from a novel approach (Riehle and Franceschini 1982, 1984; Franceschini 1985; Franceschini et al. 1986). The idea was to use H1 as a probe and to selectively excite a single EMD, with the hope that its signature-response would show up in the electrical activity of H1. This type of experiment is schematized in Fig. 4. It relies on the sequential activation of two adjacent cartridges (dotted rectangles in Fig. 4a), each via one of its receptor inputs. Under these conditions H1 expresses the response of only one single EMD because the multitude of other feeding units remain quiescent.

As shown in Fig. 5, optical microstimulation of two receptor cells, R1 and R6, within the same ommatidium (Fig. 5a) is indeed one way – among others – of activating two adjacent cartridges (Fig. 5c). This possibility is due to the remarkable

368   Nicolas Franceschini et al.

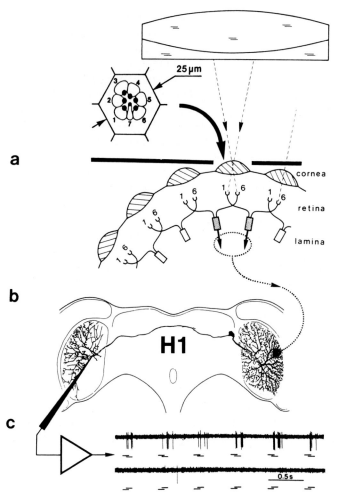

Fig. 4a-c. By recording the activity of the DS motion sensitive neuron H1 during optical micro-stimulation of only two photoreceptor cells, it is possible to isolate the response of a single elementary movement detector (EMD). The scheme (not to scale) illustrates the information flow from the objective of the "microscope-telescope" (photographed in Fig. 6, left) down to the H1-neuron, the spike activity of which is recorded contralaterally. *a* Schematic longitudinal section through the retina and lamina along a horizontal direction passing approximately through the two excited receptors R1 and R6 in Fig. 5a. Selective stimulation of the two chosen receptors is achieved by: (i) optically projecting a mask onto a given facet (the result of this projection, which is schematized by the *thick horizontal line* in *a*, is shown in Fig. 7b), (ii) selecting two appropriate directions (*dotted rays with arrows* in *a*) to illuminate the chosen receptors. During the latter process, the whole receptor pattern of the selected ommatidium (*a inset*) is under visual control, with the facet lens itself acting as a "microscope objective". *b* Heterolateral, DS neuron H1 (from an intracellular injection by Hausen 1976) the spike discharge of which is recorded extracellularly with a plastic-coated tungsten microelectrode. *c* Sequentially presenting two flashes (each with a duration of 100 ms) to the pair of receptors, R1 and R6, within the selected ommatidium consistently induces a prominent spike discharge in H1 provided the temporal interval between the two flashes is such as to simulate a micro-motion in the preferred direction (*upper trace* 50 ms lag between R1 and R6). Note the absence of response with the opposite sequence (*bottom trace* 50 ms lag between R6 and R1), which simulates motion in the nonpreferred direction. (After Franceschini 1985)

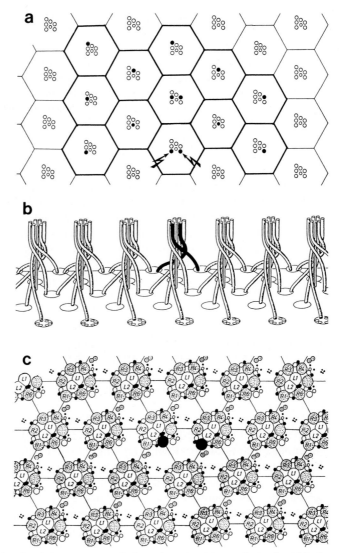

Fig. 5a-c. Axonal projection (*b*) of photoreceptor cells from the retina (*a*) onto the lamina (*c*) in the "neural superposition eye" of the fly (rev. Kirschfeld 1969; Trujillo-Cenóz 1972; Braitenberg and Strausfeld 1973; drawn after Strausfeld and Nässel 1980). The known pattern of axonal decussation illustrates the point that optical microstimulation of the two receptor cells R1 and R6 considered in Fig. 4a activates two different, immediately neighboring cartridges of the lamina, which themselves drive an elementary movement detector. (After Franceschini 1985)

wiring of photoreceptor axons (Fig. 5b), which has been worked out in great detail, and has led to the discovery of the ingenious principle of "neural superposition" in the fly's eye (rev. Kirschfeld 1969, 1973; Trujillo-Cenóz 1972; Braitenberg and Strausfeld 1973; Franceschini 1985; Nillson this Vol.). According to this principle, each of the 3000 visual sampling directions of the eye is shared by eight photo-receptor cells located in neighboring ommatidia. Six of these cells (those of the R1–R6 type) send their axons to the same cartridge of the lamina. The conveyance of six (independent) receptor signals onto the same set of second-order neurons has been interpreted as a means of enhancing the signal-to-noise ratio.

A special optical instrument — a hybrid between a microscope and a telescope — was built for performing optical microstimulation of single photoreceptor cells combined with single neuron recordings (Fig. 6, Franceschini unpublished). The design of this instrument was based on the thorough knowledge now available about the optics of the fly's eye (rev. Kirschfeld 1969, 1972; Franceschini 1972, 1975; Stavenga 1975). The instrument can be moved in azimuth and elevation around the insect so as to select an ommatidium in the frontal part of the eye (region of highest H1 sensitivity). The selected facet (Fig. 7b) itself acts as the "primary objective lens" (diameter 25 μm, focal length 50 μm) of the instrument. This lenslet ultimately focuses two microspots of light (diameter 1 μm) onto the chosen pair of rhabdomeres (diameter 1 μm), e.g., R1 and R6 that are approximately 3.5 μm apart (Fig. 4a, inset). Precise positioning of these microspots onto the rhabdomeres is achieved by turning on a third incident beam of the instrument. The latter floods all the receptor cells of the ommatidium, thereby inducing pigment granule migration in their cytoplasm, which is manifested by a distinct, yellow reflection from the rhabdomere tips (Kirschfeld and Franceschini 1969; rev. Franceschini 1972).

The analysis carried out under single photoreceptor stimulation was aimed at deciphering the logic underlying the operation of an EMD microcircuit — a single one among the thousands of EMD's feeding H1 — and defining the individual steps of the computation. By deliberately stimulating a single EMD, it is possible to avoid making the somewhat dubious assumption implicit in any study using large field moving patterns, namely that all stimulated EMD's at any time respond in a qualitatively similar fashion and add their responses linearly in such a way that the compound response faithfully reflects the properties of an individual EMD.

The type of stimulus presented to the two receptors under consideration con-sisted of various sequences of light pulses and/or steps. This kind of stimulus was suitable for three reasons. First, this "Exner-type" sequential flashing more or less simulates the "natural stimulus" experienced by an EMD when an object moves across its receptive field. Secondly, this kind of stimulus is known to be particularly useful when the aim is to disentangle and identify the various linear and nonlinear parts involved in a two-input system (see e.g., Varju 1977). Thirdly, this kind of stimulus is also suitable for analyzing nonstationary systems, and recent pieces of evidence suggest that a fly EMD is a complex, adaptive system (Mastebroek et al. 1982; Zaagman et al. 1983; Schuling and Mastebroek 1984; Maddess and Laughlin 1985; de Ruyter van Steveninck et al. 1986; Borst and Egelhaaf 1987).

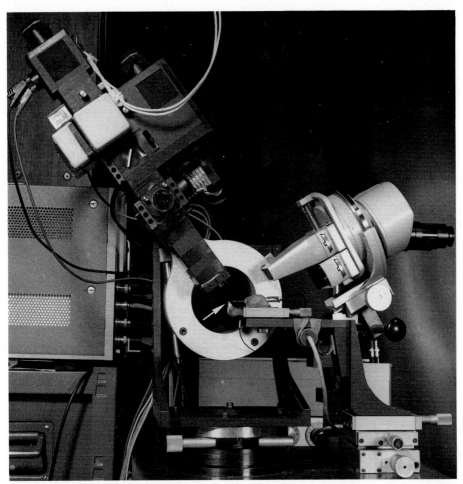

Fig. 6. "Microscope-telescope" (*left*) pointing at the fly's eye (*arrow*). This triple-beam stimulating instrument – which is secured to the ceiling of the laboratory with the halyard seen at top – can be moved in azimuth and elevation around the eye so that a given ommatidium can be selected (Fig. 7b). In this ommatidium, two receptor cells are illuminated in sequence to produce an elementary "apparent movement". The selected facet is an integral part of the instrument and acts as its primary "objective lens". It ultimately projects upon the two selected receptors R1 and R6 (Fig. 4a) the micrometer-sized images of two pinholes which are independently lit (6 V, 20 W halogen lamp with Kodak green filter No. 61) and independently movable, each being in addition equipped with its own shutter. The instrument is prefocused onto the point of intersection between its vertical and horizontal axes of rotation. The fly is brought to that point via an X-Y-Z micromanipulator (*bottom right*) that also moves the micromanipulator supporting the microelectrode. The latter is then finely advanced under binocular control (*right*) into the lobula plate through a hole cut into the back part of the head capsule. The whole apparatus is mounted on a vibration-free table, and all rotatory movements of the optical stimulator are smooth enough to allow a new facet to be optically selected without jeopardizing the electrical recording from the H1 neuron

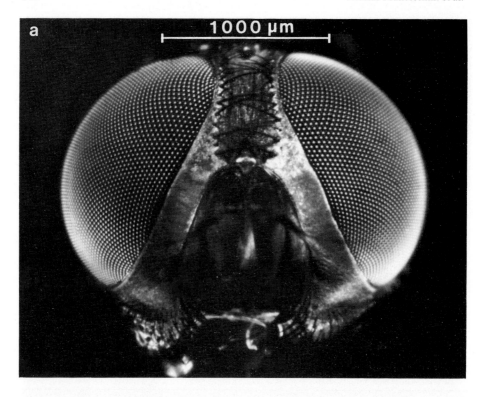

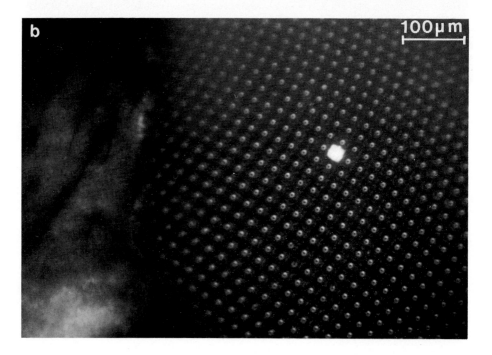

4.2 Sequence Discrimination by an EMD

Sequentially flashing the two horizontally aligned receptor cells R1 and R6 (Fig. 4a) induces a vigorous spike discharge in H1 when the sequence simulates motion in the preferred direction (R1→R6) as illustrated in Fig. 4c (upper trace). By contrast, the reverse sequence (R6→R1), which simulates motion in the null-direction, keeps H1 remarkably silent (Fig. 4c, lower trace). If the rate of the sequence repetition is low (2 Hz in Fig. 4c) there is no sign of "fatigue" in the underlying neural machinery, which discharges a burst of spikes whenever the sequence recommences.

The characteristic features of this sequence-discrimination are best demonstrated when the spiking response of H1 is evaluated in terms of instantaneous frequency, averaged over 100 stimulus presentations, and all the responses shown below were processed in this way. The series of responses in Fig. 8 shows that a 100-ms rectangular pulse of light presented to each receptor separately (Fig. 8a,b) or to both receptors simultaneously (Fig. 8c) elicits hardly any response. But as soon as a temporal interval having the correct sign is introduced between the two flashes, a conspicuous response occurs (Fig. 8d), which contrasts with the total absence of response to the opposite sequence (Fig. 8e).

This experiment illustrates the biologically relevant fact that an EMD of this kind is smart enough not to be fooled by simultaneous excitation of its two inputs (Fig. 8c) and is therefore responsive to a specific "trigger feature": movement in that particular direction simulated by the preferred sequence. The "preferred sequence" R1→R6 is indeed consistent with the "preferred direction" of H1 (Fig. 3), if one takes into account the fact that the image is reversed by the facet lens (Fig. 4a). Furthermore, this experiment illustrates the striking nonlinearity of EMD responses: the principle of superposition, which characterizes linear systems, is violated here because in no way can the vigorous response to two successive flashes (Fig. 8d) be predicted from the responses to each flash presented separately (Fig. 8a and b).

4.3 The Transient, ON-OFF Nature of the EMD Response

As expressed in the extracellularly recorded activity of the H1 "probe", the response of a single EMD to a sequence of two flashes displays characteristic features:

Fig. 7. *a* Head of the fly *Musca domestica* (male) showing the prominent compound eyes with their facetted cornea. The 3000 ommatidia of each eye define 3000 sampling directions (see Fig. 1, left) that are separated from each other by an angle $\theta \approx 1°-2°$. This photograph was taken with an electric torch converted into a Lieberkühn microscope, after precise orientation of the head on a goniometer so as to reveal two symmetrical, median *deep pseudopupils* (Franceschini 1975). *b* Frontal-dorsal part of the eye (*Musca domestica*, female) showing the intense reflection from the single illuminated ommatidium within which the two receptor cells R1 and R6 (see inset of Fig. 4a and Fig. 5a) were sequentially flashed while recording from the DS neuron H1 (Fig. 4b,c). Here a fourth incident beam from the optical instrument served to illuminate the whole eye surface, thus making it possible to precisely select a facet under good viewing conditions. (Franceschini 1985)

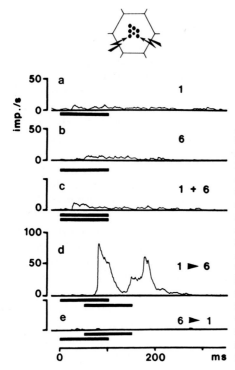

Fig. 8. Sequence discrimination by a fly elementary motion detector (EMD) that drives the H1 neuron. Response of H1 to the presentation of two light flashes (duration 100 ms each) onto receptor cells R1 and/or R6 within a single, frontal ommatidium (see *inset* at the top). The ordinate represents the instantaneous frequency of the spike discharge, averaged over 100 stimulus presentations (time resolution ≈ 5 ms; repetition frequency 2 Hz). Note that hardly any response occurs when a single receptor is stimulated (*a,b*), or when the two receptors are excited synchronously (*c*) or when the sequence of excitation mimics motion in the null direction (*e*). The prominent response, basically consisting of two transients, which is observed (*d*) when the flash sequence mimics motion in the preferred direction (R1→R6) attests to the *facilitatory* effect of the first flash. (After Riehle and Franceschini 1984)

1. The response to a sequence of two 100-ms flashes, with a time lag of 50 ms between their onsets, is transient and essentially consists of two prominent peaks in spike frequency (Fig. 8d). This response can be described as an "ON-OFF response" to the *second* flash because its ON-component and its OFF-component appear to be time-locked (after a latency of 15–30 ms) to the onset and offset of the *second* stimulus, respectively.

2. Both response transients are again observed when the two 100-ms flashes do not overlap in time, and Fig. 9 suggests that the OFF-transient may simply add to the declining phase of the ON-transient (dotted line), to form the characteristic ON-OFF response of the EMD.

3. The ON-transient can be isolated if, instead of a succession of 100 ms light pulses, a sequence of "light steps" is presented to the receptor pair (Fig. 15a). The characteristic, exponential-like decay of this step response to the second stimulus seems to exhibit the properties of a first order linear high-pass filter the time constant of which here is in the order of 100 ms.

4. Likewise, the OFF-transient can be studied separately if a sequence of "dark steps" is presented to the same receptor pair (Fig. 15c). The pure "OFF-response" resulting from such a sequence often exhibits a conspicuous rebound effect in its declining phase (Fig. 15c).

Hence, a major feature of the EMD response observed in the H1 collector neuron, is its transient nature. The second feature is that both a succession of light

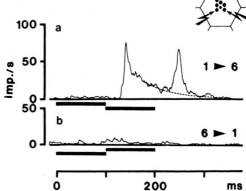

Fig. 9. Response of H1 when the two 100-ms flashes presented to R1 and R6 (*inset*) are timed so that there is no overlap. It is argued in the text (Sect. 4.7) that the dual-peak response observed here (see also Fig. 8d) results from the superimposition of responses to the two sequences ON-ON (cf. Fig. 15a) and OFF-OFF (cf. Fig. 15c) that are computed in parallel by two different microcircuits: an ON-EMD and an OFF-EMD (cf. Fig. 16)

steps and a succession of dark steps are efficient stimuli and elicit responses with the same polarity in H1. If one considers that the two types of sequences correspond to the movement of the leading edge and the trailing edge of a bright object, respectively, it certainly makes sense that an EMD microcircuit should respond to both kinds of events occurring in its receptive field and consequently produce the typical ON-OFF response observed in Figs. 8d and 9.

4.4 Evidence for a Facilitatory Control

An essential feature of EMD behavior, which can be seen in Figs. 8 and 9, is the conspicuous absence of response to the first flash of the sequence. This feature can be again observed in Fig. 10a, which shows the response to a sequence of two brief (10 ms) flashes presented to the pair of photoreceptor cells R1 and R6.

There can be no doubt that the first flash elicits a signal somewhere in the EMD circuit. However, this signal does not get through to H1 and its covert action must

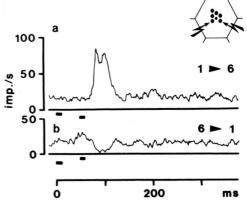

Fig. 10. Response of the H1-neuron to sequential microstimulation of two receptor cells R1 and R6 (*inset*) with brief light flashes of 10 ms duration separated by 50 ms (*a,b* averages over 100 sequences presented at 2 Hz repetition frequency). Note that in addition to the "positive" response (*a* increase in spike rate) recorded with the sequence that simulates motion in the preferred direction (R1→R6) a "negative" response (*b* inhibition of the resting discharge) is observed with the reverse sequence. The latter response is noticeable because the resting discharge of H1 in this example is relatively high

consist of dynamically adjusting some internal parameter that ultimately facilitates
the response to the second flash. The conclusion that this "parametric control",
which is exerted via the lateral branch in the functional diagram of Fig. 11, is of a
facilitatory nature seems obvious if one compares Fig. 8b and 8d — two variations
of the response to a similar flash on the same receptor (R6). If instead of remaining
"flat" as in Fig. 8b, the response to R6 is "fully developed" in Fig. 8d, this can only
be due to the fact that the priming flash applied to R1 has exerted a facilitatory
influence.

The anatomical location of this facilitatory control, which might be carried out,
for example, by a modifiable synapse, is not known. But even in functional terms
it is still questionable whether it acts as a *gain control* (which would correspond to
the "true multiplier" of Reichardt's correlation model) or as a *threshold control*. In
a threshold control circuit, transmission of a signal down through a channel
encounters a threshold the value of which is made to vary by an external control

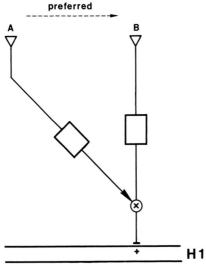

Fig. 11. Basic block diagram of an elementary motion detector (EMD) as originally inferred by
Hassenstein and Reichardt (1956) from behavioral experiments on the beetle *Chlorophanus*. A similar
functional structure can explain the response of the DS neuron H1 in the fly visual system upon precise
stimulation of a single EMD. Here, however, inputs A and B explicitly mean those two neighboring
cartridges (Fig. 5c) which were selectively activated upon optical microstimulation of the two photoreceptor cells R1 and R6 within a single ommatidium (Fig. 5a). It is argued here that the "lateral" branch
of this EMD mediates a long-lived facilitation and contains — to a first approximation — a low-pass filter
of higher order, the impulse response of which is shown in Fig. 13. By contrast the "direct" branch of
the EMD essentially contains a first-order high-pass filter, whose step response is shown in Fig. 15a. The
non-linear interaction between the two filtered signals was inferred by Hassenstein and Reichardt to be
a multiplier, but other kinds of parametric controls, e.g., a threshold control (Fig. 12b) could be implied.
This EMD is assumed to sense motion in the H1 preferred direction and consequently drives H1 with
positive polarity (see *bottom part* of the figure). Our experiments, however (e.g., Fig. 10b, see also Fig.
2G in Riehle and Franceschini 1984), suggest that the same two cartridges *A* and *B* drive a companion
EMD (not shown here) that actively responds to the opposite sequence (i.e., to the nonpreferred
direction) and drives H1 with opposite polarity. (cf. Fig. 14c)

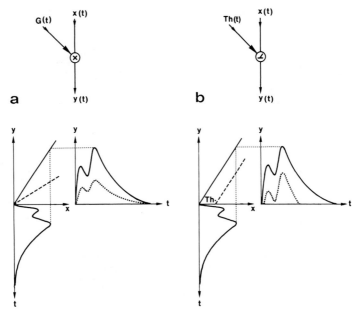

Fig. 12a,b. Gain control *(a)* and threshold control *(b)* are two kinds of parametric controls that could account for the *facilitatory* influence of the first stimulus upon the second stimulus in a fly EMD (Figs. 8d, 9a, 10). The *bottom diagrams* illustrate the fate of an input signal x(t) (arbitrarily chosen with two maxima) as it passes through a gain control *(a)* or a threshold control *(b)*. The curves show how y(t) results from x(t) by reflection onto the characteristic curve y(x) which is itself voltage-controlled. The lateral branch may change either the slope of the characteristic curve (a) or displace the latter parallel to itself (b). Note that both a gain increase and a threshold decrease lead to facilitation of the response y(t) *(dotted curves → continuous curves)*. In the latter case, however, the output y(t) does not grow as a scaled version of itself. The precise anatomical location of the parametric control is unknown at present. Presynaptic facilitation is a plausible candidate for its neuronal implementation

voltage (Fig. 12b). This kind of (nonlinear) control can be implemented simply with elements that are well within the scope of neuronal hardware (see, e.g., Calvin and Graubard 1979; Nicoll 1982).

4.5 The Impulse Response of the Facilitatory Control System

In order to estimate the time course of the facilitatory control, a series of experiments of the type depicted in Fig. 10a was carried out with varying interstimulus intervals. Figure 13 is a plot of the peak response amplitude to the second flash with respect to the lag between the onsets of the "conditioning" and "test" flashes. Figure 13 teaches us two things about the underlying system:

1. The facilitation brought about by the first flash is graded over time and long-lived. However, it does not start immediately on presentation of the first ("conditioning") flash but rather at t = 10 ms. On the other hand, the facilitation does not last indefinitely but has decayed here by t = 230 ms. This time span over

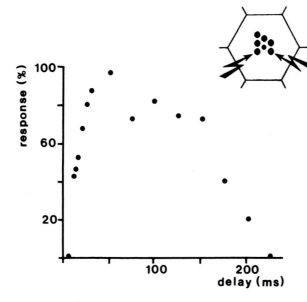

Fig. 13. Relative increase in firing rate of H1 in response to a brief (10 ms) flash on R6 presented after a similar flash on R1. The interval between the onsets of the two flashes is given on the abscissa. Average of 9 experiments (9 H1 neurons, 9 flies). Experimental conditions as in Fig. 10a. The response profile obtained describes the impulse response of the facilitatory control exerted by the lateral channel over the other (Fig. 11) in the preferred direction

which the facilitatory control is "launched" defines a characteristic "temporal window" which is of basic importance in that it governs the speed range of the EMD. Taking into account the angular separation between the lines of sight of the two receptors R1 and R6 (3.6° according to *in vivo* measurements by Kirschfeld and Franceschini 1968), we calculated that the "temporal window" described above corresponds to a range of angular speeds of 16°/s to 360°/s with the EMD under study (Franceschini 1985).

Note that maximum facilitation occurs near $t = 50$ ms, corresponding to an "optimum" angular speed of 72°/s. Note also that on each side of this maximum (Fig. 13) two different interstimulus intervals yield the same "reading" in impulses/s at the sensor output, thereby creating that very ambiguity of response versus speed which was mentioned earlier in connection with the correlation model (Sect. 2).

2. The second flash applied to receptor R6 in Fig. 10a can be said to be a test-input that in Fig. 13 sweeps across the impulse response of the facilitatory control set off by the first flash on R1. Figure 13 can therefore be said to be a replica of the impulse response of the lateral branch that brings about facilitation (Fig. 11). The sluggish time course of this response (Fig. 13) is reminiscent of the impulse response of a linear, higher order low-pass filter. This type of experiment thus gives access to the temporal properties of the low-pass filter drawn in the lateral branch of the EMD block-diagram (Fig. 11).

4.6 Two Unidirectional EMD's with Opposite Preferred Directions

In those cases where the resting discharge of the H1 neuron is not as negligible as in Fig. 8, presentation of a flash sequence corresponding to simulation of movement in the nonpreferred direction (R6→R1) rather than eliciting an increase in spike frequency elicits a suppression of activity, as shown in Fig. 10b. Again, this response appears to be locked to the second stimulus of the sequence. When under similar conditions a sequence of longer (100 ms) light pulses was presented, two transient phases of suppression were observed, which were again locked to the onset and offset of the second stimulus, respectively (Riehle and Franceschini 1984).

The consistent finding that the firing rate of H1 in response to a sequence of light stimuli on a single pair of photoreceptor cells, R1 and R6, increases or decreases depending on the temporal order within the sequence has led us to the conclusion that at nearby sites on its dendritic arbor, H1 receives antagonistic signals from two co-existing EMD's having opposite preferred directions. The results of all the experiments we have conducted so far are consistent with the hypothesis that these two sensors both rely on a facilitatory control mechanism such that their neural implementation may be similar, their specificity being expressed simply by the sign of their postsynaptic effect on H1.

Figure 14 is a schematic functional representation of these ideas. The basic module is a puppet (Fig. 14a), whose head collects the receptor signal corresponding to one direction in space, whose hands control a facilitatory mechanism acting on the right and left neighbors, whose knees contain its own facilitatory mechanisms and whose feet convey a positive (excitatory) or negative (inhibitory) signal to H1. The four limbs are assumed to convey adequately filtered signals, as shown in Fig. 11 (low-pass for the arms, high-pass for the legs).

If the head of this puppet receives a stimulus, it will transmit a positive or negative kick (Fig. 10) down to the H1 cell it is standing on, depending on which knee has been facilitated, i.e., depending on which neighbor has been activated a short time before (Fig. 14b). If no neighbor has been activated, no signal gets through to H1 because neither of the knees is facilitated. If two knees become facilitated at the same time (as may occur for example under synchronous flicker on two inputs), H1 should receive a mixture of reciprocally counteracting excitation and inhibition (Fig. 14c).

The consistency of the responses produced by the animals, choosing a different facet each time at random, in addition to the highly repetitive structure of the neural mosaic, suggests that this same functional structure is regularly replicated beneath the ommatidial mosaic. Figure 14d shows an array of such puppets, their hands on their neighbors' knees, standing side by side on an H1 dendritic branch.

4.7 The Splitting of an EMD into an ON-EMD and an OFF-EMD

Judging from the characteristic ON-OFF response of an EMD to a sequence of 100 ms light pulses (Fig. 9), we originally put forward the hypothesis that an EMD basically relies on a sluggish facilitatory channel (lateral branch in Fig. 11) which lets through an ON-OFF signal from the other channel down to the H1 neuron (Riehle and Franceschini 1984). Recent experiments suggest that this process is split into two parts, whereby the ON-component and the OFF-component are facilitated separately and independently (Franceschini et al. 1986).

As illustrated in Fig. 15, a sequence of light onsets gives rise to a vigorous ON-response (Fig. 15a), whereas no response is observed when the same light onset

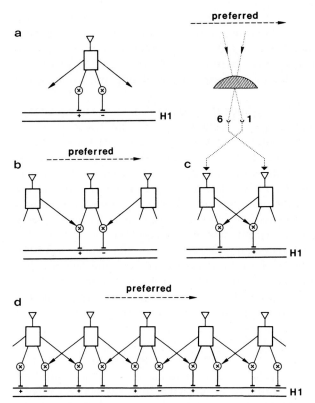

Fig. 14a-d. Basic topology of the lateral interactions leading to DS motion detection in the H1 neuron, as inferred from experiments using microstimulation of single photoreceptor cells with the "microscope-telescope" described in Fig. 6. *a* The basic functional structure describing the information flow from one lamina cartridge down to the H1 neuron is a puppet with two "feet" driving H1 with opposite signs (excitation and inhibition). Two identical *facilitatory controls* whose parameter (gain, threshold . . . cf. Fig. 12) is adjustable by an external signal form the knees of the puppet. The two arms transmit a signal to the next neighbor's knees, thereby dynamically adjusting a facilitatory control on each side. *b* Whether activation of the central puppet results in excitation or inhibition of the H1 spike discharge depends on which knee is being facilitated, i.e., which neighbor has been activated a short time before to create the long-lived facilitation whose kinetics is described in Fig. 13 (impulse response). *c* Results such as those described in Fig. 10 show that two companion EMD's of the type described in Fig. 11 are accessed by microstimulation of R1 and R6 within a single ommatidium. Both are assumed to be built similarly but detect motion in opposite directions and drive H1 with opposite signs. *d* The consistency of the responses obtained when sequentially flashing R1 and R6 in any ommatidium within the frontal part of the eye (Fig. 7b) suggests that the same microsystem for DS motion detection is replicated beneath the lamina cartridges, giving rise to a "puppet array" for each horizontal row of cartridges (Fig. 5c). The qualitatively similar DS responses obtained when sequentially flashing an oblique receptor pair (e.g., R2/R4, see inset Fig. 4a) suggest that a similar puppet array underlies oblique rows of cartidges as well

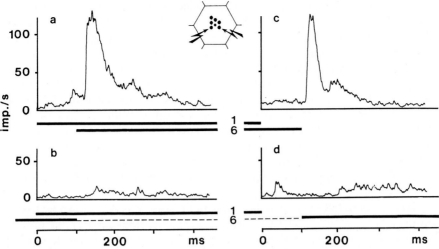

Fig. 15a-d. Basic experiment demonstrating how EMD's contribute to solving the "correspondence problem" (Ullman 1981) in motion detection. EMD's appear to respond exclusively to light sequences that are made up of contrast transitions with like signs (steps of light in *a*, steps of darkness in *c*) and ignore those sequences in which contrast transitions have opposite signs (*b*, *d*). *a* can be said to be the step response of the high-pass filter located in the direct branch of the EMD block diagram (Fig. 11). Among the eight step sequences that can be performed by manipulating the sign and the delay of the contrast transitions on R1 and R6, only the two sequences shown in *a*, *c* lead to a prominent response in H1. Note that these two sequences to a first approximation simulate the motion of the leading edge (*a*) and trailing edge (*c*) of a broad and bright object in the preferred direction (average responses over 100 sequence presentations; repetition rate: 0,5 Hz; stimulation of R6 in *b* and R1 in *d* started at t = –1000 ms)

on R1 is followed by a light offset on R6 (Fig. 15b). Likewise, the correct sequence of light offsets generates a vigorous response (Fig. 15c), whereas no response occurs when a light offset on R1 is followed by a light onset on R6 (Fig. 15d). It thus appears likely that the first contrast change is able to facilitate the response to the second contrast change only if it has the same polarity.

These response characteristics suggest that the ON- and OFF- components of the ON-OFF response observed (e.g., Figs. 8d and 9) are conveyed separately down to H1 and that the EMD responsible for sensing motion in the preferred direction actually consists of two independent EMD's: an ON-EMD sensing the motion of light edges, and an OFF-EMD sensing the motion of dark edges. Figure 16 schematizes this splitting of the EMD into two parts. If we assume that the EMD responsible for motion in the non-preferred direction (Sect. 4.6) is likewise split into two independent systems, each of the above puppets (Fig. 4a) should in fact be drawn with four arms and four legs.

5 Conclusion

The detailed knowledge acquired over the last two decades about the optics of the fly's eye (rev. Kirschfeld 1969; Franceschini 1972, 1975; Stavenga 1975), the neuronal wiring of its receptor cells (rev. Trujillo-Cenóz 1972; Braitenberg and

Strausfeld 1973; Kirschfeld 1973) and the organization of its receptor mosaic (rev. Franceschini 1984, 1985; Hardie 1985, 1986), in addition to the thorough knowledge now available on motion-sensitive neurons in the lobula plate (rev. Hausen 1981, 1984), have made this compound eye a unique model for studying the elementary process which takes place in a DS motion-detecting neuron. The present report has stressed in particular the novel possibilities opened by optical microstimulations of identified photoreceptor cells, associated with electrophysiological recording from an identifiable DS neuron. With this approach, the receptor mosaic, which can be visualised *in vivo* with a variety of techniques (rev. Franceschini 1975, 1983), is thought of as a huge "keyboard" controlling the widefield DS neuron. The strategy consists of operating only two selected "keys" in sequence, thereby inducing a directional response liable to reveal some functional properties of the underlying "integrated circuit".

It emerges that the same kind of sequential flashing as that shown by Exner a century ago to cause an illusory impression of movement ("apparent motion") in a human observer, when confined to a single pair of identified photoreceptor cells, has helped us to elucidate the intrinsic response of an elementary motion detector (Figs. 8, 9, 10, 15). It is fortunate that microstimulation of only 2 out of the 24,000 receptor cells of the eye suffices to elicit a response in the giant H1 neuron. This is all the more surprising, since under these artificial conditions of stimulation, each of the two activated lamina cartridges is running on only one out of six inputs (Fig. 5c). It has recently been demonstrated, however, that stimulation of a single receptor cell of the R1–6 type does suffice to drive second order neurons in its target cartridge (van Hateren 1987).

In contrast with the situation in rabbit retinal ganglion cells, in which motion detection has long been described as relying mainly upon a strong inhibitory process vetoing the response to null sequences (Barlow and Levick 1965), our present view suggests that in the fly visual system an elementary DS subunit (EMD) relies on a strong facilitatory interaction in the preferred direction. A similar kind of facilitatory interaction is suggested for the reverse direction, whereby in this case the facilitated signal drives H1 with the opposite polarity (Fig. 14). It should be noted that in some mammalian visual systems cases of facilitation in the preferred direction have been observed in DS neurons, in particular in the rat's lateral geniculate body (Montero and Brugge 1969), in the cat striate cortex (Emerson and Gerstein 1977; Movshon et al. 1978; Vitanova et al. 1985; Emerson et al. 1987) and in the macaque medio-temporal visual area (MT: Mikami et al. 1986).

An interesting aspect of the functional diagram established on the basis of our investigations is the apparent splitting of an EMD into two parts, the one responsible for sensing the motion of the leading edge, and the other for sensing the motion of the trailing edge of an object (Fig. 16). Separate evaluation of these two pieces of information by two separate circuits (that ultimately drive H1 with the same polarity) ingeniously contribute to solving the "correspondence problem" (Ullman 1981) in motion detection, since only contrast borders with like signs are brought into "correspondence". This feature, which points to an unexpectedly high degree of parallel processing in motion detection, entails the advantage of providing H1 with twice as much information concerning the motion of an object, even if the angular size of the latter is much smaller than the sampling basis of the EMD.

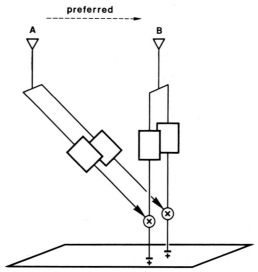

Fig. 16. A single EMD of the type described in Fig. 11 cannot account for the results shown in Fig. 15 even if full-wave rectification (ON-OFF type of response) is assumed to occur at an early stage in the processing. This suggests that the process of DS motion detection is split into two parts, whereby the ON response to a light step (Fig. 15a) and the OFF response to a dark step on R6 (Fig. 15c) are facilitated separately and independently. This novel block diagram describes the parallel processing achieved by an ON-EMD and an OFF-EMD that are responsible for sensing the motion of light edges and dark edges, respectively. These twin EMD's here respond to the preferred direction and are both excitatory to H1. Motion sensing in the nonpreferred direction is similarly assumed to be represented by an ON-EMD and an OFF-EMD (both inhibitory to H1), so that a total of four EMD's are accessed by microstimulation of R1 and R6

The results reported here relate exclusively to the type of EMD activated by a receptor pair R1-R6 in an ommatidium. Although R1 and R6 are not immediate neighbors (Fig. 5a), their target cartridges are (Fig. 5c). This is due to the conspicuous anamorphism of the fly retina-lamina projection (Braitenberg 1970), a direct demonstration of which has recently been provided by the "light-induced dye uptake" technique (Wilcox and Franceschini 1984). Preliminary results obtained by sequentially flashing other pairs of receptors within the same ommatidium suggest that there may exist other EMD's, in particular one driven by two obliquely placed, nearest neighboring receptors such as R2-R4. A similar oblique interaction with small sampling basis was previously inferred from several electrophysiological studies on H1 (Zaagman et al. 1977; Mastebroek et al. 1980; Srinivasan and Dvorak 1980; Lenting 1985), and optomotor studies on *Drosophila* (Buchner 1976) and *Musca* (Pick and Buchner 1979; see, however, Kirschfeld 1972).

The "reductionist" approach that consisted of isolating the single EMD electrical response may now shed a new light upon the "classical" response of H1 to a moving striped pattern, as shown in Fig. 3. Since this moving pattern activates a multitude of spatially distributed EMD's, we may, to a first approximation, interpret the sustained response of H1 (Fig. 3) as resulting from the weighted sum of many time-displaced ON-transients and OFF-transients that are (asynchronously) produced by each EMD, whenever the light edge or the dark edge of a stripe passes across its receptive field. Note, however, that a sequence of light flashes such as those presented in Fig. 8 would be a fair simulation of the actual sequence of excitations on a receptor pair looking at a moving bright stripe only if each single receptor had a needle-shaped visual field in space, instead of its well-known, gaussian-shaped acceptance curve. This in fact brings us to a further interesting aspect of Exner-type sequential flashing of two stationary retinal sites as a means of analysing motion detectors: the spatial low-pass filtering process occurring distally to phototransduction can be bypassed.

Activation of a single EMD can elicit in H1 a response of surprisingly high amplitude (peak discharge rate greater than 100 impulses/s in Fig. 15a, c). This attests to the presence of a dramatic nonlinearity in the "summation process" of EMD inputs to H1 and bears upon the general problem of widefield integration of local motion signals by the lobula plate DS "ganglion cells", studied earlier with moving patterns of variable sizes (rev. Hausen 1981, 1984). Whether this nonlinear summation involves a specific gain control circuit, as inferred by Poggio et al. (1981), or more simply relies on synaptic saturation on the DS cell dendrites (Hengstenberg 1982), is a question that single receptor stimulation might help to answer. Also, the same optical microstimulation technique, once combined with intracellular recordings from H1 or from other DS lobula plate neurons, should reveal details about the antagonistic synaptic drives that are mediated locally by the two feet of the postulated puppet (Fig. 14a).

As recognized early in this field (Hassenstein and Reichardt 1956), a block diagram established on the basis of input-output analyses does not claim to constitute a wiring diagram of the actual neurons. It is rather a topological structure, the main purpose of which is to extract a functional principle from the hardware, e.g., by locating and identifying the nonlinearities within the causal chain and by describing the kinetics of the various parts. This functional description is valuable not only because it raises relevant questions in connection with the neural hardware but also because it is the level par excellence at which a system can be transcribed in terms of another, man-made technology (e.g., Franceschini et al. 1985).

Acknowledgments. This research was supported by the Max Planck Gesellschaft (W-Germany), the Centre National de la Recherche Scientifique, the Ministère de la Recherche et de la Technologie, the Direction des Recherches, Etudes et Techniques, the Fondation pour la Recherche Médicale (France) and the European C.O.D.E.S.T. We are indebted to K. Hausen, S. Picaud, J.M. Pichon and C. Wehrhahn for stimulating comments and we thank Mrs M. Andre, Mrs J. Blanc, Mr R. Chagneux, Mr R. Fayolle, Mr J. Roumieu and Mrs A. Totin for expert technical assistance. A. Riehle was supported by a post-doctoral fellowship from C.N.R.S. (France) and D.F.G. (W-Germany). A. Le Nestour was supported by a predoctoral fellowship from the Fondation de France.

References

Adelson EH, Bergen JR (1985) Spatio-temporal energy models for the perception of motion. J Opt Soc Am 2:284–299

Amthor FR, Oyster CW, Takahashi ES (1984) Morphology of on-off direction-selective ganglion cells in the rabbit retina. Brain Res 298:187–190

Ariel M, Daw N (1982) Pharmacological analysis of directionally sensitive rabbit retinal ganglion cells. J Physiol 324:161–185

Barlow HB, Levick WR (1965) The mechanism of directionally selective units in rabbit's retina. J Physiol 178:477–504

Bishop LG, Keehn DG (1966) Two types of neurons sensitive to motion in the optic lobe of the fly. Nature (London) 212:1374–1376

Bishop LG, Keehn DG, McCann G (1967) Neural correlates of the optomotor response in the fly. Kybernetik 3:288–295

Bishop LG, Keehn DG, McCann GD (1968) Motion detection by interneurons of optic lobes and brain of the flies *Calliphora phaenicia* and *Musca domestica.* J Neurophysiol 31:509–525

Borst A, Egelhaaf M (1987) Temporal modulation of luminance adapts time constant of fly movement detectors. Biol Cybernet 56:209–215

Braitenberg V (1970) Ordnung und Orientierung der Elemente im Sehsystem der Fliege. Kybernetik 7:235–242

Braitenberg V, Strausfeld NJ (1973) Principles of the mosaic organization in the visual system's neuropil of *Musca domestica*. In: Handbook of sensory physiology, vol VII 3A. Jung R (ed) Springer, Berlin Heidelberg, New York, pp 631–659

Buchner E (1976) Elementary movement detectors in an insect visual system. Biol Cybernet 24:85–101

Buchner E (1984) Behavioral analysis of spatial vision in insects. In: Ali MA (ed) Photoreception and vision in invertebrates. Plenum, New York, pp. 561–621

Bülthoff H, Götz KG (1979) Analogous motion illusion in man and fly. Nature 278:636–638

Bülthoff H, Schmid A (1983) Neuropharmakologische Untersuchungen bewegungsempfindlicher Interneurone in der Lobula Platte der Fliege. Verh Dtsch Zool Ges, Gustav Fischer, Stuttgart, p 273

Calvin WH, Graubard K (1979) Styles of neuronal computation. In: Schmitt FO, Worden FG (eds) The neurosciences, 4th study program. MIT, Cambridge, pp 513–524

Collett T, Blest DA (1966) Binocular, directionally-selective neurons, possibly involved in the optomotor response of insects. Nature (London) 212:1330–1333

Collett T, Harkness LIK (1982) Depth vision in animals. In: Ingle DJ, Goodale MA, Mansfield RJW (eds) Analysis of visual behavior. MIT, Cambridge London, pp 111–176

Collett T, King AJ (1975) Vision during flight. In: Horridge GA (ed) The compound eye and vision of insects. Clarendon, Oxford, pp 437–466

de Ruyter van Steveninck RR, Zaagman WH, Mastebroek HAK (1986) Adaptation of transient responses of a movement-sensitive neuron in the visual system of the blowfly *Calliphora erythrocephala*. Biol Cybernet 54:223–236

DeVoe RD (1980) Movement sensitivities of cells in the fly's medulla. J Comp Physiol A 138:93–119

DeVoe RD, Ockleford EM (1976) Intracellular responses from cells of the medulla of the fly, *Calliphora erythrocephala*. Biol Cybernet 23:13–24

DeVoe RD, Kaiser W, Ohm J, Stone LS (1982) Horizontal movement detectors of honeybees: directionally-selective visual neurons in the lobula and brain. J Comp Physiol A 147:155–170

Doorn AJ van, Koenderink JJ (1976) A directionally sensitive network. Biol Cybernet 2:161–170

Dvorak DR, Bishop LG, Eckert HE (1975) On the identification of movement detectors in the optic lobe. J Comp Physiol 100:5–23

Dvorak D, Srinivasan MV, French AS (1980) The contrast sensitivity of fly movement-detecting neurons. Vision Res 20:397–407

Eckert H (1971) Die Spektralempfindlichkeit des Komplexauges von *Musca*. Kybernetik 9:145–156

Eckert H (1980) Functional properties of the H1-neuron in the third optic ganglion of the blowfly, *Phaenicia*. J Comp Physiol A 135:29–39

Eckert H (1982) The vertical-horizontal neuron (VH) in the lobula plate of the blowfly, *Phaenicia*. J Comp Physiol A 149:195–205

Egelhaaf H (1985) On the neuronal basis of figure-ground discrimination by relative motion in the visual system of the fly. II. Figure-detection cells, a new class of visual interneurons. Biol Cybernet 52:195–209

Emerson RC, Gerstein GL (1977) Simple striate neurons in the cat. II. Mechanisms underlying directional asymmetry and directional selectivity. J Neurophysiol 40:136–155

Emerson RC, Citron MC, Vaughn WJ, Klein SA (1987) Nonlinear directionally selective subunits in complex cells of cat striate cortex. J Neurophysiol 58:33–65

Erikson ES (1982) Neural responses to depth-motion stimulation in a horizontally sensitive interneuron in the optic lobe of the blowfly. J Insect Physiol 28: 631–639

Exner S (1875a) Experimentelle Untersuchung der einfachsten psychischen Processe. Pflüger's Arch Physiol 11:403–432

Exner S (1875b) Über das Sehen von Bewegungen und die Theorie des zusammengesetzten Auges. Sitzungsber Akad Wiss Wien Abt III 72:156–190

Exner S (1891) Die Physiologie der facettierten Augen von Krebsen und Insekten. Deuticke, Leipzig

Exner S (1894) Entwurf zu einer physiologischen Erklärung der psychischen Erscheinungen, 1. Teil. Deuticke, Leipzig, pp 37–140

Fennema CL, Thompson WB (1979) Velocity determination in scenes containing several moving objects. Comput Graph Image Proc 9:301–315

Fermi G, Reichardt W (1963) Optomotorische Reaktion der Fliege *Musca domestica*. Kybernetik 2:15–28

Foster DH (1971) A model of the human visual system in its response to certain classes of moving stimuli. Kybernetik 8:69–84

Franceschini N (1972) Pupil and pseudopupil in the compound eye of *Drosophila*. In: Wehner R (ed) Information processing in the visual systems of arthropods. Springer, Berlin Heidelberg New York, pp 75–82

Franceschini N (1975) Sampling of the visual environment by the compound eye of the fly: fundamentals and applications. In: Snyder AW, Menzel R (eds) Photoreceptor optics. Springer, Berlin Heidelberg New York, pp 98–125

Franceschini N (1983) In vivo microspectrofluorimetry of visual pigments. In: Cosens DJ (ed) The biology of photoreceptors. Univ Press, Cambridge, pp 53–85

Franceschini N (1984) Chromatic organization and sexual dimorphism of the fly retinal mosaic. In: Borsellino A, Cervetto L (eds) Photoreceptors. Plenum, New York, pp 319–350

Franceschini N (1985) Early processing of colour and motion in a mosaic visual system. Neurosci Res Suppl 2:17–49

Franceschini N, Blanes C, Oufar L (1985) Appareil de mesure passif et sans contact de la vitesse d'un objet. Patent ANVAR N°51549, Paris

Franceschini N, Riehle A, Le Nestour A (1986) Properties of the integrated circuit mediating directional selectivity in a movement sensitive neuron. Soc Neurosci Abstr 12:859

Gibson JJ (1958) Visually controlled locomotion and visual orientation in animals. Br J Psychol 49:182–194

Götz KG (1964) Optomotorische Untersuchung des visuellen Systems einiger Augenmutanten der Fruchtfliege *Drosophila*. Kybernetik 2:77–92

Götz KG (1965) Behavioral analysis of the visual system of the fruitfly *Drosophila*. Proc Symp Information processing in sight sensory systems, Caltech, Pasadena, pp 85–100

Götz KG (1968) Flight control in *Drosophila* by visual perception of motion. Kybernetik 4:199–208

Götz KG (1972) Principles of optomotor reactions in insects. Bibl Ophthalmol 82:251–259

Grind WA van de, Koenderink JJ, Doorn AJ van (1986) The distribution of human motion detector properties in the monocular visual field. Vis Res 26:797–810

Grzywacz NM, Koch C (1987) Functional properties of models for direction selectivity in the retina. Synapse 1:417–434

Hardie RC (1985) Functional organization of the fly retina. In: Ottoson D (ed) Progress in sensory physiology, vol 5. Springer, Berlin Heidelberg New York Tokyo, pp 1–79

Hardie RC (1986) The photoreceptor array of the dipteran retina. Trends Neurosci 9:419–423

Hassenstein B (1951) Ommatidienraster und afferente Bewegungsintegration. Z Vergl Physiol 33:301–326

Hassenstein B (1958) Über die Wahrnehmung der Bewegung von Figuren und unregelmässigen Helligkeitsmustern. Z Vergl Physiol 40:556–592

Hassenstein B, Reichardt W (1956) Systemtheoretische Analyse der Zeit-, Reihenfolgen- und Vorzeichenauswertung bei der Bewegungsperzeption des Rüsselkäfers *Chlorophanus*. Z Naturforsch 11b:513–524

Hateren JH van (1987) Neural superposition and oscillations in the eye of the blowfly. J Comp Physiol A 161:849–855

Hausen K (1976) Functional characterization and anatomical identification of motion sensitive neurons in the lobula plate of the blowfly *Calliphora erythrocephala*. Z Naturforsch 31c:629–633

Hausen K (1981) Monokulare und binokulare Bewegungsauswertung in der Lobula-Platte der Fliege. Verh Dtsch Zool Ges 74:49–70

Hausen K (1984) The lobula-complex of the fly: structure, function and significance in visual behavior. In: Ali MA (ed) Photoreception and vision in invertebrates. Plenum, New York, pp 523–559

Heisenberg M, Buchner E (1977) The role of retinula cell types in visual behavior of *Drosophila melanogaster*. J Comp Physiol A 117:127–162

Helmholtz H von (1867) Handbuch der physiologischen Optik. Voss, Leipzig

Hengstenberg R (1982) Common visual response properties of giant vertical cells in the lobula plate of the blowfly *Calliphora*. J Comp Physiol A 149:179–193

Hengstenberg R, Hausen K, Hengstenberg B (1982) The number and structure of giant vertical cells (VS) in the lobula plate of the blowfly *Calliphora erythrocephala*. J Comp Physiol A 149:163–177

Hertel H, Maronde U (1987) The physiology and morphology of centrally projecting visual inter-neurons in the honeybee brain. J Exp Biol 133:301–315

Hertel H, Schaefer S, Maronde U (1987) The physiology and morphology of visual commissures in the honeybee brain. J Exp Biol 133:283–300

Hildreth EC, Koch C (1987) The analysis of visual motion: from computational theory to neuronal mechanisms. Annu Rev Neurosci 10:477–533

Horridge GA (1986) A theory of insect vision: velocity parallax. Proc R Soc London Ser B 229:13–27

Horridge GA, Scholes JH, Shaw S, Tunstall J (1965) Extracellular recordings from single neurons in the optic lobe and brain of the locust. In: Treherne JE, Beament JWL (eds) The physiology of the insect central nervous system. Academic Press, New York London, pp 165–202

Hubel D, Wiesel T (1959) Receptive fields of single neurons in the cat's striate cortex. J Physiol 148:574–591

Jensen RJ, DeVoe RD (1983) Comparisons of directionally selective with other ganglion cells of the turtle retina: intracellular recording and staining. J Comp Neurol 217:271–287

Kirschfeld K (1969) Optics of the compound eye. In: Reichardt W (ed) Processing of optical data by organisms and by machines. Academic Press, New York London, pp 144–166

Kirschfeld K (1972) The visual system of *Musca*: studies on optics, structure and function. In: Wehner R (ed) Information processing in the visual system of arthropods. Springer, Berlin Heidelberg New York, pp 61–74

Kirschfeld K (1973) Das neurale Superpositionsauge. Fortschr Zool 21:229–257

Kirschfeld K, Franceschini N (1968) Optische Eigenschaften der Ommatidien im Komplexauge von *Musca*. Kybernetik 5:47–52

Kirschfeld K, Franceschini N (1969) Ein Mechanismus zur Steuerung des Lichtflusses in den Rhab-domeren des Komplexauges von *Musca*. Kybernetik 6:13–22

Kirschfeld K, Lutz B (1974) Lateral inhibition in the compound eye of the fly, *Musca*. Z Naturforsch 29c:95–97

Koch C, Poggio T, Torre V (1983) Nonlinear interaction in a dendritic tree: localization, timing and role in information processing. Proc Natl Acad Sci USA 80:2799–2802

Lenting BP (1985) Functional characteristics of a wide-field movement processing neuron in the blowfly visual system. Thesis, Groningen

Lettvin JY, Maturana HR, McCulloch WS, Pitts WH (1959) What the frog's eye tells the frog's brain. Proc Inst Radio Eng NY 47:1940–1951

Levick WR, Barlow HB, Hill RM (1969) Retinal mechanisms for the perception of movement in rabbits. In: The physiological basis for form discrimination. NIH, pp 97–105

Lillywhite PG, Dvorak DR (1981) Responses to single photons in a fly optomotor neuron. Vision Res 21:279–290

Limb JO, Murphy JA (1975) Estimating the velocity of moving images in television signals. Comp Graph Im Process 4:311–327

Maddess T (1986) Afterimage-like effects in the motion-sensitive neuron H1. Proc R Soc London Ser B 228:433–459

Maddess T, Laughlin SB (1985) Adaptation of the motion-sensitive neuron H1 is generated locally and governed by contrast frequency. Proc R Soc London Ser B 225:251–275

Marchiafava PL (1979) The response of retinal ganglion cells to stationary and moving visual stimuli. Vision Res 19:1203–1211

Marr D (1982) Vision. Freeman, San Francisco

Marr D, Poggio T (1977) From understanding computation to understanding neural circuitry. Neurosci Res Progr Bull 15:470–488

Marr D, Ullman S (1981) Directional selectivity and its use in early visual processing. Proc R Soc London Ser B 211:151–180

Masland RH, Tauchi M (1986) The cholinergic amacrine cell. Trends Neurosci 9:218–223

Mastebroek HAK, Zaagman WH, Lenting BPM (1980) Movement detection: performance of a wide-field element in the visual system of the blowfly. Vision Res 20:467–474

Mastebroek HAK, Zaagman WH, Lenting BPM (1982) Memory-like effects in fly vision: spatio-temporal interactions in a wide-field neuron. Biol Cybernet 43:147–155

McCann GD (1973) The fundamental mechanism of motion detection in the insect visual system. Kybernetik 12:64–73

McCann GD, Arnett DW (1972) Spectral and polarization sensitivity of the dipteran visual system. J Gen Physiol 59:534–558

McCann GD, Dill JC (1969) Fundamental properties of intensity, form and motion perception in the visual nervous system of *Calliphora phaenicia* and *Musca domestica*. J Gen Physiol 53:385–413

McCann GD, Foster SF (1971) Binocular interactions of motion detection fibers in the optic lobes of flies. Kybernetik 8:193–203

McCann GD, MacGinitie GF (1965) Optomotor response studies of insect vision. Proc R Soc London Ser B 163:369–401

Mikami A, Newsome WT, Wurtz RH (1986) Motion selectivity in macaque visual cortex. I. Mechanisms of direction and speed selectivity in extrastriate area MT. J Neurophysiol 55:1308–1327

Miller RF (1979) The neuronal basis of ganglion cell receptive field organization and the physiology of amacrine cells. In: Schmitt FO and Worden FG (eds) The Neurosciences, 4 study program. MIT, Cambridge, pp 247–245

Mimura K (1971) Movement discrimination by the visual system of flies. Z Vergl Physiol 73:105–138

Mimura K (1972) Neural mechanisms, subserving directional selectivity of movement in the optic lobe of the fly. J Comp Physiol 80:409–437

Montero VM, Brugge JF (1969) Direction of movement as the significant stimulus parameter for some lateral geniculate cells in the rat. Vision Res 9:71–88

Movshon JA, Thompson ID, Tolhurst DJ (1978) Spatial summation in the receptive fields of simple cells in the cat's striate cortex. J Physiol 283:53–77

Nakayama K (1985) Biological image motion processing: a review. Vision Res 25:625–660

Nicoll RA (1982) Transmitters can say more than just "yes" or "no". Trends Neurosci 5:369–374

O'Shea M, Rowell CHF (1977) Complex neural integration and identified interneurons in the locust brain. In: Hoyle G (ed) Identified neurons and behavior of arthropods. Plenum, New York, pp 307–328

Osorio D (1986) Directionally selective cells in the locust medulla. J Comp Physiol A 159:841–847

Pick B, Buchner E (1979) Visual movement detection under light- and dark-adaptation in the fly, *Musca domestica*. J Comp Physiol 134:45–54

Pierantoni R (1976) A look into the cock-pit of the fly. The architecture of the lobula plate. Cell Tissue Res 171:101–122

Poggio T, Reichardt W (1976) Visual control of orientation behavior in the fly. Part II. Towards the underlying neural interactions. Q Rev Biophys 9:377–438

Poggio T, Reichardt W, Hausen K (1981) A neuronal circuitry for relative movement discrimination by the visual system of the fly. Naturwissenschaften 68:443–446

Reichardt W (1957) Autokorrelations-Auswertung als Funktionsprinzip des Zentralnervensystems. Z Naturforsch 12b:448–457

Reichardt W (1961) Autocorrelation: a principle for evaluation of sensory information by the central nervous system. In: Rosenblith WA (ed) Principles of sensory communications. John Wiley & Sons, New York, pp 303–317

Reichardt W (1969) Movement perception in insects. In: Reichardt W (ed) Processing of optical data by organisms and machines. Academic Press, New York London, pp 465–493

Reichardt W (1987) Evaluation of optical motion information by movement detectors. J Comp Physiol A 161:533–547

Reichardt W, Poggio T (1976) Visual control of orientation behavior in the fly. Pt 1. A quantitative analysis. Q Rev Biophys 9:311–375

Riehle A, Franceschini N (1982) Response of a directionally-selective, movement detecting neuron under precise stimulation of two identified photoreceptors cells. Neurosci Lett Suppl 10:S411–5412

Riehle A, Franceschini N (1984) Motion detection in flies: parametric control over ON-OFF pathways. Exp Brain Res 54:390–394

Rind FC (1983) A directionally sensitive motion detecting neuron in the brain of a moth. J Exp Biol 102:253–271

Rind FC (1987) Non-directional movement sensitive neuron of the locust optic lobe. J Comp Physiol A 161:477–494

Santen JPH van, Sperling G (1984) Temporal covariance model of human motion perception J Optic Soc Am A 1:451–473

Santen JPH van, Sperling G (1985) Elaborated Reichardt detectors J Opt Soc Am A 2:300–321

Schouten JF (1967) Subjective stroboscopy and a model of visual movement detectors. In: Wathen-Dunn W (ed) Models for the perception of speech and visual form. MIT, Cambridge, pp 44–55

Schuling FH, Mastebroek HAK (1984) Modeling of adaptive image processing strategies in fly vision. Proc Int AMSE Conf Modelling and simulation, Athens, June 27–29, vol 4.2, pp 117–140

Shaw S (1984) Early visual processing in insects. J Exp Biol 112:225–251

Solodovnikov VV (1965) Dynamique statistique des systèmes linéaires de commande automatique. Dunod, Paris

Srinivasan MV (1983) The impulse response of a movement detecting neuron and its interpretation. Vision Res 23:659–663

Srinivasan MV, Dvorak DR (1980) Spatial processing of visual information in the movement-detecting pathway of the fly. J Comp Physiol 140:1–23

Stavenga DG (1975) Optical qualities of the fly eye – an approach from the side of geometrical, physical and waveguide optics. In: Snyder AW, Menzel R (eds) Photoreceptor optics. Springer, Berlin Heidelberg New York, pp 126–144

Strausfeld NJ (1976) Atlas of an insect brain. Springer, Berlin Heidelberg New York

Strausfeld NJ (1984) Functional neuroanatomy of the blowfly visual system. In: Ali MA (ed) Photoreception and vision in invertebrates. Plenum, New York, pp 481–522

Strausfeld NJ, Hausen K (1977) The resolution of neuronal assemblies after cobalt injection into neuropil. Proc R Soc London Ser B 199:463–476

Strausfeld NJ, Nässel DR (1980) Neuroarchitecture of brain regions that subserve the compound eye of Crustacea and insects. In: Autrum H (ed) Handbook of sensory physiology, vol VII/6B. Springer, Berlin Heidelberg New York, pp 1–133

Thorson J (1966) Small signal analysis of a visual reflex in the locust. II. Frequency dependence. Kybernetik 3:53–66

Torre V, Poggio T (1978) A synaptic mechanism possibly underlying directional selectivity to motion. Proc R Soc London Ser B 202:409–416

Trujillo-Cenóz O (1972) The structural organization of the compound eye in insects. In: Fuortes MGF (ed) Handbook of sensory physiology, vol VII/2. Springer, Berlin Heidelberg New York, pp 5–62

Ullman S (1981) Analysis of visual motion by biological and computer systems. Computer 14:57–69

Ullman S (1986) Artificial intelligence and the brain: computational studies of the visual system. Annu Rev Neurosci 9:1–26

Umeda K, Tateda H (1985) Visual interneurons in the lobula complex of the fleshfly, *Boettcherisca peregrina*. J Comp Physiol A 157:831–836

Varju D (1977) Systemtheorie für Biologen und Mediziner. Springer, Berlin Heidelberg New York

Vitanova L, Glezer V, Gauselman V (1985) On the mechanisms underlying appearance of responses to movement, directional selectivity and velocity tuning of the cat's striate cortical neurons. Biol Cybernet 52:237–246

Watanabe S, Murakami M (1984) Synaptic mechanisms of directional selectivity in ganglion cells of frog retina as revealed by intracellular recordings. Jpn J Physiol 34:497–511

Waterman TH, Wiersma G, Bush BMH (1964) Afferent visual responses in the optic nerve of the crab *Podophtalmus*. J Cell Comp Physiol 63:135–155

Wehner R (1981) Spatial vision in arthropods. In: Autrum H (ed) Handbook of sensory physiology, vol VII/6C. Springer, Berlin Heidelberg New York, pp 288–616

Wehrhahn C (1984) Visual guidance of flies during flight. In: Kerkut GA, Gilbert LI (eds) Comprehensive insect physiology, biochemistry and pharmacology, vol 6. Pergamon, Oxford pp 673–684

Wertheimer M (1912) Experimentelle Studien über das Sehen von Bewegung. Z Psychol 61:161–265

Wilcox M, Franceschini N (1984) Illumination induces dye incorporation in photoreceptor cells. Science 225:851–854

Wilson HR (1985) A model for direction selectivity in threshold motion perception. Biol Cybernet 51:213–222

Zaagman WH, Mastebroek HAK, Buyse T, Kuiper JW (1977) Receptive field characteristics of a directionally selective movement detector in the visual system of the blowfly. J Comp Physiol 116:39–50

Zaagman WH, Mastebroek HAK, Kuiper JW (1978) On the correlation model: performance of a movement detecting neural element in the fly visual system. Biol Cybernet 31:163–168

Zaagman WH, Mastebroek HAK, de Ruyter van Steveninck R (1983) Adaptative strategies in fly vision: on their image-processing qualities. IEEE Trans Syst Manag Cybernet SMC 13:900–906

Chapter 18

Neural Mechanisms of Visual Course Control in Insects

KLAUS HAUSEN and MARTIN EGELHAAF, Tübingen, FRG

1 Introduction

Visual orientation and course-stabilization of flying insects rely essentially on the evaluation of the retinal motion patterns perceived by the animals during flight. Apparent motions of the entire surrounding indicate the direction and speed of self-motion in space and are used as visual feedback signals during optomotor course-control manoeuvres. Discontinuities in the motion pattern and relative motions between pattern-segments indicate the existence of stationary or moving objects and represent the basic visual cues for flight-orientation during fixation- and tracking-sequences, and possibly also for the avoidance of obstacles, and the selection of landing sites.

Neurophysiological studies, carried out in the last three decades, have revealed a large variety of functionally different motion-sensitive interneurons in the visual systems of insects (for review see Wehner 1981). The problems tackled in these studies range from the uptake and peripheral preprocessing of visual signals, and the cellular mechanism of the elementary motion detector, to the subsequent processing of motion information in higher neuronal circuits, and the functional relevance of these circuits in flight-control. It is evident that the latter questions, in particular, can only be adequately studied in conjunction with behavioral analyses.

The visual system of the fly has proven to be particularly well suited for investigations on these topics. Its compound eyes, optic lobes and visuo-motor pathways have been the subject of numerous optical, physiological, and ana- tomical studies (reviews: Strausfeld 1976; Hausen 1977; Heide 1983; Laughlin 1984; De Voe 1985; Franceschini 1985; Hardie 1985; Järvilehto 1985) and it may not be exaggerating to claim that it is the most thoroughly investigated insect visual system at present. Furthermore, visually guided behavior has been extensively studied in various species of flies under both natural and laboratory conditions (reviews: Reichardt and Poggio 1976; Land 1977; Götz 1983; Buchner 1984; Heisenberg and Wolf 1984; Wehrhahn 1985; Reichardt 1986).

The present account, which reviews neuronal mechanisms of optomotor course-control and the detection and fixation of objects, will therefore be mainly concerned with studies in flies. We will first discuss the results of recent behavioral investigations which have been of major relevance for the electrophysiological analysis of motion-computation in the fly visual system. The general architecture of visuo-motor pathways, and the anatomy, physiology and functional role of identified interneurons in the control of visual behavior are topics of the following sections. Related studies on other insect species will be considered briefly at the end of the paper.

2 Motion-Induced Behavior in Flies

2.1 Visual Course-Stabilization

Flies stabilize their flight direction by following, and hence reducing the global movements of the retinal image induced by involuntary deviations from their flight course. For example, rotatory retinal motions, which indicate rotations of the body with respect to the environment, elicit syndirectional torque responses around the three body axes. Translatory motions in vertical and horizontal directions, which indicate deviations from the flight altitude and changes of the flight velocity, induce corrective modulations of the translational flight forces (lift and thrust).

The motor activity of the flying animal during these maneuvers is complex and far from being fully understood. The generation of yaw torque, and of thrust and lift, results mainly from differential and covariant changes of the beat amplitudes of both wings, respectively (Götz 1968; Götz et al. 1979; Zanker 1987). Differential changes of wing pitch on both sides may be involved at least in the generation of roll responses (Hengstenberg et al. 1986). In addition, changes of the plane of the wing-stroke and active rudder movements of hindlegs and abdomen play a prominent role during steering maneuvres (Götz 1968; Götz et al. 1979; Zanker 1987).

The neural computations underlying these complex motor responses have been extensively investigated in the last decades. One of the first, and most important, results of these studies was the finding that the motion detectors involved in the control of optomotor responses are of the so-called correlation type (Hassenstein and Reichardt 1956; Reichardt 1957, 1961, 1987). A motion detector of this kind basically performs a multiplication of input signals received by two retinal sampling stations (Fig. 1a). Prior to multiplication, one of the signals is delayed by low-pass filtering, whereas the other remains unaltered. Due to these operations, the circuit responds preferentially to visual stimuli moving in one particular direction. By connecting two of them with opposite directional selectivities as excitatory and inhibitory elements to an integrating output stage, one gains a bidirectional elementary movement detector (EMD) having a preferred and a null direction.

The preferred direction of an EMD depends on the spatial arrangement of the sampling stations in the compound eye. In the visual system of the fly, individual EMD's receive input mainly from contiguous points of the hexagonal ommatidial lattice (Fig. 1a). This sample scheme holds true at least for *Drosophila* (Götz 1968, 1983; Buchner 1976; Götz et al. 1979), and there is some evidence for a similar organization in the larger flies *Musca* and *Calliphora* (Zaagman et al. 1977). The situation seems, however, somewhat more complex in the latter species, since

EMD's having enlarged and horizontally or vertically aligned sampling bases have been also demonstrated (Kirschfeld 1972; Riehle and Franceschini 1984).

Apart from its directional selectivity, an EMD of the correlation type is characterized by the fact that its average output depends on the contrast frequency of a motion stimulus (i.e., the ratio of angular velocity and wavelength of its spatial Fourier components) rather than its actual velocity. Behavioral studies have demonstrated that the amplitudes of the optomotor responses controlled by these motion detectors are indeed strongly dependent on the contrast frequency of motion stimuli (for a detailed discussion of this point see Buchner 1984; see also Reichardt and Guo 1986).

Two further functional aspects of the detector shall be briefly mentioned. The first one regards the time constant of the filter in the input channel, which was recently found to depend on the actual stimulus conditions. It has been shown that prolonged stimulations with high contrast frequencies can shorten the time constant considerably (Maddes and Laughlin 1985; de Ruyter van Steveninck et al. 1986; Borst and Egelhaaf 1987). The result of this adaptive process is an effective increase of the temporal operating range of the motion detector, which is reflected in the broad peak of the contrast frequency function of the optomotor system (1-10Hz) as found in behavioral investigations (Götz 1964; McCann and Mac-Ginitie 1965; Eckert 1973; Wehrhahn 1986; Borst and Bahde 1987; Hausen and Wehrhahn 1987a), and in electrophysiological studies (Mastebroek et al. 1980; Eckert 1980; Hausen 1981, 1982b; Eckert and Hamdorf 1981; Maddess and Laughlin 1985).

The second aspect concerns the role of spatial integration processes in motion computation. Studies on the dynamic properties of motion detectors have revealed that their transient responses to motion are strongly influenced by the spatial structure of the stimulus (Reichardt and Guo 1986; Egelhaaf and Reichardt 1987). For particular classes of motion stimuli, this may even lead to an inversion of the detector-response and, thus, to incorrect signals with respect to the actual direction of motion. Ambiguities of this kind are certainly fatal for a flight-control system relying essentially on motion information and can be avoided by integrating the outputs of extended arrays of EMD's. As will be discussed below, such spatial integration does indeed play an important role in the motion-detection circuits of the fly.

Whereas the above studies revealed basic functional properties of the EMD, further series of behavioral investigations elucidated the spatial organization of the EMD networks in the visual system and their functional connections to the motor system. Under natural conditions, optomotor responses are usually induced by binocular motions of the environment. Laboratory experiments have demonstrated, however, that monocular stimulation of arbitrary small areas of the retina, and even stimulation of individual EMD's is sufficient to elicit measurable motor responses (Götz 1964; Kirschfeld 1972). This shows that the entire retina is subserved by dense networks of EMD's, each ommatidium representing an input channel to the various directional types of EMD's described above.

Optomotor yaw torque responses are selectively induced by horizontal motions (Fermi and Reichardt 1963; Götz 1968; Wehrhahn 1986; Hausen and Wehrhahn 1987a) and, hence, must be controlled by EMD's having sampling bases

parallel to the horizontal axis of the eye lattice (h, Fig. 1a), or by pairs of EMD's
having symmetrically arranged sampling bases with respect to this axis (x and y,
Fig. 1a), the outputs of which are added. Monocular stimulation with progressive
(front to back) motion leads to a simultaneous decrease of the ipsilateral and an
increase of the contralateral wing beat amplitude and thus to generation of yaw
torque turning the animal in the same direction as the perceived motion. Regressive
(back to front) motion is less effective but leads also to syndirectional torque

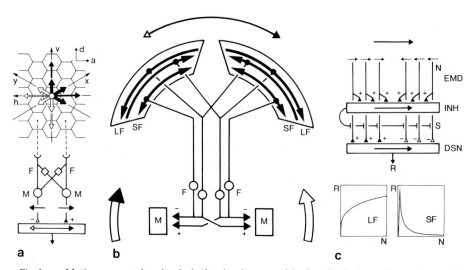

a   b   c

Fig. 1.a-c. Motion computation circuits in the visual system of the fly. *a* Retinal sampling stations and
organization of elementary motion detectors (EMD's). The wiring diagram in the lower part of the
figure shows a bidirectional EMD consisting of two subunits. Each subunit has two input channels
derived from the retina, one containing a low-pass filter (*F*), and a multiplication stage (*M*); it responds
preferentially to motion across the retinal sampling stations in the direction indicated by an *arrow* in the
output channel. The bidirectional EMD has a preferred and null direction as indicated by the *black and
white arrowhead* in the summation stage at the bottom of the circuit. The *upper graph* shows the sampling
stations of presently known EMD's in the ommatidial lattice of the compound eye (*thick black and white
arrows*). h, v, x, y: axes of the hexagonal ommatidial lattice; d,a dorsal, anterior. *b* Model of neural
circuits controlling motion-induced yaw torque responses. Horizontal motions activate two motion
sensitive systems behind each eye, which integrate the outputs of arrays of EMD's, and which are
specifically tuned to large-field motion (*LF*-system) and small-field or object motion (*SF*-system). The
LF-system is activated by ipsilateral front-to-back motion and contralateral back-to-front motion, and
induces syndirectional yaw torque responses of the fly by simultaneous excitation and inhibition of the
contralateral and ipsilateral flight motor (*M*), respectively. The SF-system is activated by ipsilateral
horizontal motion of small objects in both directions and induces turnings towards the stimulus. The
output channels of the LF- and SF-system contain different frequency-filters (*F*). The SF- and
LF-system dominate in torque-control under stimulation with high-frequency and low-frequency
oscillatory motion, respectively. *c* Model of gain-control mechanism underlying the spatial tuning of the
LF- and SF-system shown in *b*. The outputs of excitatory and inhibitory arrays of EMD's sensitive to
front-to-back motion and back-to-front motion are integrated by a direction selective element (*DSN*),
which represents an output element of the LF- or SF-system. The EMD-outputs are also integrated by
an inhibitor (*INH*) which shunts the individual EMD-channels (*S*) peripheral to the DSN. Depending
on the transfer characteristics of the output synapses of the EMD's, the output (*R*) of the DSN increases
with the number (*N*) of EMD's activated by a motion-stimulus (*LF*-tuning, *left curve*), or reaches a
maximum when only few EMD's are excited and declines subsequently (*SF*-turning, *right curve*). (*a*
After Reichardt 1961; Kirschfeld 1972; Buchner 1976; Götz et al. 1979; *b,c* after Reichardt et al. 1983;
Buchner 1984; Egelhaaf 1985c)

responses. We must hence assume functional connections between the networks of elementary motion detectors and the flight motor of the fly as sketched in Fig. 1b, which are termed large-field system (LF). Corresponding systems, which shall not be discussed in detail, control roll and pitch responses (Götz 1968; Wehrhahn and Reichardt 1975; Srinivasan 1977; Götz and Buchner 1978; Buchner et al. 1978; Wehrhahn 1978; Götz et al. 1979; Blondeau and Heisenberg 1982; Zanker 1987).

2.2 Discrimination and Fixation of Objects

Freely flying flies frequently display a pursuit behavior, in which two flies follow each other (Land and Collett 1974; Wehrhahn 1979; Wagner 1986a). This demonstrates that the animals are able to detect, fixate, and track a moving object in a structured environment even when the retinal image of the environment is also moving. Behavioral and electrophysiological studies have revealed that the evaluation of relative motion between object and background plays a key role in the neural control of this behavior, and that the cellular circuits performing this computation are closely interrelated with those controlling optomotor course stabilization.

In the early experimental investigations of fixation behavior (Reichardt 1973; Reichardt and Poggio 1976; Heisenberg and Wolf 1984), tethered flying flies were positioned in the centre of a bright panorama and were stimulated with a single dark stripe. The stripe was oscillated horizontally around arbitrary positions and the yaw torque responses of the flies were recorded. The measurements revealed that the average torque responses were consistently directed towards the moving stimulus, indicating that the flies attempted to fixate it. The amplitudes of these responses were found to be strongly dependent on the stimulus position, being largest for frontal stimuli.

The mechanism underlying fixation behavior was originally thought to rely on the fact that yaw torque responses induced by progressive motion are not only of opposite polarity but also larger than those induced by regressive motion (see above). An oscillating stimulus will consequently elicit a net response towards the target and thus lead to the observed fixation behavior (Reichardt 1973; Wehrhahn and Hausen 1980). In an alternative hypothesis, it was proposed that fixation responses are controlled by flicker detectors in the visual system rather than by motion detectors (Pick 1974).

In subsequent studies (Virsik and Reichardt 1976; Reichardt and Poggio 1979; Poggio et al. 1981; Reichardt et al. 1983; Egelhaaf 1985a,b,c), the yaw torque responses of tethered flies were measured under stimulation-conditions more closely resembling the complex retinal motion pattern encountered by freely moving animals. In these experiments the flies were stimulated by a textured stripe (figure or object) moving in front of a textured moving panorama (ground). Figure and ground had the same texture and were oscillated horizontally with the same frequency and amplitude, but with a certain phase difference. Hence, the only visual cue for the discrimination of the figure was the relative motion between both patterns. The experiments clearly demonstrated that flies are able to detect and fixate a figure under these conditions, and that, hence, motion detection and not flicker detection is the essential process underlying this behavior. However, the hypothesis that fixation is simply caused by the response asymmetry for progressive and regressive motion turned out to be insufficient to explain the dynamic properties of the torque responses recorded in these experiments.

The analysis of these transient yaw torque components and related electro-physiological studies of the motion computation circuits in the optic lobes of the fly revealed that optomotor course stabilization and fixation behavior are jointly governed by two control systems (Reichardt et al. 1983; Egelhaaf 1985a,b,c, 1987). The first one is the aforementioned LF system, which is sensitive to motion of the entire environment, and which induces syndirectional torque responses. The second system, which will be called small-field (SF) system (Fig. 1b), evaluates horizontal motion of small objects. It induces turning responses towards an object irrespective of its actual direction of motion and is inhibited, when the size of the object increases, or when the background moves in the same direction as the object. Due to these characteristics, the relative influence of both systems on torque generation depends critically on the actual stimulus condition: e.g., oscillatory motions of the entire surrounding, which greatly inactivate the SF-system elicit syndirectional optomotor following responses, whereas oscillatory object-motions activate predominantly the SF-system and hence will lead to torque oscillations with a strong component directed towards the target. The contributions of both control systems to the final motor output vary with the oscillation frequency, the LF-system dominating at low and the SF-system at high frequencies. It is assumed that these temporal characteristics are due to different frequency filters in the output channels of the two control systems (Egelhaaf 1987).

The different spatial sensitivities of the LF- and SF-system can both be modelled by the circuitry shown in Fig. 1c (Poggio et al. 1981, Reichardt et al. 1983; Egelhaaf 1985a,c). It consists of two arrays of EMD's sensitive to regressive and progressive motion, an integrative inhibitory element (INH) and an integrative directionally selective output element (DSN). The EMD's represent the entire arrays of motion detectors sensitive to horizontal motion behind one eye and are for clarity separated into two groups according to their directional selectivity. The output element DSN may be regarded as one of the output elements of the LF-system shown in Fig. 1b, or as one of several directionally selective output elements of the SF-system having opposite preferred directions. The latter are shown as a single bidirectional element in Fig. 1b, since they induce turning tendencies into the same direction. Under stimulation with motion, the inhibitory element integrates the signals of the EMD's and inhibits them individually in a shunting-operation (S). The strength of this inhibition depends thus on the number (N) of EMD's stimulated and hence on the size of the stimulus. The output element subsequently integrates the reduced signals of the EMD's. Depending on the transfer characteristics of the output synapses of the EMD's, stimulation with a motion stimulus of increasing size will either lead to an increasing signal amplitude in the output element, as is characteristic for the LF-system, or to a peak response at small stimuli and a subsequent response decay, as is found in the SF-system. Hence, the principle difference between both systems lies in the synaptic efficiencies of the EMD-terminals on the integrating output elements. Further differences concern the actual neuronal realization of these circuits in the optic lobe of the fly, which will be discussed below. The computations performed by this inhibitory network have been mathematically formulated and investigated in computer simulations (Reichardt et al. 1983). The graphs in Fig. 1c show the simulated responses of an LF-and an SF-output element as function of the number of stimulated channels and illustrate the spatial integration properties of the two systems.

The interplay of the two systems in flight control is illustrated in Fig. 2a which shows the yaw torque response of a fly to stimulation with motion of a textured figure and background. As indicated in the stimulus trace at the bottom, the

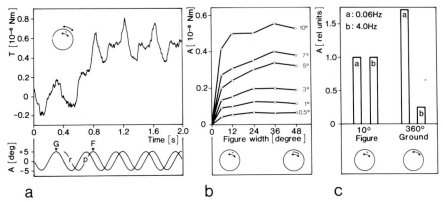

a   b   c

Fig. 2.a-c. Yaw-torque responses of a tethered flying housefly (*Musca*) to horizontal motion. *a* Averaged transient responses T recorded under stimulation with horizontal motion of a narrow stripe (figure *F*) in front of a moving panorama (ground *G*). The stimulus situation is shown in the inset, and the motion of figure and ground are indicated at the bottom of the figure. The figure is positioned in the equatorial, frontolateral visual field of the right eye and has an angular width of 12°. Progressive and regressive motion of the figure are marked with *p* and *r*. Figure and ground have identical textures (random dot patterns) and are oscillated with the same amplitude (10°) and frequency (2.5 Hz); both are first moved in synchrony and later with a phase difference of 90°. Synchronous motion induces zero-symmetric optomotor yaw torque fluctuations, which are syndirectional to the stimulus motion. Relative motion between figure and ground elicits strong torque responses towards the figure, indicating that the fly is attempting to fixate it. The complicated transient response pattern reflects the influence of two torque control systems on the motor output: a large-field (LF) system, which is tuned to large motion-stimuli, and a small-field (SF) system, which responds selectively to small motion stimuli (see text). *b* Response dependency on the size of the stimulus. The curves show the amplitudes *A* of yaw torque responses of the fly as function of the width of a figure, oscillating horizontally with a frequency of 2.5 Hz within a stationary panorama (see *insets*). Parameter in the experiment is the oscillation amplitude of the figure. Each value plotted is derived from 100 measurements. Under all conditions the response amplitudes increase first steeply and reach later a plateau. This is compatible with the hypothesis that torque generation is jointly controlled by the SF- and the LF-system. *c* Response dependency on the stimulus size under different oscillation frequencies. The histogram shows the amplitudes A of yaw-torque responses to oscillations of a small figure (width: 10°) and the entire ground (360°). Responses to both stimuli were measured at two oscillation frequencies (*a* 0.06 Hz; *b* 4 Hz). The values plotted were obtained from 32 (0.06 Hz) and 52 (4 Hz) flies; they were normalized using the response to figure motion under each condition as reference. At low oscillation frequencies the responses to motion of the ground are large as compared to those induced by the figure. The opposite situation is found under at high frequencies. This shows that the influence of the LF-system, dominating in torque control under stimulation with large patterns, decreases with increasing oscillation frequencies, and indicates different frequency-filtering in the output channel of the LF- and SF-system (*a,b* From Reichardt et al. 1983; *c* from Egelhaaf 1987)

experiment starts with synchronous oscillations of figure and ground. The fly shows a normal corrective optomotor response: clockwise motion (from $A = -5°$ to $A = +5°$) induces a positive torque response which indicates intended clockwise turning of the fly, whereas counterclockwise motion leads to a negative response. Obviously, the figure is not detected under this condition. When the figure starts moving relative to the background, the response of the fly changes considerably. It shifts to positive values and shows conspicuous peaks. The positive response level indicates that the figure is now detected and that the fly attempts to fixate it. The ongoing torque fluctuations reflect the activity of both control systems. Large response peaks occur, when the figure moves with maximal speed in progressive direction, while the ground velocity is transiently zero. In these instants, the SF-system is maximally stimulated by progressive motion of the figure and is simultaneously released from the inhibition caused by background-motion.

Figure 2b shows the dependency of the torque amplitudes induced by oscillatory figure motion as a function of the size of the figure, while the ground is held stationary (see Reichardt et al. 1983). The oscillation frequency is chosen such that the LF- and SF-system are about equally effective in torque control. Parameter in this experiment is the oscillation amplitude of the figure. In all cases the torque responses increase steeply initially and remain then rather independent of the stimulus size. This is in accordance with the concept that the response consists of an increasing LF-controlled component and a simultaneously decreasing SF-component (compare with Fig. 1c).

The relative influence of both systems under different oscillation frequencies is illustrated in Fig. 2c, which shows the responses to oscillation of the figure and the ground, measured at a very low and a high oscillation frequency. It can be seen that in agreement with the proposed filter characteristics in the output channels of the LF- and the SF-system, the responses to ground motion become small compared to the figure-induced responses at high oscillation frequencies and large at low frequencies (Egelhaaf 1987).

Thus, the behavioral data compiled in Fig. 2 can at least qualitatively be interpreted in terms of the two control systems proposed in the model. More convincing are simulations of the model output which show a high similarity to the behavioral data. We will consider these simulations later and discuss first the basic neural architecture of the visual system of the fly and the neuronal motion computation circuits in the optic lobe.

3 Visuo-Motor Pathways in the Nervous System of the Fly

The visual system of the fly consists of the ocelli, the compound eyes, the optic lobes, and the visual tracts and projection centres in the brain (Strausfeld 1976). Each compound eye scans about one hemisphere of the environment and samples the light-intensity distribution in the optical axes of the ommatidia. The signals of the compound eye are conveyed by receptor axons into the optic lobe, which consists of successive visual neuropils (lamina, medulla, lobula, lobula plate) and two chiasms. The basic blueprint of all visual neuropils is similar: each is composed of retinotopically arranged columns and superimposed layers. Columns are built up by parallel centripetal and centrifugal small-field neurons (columnar cells), the axons of which project through the chiasms and establish the retinotopic connections between the neuropils. The columnar cells mediate, in addition, local interactions between adjacent columns. In contrast, layers are dense synaptic regions within the neuropils, which are oriented orthogonally to the columns, and which contain the arborizations of large interneurons (tangential and amacrine cells). The latter elements are the structural substrate of long-range interactions and spatial integration processes within the columnar array. Output connections from the optic lobe to the brain are established by both columnar and tangential cells (Fig. 3). Columnar cells leave the highest-order visual neuropils lobula and lobula plate as dense bundles and terminate in visual centers of the ventrolateral brain termed optic foci. Tangential output neurons of the optic lobe originate from all neuropils except the lamina. Some of them connect both optic lobes, others project also into the optic foci.

The optic foci can be regarded as major sensory integration areas for the motor control in the brain. Apart from the optic lobes, they receive visual input from the ocelli, and mechanosensory inputs from antennae and halteres (Strausfeld and Bacon 1983; Strausfeld et al. 1984). The output elements of these areas are descending neurons, which pass through the cervical connective and terminate in the motor neuropils of the thoracic ganglion. Intensive path-tracing studies employing transsynaptic cobalt stainings have revealed evidence for connections between output neurons of the optic lobe, descending neurons and motor neurons of the neck muscles (Strausfeld and Seyan 1985; Strausfeld et al. 1987; Milde et al. 1987), the leg musculature (Strausfeld and Bassemir 1985a), and the indirect flight muscles, which play a major role in torque generation during steering maneuvres (Hausen and Hengstenberg 1987).

In short, the major visuo-motor pathways between compound eye and motor system consists of a sequence of retinotopic visual neuropils, the output elements of which converge together with other sensory tracts onto descending neurons. The latter represent the bottle neck of the pathway and in the thoracic ganglion establish divergent connections to the motoneurons of various groups of muscles. The direct neuronal chain between eye and, e.g., the flight muscles consists of about six to seven cells.

The main motion computation centre in the whole pathway is the lobula plate, the structural organization of which shall be described in some detail.

4 Cellular Architecture of the Lobula Plate

The lobula plate receives input from columnar cells derived from the medulla and lobula. The retinotopic order of these elements is such that the lateral and medial columns of the neuropil subserve the frontal and caudal parts of the ipsilateral visual field, whereas the dorsal and ventral columns subserve the dorsal and ventral regions, respectively. The complicated network of columnar inputs to the lobula plate is not yet fully analyzed. Best investigated are three classes of neurons, namely

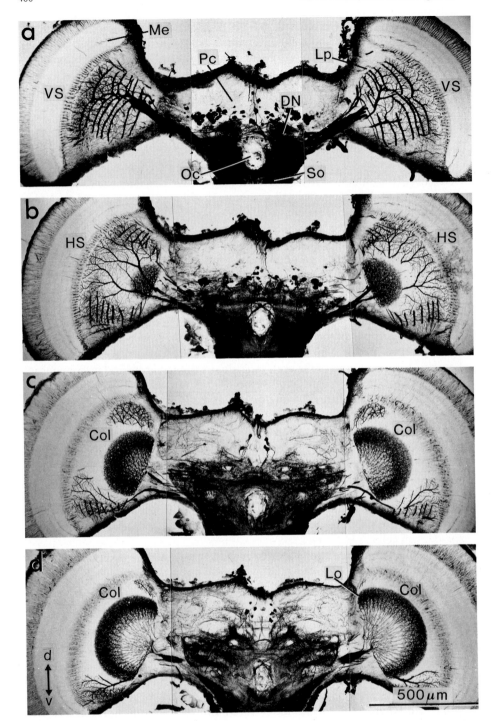

the Y-cells, which are putative GABA-ergic elements (Meyer et al. 1986) originating from the medulla and terminating in both lobula and lobula plate; the T4-cells, which occur as twin elements in each medulla column and terminate at two different levels in the lobula plate; and the T5-cells of the lobula, which are also columnar twin elements showing a similar terminal organization in the lobula plate. Since the terminal of each T4- and T5-cell is bilayered, one finds four discrete input layers in the lobula plate (Strausfeld 1984).

The large interneurons of the lobula plate have been the subject of numerous anatomical studies (Dvorak et al. 1975; Hausen 1976, 1981, 1982a,b, 1984, 1987; Pierantoni 1976; Eckert and Bishop 1978; Hausen et al. 1980; Eckert 1981; Bishop and Bishop 1981; Hengstenberg 1982; Hengstenberg et al. 1982; Egelhaaf 1985b; Strausfeld and Bassemir 1985a,b). A recent account on the structural organization of this network lists about 50 identified neurons (Hausen 1987). The individual cells can be classified anatomically as anaxonal amacrine cells, and as tangential cells having a separate input and output region connected by an axon. The latter group of cells can be further subdivided into centripetal output cells of the lobula plate projecting into the brain, centrifugal feedback cells terminating in the lobula plate and heterolateral connection elements between the lobula plate and neuropils of the contralateral optic lobe. Examples of these cell types are compiled in Fig. 4.

Typical output elements of the lobula plate are two classes of giant neurons termed the horizontal system and the vertical system (HS, VS; see also Fig. 3). The horizontal system consists of three elements, the dendritic arbors of which are located near the anterior surface of the neuropil and cover the dorsal, medial and ventral region of the neuropil. According to these dendritic locations the three cells are termed north, equatorial, and south horizontal cell (HSN, HSE, HSS). The axons of the HS-cells project into the ipsilateral ventrolateral brain. Dye or cobalt injections into HS-cells lead to transneuronal lalls lead to transneuronal labeling of descending neurons, indicating synaptic couplings between both types of cells (Hausen 1987). Two of these descending neurons are shown together with the reconstruction of the horizontal system in Fig. 4. Further synaptic contacts of horizontal cells to a bundle of descending elements have recently been demonstrated (Strausfeld and Bassemir 1985b).

The vertical cells are a class of 11 output neurons (VS1–11) which lie serially arranged at the posterior surface of the lobula plate and which have narrow vertical dendritic domains. The axons of the VS-cells project centrally and terminate near the oesophageal channel at the posterior surface of the brain. Like those of the HS-cells, the axon-terminals of the VS-cells show numerous presynaptic

◄ ───

Fig. 3. a-d. Cobalt-stained output neurons in the lobula complex and descending neurons of the brain in the blowfly (*Calliphora*). The figure shows serial sections (30 μm thick) in a sequence from posterior (*a*) to anterior (*d*) through the brain and the proximal parts of the optic lobes. Giant output elements of the optic lobe are the cells of the vertical system (*VS*) and the horizontal system (*HS*), showing large dendrites at the posterior and anterior surface of the lobula plate, respectively. Columnar output elements of the optic lobe are the Col A-cells (*Col*) of the lobula. The axons of the cells project into the ventrolateral brain, which is heavily invested by dendrites of descending neurons (*DN*). The latter project through the cervical connective into the motor centres of the thoracic ganglion. *Me* medulla; *Lo* lobula; *Lp* lobula plate; *Pc* protocerebrum; *So* suboesophageal ganglion; *Oc* oesophageal channel; *d,v* dorsal, ventral

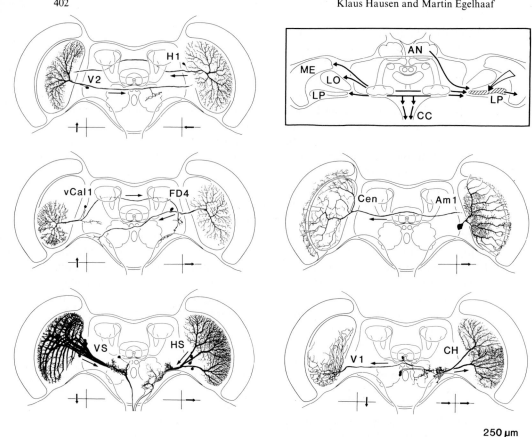

Fig. 4. Tangential and amacrine neurons of the lobula plate (*Calliphora*). The figure shows examples of the different anatomical types of cells in the lobula plate reconstructed after cobalt or Lucifer stainings from frontal serial sections as shown in Fig. 3. Homo- and heterolateral output elements of the lobula plate are the HS- and VS-cells, and the cells vCal1 and FD4, respectively. All cells of this type show large dendrites in the lobula plate and project into the ventrolateral brain. Examples of heterolateral connection-elements between both lobula plates are the cells V2 and H1. The CH-cells and the V1-cell represent centrifugal elements, which originate in the ventrolateral brain and give rise to extended terminal arborizations in the lobula plate. Further centrifugal connections to the medulla are established by the amacrine cell Am 1 and the cell Cen. The polarities of the individual neurons are indicated by arrows at the axons. All neurons shown are motion-sensitive elements, the directional selectivities of which are indicated by arrows in cross-diagrams representing the left and right, and the dorsal and ventral hemispheres of the visual field. The inset shows a summary diagram of the axonal pathways of lobula plate cells in the brain. AN: antennal lobe, *CC* cervical connective; *ME* medulla; *LO* lobula; *LP* lobula plate. (From Hausen 1987; reconstruction of the FD4-cell from Egelhaaf 1985b)

specializations (Hausen et al. 1980; Bishop and Bishop 1981) and are coupled to descending neurons (Hengstenberg et al. 1982). A detailed investigation of the synaptic connectivity in this region has so far revealed two major pathways between VS-cells and the motor system (Strausfeld and Bassemir 1985b): the cells VS2–3 are directly coupled to motoneurons of particular neck muscles , whereas the cells VS4–9 are coupled to descending neurons, which project into the thoracic ganglion. One of these descending neurons is shown in the VS-reconstruction of Fig. 4.

Whereas the HS- and VS-cells are examples of output neurons of the lobula plate terminating in the ipsilateral part of the brain, a second group of output neurons projects into the contralateral part and terminates in close vicinity to the axon terminals of the contralateral HS-cells. Examples of such heterolateral elements are the cells vCal1 and FD4 shown in Fig. 4. Although detailed connectivity studies are still lacking, it is likely that these cells are also synaptically coupled to descending neurons.

Centrifugal cells project from the brain back into the lobula plate. Homolateral neurons of this kind are the dorsal and ventral centrifugal horizontal cell (CH), which show small dendritic arborizations at the HS-terminals and project into the dorsal and ventral half of the ipsilateral lobula plate, respectively. The V1-cell is a heterolateral centrifugal element arising from the termination area of the VS-cells and projecting into the contralateral lobula plate. More complex centrifugal connections are established by elements like the amacrine cell Am1 and the tangential cell Cen, which project from the lobula plate back into the medulla.

The last group to be mentioned are the heterolateral connection elements, which establish direct pathways between the lobula plate and the contralateral optic lobe. Figure 4 shows two cells of this kind, the H1 and the V2, both having large dendritic domains in one lobula plate and large terminals (not shown) in the contralateral lobula plate.

The inset in Fig. 4 summarizes schematically all connections established by the groups of cells discussed above. The open arrow in the right internal chiasm represents the retinotopic visual input of the lobula plate. Thin arrows indicate the bidirectional pathways between lobula plate and the ipsi- and contralateral projection areas in the ventrolateral brain and the visual neuropils of the contralateral optic lobe. Unidirectional pathways link the lobula plate to the ipsilateral antennal lobe and the ipsilateral medulla.

Finally, it should be mentioned that the tangential cells of the lobula plate show extraordinary structural constancy: the position, size, and layer of their arborizations within the lobula plate, as well as their axonal pathways, are nearly identical from animal to animal. In contrast to the lobula, which shows a remarkable sexual dimorphism (Hausen and Strausfeld 1980; Strausfeld 1980), this holds true for both sexes (Hausen 1984, 1987). The anatomical invariance of the tangential cells is paralleled by physiological similarity between individual cells of the same type in different animals, which has considerably facilitated the electrophysiological analysis of the lobula plate.

5 Functional Architecture of the Lobula Plate

5.1 Response Characteristics and Input Organization of the Tangential Cells

All tangential cells of the lobula plate investigated so far are motion-sensitive elements differing physiologically from each other with respect to their directional selectivities and receptive fields, and their particular response properties resulting from interactions with other tangential cells of the ipsi- and contralateral optic lobe.

The response characteristics of individual tangential cells have been reviewed in detail previously (Hausen 1981, 1984). We will therefore discuss only briefly the functional properties of the equatorial horizontal cell HSE and the V1-cell as representative examples of the homolateral output cells and the heterolateral centrifugal cells (Fig. 5).

The HSE responds selectively to progressive motion and is inhibited by motion in the opposite direction. As demonstrated by the recordings shown in Fig. 5a, the responses to motion consist of graded potentials and superimposed irregular action potentials. This type of signal is characteristic for all three horizontal cells and also for the vertical cells, which selectively respond to vertical motion. Common anatomical features of both classes of cells are their rather short and thick axons, which allow virtually decrement-free transmission of graded dendritic potentials into their terminals (Eckert 1981; Hausen 1982b). The significance of the action potentials in the signal transmission of these cells is still an open question (for discussion of this point see Hausen 1982b). Therefore the graded responses are used to characterize their functional properties. A quantitative evaluation of the directional selectivity of the HSE (Fig. 5b) demonstrates that the graded response amplitudes decrease in cosine-like fashion from the peak values obtained under motion in the preferred and null direction (P,N). The latter are opposite to each other and are parallel to the horizontal lattice axis h of the ommatidial mosaic (see Hausen 1982b). The measurements show further that the cell does not respond to vertical motion. The dependence of the HSE's response on the contrast frequency of a moving periodic stimulus is shown in Fig. 5c. The lower and upper response thresholds lie at about 0.01 Hz and 50 Hz, respectively, and the response optimum is located between 1–10 Hz. More detailed measurements have revealed that the response curve is nearly flat in the peak range (Hausen 1982b), which can be explained by the adaptation of the time constants in the motion detectors (see Sect. 2.1). These basic response characteristics are identical in all three horizontal cells.

Fig. 5a-f. Response characteristics of the HSE-cell (_a-c_) and the V1-cell (_d-f_) of the lobula plate (_Calliphora_). The cells were recorded intracellulary in the right optic lobe, and were stimulated with moving periodic gratings of dark and bright stripes placed in the equatorial plane of the right and left visual field at $\psi = +40°$ and $\psi = -50°$ (see insets; ψ and θ denote azimuth and elevation in the spherical visual field of the fly). The gratings had diameters of 39° and were moved with a contrast frequency of 1.5 Hz in _a,b_ and _d,e. a_ Signals of the HSE. The HSE shows graded responses to motion. It is inactivated (hyperpolarized) by regressive motion and activated (depolarized) by progressive motion in the ipsilateral visual field. The depolarizing graded responses are accompanied by irregular spike activity. _b_ Response dependency on the direction of motion α. The graded response amplitudes of the HSE depend in cosine-like fashion on the motion-direction. Preferred (_P_) and null (_N_) direction are antiparallel. The data are normalized values from N = 15 measurements. _c_ Response dependency on the contrast frequency of the grating moving in preferred direction (progressive). The graded response amplitudes are maximal under stimulation with contrast frequencies of 1–10 Hz. N = 3 measurements. _d_ Signals of V1. The cell responds to motion in the left visual field with modulation of its axonal spike frequency. The cell is inhibited by upward motion and excited by motion in the opposite direction. _e_ Response dependency on the direction of motion α. As in the HSE, the response curve shows a cosine-like slope; the modulation under stimulation with motion in null-direction (_N_) is small because of the low spontaneous activity (_SA_) of the cell. Null and preferred direction (_N,P_) are antiparallel. N = 3 measurements. _f_ Response dependency on the contrast frequency of motion in preferred direction. The responses of the V1 are almost identical to those of the HSE. N = 3 measurements. (From Hausen 1982b, 1984)

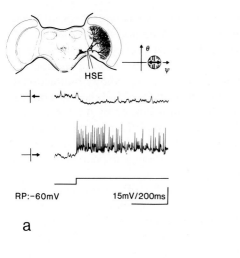

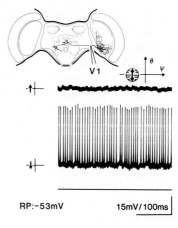

a

d

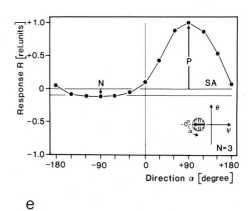

b

e

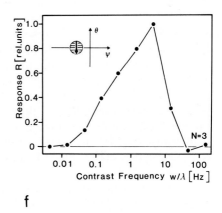

c

f

Responses of the V1-cell are shown in Fig. 5d-f. As is typical for all hete-rolateral tangential cells of the lobula plate having long and thin axons, the V1 transmits information by modulating its axonal spike frequency. The cell responds preferentially to downward motion and is inactivated by upward motion. In contrast to the graded response pattern of the HSE, which allows bidirectional potential modulation around the resting value, the response-modulation of the V1 under stimulation with motion in the null direction is limited by the low resting activity of the cell (about 10 spike/s). The orientation of the preferred and null direction of the V1-cell are, again, antiparallel, but aligned in this case with the v-axis of the ommatidial lattice. The contrast frequency dependence of the V1-responses resembles closely that of the HSE.

The response properties of the two cells are typical for all tangential cells of the lobula plate so far investigated: all of them show the same velocity characteristics and have antiparallel preferred and null directions, which are aligned either with the h-axis or the v-axis of the ommatidial lattice. Exceptions from this latter characteristic are found in only a few cells (e.g., the VS1 and VS7-9), the receptive fields of which are subdivided into compartments with different preferred directions (Hengstenberg 1982). Some elements show also two preferred directions within the same area of the receptive field (Hausen 1981).

These common characteristics indicate that all tangential cells receive excitatory and inhibitory input from arrays of functionally identical elementary motion detectors of the correlation type, which differ only with respect to the orientation of their sampling bases in the retina. The location and cellular structure of these EMD's has remained an unsolved problem so far. There is physiological evidence that the tangential cells receive input from individual EMD's rather than peripheral integrating elements, and that the actual process of motion detection does not take place directly at their dendrites (Hausen 1982b; see also Hausen et al. 1980). It cannot be decided, however, whether the EMD's are located in the lobula plate or in the medulla. The most plausible guess which can be made at present is that the columnar input elements of the lobula plate, the T4-, T5- and Y-cells, are parts of the motion-detection circuits. Direct synaptic connections betweeen T4-terminals and tangential cells have been demonstrated electron microscopically (Strausfeld 1984). For simplicity, we will assume in the following that there are eight arrays of retinotopic EMD's, two for each of the four preferred directions, which feed as excitatory and inhibitory input elements into the assembly of tangential cells. Deoxyglucose studies, in which motion-induced activity in the lobula plate was visualized by radioactive labeling (Buchner et al. 1979, 1984), and combined electrophysiological and light microscopical investigations (Hausen 1987) have revealed that the neuropil of the lobula plate is composed of four directionality layers, which contain (in a sequence from anterior to posterior) the tangential cells responding to progressive, regressive, upward and downward motion. It is tempting to assume that these four layers represent the terminal areas of the arrays of EMD's having the respective preferred directions. Interestingly, the four terminal-strata of the T4- and T5-cells seem to accord to these layers (Strausfeld 1984).

Comparisons of the location and size of the receptive fields of individual tangential cells and their dendritic organizations show that the receptive fields are directly determined by the position and extent of their dendritic domains within the retinotopic columnar array of the lobula plate. Furthermore, the spatial sensitivity distribution in the receptive fields of the cells can be correlated with local variations of the arborization densities in their dendritic domains (Hausen 1981, 1982a,b). This indicates again that the dendrites of the tangential cells receive input from individual retinotopic EMD's, and suggests, in addition, that the local synaptic density accounts for the local gain of signal transmission between EMD's and tangential cells.

In summary, motion is evaluated in the visual system of the fly by retinotopic arrays of EMD's, which have different preferred directions and seem to terminate in the four directionality layers of the lobula plate. The tangential cells gain their individual directional selectivity by investing a particular layer in the lobula plate and contacting excitatory and inhibitory EMD's having opposite preferred directions. The position, size and density of the dendritic arborization within the layer determines the receptive field of a cell and its local sensitivity to motion.

5.2 Gain Control Mechanisms in Tangential Cells

From the foregoing it is evident that the basic operation performed by the tangential cells of the lobula plate is the spatial integration of local motion information. There are two functional categories of tangential cells, the spatial integration properties of which differ significantly: cells of the first group respond to stimulation with moving patterns of increasing size with increasing response amplitudes (Hausen 1981, 1982b; Egelhaaf 1985a), whereas cells of the second group show peak responses under stimulation with small patterns (Egelhaaf 1985b). The measurements on the HSE and the so-called figure-detection cell FD1 shown in Fig. 6 illustrate these response characteristics.

Figure 6a shows the response amplitudes of a HSE as function of the angular extent of a progressively moving periodic pattern. Parameters in the experiment were the mean luminance of the pattern (compare curves 1, 3 and 4) and the contrast frequency of motion (curves 2, 4). The data demonstrate that the response amplitudes of the cell increase under all conditions with increasing stimulus size. The same behavior is observed under a different experimental condition (Fig. 6b), in which the response dependency on the width of a textured figure oscillating horizontally around the animal was evaluated. In contrast, the FD1-cell (Fig. 6b) responds selectively to a small pattern and becomes almost insensitive to motion when the pattern grows large.

These spatial integration properties of the two cells can be fairly well modeled by the inhibitory gain control mechanism for the LF- and SF-system outlined already in Sect. 2.2 and shown in more detail in Fig. 7. Experimental evidence, which will not be discussed here (see Reichardt et al. 1983), indicates that the inhibitory element INH in the gain control circuitry of the HSE-cell is activated by progressive and regressive motion in the ipsilateral visual field. In case of the FD1-cell, the inhibitory element must be assumed to be sensitive to contralateral regressive and ipsilateral progressive motion (Egelhaaf 1985b,c). If these inhibitory elements are implemented in the model circuits and the respective synaptic

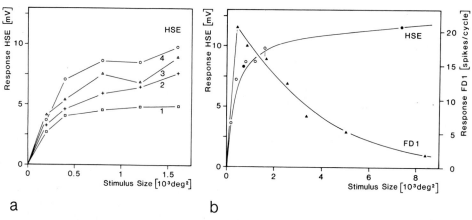

Fig. 6a,b. Spatial integration characteristics of the HSE-cell and the FD1-cell (*Calliphora*). *a* Response amplitudes of the HSE as function of the angular size of a periodic grating moving in progressive direction. Parameters in the experiment were the contrast frequency and the mean luminance of the pattern. In experiments 1,3, and 4 the contrast frequency was kept constant (1.5 Hz) and the mean luminance was varied (1:7 cd/m², 3:20 cd/m², 4:70 cd/m²), in experiments 2 and 4 the mean luminance was constant (70 cd/m²) and the contrast frequency was changed (2:0.45 Hz, 4:1.5 Hz). The values plotted are means of 3 measurements. The cell responds preferentially to large stimuli in all cases. In addition, even under large-field stimulation complete saturation does not occur, since different response levels are reached under the different stimulus conditions. This indicates the existence of a gain control mechanism. *b* Response amplitudes of the HSE and the FD1 as function of the size of a textured figure oscillating horizontally in a panorama. Maximal width of figure: whole panorama, oscillation frequency: 2.5 Hz, oscillation amplitude: 10°. Values plotted are averages of 20 measurements (HSE *filled circles; open circles* show the values of curve 4 in *a* for comparison), and 24 measurements (FD1). The *curves* demonstrate that the HSE and FD1 are tuned to large and small stimuli, respectively. (*a* From Hausen 1982b; *b* after Hausen 1982b; Reichardt et al. 1983; Egelhaaf 1985c)

efficiencies of the EMD's are chosen appropriately (see Reichardt et al. 1983; Egelhaaf 1985a,c), the experimental data of Fig. 6 are closely fitted by the corresponding model simulations shown in Fig. 7.

The different spatial integration properties of the HSE and FD1 are representative for a number of further tangential cells investigated in this respect. Large-field characteristics were also found in the cells of the VS-system (Hengstenberg 1982) and in the heterolateral connection element H1 (Egelhaaf 1985a); further small-field elements are the cells FD2–4 (Egelhaaf 1985b). Thus, the experimental evidence available so far strongly suggests that the assembly of tangential cells in the lobula plate is subdivided into two groups with LF- and SF-tuning.

5.3 Synaptic Interactions Between Tangential Cells

Although synaptic interactions in the lobula plate have been thoroughly studied (Hausen 1981) the inhibitory circuits mediating gain control in the cells sensitive to large-field motion are not yet identified. There are several possible explanations for this. First of all, it cannot be excluded that the gain control operation takes place

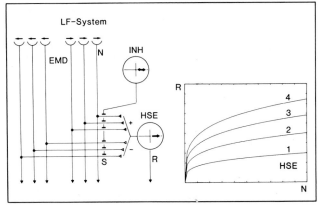

a

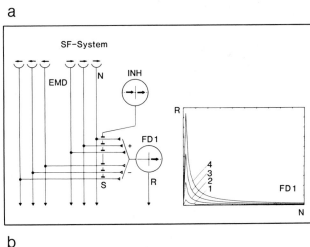

b

Fig. 7a,b. Models of the gain control mechanism in the HSE-cell and the FD1-cell. The inhibitory circuits shown in *a* and *b* are basically identical to the gain control circuit of the LF- and SF-system shown in Fig. 1c. The HSE and FD1 integrate the outputs of excitatory EMD's sensitive to progressive motion and of inhibitory EMD's sensitive to regressive motion. The individual EMD's are shunted by an inhibitory element INH. There is experimental evidence that in case of the HSE, the inhibitor shows monocular sensitivity to horizontal motion in both directions, whereas the inhibitor in the input network of the FD1 shows binocular sensitivity to clockwise horizontal motion. Implementation of these different inhibitors into the circuits leads to spatial integration characteristics in the output elements (HSE, FD1) which are in close agreement with the measurements shown in the previous figure. The curves show the simulated responses (*R*) of the two output elements as function of the number (*N*) of EMD's stimulated. Parameter in the simulations are the signal amplitudes of the EMD's, which depend on stimulus properties like contrast frequency or mean luminance. (After Reichardt et al. 1983 and Egelhaaf 1985a,c)

peripheral to the lobula plate in the medulla, which has not been investigated in this respect. Alternatively, elements of the model circuitry like, for example, the bidirectional inhibitor shown in Fig. 7a, may in fact consist of two or more unidirectional cells, having opposite preferred directions. Various elements of this kind are known; it is still unclear, however, whether they interact with EMD's. These problems deserve further investigation.

The large-field cells of the lobula plate exhibit another type of synaptic interaction, which is of major significance for their functional role in the control of motion-induced behavior. These interactions are based on direct synaptic contacts between tangential cells of both lobula plates and lead to enhanced sensitivities of the cells to particular binocular motion stimuli. Examples of these synaptic connections are compiled in Fig. 8.

The HS-cells (except for the South Horizontal Cell) are postsynaptic to the H2-neuron of the contralateral lobula plate, which responds selectively to regressive motion in its receptive field (Hausen 1976, 1981; Eckert 1981). Due to this excitatory synaptic input the HS-cells gain binocular sensitivity and respond selectively to horizontal rotatory motion (Fig. 8, yaw). Under natural conditions, rotatory retinal motion patterns arise from self-rotations of the animal in space, and one can thus interpret the HS-system as a visual yaw-monitor, specifically designed for the control of course-stabilizing yaw torque generation by the motor system.

A similar situation is found in the VS-system (Fig. 8, roll). The medial vertical cells VS2–6 respond to downward motion within their receptive fields, which are located in the frontolateral part of the ipsilateral visual field. Some, if not all of them, respond additionally to upward motion in the lateral part of the contralateral visual field (Hengstenberg 1981, 1982). It is most likely that this contralateral input is mediated by the V2-cell (Hausen 1981). The resulting rotational sensitivity of

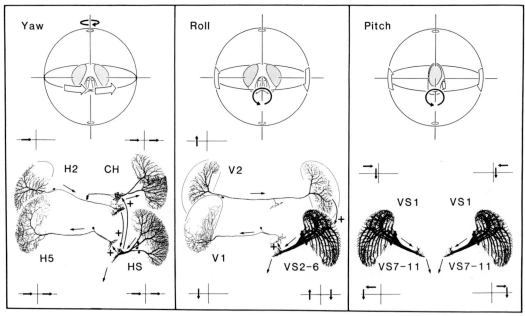

Fig. 8. Binocular interactions between tangential cells of the lobula plate showing large-field tuning (*Calliphora*). The monocular and binocular directional selectivities of the cells are indicated by arrows. The excitatory synaptic interactions indicated lead to high sensitivities of the cells to rotatory retinal motion patterns. Under natural conditions, such patterns result from self-rotations of the animal in space (*upper graphs*). The cells thus code the information necessary to control optomotor course-stabilization maneuvres. (After Hausen 1981, 1987)

the VS-cells indicates, again, that they are particularly well suited to signal self-rotations of the animal with respect to the environment, in this case around the longitudinal body axis.

A remarkable combination of directional sensitivities is found in the most lateral VS-cell VS1, which scans the dorsal and a narrow frontal region of the ipsilateral visual field, and in VS7-9, which also scan the dorsal and, in addition, the caudal region of the visual field. VS1 and VS7-9 respond to regressive and progressive motion in their dorsal fields, respectively, and to downward motion in the other parts of their receptive fields (Hengstenberg 1981). This particular combination of preferred directions leads to specific sensitivities of these cells to pitch movements of the animal (Fig. 8, pitch). It should be noted that, in contrast to the situation during yaw and roll, visual detection of pitch movements does not require binocular information. This may be the reason why these cells show purely monocular responses.

In conclusion, a number of large-field output cells of the lobula plate are particularly sensitive to the binocular and monocular retinal motion patterns arising during rotations of the animal around the three body axes. They thus code the information necessary to control course-stabilizing torque responses.

We will now consider the response behavior of the HS- and FD-cells under stimulation with relative motion between a stripe and a panorama as used in the behavioral experiments on object fixation (Sect. 2.2). Experiments of this kind were performed on the HSE-cell and the FD-cells in order to investigate their potential role in fixation and tracking behavior (Reichardt et al. 1983; Egelhaaf 1985a,b,c).

Figure 9 shows the responses of HSE- and FD-cells to three characteristic stimuli of this kind: progressive motion of a single figure (A), and simultaneous motion of figure and ground in the same and in opposite direction (B,C). In the HSE, the largest response is found under ipsilateral progressive and contralateral regressive motion (B). There is also a considerable response of the cell to progressive motion of the small figure, which is compatible with the steep increase of the response curves shown in Fig. 6. The response to antiparallel motion of figure and ground (C) results from the simultaneous excitatory and inhibitory inputs to the cell activated by the progressive figure-motion and the regressive ground-motion, respectively. The response amplitude under this stimulus condition depends critically on the relative weight of both inputs, and thus on parameters like size and velocity of the stimuli.

A comparison of this response pattern to that of FD-cells demonstrates again the basic functional difference between the two types of cells. Both FD-cells shown in Fig. 9 respond to progressive motion within their receptive fields but are subject to different inhibitory influences. As already discussed, the inhibitory element in the gain control circuitry of the FD1 is sensitive to binocular horizontal rotatory motion. In contrast, the inhibitor of the FD4 shows bidiretional sensitivity in the contralateral visual field (see, however, Egelhaaf 1985b,c). The response histograms of the two cells demonstrate clearly the preferential sensitivity of both cells to the figure motion and the response inhibition during simultaneous motion of the large, binocular ground. Whereas the response depression under stimulus B results evidently from the activation of the inhibitory inputs to the cells, the reason for the

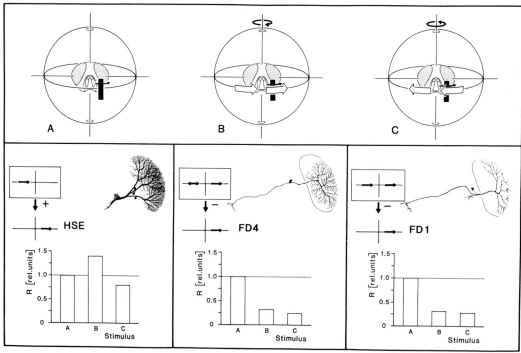

Fig. 9A-C. Responses of the HSE-cell and FD-cell° under stimulation conditions which represent object motion (*A*) and simultaneous self and object motion (*B,C*). The experimental data shown in the three histograms were obtained from recordings of the cells in tethered animals (*Calliphora*), which were stimulated by a texture stripe (figure) oscillating horizontally in a textured panorama (width of stripe: 10° (HSE), 6° (FD1),24° (FD4), oscillation frequency: 2.5 Hz). Response-values *A,B,C* are normalized average responses measured during progressive motion of the figure, and during syndirectional and antidirectional motion of figure and binocular ground, respectively. The responses of the cells reflect the illustrated binocular excitatory and inhibitory interactions within their input circuitries. The HSE shows maximal responses under syndirectional figure and ground motion in the preferred direction of the cell (*B*). Both FD-cells are maximally excited by the motion of the figure (*A*), and are inhibited during motion of figure and background in the same and in opposite directions. The latter two stimuli arise, under natural conditions, during turnings of the animal towards or against a moving object. FD-cells are hence inactivated in both situations. (After Reichardt et al. 1983 and Egelhaaf 1985b)

low response amplitudes measured under stimulus C is less obvious. They are conceivable, however, if one takes into account that the cells, like the HSE, receive direct input from excitatory and inhibitory EMD's when figure and ground move in opposite direction.

Stimulation conditions B and C resemble the motion patterns seen by a fly turning away from or towards an object, which moves independently from front to back through the visual field. In particular the latter situation (C) is frequently encountered by the animals during the initial phase of object fixation and tracking in free light. The response histograms demonstrate that the FD-cells, although highly sensitive to motion of small objects, can be strongly inhibited during such self-motions. This appears to be an undesired but inevitable consequence of the small-field tuning of the FD-cells. There is no other mechanism preventing these cells from firing in response to extended moving patterns which can be as easily implemented as their inhibition by a directionally selective element sensitive to large-field motion.

The cellular components of the gain control circuitry of FD-cells have also not yet been identified. However, the CH-cells (see Fig. 8) are likely candidates for the inhibitor elements of the FD1 and FD4. This is suggested by anatomical evidence, as well as the directional selectivity of these cells shown in Fig. 8, which fits to the directional selectivity of the inhibitor at least in the case of the FD1. The finding that the CH-cells may be GABA-ergic (Meyer et al. 1986), and thus very probably inhibitory elements, supports this view. The other FD-cells known so far (FD2, FD3), which are sensitive to regressive motion might be inhibited by the H5-cell, which has the opposite directional selectivity to the CH-cells (Fig. 8).

The discussion in the last two sections has demonstrated that the large-field cells and the FD-cells of the lobula plate show the functional characteristics of the LF-system and the SF-system deduced originally from the results of behavioral analyses: the spatial integration properties of both groups of cells can be simulated with the proposed gain control circuitry, and the particular combinations of binocular directional selectivities found in the different cells suggest functional significance of the elements in the control of torque generation about the different body axes. We will now consider the transient responses of tangential cells to stimulation with motion and compare them to respective computer simulations performed on the basis of the proposed model network.

Figure 10a shows the signals of the HSE and the FD4 recorded under oscillatory horizontal figure motion, ground motion and relative motion between figure and ground (phase difference:90°). The responses obtained under the first two stimulus conditions reflect characteristics discussed above: both cells are selectively activated by progressive motion but respond differentially to the stimulus size. In contrast, the responses recorded to simultaneous motion of figure and background are rather complex. The FD4-signals obtained under this stimulation are particularly interesting. The cell generates large response-peaks which occur, after a certain response delay, in those instants, in which the figure moves in progressive direction while the ground motion is transiently zero.

Figure 10b shows computer simulations of the output signals of LF-and SF-elements of the model circuitry, in which the particular excitatory and in-hibitory interactions of the HSE and FD4 as shown in Fig. 9 were implemented. The simulations demonstrate that the model describes not only the spatial inte-gration properties but also the complex transient response behavior of the HS-cells and the FD-cells under stimulation with relative motion.

6 Motor Control by the Lobula Plate

The electrophysiological data and the computer simulations discussed in the last section suggest strongly that the HS-cells and the FD-cells are the neural correlates of LF-and SF-system controlling motion-induced yaw torque generation. When implementing the particular binocular input organization of both groups of cells into the model-circuitry of Fig. 1b it is indeed possible to simulate the complex transient yaw torque responses which are generated by tethered flying flies under

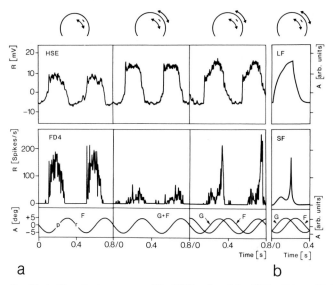

a   b

Fig. 10a,b. Transient responses of the HSE-cell and the FD4-cell to oscillatory horizontal motion of a textured stripe (figure *F*) and a textured panorama (ground *G*), and computer simulations of the response patterns of the two cells. *a* The response traces show the motion-induced graded responses (R) of the HSE and PST-histograms of the responses of the FD4 (*Calliphora*). Stimulus conditions are indicated by insets and the stimulus traces. Textures of figure and ground random dot pattern; width of figure: 10° (*HSE*), 24° (*FD4*); oscillation-frequency: 2.5 Hz; oscillation-amplitude A: 10°; ρ, r: progressive and regressive motion of the figure, placed in the fronto-lateral part of the right visual field of the test-fly. The data are averages of 29 (HSE) and 16 (FD4) measurements. The three recordings in each cell show the responses to figure motion in front of the stationary ground panorama, to syndirectional motion of figure and ground, and to relative motion of both stimuli (phase difference: 90°). The responses of the two cells under the first two conditions reflect the preferred directions and the spatial tuning of the two cells: the HSE shows large response-amplitudes under clockwise synchronous motion of figure and ground whereas the FD4 responds preferentially to progressive motion of the figure. The response patterns of the cells under stimulation with relative motion of figure and ground is complex and results from superposition of excitatory and inhibitory response components. The response of the FD4 under this condition shows characteristic peaks, which occur when the figure moves progressively while the ground motion stops (note that there is a certain response delay of the cell. *b* Computer-simulations of the responses of the HSE and FD4 to relative motion of figure and ground (phase difference: 90°)) performed with the gain control circuits of the LF- and SF-system shown in Fig. 7. For better comparison, the simulations were shifted on the time-axis according to the response-delays of the cells. The simulations are in close agreement with the experimental data. (From Reichardt et al. 1983 and Egelhaaf 1985b,c)

stimulation with horizontal motion of a ground and a small figure (Fig. 11; Egelhaaf 1985c). The close agreement between the simulation and the torque response is indirect, but rather compelling evidence that the HS- and FD-cells are in fact the major control elements for torque generation under this stimulation, and that the model is a basically correct representation of the neural computations underlying course stabilization and object fixation in flies.

More direct evidence for the functional role of the cells in torque control was gained by lesion experiments, in which the axonal pathways of HS- and FD-cells of flies were microsurgically cut and the torque responses of the treated animals were subsequently measured (Hausen and Wehrhahn 1983, 1987b). Results of

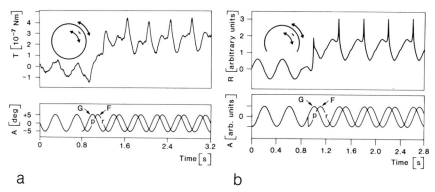

Fig. 11a,b. Transient yaw torque responses of a fly (*Calliphora*) to stimulation with oscillatory horizontal motion of figure (F) and background panorama (G), and computer simulation of the output of the LF- and SF-system. *a* Torque response-pattern of a tethered flying fly positioned in the center of the experimental setup sketched in the inset. The fly detects and attempts to fixate the figure, when it is moved relative to the background. Stimulus condition and response pattern are virtually identical to those shown in Fig. 2a. The response trace represents the average of 100 sweeps. *b* Simulation of the yaw torque responses to stimulation with oscillatory motion of figure and ground. The simulation was performed on the basis of the torque control model shown in Fig. 1b. It was assumed that the LF-system consisted of the HSE-cell, and that the SF consisted of the cells FD1–4. The particular binocular excitatory and inhibitory interactions within the input circuitries of the individual cells were implemented in the model. (From Egelhaaf 1985a,c)

such experiments are shown in Fig. 12. The first graph shows the locations and the effects of the lesions: the terminals of the right HS-cells and the left FD-cells in the right ventrolateral brain were destroyed by lesion L1; the axonal pathways of the FD-cells and the H2-cells in the central protocerebrum were cut by lesion L2. Torque responses of the operated animals were recorded under monocular and binocular stimulation with horizontally moving gratings, placed in the frontolateral visual fields of the eyes. A stimulus size was chosen which elicited comparable responses in the FD-cells and HS-cells. The stimulus combinations in the different experiments are indicated below the torque traces.

The traces in the first row of Fig. 12b and c show the response behavior of intact control flies. Monocular stimulations with progressive motion in front of the left and right eye induce strong negative and positive torque responses, which represent intended turnings of the flies to the left and right side, respectively. Under monocular stimulation with regressive motion the measured torque responses are only weak or even absent. This response difference can be understood on the basis of the differing output connections of the HS- and FD-cells as shown in Fig. 12a; the torque components induced by the HS- and FD-cells are of the same sign only under stimulation with progressive motion, whereas regressive motion induces torque components of opposite sign, which more or less cancel each other. Under binocular stimulation, significant torque responses are only induced by rotatory motions, whereas the response components induced by translatory motions cancel each other for symmetry reasons.

The response pattern of lesioned animals shows a number of characteristic differences, only two of which will be discussed here. We will first consider the response to monocular stimulation with progressive motion in the right visual field

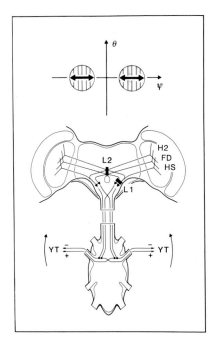

a

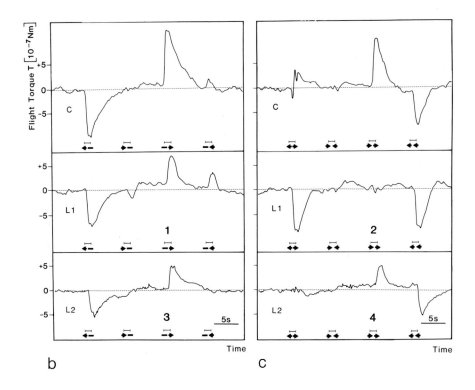

b

c

(1) in animals with lesion L1. In this case, the right HS-cells are cut and the recorded positive response can be induced only by the right FD-cells sensitive to progressive motion. When displaying simultaneously regressive motion as a second stimulus in the left visual field (2), the response vanishes completely. This is consistent with the fact that FD-cells are inhibited by contralateral motion. Additional response components induced under this stimulus condition by the regressive-sensitive FD-cells of the left lobula plate do not occur, since their terminals are destroyed also by the lesion L1.

In contrast to the situation in (1), the positive response to monocular progressive motion recorded in animals with lesion L2 (3) reflects the activity of the right HS-cells, since in this case only the central axonal pathway of the FD-cells is cut. Accordingly, the response is not suppressed when regressive motion is simultaneously displayed in the contralateral visual field (4). These examples illustrate that the specific alterations of the response patterns in lesioned animals are fully consistent with the concept of torque control by HS- and FD-cells illustrated in Fig. 12a, which is equivalent to the initial model circuitry shown in Fig. 1b.

Further evidence for a joint control of torque generation by two separate neural systems showing LF- and SF-tuning was gained in behavioral investigations on flies, the HS-cells of which were eliminated by laser-ablation of their precursor cells in an early larval stage (Geiger and Nässel 1982), and also in detailed behavioral studies on *Drosophila* mutants, which lack the HS- and VS-cells of the lobula plate (Heisenberg et al. 1978; Heisenberg and Wolf 1984; Bausenwein et al. 1986). In all cases the torque responses to motion of large patterns were found to be strongly impaired, whereas responses to small moving patterns or objects could still be observed.

In summary, there are several lines of evidence which demonstrate convincingly that yaw torque generation during course stabilization maneuvers and fixation sequences is controlled by the HS- and FD-cells, which gain their particular spatial tunings and binocular sensitivities from inhibitory and excitatory interactions within their input circuitries. Although the cellular components of

◀ ───

Fig. 12a-c. Lesion experiments, demonstrating the importance of the HS- and FD-cells in yaw torque control (*Calliphora*). *a* Flies were operated at sites in the brain marked lesion L1 and lesion L2. The terminals of the right HS-cells and the terminals of their presynaptic elements H2 originating from the left lobula plate were destroyed by L1 in a first group of animals. The central axonal pathway of the FD-and the H2-cells was cut by L2 in a second group of animals. The sketch of the output connections of the HS- and FD-cells to the motor system (YT: yaw torque) accords with the model circuitry of Fig. 1b and is in agreement with the results of the experimental data shown in *b* and *c*. Operated and normal control animals were stimulated with periodic gratings placed in the equatorial plane of the left and right frontolateral visual field, and moving horizontally with a contrast frequency of 1.5 Hz. The size of the patterns was such that the HS-cells and the FD-cells were stimulated about equally strongly. *b* Yaw torque responses of control animals (C) and operated animals (L1, L2) to monocular stimulation with motion as indicated below the stimulus traces. Positive and negative deflections in the response traces indicate turning tendencies to the right and left side, respectively. *c* Yaw torque responses to binocular stimulation with motion. All response traces are averages obtained from at least 200 stimulation sequences in 10 flies. The data show characteristic differences between the response behaviors of the operated and normal flies, which demonstrate directly the functional role of the HS- and FD-cells in torque-control. See text for further explanation. (Hausen and Wehrhahn 1987b)

these circuits are not yet completely identified, it is evident that they are at least in part constituted by the homo- and heterolateral centrifugal cells and the heterolateral connection cells between both lobula plates, which are sensitive to horizontal motion. It seems likely that respective bilateral gain-control circuits exist between the tangential cells sensitive to vertical motion.

Two final aspects of this control system warrant discussion. As shown in the wiring diagram of Fig. 1a, the HS- and FD-cells elicit torque components of the same sign under stimulation with monocular progressive motion, but are assumed to have antagonistic effects on the motor system under stimulation with regressive motion (see Egelhaaf 1985c). This has the consequence that selective stimulation of the horizontal system by horizontal large-field motions in both directions (or by respective self-motions of the animal in space) leads to syndirectional turning responses and, thus, to a course-stabilization maneuver. In contrast, selective stimulation of the FD-cells by horizontal object motions in both directions might lead to turnings towards the stimulus and, thus, to a fixation response. Another interesting point concerns the different efficiencies of the two systems in torque control under stimulus oscillations at low and high frequencies, as discussed above. Under free-flight conditions, slow changes of the direction of the perceived motion can arise during passive course deviations due to turbulences in the air or can result from asymmetries in the flight motor; they are compensated by corrective steering maneuvers dominated by the HS-system. On the other hand, houseflies do not change course smoothly in free flight, but in sequences of rapid turns which inevitably lead to retinal large-field motion of continually changing sign (Wagner 1986b). Under this condition, the LF-system contributes to the turning response with only a relatively small gain and thus does not counteract the active turns, while the SF-system is still operational (Egelhaaf 1987). Both aspects illustrate the high adaptation of the torque-control system to the particular demands of course stabilization and fixation under natural conditions.

This review has been mainly concerned with the neural mechanisms underlying the visually induced yaw torque responses in the fly. The neural control of the torque responses about the other body axes and the control of the additional steering movements like abdominal and hindleg deflection are less well analyzed. There is accumulating evidence, however, that they are also governed by the lobula plate. Very recent studies have revealed, in addition, that the tangential cells of the lobula plate are the major sensory elements controlling motion-induced head movements (Hengstenberg 1984; Milde and Strausfeld 1986; Milde et al. 1987; Strausfeld et al. 1987).

7 Motion Computation in the Visual System of Other Insects

The computation of motion is evidently not a peculiarity of the fly visual system, but appears to be of widespread importance throughout the animal kingdom and, in particular, in the various insect groups (for an exhaustive literature review see Wehner 1981). However, it is hard to gain a coherent view on motion computation in insects in general, due to paucity of evidence.

Nevertheless, the existence of control systems mediating optomotor course stabilization is vital for all insects and, particularly, for the fast-flying ones. Accordingly, directionally selective units sensitive to large-field motion have also been found in different nondipterous insect groups like *Odonata* (Olberg 1981a,b), *Orthoptera* (Burtt and Catton 1954; Horridge et al. 1965; Kien 1974, 1975), *Lepidoptera* (Collett and Blest 1966; Collett 1970, 1971a; Rind 1983), and *Hymenoptera* (Kaiser and Bishop 1970; DeVoe et al. 1982). The fact that at least some of these neurons receive input from both eyes and are sensitive to rotations of the entire surrounding underlines their potential involvement in the optomotor turning response.

There is much less uniformity among the different insect species with respect to the functional properties of neurons which are selectively tuned to retinal small-field motion (Palka 1969, 1972; Collett 1971b, 1972; Collett and King 1975; O'Shea and Rowell 1975; Frantsevich and Mokrushov 1977; Rowell et al. 1977; Pinter 1977; Olberg 1981a, 1986). Although in most of these examples there is no experimental evidence on the functional significance of these units in orientation behavior, it can be speculated that this diversity might be due to the different behavioral contexts in which small objects moving relative to the eye have to be detected. For instance, the stimulus conditions in flying and in walking or even stationary animals differ considerably and, consequently, impose quite different constraints on the neuronal mechanisms underlying figure-ground discrimination (e.g., compare Collett 1971a; Egelhaaf 1985b,c and O'Shea and Rowell 1975; Rowell et al. 1977). Moreover, even in flying insects, different pursuit strategies inevitably lead to different temporal characteristics of retinal image motion and are likely to be reflected in the functional properties of the respective control systems (compare, e.g., syrphid flies, Collett 1980, and the houseflies, Wagner 1986a,b). Without thorough analyses at both the behavioral and the neuronal level it is not possible, however, to relate — in more than a superficial way — the electrophysiological findings in other insect species to what is known in flies on the detection and fixation of objects and the underlying neuronal circuits.

8 Conclusions

The results discussed here have clearly revealed that in the fly the visual stabilization of the flight course and the detection and fixation of objects essentially depend on the evaluation of motion and relative motion from the retinal image flow. This is mainly achieved by two control systems which are selectively tuned to large-field and small-field motion, respectively. They reside in the lobula plate which, therefore, represents a prominent motion computation center in the fly's brain. Although there remain many problems to be solved in this context, the principle mechanisms underlying these tasks can now be understood, at least in outline. These mechanisms might be of importance beyond the detection and fixation of stationary and moving objects as measured in our behavioral paradigms. Relative motion between objects and their background might also be the decisive cue in other tasks which require knowledge on the three-dimensional layout of the visual surround, such as the avoidance of obstacles and predators, or the control of landing.

Acknowledgments. We are indebted to W. Reichardt and C. Wehrhahn for numerous stimulating discussions and for critically reading the paper. T. Wiegand drew the figures and I. Geiss typed the manuscript. We would like to thank them all.

References

Bausenwein B, Wolf R, Heisenberg M (1986) Genetic dissection of optomotor behavior in *Drosophila melanogaster*. Studies on wild-type and the mutant optomotor-blind[H31]. J Neurogenet 3:87–109

Bishop CA, Bishop LG (1981) Vertical motion detectors and their synaptic relations in the third optic lobe of the fly. J Neurobiol 12:281–296

Blondeau J, Heisenberg M (1982) The three-dimensional torque system of *Drosophila melanogaster*. Studies on wildtype and the mutant optomotor-blind[H31]. J Comp Physiol A 145:321–329

Borst A, Bahde S (1987) Comparison between the movement detection systems underlying the optomotor and the landing response in the housefly. Biol Cybernet 56:217–224

Borst A, Egelhaaf M (1987) Temporal modulation of luminance adapts time constant of fly movement detectors. Biol Cybernet 56:209–215

Buchner E (1976) Elementary movement detectors in an insect visual system. Biol Cybernet 24:85–101

Buchner E (1984) Behavioral analysis of spatial vision in insects. In: Ali MA (ed) Photoreception and vision in invertebrates. Plenum, New York London, pp 561–621

Buchner E, Götz KG, Straub C (1978) Elementary detectors for vertical movement in the visual system of *Drosophila*. Biol Cybernet 31:235–242

Buchner E, Buchner S, Hengstenberg R (1979) 2-Deoxy-D-glucose maps movement-specific nervous activity in the second visual ganglion of *Drosophila*. Science 205:687–688

Buchner E, Buchner S, Bülthoff I (1984) Deoxyglucose mapping of nervous activity induced in *Drosophila* brain by visual movement. I. Wildtype. J Comp Physiol A 155:471–483

Burtt ET, Catton WT (1954) Visual perception of movement in the locust. J Physiol 125:566–580

Collett TS (1970) Centripetal and centrifugal visual cells in medulla of the insect optic lobe. J Neurophysiol 33:239–256

Collet TS (1971a) Visual neurons for tracking moving targets. Nature (London) 232:127–130

Collett TS (1971b) Connections between wide-field monocular and binocular movement detectors in the brain of a hawk moth. Z Vergl Physiol 75:1–31

Collett TS (1972) Visual interneurons in the anterior optic tract of the privet hawk moth. J Comp Physiol 78:396–433

Collett TS (1980) Angular tracking and the optomotor response. An analysis of visual reflex interaction in a hoverfly. J Comp Physiol A 140:145–158

Collett TS, Blest AD (1966) Binocular, directionally selective neurons, possibly involved in the optomotor response of insects. Nature (London) 212:1330–1333

Collett TS, King AJ (1975) Vision during flight. In: Horridge GA (ed) The compound eye and vision of insects. Clarendon, Oxford, pp 437–466

de Ruyter van Steveninck RR, Zaagman WH, Mastebroek HAK (1986) Adaptation of transient responses of a movement-sensitive neuron in the visual system of the blowfly *Calliphora erythrocephala*. Biol Cybernet 54:223–236

DeVoe R (1985) The eye: electrical activity. In: Kerkut GA, Gilbert LI (eds) Comprehensive insect physiology, biochemistry and pharmacology, vol 4. Nervous system: sensory. Pergamon, Oxford, pp 277–354

DeVoe R, Kaiser W, Ohm J, Stone LS (1982) Horizontal movement detectors of honeybees: directionally-selective visual neurons in the lobula and brain. J Comp Physiol A 147:155–170

Dvorak DR, Bishop LG, Eckert HE (1975) On the identification of movement detectors in the fly optic lobe. J Comp Physiol A 100:5–23

Eckert H (1973) Optomotorische Untersuchungen am visuellen System der Stubenfliege *Musca domestica* L. Kybernetik 14:1–23

Eckert H (1980) Functional properties of the H1-neuron in the third optic ganglion of the blowfly, *Phaenicia*. J Comp Physiol A 135:29–39

Eckert H (1981) The horizontal cells in the lobula plate of the blowfly, *Phaenicia sericata*. J Comp Physiol A 143:511–526

Eckert H, Bishop LG (1978) Anatomical and physiological properties of the vertical cells in the third optic ganglion of *Phaenicia sericata* (Diptera, Calliphoridae). J Comp Physiol A 126:57–86

Eckert H, Hamdorf K (1981) The contrast frequency-dependence: A criterion for judging the non-participation of neurons in the control of behavioral response. J Comp Physiol A 145:241–247

Egelhaaf M (1985a) On the neuronal basis of figure-ground discrimination by relative motion in the visual system of the fly. I: Behavioral constraints imposed on the neuronal network and the role of the optomotor system. Biol Cybernet 52:123–140

Egelhaaf M (1985b) On the neuronal basis of figure-ground discrimination by relative motion in the visual system of the fly. II: Figure-detection cells, a new class of visual interneurons. Biol Cybernet 52:195–208

Egelhaaf M (1985c) On the neuronal basis of figure-ground discrimination by relative motion in the visual system of the fly. III: Possible input circuitries and behavioral significance of the FD-cells. Biol Cybernet 52:267–280

Egelhaaf M (1987) Dynamic properties of two control systems underlying visually guided turning in house-flies. J Comp Physiol A 161:777–783

Egelhaaf M, Reichardt W (1987) Dynamic response properties of movement detectors: theoretical analysis and electrophysiological investigation in the visual system of the fly. Biol Cybernet 56:69–87

Fermi G, Reichardt W (1963) Optomotorische Reaktionen der Fliege *Musca domestica*. Abhängigkeit der Reaktion von der Wellenlänge, der Geschwindigkeit, dem Kontrast und der mittleren Leuchtdichte bewegter periodischer Muster. Kybernetik 2:15–28

Franceschini N (1985) Early processing of color and motion in a mosaic visual system. Neurosci Res Suppl 2:17–49

Frantsevich LI, Mokrushov PA (1977) Jittery movement fibres (JMF) in dragonfly nymphs: Stimulus surround interaction. J Comp Physiol A 120:203–214

Geiger G, Nässel DR (1982) Visual processing of moving single objects and wide-field patterns in flies: Behavioral analysis after laser-surgical removal of interneurons. Biol Cybernet 44:141–149

Götz KG (1964) Optomotorische Untersuchung des visuellen Systems einiger Augenmutanten der Fruchtfliege *Drosophila*. Kybernetik 2:77–92

Götz KG (1968) Flight control in *Drosophila* by visual perception of motion. Kybernetik 4:199–208

Götz KG (1983) Bewegungssehen und Flugsteuerung bei der Fliege *Drosophila*. In: Nachtigall W (ed) Biona-Report 2. Fischer, Stuttgart New York, pp 21–33

Götz KG, Buchner E (1978) Evidence for one-way movement detection in the visual system of *Drosophila*. Biol Cybernet 31:243–248

Götz KG, Hengstenberg B, Biesinger R (1979) Optomotor control of wing beat and body posture in *Drosophila*. Biol Cybernet 35:101–112

Hardie RC (1985) Functional organization of the fly retina. In: Ottoson D (ed) Progress in sensory physiology, vol 5. Springer, Berlin Heidelberg New York Toronto, pp 1–79

Hassenstein B, Reichardt W (1956) Systemtheoretische Analyse der Zeit-, Reihenfolgen- und Vorzeichenauswertung bei der Bewegungsperzeption des Rüsselkäfers *Chlorophanus*. Z Naturforsch 11b:513–524

Hausen K (1976) Functional characterization and anatomical identification of motion sensitive neurons in the lobula plate of the blowfly *Calliphora erythrocephala*. Z Naturforsch 31c:629–633

Hausen K (1977) Signal processing in the insect eye. In: Stent GS (ed) Function and formation of neural systems. Dahlem Konf, Berlin, pp 81–110

Hausen K (1981) Monocular and binocular computation of motion in the lobula plate of the fly. Verh Dtsch Zool Ges 1981:49–70

Hausen K (1982a) Motion sensitive interneurons in the optomotor system of the fly. I. The horizontal cells: Structure and signals. Biol Cybernet 45:143–156

Hausen K (1982b) Motion sensitive interneurons in the optomotor system of the fly. II. The horizontal cells: Receptive field organization and response characteristics. Biol Cybernet 46:67–79

Hausen K (1984) The lobula-complex of the fly: Structure, function and significance in visual behavior. In: Ali MA (ed) Photoreception and vision in invertebrates. Plenum, New York London, pp 523–559

Hausen K (1987) The neural architecture of the lobula plate of the blowfly, *Calliphora erythrocephala*. Cell Tissue Res (submitted)

Hausen K, Hengstenberg R (1987): Multimodal convergence of sensory pathways on motoneurons of flight muscles in the fly (*Calliphora*). Soc Neurosci Abstr 13:1059

Hausen K, Strausfeld NJ (1980) Sexually dimorphic interneuron arrangements in the fly visual system. Proc R Soc London Ser B 208:57–71

Hausen K, Wehrhahn C (1983) Microsurgical lesion of horizontal cells changes optomotor yaw responses in the blowfly *Calliphora erythrocephala*. Proc R Soc London Ser B 219:211–216

Hausen K, Wehrhahn C (1987a) Neural control of flight-torque during visual orientation in flies. 1. Functional characteristics of visual interneurons and the yaw torque generating motor system. (in preparation)

Hausen K, Wehrhahn C (1987b) Neural control of flight-torque during visual orientation in flies. 2. Separation of two control systems by selective lesions of visual pathways in the brain (in preparation)

Hausen K, Wolburg-Buchholz K, Ribi WA (1980) The synaptic organization of visual interneurons in the lobula complex of flies. Cell Tissue Res 208:371–387

Heide G (1983) Neural mechanisms of flight control in Diptera. In: Nachtigall W (ed) Biona-Report 2. Fischer, Stuttgart New York, pp 35–52

Heisenberg M, Wolf R (1984) Vision in *Drosophila*. Springer, Berlin Heidelberg New York

Heisenberg M, Wonneberger R, Wolf R (1978) Optomotor-blind[H31]-a *Drosophila* mutant of the lobula plate giant neurons. J Comp Physiol 124:287–296

Hengstenberg R (1981) Visuelle Drehreaktionen von Vertikalzellen in der Lobula-Platte von *Calliphora*. Verh Dtsch Zool Ges 1981:180

Hengstenberg R (1982) Common visual response properties of giant vertical cells in the lobula plate of the blowfly *Calliphora*. J Comp Physiol A 149:179–193

Hengstenberg R (1984) Roll stabilization during flight of the blowfly's head and body by mechanical and visual cues. In: Varjú D, Schnitzler HU (eds) Localization and orientation in biology and engineering. Springer, Berlin Heidelberg New York, pp 121–134

Hengstenberg R, Hausen K, Hengstenberg B (1982) The number and structure of giant vertical cells (VS) in the lobula plate of the blowfly *Calliphora erythrocephala*. J Comp Physiol A 149:163–177

Hengstenberg R, Sandeman DC, Hengstenberg B (1986) Compensatory head roll in the blowfly *Calliphora* during flight. Proc R Soc London Ser B 227:455–482

Horridge GA, Scholes JH, Shaw S, Tunstall J (1965) Extracellular recordings from single neurons in the optic lobe and brain of locust. In: Treherne JE, Beaument JWL (eds) Pap 12th Int Congr Entomology. Academic Press, London New York, pp 165–202

Järvilehto M (1985) The eye: vision and perception. In: Kerkut GA, Gilbert LI (eds) Comprehensive insect physiology, biochemistry, and pharmacology, vol 6. Nervous system: sensory. Pergamon, Oxford, pp 355–429

Kaiser W, Bishop LG (1970) Directionally selective motion detecting units in the optic lobe of the honey bee. Z Vergl Physiol 67:403–413

Kien J (1974) Sensory integration in the locust optomotor system. II. Direction selective neurons in the circumoesophageal connectives and the optic lobes. Vision Res 14:1255–1268

Kien J (1975) Neuronal mechanisms subserving directional selectivity in the locust optomotor system. J Comp Physiol A 102:337–355

Kirschfeld K (1972) The visual system of *Musca*: Studies on optics, structure and function. In: Wehner R (ed) Information processing in the visual system of arthropods. Springer, Berlin Heidelberg New York, pp 63–74

Land MF (1977) Visually guided movements in invertebrates. In: Stent GS (ed) Function and formation of neural systems. Dahlem Konf, pp 161–177

Land MF, Collett TS (1974) Chasing behavior of houseflies (*Fannia canicularis*). A description and analysis. J Comp Physiol 89:331–357

Laughlin SB (1984) The roles of parallel channels in early visual processing by the arthropod compound eye. In: Ali MA (ed) Photoreception and vision in invertebrates. Plenum, New York London, pp 457–481

Maddess T, Laughlin SB (1985) Adaptation of the motion-sensitive neuron H1 is generated locally and governed by contrast frequency. Proc R Soc London Ser B 225:251–275

Mastebroek HAK, Zaagman WH, Lenting BPM (1980) Movement detection: Performance of a wide-field element in the visual system of the blowfly. Vision Res 20:467–474

McCann GD, MacGinitie GF (1965) Optomotor response studies of insect vision. Proc R Soc London Ser B 163:369–401

Meyer EP, Matute C, Streit P, Nässel DR (1986) Insect optic lobe neurons identifiable with monoclonal antibodies to GABA. Histochemistry 84:207–216

Milde J, Strausfeld NJ (1986) Visuo-motor pathways in arthropods. Naturwissenschaften 73:151–154

Milde J, Seyan HS, Strausfeld NJ (1987) The neck motor system of the fly *Calliphora erythrocephala*. II. Sensory organization. J Comp Physiol A 160:225–238

Olberg RM (1981a) Object- and self-movement detectors in the ventral nerve cord of the dragonfly. J Comp Physiol A 141:327–334

Olberg RM (1981b) Parallel encoding of direction of wind, head, abdomen, and visual pattern movement by single interneurons in the dragonfly. J Comp Physiol A 142:27–41

Olberg RM (1986) Identified target-selective visual interneurons descending from the dragonfly brain. J Comp Physiol A 159:827–840

O'Shea M, Rowell CHF (1975) Protection from habituation by lateral inhibition. Nature (London) 254:53–55

Palka J (1969) Discrimination between movements of eye and object by visual interneurons of crickets. J Exp Biol 50:723–732

Palka J (1972) Moving movement detectors. Am Zool 12:497–505

Pick B (1974) Visual flicker induces orientation behavior in the fly *Musca*. Z Naturforsch 29c:310–312

Pick B (1976) Visual pattern discrimination as an element of the fly's orientation behavior. Biol Cybernet 23:171–180

Pierantoni R (1976) A look into the cock-pit of the fly. The architecture of the lobula plate. Cell Tissue Res 171:101–122

Pinter RB (1977) Visual discrimination between small objects and large textured backgrounds. Nature (London) 270:429–431

Poggio T, Reichardt W, Hausen K (1981) A neuronal circuitry for relative movement discrimination by the visual system of the fly. Naturwissenschaften 68:443–446

Reichardt W (1957) Autokorrelationsauswertung als Funktionsprinzip des Nervensystems. Z Naturforsch 12b:418–457

Reichardt W (1961) Autocorrelation, a principle for evaluation of sensory information by the central nervous system. In: Rosenblith WA (ed) Principles of sensory communication. John Wiley & Sons, New York, pp 303–317

Reichardt W (1973) Musterinduzierte Flugorientierung: Verhaltens-Versuche an der Fliege *Musca domestica*. Naturwissenschaften 60:122–138

Reichardt W (1986) Processing of optical information by the visual system of the fly. Vision Res 26:113–126

Reichardt W (1987) Computation of optical motion by movement detectors. Biophys Chem 26:263–278

Reichardt W, Guo A (1986) Elementary pattern discrimination (behavioral experiments with the fly *Musca domestica*). Biol Cybernet 53:285–306

Reichardt W, Poggio T (1976) Visual control of orientation behavior in the fly. Part I. A quantitative analysis. Q Rev Biophys 9:311–375

Reichardt W, Poggio T (1979) Figure-ground discrimination by relative movement in the visual system of the fly. Part I: Experimental results. Biol Cybernet 35:81–100

Reichardt W, Poggio T, Hausen K (1983) Figure-ground discrimination by relative movement in the visual system of the fly. Part II: Towards the neural circuitry. Biol Cybernet 46 (Suppl):1–30

Riehle A, Franceschini N (1984) Motion detection in flies: Parametric control over on-off pathways. Exp Brain Res 54:390–394

Rind C (1983) A directionally sensitive motion detecting neuron in the brain of a moth. J Exp Biol 102:253–271

Rowell CHF, O'Shea M, Williams JLD (1977) The neuronal basis of a sensory analyser, the acridid movement detector system. IV. The preference for small field stimuli. J Exp Biol 68:157–185

Srinivasan MV (1977) A visually-evoked roll response in the housefly. Open-loop and closed-loop studies. J Comp Physiol A 119:1–14

Strausfeld NJ (1976) Atlas of an insect brain. Springer, Berlin Heidelberg New York

Strausfeld NJ (1980) Male and female visual neurons in dipterous insects. Nature (London) 283:381–383

Strausfeld NJ (1984) Functional neuroanatomy of the blowfly's visual system. In: Ali MA (ed) Photoreception and vision in invertebrates. Plenum, New York London, pp 483–522

Strausfeld NJ, Bacon JP (1983) Multimodal convergence in the central nervous system of insects. In: Horn E (ed) Multimodal convergence in sensory systems. Fortschr Zool 28. Fischer, Stuttgart, pp 47–76

Strausfeld NJ, Bassemir UK (1985a) Lobula plate and ocellar interneurons converge onto a cluster of descending neurons leading to leg and neck motor neuropil in *Calliphora erythrocephala*. Cell Tissue Res 240:617–640

Strausfeld NJ, Bassemir UK (1985b) The organization of giant horizontal-motion-sensitive neurons and their synaptic relationships in the lateral deutocerebrum of *Calliphora erythrocephala* and *Musca domestica*. Cell Tissue Res 242:531–550

Strausfeld NJ, Seyan HS (1985) Convergence of visual, haltere and prosternal inputs at neck motor neurons of *Calliphora erythrocephala*. Cell Tissue Res 240:601–615

Strausfeld NJ, Bassemir U, Singh RN, Bacon JP (1984) Organizational principles of outputs from dipteran brains. J Insect Physiol 30:73–39

Strausfeld NJ, Seyan HS, Milde JJ (1987) The neck motor system of the fly *Calliphora erythocephala*. I. Muscles and motor neurons. J Comp Physiol A 160:205–224

Virsik RP, Reichardt W (1976) Detection and tracking of moving objects by the fly *Musca domestica*. Biol Cybernet 23:83–98

Wagner H (1986a) Flight performance and visual control of flight of the free-flying housefly (*Musca domestica* L.), II. Pursuit of targets. Philos Trans R Soc London Ser B 312:553–579

Wagner H (1986b) Flight performance and visual control of flight of the free-flying housefly (*Musca domestica* L.) III. Interactions between angular movement induced by wide- and smallfield stimuli. Philos Trans R Soc London Ser B 312:581–595

Wehner R (1981) Spatial vision in insects. In: Autrum H (ed) Handbook of sensory physiology, vol VII/6C. Springer, Berlin Heidelberg New York, pp 287–616

Wehrhahn C (1978) The angular orientation of the movement detectors acting on the flight lift responses in flies. Biol Cybernet 31:169–173

Wehrhahn C (1979) Sex-specific differences in the chasing behavior of houseflies (*Musca*). Biol Cybernet 32:239–241

Wehrhahn C (1985) Visual guidance of flies during flight. In: Kerkut GA, Gilbert LI (eds) Comprehensive insect physiology, biochemistry and pharmacology, vol 6. Nervous sytems: sensory, Pergamon, Oxford, pp 673–684

Wehrhahn C (1986) Motion sensitive yaw torque responses of the housefly *Musca*: A quantitative study. Biol Cybernet 55:275–280

Wehrhahn C, Hausen K (1980) How is tracking and fixation accomplished in the nervous system of the fly? A behavioral analysis based on short time stimulation. Biol Cybernet 38:179–186

Wehrhahn C, Reichardt W (1975) Visually induced height orientation of the fly *Musca domestica*. Biol Cybernet 20:37–50

Zaagman WH, Mastebroek HAK, Buyse T, Kuiper JW (1977) Receptive field characteristics of a directionally selective movement detector in the visual system of the blowfly. J Comp Physiol A 116:39–50

Zanker JM (1987) Uber die Flugkrafterzeugung und Flugkraftsteuerung der Fruchtfliege *Drosophila melanogaster*. Diss, Eberhard-Karls-Univ, Tübingen

Chapter 19

Size and Distance Perception in Compound Eyes

RUDOLF SCHWIND, Regensburg, FRG

1 Introduction

Several arthropods show by their behavior that they are able to measure the absolute distance or to gauge the absolute size of a nearby object, be it a prey, a mate or anything onto which the animal is going to jump (revs. Wehner 1981; Collett and Harkness 1982; Collett 1987).

As Exner (1891) pointed out, vergence and accommodation as cues for absolute distance can be ruled out for most of the arthropods. Accommodation mechanisms are not available in compound eyes and would be of little use anyhow, since due to the short focal length of the dioptric apparatus, all objects farther away than several millimeters are in practice at infinity for a single ommatidium. Vergence as a mechanism for absolute distance measurement can be ruled out for insects, since their eyes are fixed, and until now there are no hints that crayfish with mobile eyestalks use such a mechanism.

Exner (1918) supposed that arthropods could measure distances binocularly using disparity, and monocularly by the use of motion parallax.

Although results of Baldus (1926), Hoppenheit (1964) and other authors provided strong evidence that some predatory insects use binocular cues to measure absolute distances, the idea that arthropods could measure distances binocularly was generally not accepted, until Rossel (1983, 1986) demonstrated stereopsis in mantids (cf. Sect. 4).

Use of motion parallax as a possible means for distance estimation was favored by many authors (e.g., Horridge 1977; Wehner 1981). The importance of motion parallax for the perception of relative distances or even for a step in vision which precedes any perception of form and which allows foreground objects to be distinguished from background, was recently stressed again (Egelhaaf 1985, Horridge 1986, 1987).

Motion parallax alone, however, cannot in each case explain the ability of certain insects or crayfish to measure absolute distances or sizes of objects (cf. Sects. 2–5). What cues are used by arthropods to fulfill this task?

2 Measuring Absolute Distances by Motion Parallax

Translatory movement of an animal leads to retinal image displacements, the speed (or amplitude) of which depends on distance. In freely moving animals, relative distances of objects could be derived in principle by the differences of speed (or amplitude) of the retinal image displacements (rev. Horridge 1986).

For absolute distance measurements by motion parallax, the exact speed (or amplitude) of the motion performed by the animal has to be known, which is a difficult task, at least for flying insects, since wind speed cannot be readily calculated (cf. Heran and Lindauer 1963).

Self-induced retinal image displacements which can be utilized to gauge absolute distances include those provided by peering movements, which are sideward-swaying movements usually caused by leg movements. Zänkert (1938) described these movements in mantids and proposed that the mantis would thus scrutinize an object by looking at it from several sides. Peering movements were studied in more detail by Wallace (1959) and Collett (1978) in locust nymphs, by Goulet et al. (1981) in wood crickets, and by Eriksson (1980) in the grasshopper *Phaulacridium vittatum*. When a locust nymph is viewing a stationary object, it may perform sideward-swaying body movements, thereby counterrotating the head so that the head does not rotate relative to external coordinates (Fig. 1). This lateral displacement leads to a retinal image displacement, the amplitude (or speed) of which depends on distance. Wallace in fact showed that a locust nymph can measure absolute distances in this way: when he moved an object on to which a locust was about to jump horizontally in synchrony with peering so that the retinal image displacements were increased in amplitude (or speed), the locust jumped short.

This method of distance measuring by motion parallax, however, does not work if the object moves, since the extent of the target movements cannot be known by the animal and it should be noticed that whilst mantids perform peering movements when viewing stationary objects such as a twig, distance estimation by self-induced retinal image displacements is not how they gauge distance of a moving prey (see Sect. 4.2).

3 Use of Image Size

Image size is a well-known monocular cue for absolute distance measurements in man. If an object has been identified, distance can be gauged by image size: the smaller the retinal distance of an identified object, the farther is the perceived distance. Due to the poor resolution of compound eyes, identification of objects is rarely possible; however, in some special cases image size can nevertheless be used to estimate absolute distance. For example, male *Syritta* track other flies to find a female. If it has the good fortune to be following a conspecific female, then a preprogrammed distance measuring mechanism works (Collett and Land 1975): the male keeps a relatively constant distance of about 10 cm (Fig. 2) by adjusting the vertical extent of the target image to a certain preprogrammed range.

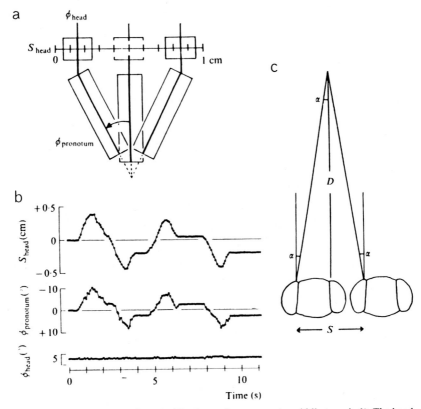

a

ϕ_{head}

S_{head} 0   1 cm

$\phi_{pronotum}$

c

α

D

α α

← S →

b

S_{head} (cm)

+0·5

0

−0·5

$\phi_{pronotum}$ (°)

−10

0

+10

ϕ_{head} (°)

5

0   5   10

Time (s)

Fig. 1a-c. Locust peering: pronotum is rotated by the angle $\phi_{pronotum}$ (*a*, middle trace in *b*). The head is counterrotated in such a way that it does not rotate relative to external coordinates (lower trace in *b*). Object distance *D* can be calculated by lateral displacement of head *S* and image displacement 2α (*c*): $D = S/2\tan\alpha$. (After Collett 1978)

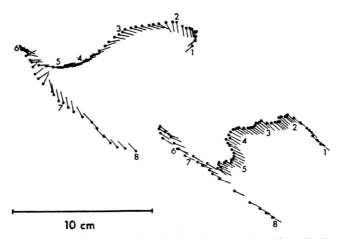

Fig. 2. A male *Syritta* (*right*) tracking a female (*left*), thereby adjusting distance to about 10 cm. Head position shown by *solid circle*, position plotted every 40 ms. (Collett and Land 1975)

10 cm

Of course this mechanism can only work if the size of the desired object is known, and such a mechanism is not suitable, for example, for capturing different-sized prey.

4 Distance Estimation by Binocular Cues

If the optical axes of the eyes can be tilted so that the image of a nearby target can be fixed on a particular region of the retina, the absolute distance of the object can be derived binocularly by the vergence angle, i.e., the difference in the position of two optical axes (e.g., Ogle 1962; Graham 1965; Collett and Harkness 1982). Until now it is not known whether or not crabs with their mobile eyestalks use such a mechanism for range finding. As insect eyes are generally immobile, this method cannot be employed for distance measuring. It should be noted, however, that in flies, eye muscles may cause a slight displacement of the retina and with this a slight tilt of the optical axes (Hengstenberg 1971). Whether or not the functional significance of this is a distance measurement is not yet clear. With fixed eyes, absolute distance can, however, be gauged by the disparity of the two retinal images. In a given direction a remote object point is seen by corresponding ommatidia, the optical axes of which point in parallel (Fig. 3). Within a certain range a nearby object point will be seen by disparate ommatidia, the optical axes of which deviate from those of the corresponding ommatidia by the angles α_L and α_R (Fig. 3). Disparity α is given by the sum $\alpha_L + \alpha_R$.

Due to the small base, the range within which distances may be measured in insects is relatively short and depends on the interocular distance and the inter-ommatidial angles ($\Delta\phi$). In the schematic drawing in Fig. 4, the hatched areas show the visual fields of two forward-looking ommatidia of the right (R) and the left (L) eye respectively. An object point which is farther than the "infinite distance", E_∞, from the eye will be seen as at infinity. The range within which distance estimation is possible, is given by the simple formula $E_\infty = b/2 \times \cot \Delta\phi/2$ (Burkhardt et al. 1973).

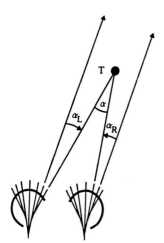

Fig. 3. A nearby object T is seen by two groups of ommatidia which are different from those viewing a remote object. Disparity α is given by $\alpha_L + \alpha_R$. (After Rossel 1986)

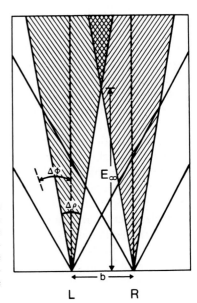

Fig. 4. Definition of E_∞ for idealized compound eyes. *R* and *L* indicate the centers of each eye, where the visual axes of neighbouring ommatidia intersect, *b* the distance between these points. For each eye visual fields of three ommatidia are shown. Optical axes (*dashed lines*) of ommatidia with hatched visual fields are parallel to the median plane. Acceptance angles $\Delta\rho$ are equal to interommatidial angles $\Delta\Phi$. The infinite distance $E_\infty = b/2 \times \cot\Delta\Phi/2$. (After Burkhardt et al. 1973)

Table 1 gives some examples, especially for predatory insects.

Whatever the mechanism of binocular distance estimation may be, the accuracy depends on the intersection pattern of the optical axes of the ommatidia (Burkhardt et al. 1973, cf. Figs. 7 and 10). Distance estimation by disparity is theoretically imprecise at distances in the order of E_∞, since here the grain of the overlapping fields of ommatidia is particularly coarse.

If any insect could localize the position of an object point by determining in which of the overlapping fields it falls, then according to Burkhardt et al. (1973), localization would be most precise in front of the animal at a distance half of the interocular distance. Assuming homogeneous $\Delta\phi$ values, the grain of the overlapping visual fields is finest at that position.

Table 1. "Infinite distance" E_∞ for various species

	Horizontal divergence angle	Interocular distance (mm)	E_∞ (mm)
Cyrtodiopsis whitei (stalk-eyed fly) a diopsid	1°	10	572 (Burkhardt and de la Motte 1983)
Tenodera australasiae (mantis)	0.6°	4.8	458 (Rossel 1986)
Aeshna cyanea larva (dragonfly)	1.2°	5.6	267 (Baldus 1926)
Notiophilus biguttatus (beetle)	2.2°	1.64	42.6 (Bauer 1981)
Anax junius (dragonfly)	0.33°	0	— (Sherk 1978)

Exner (1891) thought that insects and crayfish indeed use binocular cues for distance estimation. He states: "There is no reason, however, to suppose that the former factor (the different images of an object at the retina of the two eyes, due to disparity) does not play the same role in facetted eyes as it does in the eyes of vertebrates. Binocular depth perception in vertebrates is more precise, the greater the distance between the two eyes . . . This distance is small in insects. . . . but from this it does not follow that this sort of distance perception is beyond them. . . At . close quarters this principle of depth perception can function exactly as in vertebrates . . . . depth perception will be most acute at this distance (of a few centimeters); and this may be more advantageous for the insect than a precise estimation of greater distances".

Certainly depth perception by stereopsis as proposed by Exner cannot be a general principle in insects. This is because not all possess an extensive zone of binocular overlap and an appropriate interocular distance between the forward-looking facets (Bauers 1953; Frantsevich and Pichka 1976; Horridge 1977, 1978). Frantsevich and Pichka, who measured the zones of binocular overlap in various predatory and anthophilous insects, grouped the insects studied into distinct groups according to features of their field of binocular vision and to behavioral features:

1. Insects which intercept a point target against the sky in rapid flight — be it prey as for adult dragonflies or robber flies or a conspecific as for drone bees or male syrphids. In this group the general visual field is huge and the dorsal parts may be specialized for the detection of prey (Pritchard 1966; Wehner 1981) or females (Dietrich 1909; Zeil 1983; Land this Vol.). The binocular zone, however, is minimal or lacking. In some dragonflies, in syrphid males (Collett and Land 1975) and in drone bees the eyes meet in the median plane so that the distance between forward-looking facets approaches zero. Distance measurements by binocular cues is unlikely in these animals. They probably learn the approach of a target or an obstacle by monocular cues, such as motion parallax (Horridge 1977, 1986), increase of angular size, or increase of contrast.

2. In anthophilous insects which hover over flowers (sphingids, bee flies) or which alight on them (syrphid flies, bees, diurnal butterflies) there is usually a medium-sized zone of binocular overlap pointing in the landing direction or to the tip of the proboscis. Demoll (1909) showed that in many butterflies the optical axes of the most median ommatidia of both eyes intersect near the tip of the protruded proboscis. He presumed that butterflies could measure the distance of a nearby food source stereoscopically; however, up to now no experiments have been performed to prove this.

3. Binocular zones of considerable extent were found in a third group: in predatory insects which hunt amidst vegetation, for example, larval dragonflies (in *Aeshna cyanea* larvae, 76% of the ommatidia contribute to the zone of binocular vision), cicindelids (*Cicindela hybrida*, 54%), katydids (*Tettigonia viridissima*, 47%), mantids (*Bolivaria xanthoptera*, 46%) and water bugs (*Ranatra linearis*), 39% binocular overlap; after Frantsevich and Pichka 1976, see also Cloarec 1986).

Even for predators belonging to the latter group of insects, the presence of an extensive binocular field need not mean that the insect is able to measure absolute distances by stereopsis. An extensive binocular field could be the result of certain morphological necessities: if, for example, the eyes are separated because the space between them is needed for prognathan mouth parts, as in carabids, then there must be a zone of binocular vision, unless there is a blind zone ahead of the animal. In this case, binocular vision might indeed "just double the amount of input rather than providing an exotic third dimension", as concluded by Via (1977), who carried out experiments with the bulldog ant *Myrmecia gulosa*, in order to test the distance perception abilities of this aggressive insect.

4.1 Size-Distance Ambiguity in an Insect with Binocular Vision

The Australian bulldog ant is an example of an animal with a wide binocular field (60% overlap), which appears to cope without true distance perception (Via 1977). In experiments performed with isolated heads, approaching targets (cardboard rectangles) induced responses in a predictable way: the mandibles opened whenever the vertical edges of the target reached the visual fields of two contralaterally viewing groups of ommatidia ("trigger zones") irrespective of target distance. Mandible closure was elicited when the target moved ever closer, entering the visual fields of a second pair of trigger zones (Fig. 5). Although similar results were obtained in monocular animals, only 14% of the animals continued to snap when one eye was occluded.

If intact animals really approached the center of the prey, as was the case in the experiments, then the animal would confuse size and distance. It is, however, quite conceivable that unrestrained animals do not fixate the center of their victims, but an edge instead. In this case the jaws would snap at the right moment, irrespective of object size, and the underlying mechanism would be a simple form of binocular distance measurement by triangulation.

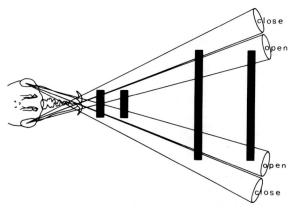

Fig. 5. Size-distance ambiguity in the bulldog ant. Mandibles open when the edges of an approaching object are seen by a "trigger zone" in each eye, and are closed again when the edges are seen by a second pair of trigger zones. Distances at which reactions are elicited depend on object size. (After Via 1977)

4.2 Insects Able to Judge Distance by Binocular Cues

While in the bulldog ant and in similar hunting forms, for example several ground beetles (e.g., Bauer 1985a,b), binocular distance vision may be absent or may be poor and only assist other nonvisual cues in prey localization, other predators use strategies to seize alert prey which work only if there is a mechanism of absolute distance estimation. Rather a striking example is shown in Fig. 6. The staphylinid beetle *Stenus* stalks its prey up to a distance of ca. 3 mm and then seizes the prey by a sudden protrusion of the eulabrium, which is accompanied by a slight forward movement of the whole body (Weinreich 1968). The beetle does not perform sideward-swaying movements, which would allow distance judgement by motion parallax, so it is natural to assume that the precise distance estimation required for its high rate of strike success is due to a binocular mechanism.

Friederichs (1931) suggested that ground beetles (cicindelids) could estimate distance by binocular cues, but this was merely based on the observation that monocular beetles perform worse in prey-catching than binocular ones. This is not convincing, since "it is difficult to assess what is impaired when one eye is occluded" (Rossel 1983). There is, however, a more convincing example for distance measurement in a carabid ground beetle (*Notiophilus biguttatus*) which hunts fast-fleeing springtails. *Notiophilus* is more successful in hunting this sort of prey than

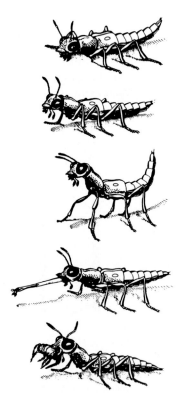

Fig. 6. Prey capture in the staphylinid beetle *Stenus bipunctatus.* Labium is protruded by enhancing pressure of hemolymph. (Weinreich 1968)

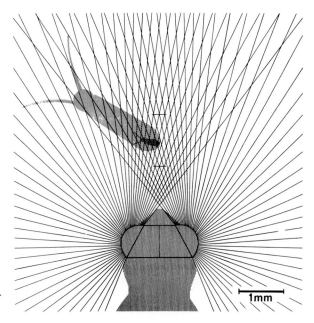

Fig. 7. Notiophilus biguttatus: Structure of visual field in a horizontal plane, range of prey size and catching distance is shown. The beetle fixates the hind end of its prey. (Bauer 1981)

other carabid beetles because it is able to gauge the distance of the prey accurately before the final attack. *Notiophilus* approaches its prey by brief movements until the catching distance is reached (1.1–2.3 mm, Fig. 7). In the terminal phase the beetle fixates an edge of the prey. A movement of the prey finally causes the beetle to perform a sudden jerk with wide-open mandibles towards its prey and in 50% of the attacks the springtail is grasped. The extent of the final jerk does not depend on the size of the prey, but is linearly related to the distance of the prey (Bauer 1981). As there is no size-distance ambiguity and as motion parallax is not available, the best explanation for the remarkable precision of the final attack is that *Notiophilus* uses binocular cues for distance measurement.

The use of binocular cues for absolute distance measurement is even better documented in visually hunting dragonfly larvae. The *Aeshna* larva captures prey in a similar way to *Stenus*: it fixates the prey, approaches until the prey is in reach of the labium, and if the prey moves, shoots out the labium which lies folded beneath the head (Baldus 1926). Baldus clearly states that the animals will never snap at prey which is out of reach of the labium. Prey size may vary by more than a factor of two — hence there is no size-distance ambiguity. Only monocular animals shoot at big objects which are out of range, confusing size and distance. Normal animals approach an oversized object as near as several centimeters and then flee. Shortly before the attack, motion of the prey is essential to elicit the strike, but the dragonfly larva itself does not move (Baldus 1926; Hoppenheit 1964; Vogt 1964; Etienne 1969); hence motion parallax cannot be the cue for distance measurement. Baldus concluded that the dragonfly larvae use a binocular mechanism to estimate distance. The snapping response will be triggered, if some features, for example the center of the prey, fall within the visual field of two symmetrical groups of ommatidia, the optical axes of which intersect at the midline near the tip of the unfolded labium (B in Fig. 8, left drawing).

Hoppenheit (1964) mentioned an observation which confirmed Baldus' idea and which must be regarded as an elegant proof for stereopsis in insects: *Aeshna* larvae can be deceived by two identical remote dummies (A und B in Fig. 8, right drawing) presented symmetrically to the midline. If the visual lines of two groups of ommatidia which look at the two dummies intersected within the range of the labium as shown in the figure, the larvae would occasionally shoot to this virtual image of a nearby prey. If the intersection point was still nearer to the animal, the larvae would continue to shoot, but the labium would no longer be completely unfolded.

Probably the most familiar example for a predatory insect which is capable of accurately measuring the distance to its prey is the mantid. The mantid's method of catching prey has been frequently described (Roeder 1937, 1960; Zänkert 1938; Mittelstaedt 1957; Rilling et al. 1959; Maldonado et al. 1967). Mantids usually sit in wait to ambush passing insects. When the target comes into view, the mantis turns its head to look at it with its forward-viewing acute zones of the two eyes (Maldonado and Barrós-Pitta 1970). When the prey comes into reach of the forelegs the mantis catches it by a rapid promotion of the forelegs, which in *Hierodula* sp. lasts only about 60 ms (Roeder 1960). Normal mantids never confuse size and distance, i.e., they never strike when the prey is out of reach of their raptorial legs and, as dragonfly larvae, catching distance does not depend on prey size, which may vary by at least a factor of 2 (Rilling et al. 1959). Motion parallax as a cue for target distance is not available, since mantids do not perform sideward swaying movement prior to the strike which is elicited by motion of the prey. For these reasons and because of the fact that monocularly blinded mantids may perform fixation reactions, but can no longer estimate prey distance, it was concluded that mantids gauge the prey distance binocularly by triangulation

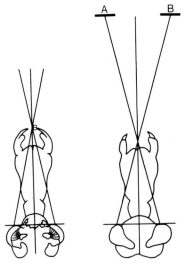

Fig. 8. Aeshna larvae which normally snap at an object within reach of the unfolded labium (*B* in the *left* drawing) can be deceived by two identical dummies (*A* and *B*, in the *right* drawing). (Redrawn and combined from Baldus 1926 and Hoppenheit 1964)

(e.g., Zänkert 1938; Maldonado and Levin 1967; Maldonado and Barrós-Pita 1970).

In an ingenious experiment, Rossel (1983, 1986) has now shown that this view was right. When the mantids *Tenodera australasiae* or *Mantis religiosa* looked at their prey through base-out prisms, the mantids struck when the apparent target position was at the strike distance (open symbols, Fig. 9), i.e., when disparity was within a certain range. Since monocular cues such as object size or motion parallax were not altered by the prisms, the experiment shows clearly that the mantids rely upon binocular disparity to measure distance. Binocular depth perception in mantids is not restricted to the frontal visual field but operates over a wide range of the binocular visual field up to an eccentricity of at least 20° (Rossel 1983).

The range at which absolute distance can be gauged goes far beyond the catching distance (30–15 mm in *T. australasiae*): At a prey distance of 100 mm or less the mantis brings its forelegs into the typical striking posture – the same response is performed at the appropriate disparity when the prey is seen through prisms.

An experiment suggesting that mantids may be able to measure even longer distances binocularly was described by Maldonado et al. (1970): *Stagmatoptera biocellata* displays a typical defensive posture when it is confronted with a potential predator such as a bird. In monocularly blinded mantids, the frequency at which this reaction was released depended on both the absolute size and the distance of the bird. Binocular mantids, however, did not confuse image size and distance. At a given distance, they responded with the same frequency to a small as well as to

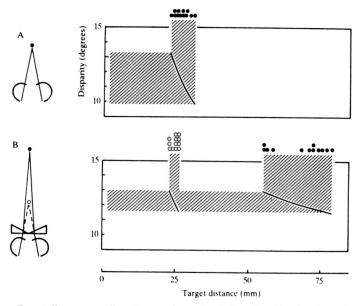

Fig. 9. Mantis religiosa: Target distance at strike release and correlated disparity. *Closed circles* real target position; *open circles* apparent target position. Mantis B looked at the target through a base-out prism. (After Rossel 1986)

a large bird. At a distance for example, as large as 450 mm, still 25% of the mantids responded to the different-sized birds. This distance is near the theoretically derived "infinite distance" E_∞ for *Tenodera* (see Table 1).

Because of the geometry of the binocular visual field, in theory tropical stalk-eyed flies should be able to perceive absolute distance (or size) even better than any predatory insect. In the diopsid *Cyrtodiopsis whitei* (Fig. 10), the males can have an eyestalk-span and with this, the base for triangulation measurements, of 10 mm or more. The binocular field is huge (Fig. 10) and comprises 70% of all ommatidia, and in a forward pointing acute zone (hexagonal grids in Fig. 10) the interommatidial angles barely exceed 1°. E_∞ is therefore more than 600 mm.

In *Cyrtodiopsis whitei*, the eyespan of males are considerably larger than those of the females. This extreme heteromorphy is explained as a result of sexual selection (Burkhardt and de la Motte 1983, 1985). But there are other diopsids

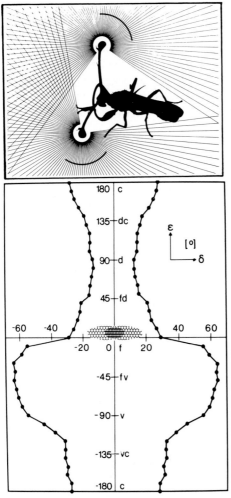

Fig. 10. *Lower part* Binocular visual field of *Cyrtodiopsis*. δ Limiting visual angles measured in planes perpendicular to the median plane: ε angle of measuring plane to the horizontal plane of the animal. *f, v* and *d, c* frontal, ventral, dorsal and caudal. *Grids* acute zones of both eyes. *Upper part* Silhouette of *Cyrtodiopsis* and lines of sight in the plane ε = −20°. (de la Motte and Burkhardt 1983)

where the males and females have modest eye-stalks of equal lengths. Therefore Burkhardt and de la Motte assume that originally the selective pressure for wide-set eyes was the need for improved spatial vision to find mates, which are difficult to identify in tropical rainforests where optical background is densely structured and species diversity is high. When *Cyrtodiopsis whitei* males encounter a conspecific, they may respond with a variety of behavioral displays depending on size and distance, and they clearly show by their reactions that they are able to identify a conspecific visually and to gauge absolute distance or size.

If, for example, a dummy of a conspecific is moved towards a *Cyrtodiopsis* male, the animal may turn to face the dummy. This fixation response was elicited when the dummy was at a distance of ca. 50 mm — irrespective of the size of the dummy. While the distance was independent of size, the number of reactions depended significantly on the size of the eyespan. If, for example, a large dummy was moved towards a large animal with eyespan of equal length, turning responses were frequently elicited while a smaller dummy was more often ignored. For a small male the smaller dummy was the more effective, and the bigger one less effective at eliciting the fixational response. The results show that the diopsids are not only able to measure the absolute distance but also to gauge the absolute size of a conspecific.

The experimental results were different when the dummy was suddenly introduced into the visual field of the animal. In this case — when the animal could not have identified the dummy as a conspecific — fixational responses could be elicited at much larger distances depending on the size of the object (Burkhardt and de la Motte 1983; Burkhardt pers. commun.).

While until recently the question of whether or not insects are able to perform absolute distance measurements by binocular cues was a matter of debate (e.g., Horridge 1977; Via 1977; Collett and Harkness 1982), there are now enough observations and experiments available confirming that Exner (1891) was right to assume that insects use disparity at least to gauge distances of nearby objects.

Of course, "stereopsis" in insects is different from that in man. In man absolute distances can be measured by vergence and accommodation, while stereopsis provides information of the depth of complex patterns relative to a fixation plane. This requires a high amount of neuronal wiring since the disparate images of many details of the three-dimensional world have to be compared (see, e.g., Bishop 1973). In insects the number of the axonal connections between the two optical centers is relatively small (e.g., Pichka 1976; Strausfeld and Nässel 1981) and in diopsids, for example, it is far less than the number of intersection points of all optical axes within the binocular visual field (Burkhardt and de la Motte 1983). It can thus be concluded that stereopsis comparable to that in man may not occur in insects. In insects, the task stereopsis has to fulfill is, however, simpler than that in man: it is not to perceive relative depth of many features of a complex surroundings, but to gauge the absolute distance of a single nearby object which might have been identified by simple cues as being of especial interest for the animal. As Rossel (1986) pointed out, this task could be performed by a relatively small set of neurons. In a first step prey or other objects could be identified monocularly: retinotopic feature detector neurons could respond to a few features which characterize the object of interest, such as angular size, characteristic movement or to a lesser extent also shape (Rilling et al. 1959). In a second step, the mean position of active feature neurons relative to the fovea could be coded by a few neurons with wide receptive fields. In principle, only four neurons in each optic lobe would be required, two for the horizontal and two for the vertical position relative to the fovea. With appropriate computation both the direction and distance could be determined by the signals of the corresponding neurons of the two optic lobes. Although this model is tentative, it seems to point the right way for further research.

5 Nonbinocular Methods of Triangulation

5.1 Use of a Flat World

In binocular vision, the distance between two symmetrical eyes serves as a base for triangulation measurements. There are a few examples in arthropods in which distances other than the interocular distance are used for triangulation. In animals which live on entirely flat ground (or beneath the water surface — see below), the vertical extent between eyes and ground may be used as a base to gauge distance or size of small objects by triangulation. The distance, and in a similar way, also the absolute size of a small object on the ground may be determined by the inclination of lines of sight which view it. The range within which such a mechanism can work is larger, the higher the eye is above the ground and the better the vertical resolving power, especially in the eye region looking to the horizon.

In the backswimmer, *Notonecta glauca*, such a mechanism serves to judge the absolute size of its prey. *Notonecta* is a predatory water bug which often hangs in its characteristic inverted posture beneath the water surface waiting for prey — small insects which have fallen onto the water surface (Fig. 11). In the *Notonecta* eye, the vertical interommatidial angles are particularly small in the eye region looking forward and parallel to the water surface (Schwind 1980). The interommatidial angles increase ventrally (upwards for the upside-down backswimmer) in such a way that, within a certain range, the vertical extent of an object floating in the plane of the water surface is always seen by the same number of ommatidia, irrespective of the distance from the backswimmer. (There is an exception to this rule: a very narrow acute zone with which, due to refraction, the animal is able to scan the water surface.)

In the optic lobes of *Notonecta*, neurons with wide receptive fields were found which respond to movement of single contrasting objects (Schwind 1978). One characteristic feature of these neurons is their different response to objects of different angular sizes. To a moving object of small angular extent (3.5°) the neurons respond maximally if the object is positioned immediately above the line of horizon (at 40° in Fig. 11). When objects of bigger angular extent were used as stimuli, this peak response shifted farther ventrally (i.e., farther above for the upside-down hanging backswimmer) the longer the angular extent of the object. This characteristic of the neurons can, in part, be explained by the geometrical properties of the eye. It leads to a "size constancy phenomenon": the neurons

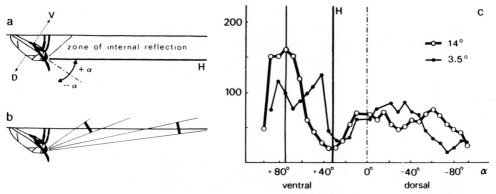

Fig. 11. a Notonecta hanging beneath the water surface. The insect's environment is divided into three parts: the zone of the transparent water surface above the animal, the zone of internal reflection and the region of deep water, below the insect's horizon (H). b and c Size constancy in Notonecta. c Responses of an optic lobe neuron to dark objects, moving in the median plane in front of the animal. Abscissa position of object in median plane; ordinate spikes per second. Solid circles response to movement of an object of small angular extent (black square angular width 3.5°). Open circles response to movement of a larger object (14°). Peak response is farther ventrally (upwards) for the larger object. This leads to a size constancy phenomenon (b): The neuron responds optimally to object of prey size within a certain range, irrespective of distance. (Schwind 1978)

respond optimally to objects of prey size moving in the plane of the water surface — in a certain range irrespective of the distance to the animal (Fig. 11b).

Recently, Zeil et al. (1986) described some peculiarities in the eye designs of different species of semi-terrestrial crabs. They show that in all probability, certain crabs living in flat habitats use a similar method either to gauge absolute size or absolute distance of an object. Most families of the brachyurans are broad-fronted with short eyestalks, positioned far apart. On the other hand, there are a few families which have long eyestalks, the basal joint of which lie close together (Fig. 12a). It has long been known that the latter type tends to occur in flat habitats such as intertidal sand- and mudflats, while the types with short eyestalks are found closer to vegetation or rocks (e.g., Crane 1975). The functional significance, however, was quite unclear. Zeil et al. (1986) found that in crabs with long vertical eyestalks, there is a horizontally pointing acute zone with extremely narrow vertical $\Delta\phi$ values, i.e., good vertical resolution. No such acute zone was found in the type with short eyestalks and in neither of the two groups was a zone for enhanced horizontal resolving power found. As in Notonecta, $\Delta\phi$ values increase ventrally in such a way that within a certain range an object on the ground is seen by the same number of ommatidia — irrespective of distance. Zeil et al. concluded that the special design of the visual system in the long eyestalk crabs does not fulfill the function of binocular depth perception, but enables the animals to gauge size or distance of an object by determining the point of intersection of optical axes with the flat ground.

a

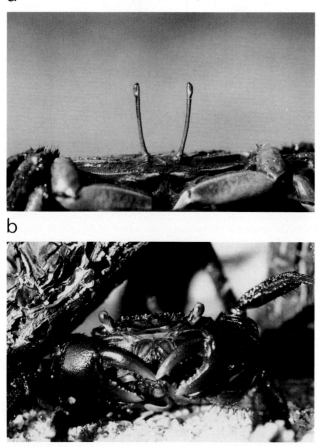

b

Fig. 12a,b. Macrophthalmus setosus (a) and Sesarma sp. (b). Two extreme representatives of two groups of brachyuran crabs, which differ according to the basic design of their visual system. (Photographs courtesy of J. Zeil)

5.2 Intraocular Mechanisms of Range Finding

Exner (1891) reported that in *Squilla mantis* two regions of one eye look at the same target. He noticed that this animal "has the advantage of seeing binocularly with each of its eyes. The superiority of binocular vision is undoubtedly distance estimation. Thus *Squilla* is able to gauge distances with each eye alone". Schaller (1953) proved this to be true, although it turned out that the ability to measure distance monocularly was restricted to a range of 4 cm or less, whereas for distances of more than 4 cm binocular vision becomes necessary. A closer look at the eyes shows that in most stomatopods there are three zones, a central band and two zones proximally and distally to this band in the cylindrical eye, the optical axes of which all point in parallel. Between these zones there are skewed ommatidia, the optical

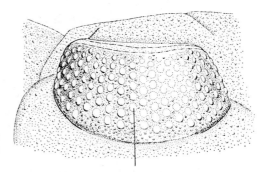

Fig. 13. Schizochroal eye of *Acaste. Line* points to a vertical file of lenses. (Stockton and Cowen 1976)

axes of which intersect with the parallel viewing axes at different distances (Schiff et al. 1985), so that a complex pattern of intersecting ommatidial visual fields arises. Schiff et al. (1986) assume that temporal excitatory patterns induced by moving stimuli which, due to the complex intersection pattern, are different for different distances, are decisive for distance estimation. One should notice that certain dynamic properties of stimuli and excitation also play a role in all other mechanisms for distance estimation discussed above, and that, on the other hand, dynamic excitatory patterns which lead to binocular depth perception are based on certain morphological prerequisites: the intersection pattern of the optical axes.

Finally, a rather uncommon type of "compound eye" should be mentioned, which probably was designed to estimate absolute distances. The extinct trilobites of the suborder *Phacopina* had schizochroal eyes in which relatively few large lenses were distributed over the eye surface. These lenses were arranged in vertical rows on the eyes (Fig. 13). The optics of the single eyes must have been of high quality (Clarkson and Levi-Setti 1975) so that in each eye a sharp image could be formed on a relatively large retina. From measurements of the optical axes, it was concluded that the visual fields of neighboring eyes on each vertical row would have overlapped greatly, whereas the receptive field of a given single eye would have only marginally overlapped with that of its next neighbor in the next vertical row (Clarkson 1966). Stockton and Cowen (1976) have proposed that due to large spacing between horizontal rows a large visual field was scanned, while the overlapping visual fields of neighboring eyes within the vertical rows might have allowed depth perception due to vertical disparity.

6 Conclusion

The examples presented in the preceding passages were selected to illustrate some principles by which absolute sizes or absolute distances of single objects can be gauged by arthropods with compound eyes. It was not intended to give a general discussion on "depth perception", which would have had to comprise a discussion on flow field features, by which relative distances of objects could be determined by freely moving animals (see, e.g., Horridge 1986).

In the different mechanisms discussed above, one feature seems to be common: motion of the retinal image, either due to movement of an object or induced by self-movement of the animal, is necessary. Apart from this, no general principle seems to be present in size and distance perception. Even if one compares only binocular methods of distance perception, one will find that they are different in species with different ways of life. The strategy employed by a particular species to gauge size or distance is adapted for survival in its specific, very narrow visual world, which is different in different species according to the ecological niches in which the animals live.

Acknowledgments. I wish to thank Liz Jerger, for help in improving the English style, Friedl Webinger for her assistance, Dr. J. Zeil for the photographs shown in Fig. 12 and Prof. D. Burkhardt and Dr. I. de la Motte for stimulating discussions.

References

Baldus K (1926) Experimentelle Untersuchungen über die Entfernungslokalisation der Libellen (*Aeschna cyanea*). Z Vergl Physiol 3:475–505

Bauer T (1981) Prey capture and structure of the visual space of an insect that hunts by light on the litter layer (*Notiophilus biguttatus* F., Carabidae, Coleoptera). Behav Ecol Sociobiol 8:91–97

Bauer T (1985a) Beetles which use a setal trap to hunt springtails: The hunting strategy and apparatus of *Leistus* (Coleoptera, Carabidae). Pedobiologia 28:275–287

Bauer T (1985b) Different adaptation to visual hunting in three ground beetle species of the same genus. J Insect Physiol 31:593–603

Bauers C (1953) Der Fixierbereich des Insektenauges. Z Vergl Physiol 34:589–605

Bishop PO (1973) Neurophysiology of binocular single vision and stereopsis. In: Jung R (ed) Handbook of sensory physiology, vol VII/3A. Springer, Berlin Heidelberg New York, pp 255–305

Burkhardt D, de la Motte I (1983) How stalk-eyed flies eye stalk-eyed flies: observations and measurements of the eyes of *Cyrtodiopsis whitei* (Diopsidae, Diptera). J Comp Physiol A 151:407–422

Burkhardt D, de la Motte I (1985) Selective pressures, variability, and sexual dimorphism in stalk-eyed flies (Diopsidae). Naturwissenschaften 72:204–206

Burkhardt D, Darnhofer-Demar B, Fischer K (1973) Zum binokularen Entfernungssehen der Insekten. I. Die Struktur des Sehraums von Synsekten. J Comp Physiol 87:165–188

Clarkson ENK (1966) Schizochroal eyes and vision in some phacopid trilobites. Palaeontology 9:464–487

Clarkson ENK, Levi-Setti R (1975) Trilobite eyes and the optics of Descartes and Huygens. Nature (London) 254:663–667

Cloarec A (1986) Distance and size discrimination in a water stick insect, *Ranatra linearis* (Heteroptera). J Exp Biol 120:59–77

Collett TS (1978) Peering – a locust behavior pattern for obtaining motion parallax information. J Exp Biol 76:237–241

Collett TS (1987) Binocular depth vision in arthropods. TINS 10:1–2

Collett TS, Harkness L (1982) Depth vision in animals. In: Ingle DJ, Goodale MA, Mansfield RJW (eds) Analysis of visual behavior. MIT, Cambridge London, pp 111–176

Collett TS, Land MF (1975) Visual control of flight behavior in the hoverfly *Syritta pipiens* L. J Comp Physiol A 99:1–66

Crane J (1975) Fiddler crabs of the world. Univ Press, Princeton, NJ

de la Motte I, Burkhardt D (1983) Portrait of an Asian stalk-eyed fly. Naturwissenschaften 70:451–461

Demoll R (1909) Über die Beziehungen zwischen der Ausdehnung des binokularen Sehraumes und dem Nahrungserwerb bei einigen Insekten. Zool Jahrb Abt Syst Geogr Biol 28:523–530

Dietrich W (1909) Die Facettaugen der Dipteren. Z Wiss Zool 92:465–539

Egelhaaf M (1985) On the neuronal basis of figure-ground discrimination by relative motion in the visual system of the fly. Biol Cybernet 52:123–280

Eriksson ES (1980) Movement parallax and distance perception in the grasshopper (*Phaulacridium vittatum* (Sjöstedt). J Exp Biol 86:337–341

Etienne AS (1969) Analyse der schlagauslösenden Bewegungsparameter einer punktförmigen Beuteatrappe bei der Aeschnalarve. Z Vergl Physiol 64:71–110

Exner S (1891) Die Physiologie der facettirten Augen von Krebsen und Insecten. Deuticke, Leipzig Wien

Frantsevich LI, Pichka VE (1976) The size of the binocular zone of the visual field in insects. J Evol Biochem Physiol (USSR) 12:461–465 (in Russian)

Friederichs HF (1931) Beiträge zur Morphologie und Physiologie der Sehorgane der Cicindeliden (Coleoptera). Z Morphol Ökol Tiere 21:1–172

Goulet M, Campman R, Lambin M (1981) The visual perception of relative distances in the woodcricket, *Nemobius sylvestris*. Physiol Entomol 6:357–367

Graham CH (1965) Visual space perception. In: Graham CH (ed) Vision and visual perception. John Wiley & Sons, New York, pp 504–547

Hengstenberg R (1971) Das Augenmuskelsystem der Stubenfiege *Musca domestica*. I. Analyse der "clock-spikes" und ihrer Quellen. Kybernetik 9:56–77

Heran H, Lindauer M (1963) Windkompensation und Seitenwindkorrektur der Bienen beim Flug über Wasser. Z Vergl Physiol 47:39–55

Hoppenheit M (1964) Beobachtungen zum Beutefangverhalten der Larve von *Aeschna cyanea* Müll. (Odonata) Zool Anz 172:216–232

Horridge GA (1977) Insects which turn and look. Endeavour N Ser 1:7–17

Horridge GA (1978) The separation of visual axes in apposition eyes. Philos Trans R Soc London Ser B 285:1–59

Horridge GA (1986) A theory of insect vision: velocity parallax. Proc R Soc London Ser B 229:13–27

Horridge GA (1987) The evolution of visual processing and the construction of seeing systems. Proc R Soc London Ser B 230:279–292

Maldonado H, Barrós-Pita JC (1970) A fovea in the praying mantis eye. I. Estimation of the catching distance. Z Vergl Physiol 67:58–78

Maldonado H, Levin L (1967) Distance estimation and the monocular cleaning reflex in praying mantis. Z Vergl Physiol 56:258–267

Maldonado H, Levin L, Barrós-Pita JC (1967) Hit distance and the predatory strike of the praying mantis. Z Vergl Physiol 56:237–257

Maldonado H, Benko M, Isern M (1970) Study of the role of binocular vision in mantids to estimate long distances, using the deimatic reaction as experimental situation. Z Vergl Physiol 68:72–83

Mittelstaedt H (1957) Prey capture in mantids. In: Scher BT (ed) Recent advances in Invertebrate Physiology. Univ Oregon Publ, pp 51–71

Ogle KN (1962) The optical space sense. In: Davson H (ed) The Eye, vol 4, Pt 2. Academic Press, New York London pp 211–417

Pichka VE (1976) Visual pathways in the protocerebrum of the dronefly *Eristalis tenax*. J Evol Biochem Physiol (USSR) 12:495–500

Pritchard G (1966) On the morphology of the compound eyes of dragonflies (Odonata, Anisoptera) with special reference to their role in prey capture. Proc R Ent Soc London Ser A 41:1–8

Rilling S, Mittstaedt M, Roeder KD (1959) Prey recognition in the praying mantis. Behavior 14:164–184

Roeder KD (1937) The control of tonus and locomotor activity in the praying mantis (*Mantis religiosa* L.). J Exp Zool 76:353–373

Roeder KD (1960) The predatory and display strikes of the praying mantis. Med Biol Illustr 10:172–178

Rossel S (1983) Binocular stereopsis in an insect. Nature (London) 302:821–822

Rossel S (1986) Binocular spatial localization in the praying mantis. J Exp Biol 120:265–281

Schaller F (1953) Verhaltens- und sinnesphysiologische Beobachtungen an *Squilla mantis* L. Z Tierpsychol 10:1–12

Schiff H, Abbott BC Manning RB (1985) Possible monocular range-finding mechanisms in stomatopods from different environmental light conditions. Comp Biochem Physiol A 80:271–280

Schiff H, d'Isep F, Candone P (1986) Superposition and scattering of visual fields in a compound, double eye. II. Stimulation sequences for different distances in a stomatopod from a bright habitat. Comp Biochem Physiol A 83:445–455

Schwind R (1978) Visual System of *Notonecta glauca*: A neuron sensitive to movement in the binocular visual field. J Comp Physiol A 123:315–328

Schwind R (1980) Geometrical optics of the *Notonecta* eye: Adaptations to optical environment and way of life. J Comp Physiol A 140:59–68

Sherk TE (1978) Development of the compound eyes of dragonflies (Odonata). III. Adult compound eyes. J Exp Zool 203:61–80

Stockton WL, Cowen R (1976) Stereoscopic vision in one eye: paleophysiology of the schizochroal eye of trilobites. Palaebiology 2:304–315

Strausfeld NJ, Nässel DR (1981) Neuroarchitecture serving compound eyes of Crustacea and insects. In: Autrum H (ed) Handbook of sensory physiology, vol VII/6B. Springer, Berlin Heidelberg New York, pp 1–132

Via S (1977) Visually mediated snapping in the bulldog ant: a perceptual ambiguity between size and distance. J Comp Physiol A 121:33–51

Vogt P (1964) Über die optischen Schlüsselreize beim Beuteerwerb der Larven der Libelle *Aeschna cyanea* Müll. Zool Jahrb Physiol 71:171–180

Wallace GK (1959) Visual scanning in the desert locust *Schistocerca gregaria* Forskål. J Exp Biol 36:512–525

Wehner R (1981) Spatial vision in arthropods. In: Autrum H (ed) Handbook of sensory physiology, vol VII/6C. Springer, Berlin Heidelberg New York, pp 287–616

Weinreich E (1968) Über den Klebefangapparat der Imagines von *Stenus* Latr. (Coleopt. Staphylinidae) mit einem Beitrag zur Kenntnis der Jugendstadien dieser Gattung. Z Morphol Tiere 62:162–210

Zänkert A (1938, 1939) Vergleichend-morphologische und physiologisch-funktionelle Untersuchungen an Augen beutefangender Insekten. Sitz Ber Ges Naturforsch Berlin 1–3:82–169

Zeil J (1983) Sexual dimorphism in the visual system of flies: The free flight behavior of male Bibionidae (Diptera). J Comp Physiol A 150:395–412

Zeil J, Nalbach G, Nalbach H-O (1986) Eyes, eye stalks and the visual world of semiterrestrial crabs. J Comp Physiol A 159:801–811

Subject Index

Abbe theory 91
aberration 38
  lens 86
  spherical 51
absorbance spectrum,
  Calliphora visual pigment 121, 154
  3-hydroxy-retinol 138, 140
  rhodopsin 119, 153
absorption, of modes 80, 81
  maxima of visual pigments 8, 118,
  153
acceptance angle 49, 262
acetylcholine 241, 248–250
Acheta 242
achromatic vision 286
acone 35
acridids 96
actin 4, 271
acute zone 91, 93
  and forward flight 96–99
  dorsal 103–107, 330
  frontal 99–103, 108, 330
  horizontal 107, 108
  male 102, 336
adaptation 43, 48–50, 181, 249, 257
  circadian control 257ff
  in H1 neuron 230
  in superposition eyes 58–60
  localization 175, 176
  neural 217–219, 222, 229, 322, 339
  role of calcium 177, 181
adenylate cyclase 272
aequorin 177, 178, 179
Aeshna 106, 148, 242, 429, 430, 433
agaristids 59
Airy disk 38, 64, 75, 78
alderflies, see Megaloptera
Aleyrodes proletella 108
aliasing 393
α-bungarotoxin 248, 249
Amegilla 96, 97
Ampelisca 60, 61
amphipods 35, 40, 45, 46, 48, 50, 60–62,
  104
amplification, in LMC's 216, 217

matched 223–225, 229
Amyelois 159
Anax junius 106, 429
Androctonus australis 265
angular magnification 51
angular sensitivity
  effect of light adaptation 85, 163, 262
  LMC 221
Anisopodidae 195
Anomura 56
antidromic illumination 80
ants 49, 164, 310, 431, see *Cataglyphis*
Apis, see bee, Hymenoptera
  acute zone 96, 98, 99
  eye pigments 158
  facets 92
  neurotransmitters 238, 241–244, 251
  polarization sensitivity 302
  sky compass 305–309
  waggle dances 298
Aplysia 238, 265
Apocephalus laceyi 108
apposition eyes 38ff, 95, 213, 214
  afocal 44, 45, 64
  beetle 40
  focal 38, 41, 44, 51
  geometry 92
  simple 38
  transparent 45–48, 64
apposition image 32
Apterygota 60, 63
arhabdomeral lobe 174
Artemia 48, 49
Ascalaphus 24, 104, 148
  chromophore 146
  visual pigment 116–118, 120
asilids 103, 189
aspartate 250, 251
Astacilla 48
Astacus 55
  rhodopsin 126
Autrum 15ff, 17, 20, 24, 186, 215, 235,
  265, 282, 318
axonal projection in dipteran lamina
  189–193

S. Exner

The Physiology of the Compound Eyes of Insects and Crustaceans

With a Foreword from K. von Frisch

Translated and Annotated by R. C. Hardie

1988. 28 figures, 76 figures in 7 plates. Approx. 180 pages. ISBN 3-540-50239-4

Contents: Physical Principles. – Dioptrics of the Compound Eye. – The Iris Pigment and its Function. – The Retinal Image of Various Insects and Crustaceans. – The Retina, its Pigment and Tapetum. – Eyes with Non-Uniform Construction. – Short Descriptions of Selected Eyes of Insects. – The Eyes of **Squilla, Phronima** and **Copilia.** – Accessory Optical Phenomena in Compound Eyes. – Vision with Compound Eyes. – Some Remarks on the Phylogeny of Compound Eyes Considered from a Functional Viewpoint. – Appendices. – Index.

Exner's classic monograph describes the basic optical mechanisms in operation in compound eyes and, despite the passage of time, still remains a definitive work. Although his findings were seriously questioned during the modern revival of interest in compound eyes, all his major discoveries have now been validated. The principle of the lens cylinder and the elucidation of the mechanics of apposition and superposition optics are amongst his outstanding contributions. It also includes a broad survey of the optics and anatomy of the eyes of many insect and crustacean species, and the first explanation for the phenomena of pseudopupils and eyeglow.
It has been faithfully translated from the original with annotations to aid the reader. The new edition, with a foreword by the late Karl von Frisch, also includes a concise illustrated appendix summarizing present knowledge of optical mechanisms in compound eyes and a useful bibliography.

Springer-Verlag Berlin
Heidelberg New York London
Paris Tokyo Hong Kong

Springer